D0605747

BLACKFLIES
The future for biological methods in integrated control

BLACKFLIES

The future for biological methods in integrated control

Editor:

MARSHALL LAIRD

The Memorial University of Newfoundland

Sponsored By:
International Development Research Centre,
60 Queen Street,
Ottawa, Canada

1981

ACADEMIC PRESS
A Subsidiary of Harcourt Brace Jovanovich, Publishers
London   New York   Toronto   Sydney   San Francisco

ACADEMIC PRESS INC. (LONDON) LTD.
24/28 Oval Road, London NW1 7DX

United States Edition published by
ACADEMIC PRESS INC.
111 Fifth Avenue, New York, New York 10003

Copyright © 1981 by
ACADEMIC PRESS INC. (LONDON) LTD.

All Rights Reserved
No part of this book may be reproduced in any form by photostat, microfilm, or any
other means without written permission from the publishers

British Library Cataloguing in Publication Data
Blackflies
  1. Simulium damnosum – Control
  I. Laird, M.
  595.77′1   QL537.555

  ISBN 0-12-434060-1

  LCCCN 81-66373

Text set in 10/11 pt Linotron 202 Palatino, printed and bound
in Great Britain at The Pitman Press, Bath

RA
641
S55
BS7
1981

Contributors

J. F. Burger Assistant Professor of Entomology, College of Life Sciences and Agriculture, University of New Hampshire, Department of Entomology, Nesmith Hall, Durham, New Hampshire, 03824, USA

G. Carlsson Drottningatan 109, 25233 Helsingborg, SWEDEN

W. A. Charnetski Crop Entomology Section, Research Station, Lethbridge, Alberta, T1J 4Bl, CANADA

B. Cliff World Health Organization, Onchocerciasis Control Programme, BP 549, Ouagadougou, UPPER VOLTA

M. H. Colbo Research Unit on Vector Pathology, Memorial University of Newfoundland, St. John's, Newfoundland, A1C 5S7, CANADA

T. L. Couch Abbott Laboratories, The Chemical and Agricultural Products Division, Department 912, T-9 North Chicago, Illinois, 60064, USA

R. W. Crosskey Department of Entomology, British Museum (Natural History), Cromwell Road, London SW7 5BD, ENGLAND

E. W. Cupp Associate Professor of Medical Entomology, Cornell University, Department of Entomology, Comstock Hall, Ithaca, New York, 14853, USA

P. T. Dang* Department of Biology, Carleton University, Ottawa, Ontario, K1S 5B6, CANADA

D. M. Davies Professor, Department of Biology, McMaster University, 1280 Main Street West, Hamilton, Ontario, L8S 4K1, CANADA

J. B. Davies World Health Organization, Onchocerciasis Control Programme, BP 549, Ouagadougou, UPPER VOLTA

A. M. Dubitskii** Chief, Laboratory of Biological Control, Deputy Director, Institute of Zoology, Kazakh Academy of Sciences, Alma-Ata, 480032, USSR

R. W. Dunbar World Health Organization, Kaduna, NIGERIA

P. Elsen Laboratoire de Zoologie Médicale, Institut de Médicine Tropicale Prince Leopald, B-2000 Antwerpen, BELGIUM

J. R. Finney Research Unit on Vector Pathology, Memorial University of Newfoundland, St. John's, Newfoundland, A1C 5S7, CANADA

J. Grunewald Universität Tübingen, Tropenmedizinisches Institut, 7400 Tübingen 1, FEDERAL REPUBLIC OF GERMANY

W. O. Haufe Research Branch, Agriculture Canada, Research Station, Lethbridge, Alberta, T1J 4B1, CANADA

H. B. N. Hynes Department of Biology, University of Waterloo, Waterloo, Ontario, N2L 3G1, CANADA

H. Jamnback Director, State Science Service, New York State Museum and Science Service, The State Education Department, Albany, New York, 12224, USA

M. Laird Director, Research Unit on Vector Pathology, Memorial University of Newfoundland, St John's, Newfoundland, A1C 5S7, CANADA

P. A. La Scala Department of Entomology, University of New Hampshire, Durham, New Hampshire, 03824, USA

* Present address: Diptera and Hemiptera Section, Research Branch, Biosystematics Research Institute, Agriculture Canada, Ottawa, Ontario, K1A 0C6, CANADA
** Present address: Division of Vector Biology and Control, World Health Organization, 1211 Geneva 27, SWITZERLAND

J. E. Mokry Research Unit on Vector Pathology, Memorial University of Newfoundland, St John's, Newfoundland, A1C 5S7, CANADA

K. Müller Department of Biology, University of Umcá, SWEDEN

R. A. Nolan Department of Biology, Memorial University of Newfoundland, St John's, Newfoundland A1B 3X9, CANADA

K. Ogata Chief, Section of Medical Zoology, Japan Environmental Sanitation Center, 10–6 Yotsuya-Kamicho, Kawasaki, JAPAN

A. L. Paul Chemical and Agricultural Products Division, Abbott Laboratories, North Chicago, Illinois, 60064, USA

B. V. Peterson Head, Diptera and Hemiptera Section, Research Branch, Biosystematics Research Institute, Agriculture Canada, Ottawa, Ontario, K1A 0C6, CANADA

G. O. Poinar Jr Division of Entomology, University of California at Berkeley, 333 Hilgard Hall, Berkeley, California, 94720, USA

J. N. Raybould Onchocerciasis Control Programme, c/o Public Health Laboratory, PO Box 88, Akosombo, GHANA

K. H. Rothfels Department of Botany, University of Toronto, Toronto, Ontario, M5S 1A1, CANADA

I. A. Rubtsov Professor, Zoological Institute, Academy of Sciences of the USSR, Leningrad 199164, USSR

M. W. Service Department of Medical Entomology, Liverpool School of Tropical Medicine, Pembroke Place, Liverpool L3 5QA, ENGLAND

B. H. Thompson Research and Assessment Branch, Department of Consumer Affairs and Environment, St John's, Newfoundland, A1C 5T7, CANADA

A. H. Undeen* Deputy Director, Research Unit on Vector Pathology, Memorial University of Newfoundland, St John's, Newfoundland, A1C 5S7, CANADA

I. C. Vajime World Health Organization, Onchocerciasis Control Programme, PO Box 211, Tamale, GHANA

R. R. Wallace Petro-Canada Exploration, P.O. Box 2844, Calgary, Alberta, T2P 2M7, CANADA

J. F. Walsh** World Health Organization, Onchocerciasis Control Programme, B.P. 549, Ouagadougou, UPPER VOLTA

J. Weiser Institute of Entomology, Czechoslovak Academy of Sciences, Flemingova Nam. 2, Prague 6, CZECHOSLOVAKIA

P. Wenk Professor, Tropenmedizinisches Institut der Universität, 74 Tübingen, Wilhelmstrasse 11, FEDERAL REPUBLIC OF GERMANY

R. S. Wotton Biological Sciences Department, University of London, Goldsmiths' College, New Cross, London, SE14 6NW, ENGLAND

* Present address: Insects Affecting Man and Animals Research Laboratory, USDA AR, PO Box 14565, Gainesville, Florida, 32604, USA
** Present address: Department of Biology, University of Salford, Salford M5 4WT, ENGLAND

Preface

Blackflies (Diptera: Simuliidae) are important pests of, and vectors of disease to, humans and both domestic and wild animals. In the former role they pester people seeking recreation and sport in river valleys from New Zealand (where they are confusingly called "sandflies", the vernacular name otherwise reserved for Phlebotomidae) to boreal regions of the USSR and Canada (where their summertime incidence can be high enough to threaten construction projects by disrupting the work force). As enormously abundant pests of domestic animals they are directly responsible for major losses to the cattle industry on Canadian prairies. They transmit filarioid worms and blood protozoans to a wide range of domestic vertebrates and wildlife. In some countries, their vectoring of haemosporidans of the genus *Leucocytozoon* causes serious losses among, for example, turkeys, ducks and geese. While the epidemiological consequences of related infections among waterfowl and other wild bird populations await elucidation, leucocytozoonosis is known to have very many avian hosts around the world.

In Africa, where it is speculated that more than 20 million humans harbour *Onchocerca volvulus* (the main vectors of which are blackflies of the *Simulium damnosum* complex), the World Health Organization's Onchocerciasis Control Programme (WHO/OCP) is now endeavouring to protect the million or so sufferers from this disease (= river blindness) in the Volta River Basin. In Central America, to which onchocerciasis is believed to have been imported long ago with West African slaves (as Arab slavers presumably spread it to Yemen), the Japan-Guatemala Onchocerciasis Pilot Project is under way against *Simulium ochraceum*.

To date, the control of Simuliidae has meant the use of appropriately formulated synthetic organic pesticides against their larvae and of space sprays, notably fogging, against adults. Although a very wide variety of pathogens, parasites and predators has been found wherever searched for in blackfly populations, none of these showed any promise of practical biocontrol effectiveness until the 1970s. At the start of this decade, the only natural population-limiting factor known for the simuliids that seemed to have any prospect of forming the basis of commercial products to supplement currently available chemicals, was a group of parasitic worms, the Mermithidae. There was no knowledge of any entomopathogenic viruses from these hosts and none of the bacteria reported from them had been cultured or had shown any control prospects.

Many of the fungi listed for simuliids appeared benign, although a single report suggesting the possible occurrence of *Coelomomyces* sp. in a Honduran blackfly, remains of interest (despite subsequent failure to confirm such an association in Central America or elsewhere) in view of the fact that the genus consists almost exclusively of mosquito pathogens that have been under study for many years as candidate biocontrol agents. Microsporidan protozoa, referable to only a few easily-recognized species, are widespread in blackflies; but the life cycles have yet to be determined, and means of manipulating them to the disadvantage of host populations remain to be devised.

There were also a few reports of parasitoid Hymenoptera (the backbone of much of "conventional" biocontrol in economic entomology) from European blackflies, somewhat surprising in view of the absence of such parasitoids from the Culicidae (there being only one doubtful record for this entire family). As regards predators, while many random observations had been published and some small-scale experiments undertaken, none of the results had led to any kind of practical control demonstration.

However, the known presence of Mermithidae in blackflies in many parts of the world, seemed of much interest in view of the fact that the warm-water mosquito parasite, *Romanomermis culicivorax*, was beginning, in the early 1970s, to show promise of a degree of effectiveness justifying commercialization. During the past decade, this question therefore received a great deal of attention. Consequently, much was learnt about host–parasite relationships between Simuliidae and their Mermithidae. It was also found that under artificial still-water conditions and exposure to extremely large numbers of pre-parasitic *R. culicivorax*, first-instar *blackfly* larvae may be fully penetrated, with a degree of development of the worm sometimes taking place. This finding eventually led to the demonstration of a few total penetrations of first-instar *Simulium damnosum s.l.* larvae in West Africa under artificial stream conditions, indicating the serviceability of the bizarre relationship for experimental modelling purposes. On the other hand, while some enhancement of larval blackfly populations in Newfoundland streams was achieved via their hand-seeding with large numbers of naturally infected larvae, by the end of the 1970s mermithids were not much nearer to forming the basis of practical blackfly biocontrol agents than they had been at the beginning of the decade. One reason for this was the lack of any self-maintaining, closed colonies of Simuliidae (as distinct from larval rearing systems dependent upon field-collected eggs) until late in 1978.

A new candidate biocontrol agent was, however, about to emerge. The discovery of *Bacillus thuringiensis* var. *israelensis* (*B.t.i.*) in an Israel mosquito population in 1976 was soon followed by demonstration of the vulnerability of larval blackflies to this dipteran serotype of a "safe" and environmentally acceptable biocontrol agent that had been in use against a wide variety of lepidopterous pests of crops and forests for many years. In 1978–1980, there were encouraging demonstrations of a high level of effectiveness (*c.* 95%) under both laboratory and field conditions in Newfoundland, Ivory Coast and Guatemala. In the two latter countries, the major vectors of onchocerciasis were the test subjects: *Simulium damnosum s.l.* in the former, and *Simulium ochraceum* in the latter.

Meanwhile, industry in the USA, France and Belgium was actively exploring commercialization, and there were plans for mass production in the Peoples' Republic of China, too. Safety testing conducted to date has shown minor adverse effects among severely stressed laboratory mammals, but there has not yet been any evidence of adverse effects upon non-target organisms (NTOs) in streams (nor for that matter upon NTOs in static mosquito larval habitats). At the end of the decade, therefore, *B.t.i.* suddenly became by far the best hope for practical biocontrol of Simuliidae.

The timeliness of this from the standpoint of vector control is self-evident. Until mid-1980, the WHO/UNDP/WB Onchocerciasis Control Programme (OCP) in the Volta River Basin, depended utterly upon the regular and repeated application of ABATE® (Temephos) to all blackfly breeding areas of streams throughout this area of approximately 654 000 km². *

All these matters are considered in detail in the following pages against the background of a sufficient outline of blackfly taxonomy, ecology, physiology, conventional control, etc., for the book to comprise an adequate single source of information relating to all relevant aspects of biocontrol practicability within the broader context of future integrated control methodologies.

Moreover, thanks to the enormous stimulus given to basic studies in these and related aspects of blackfly research by the purposeful conventional control developments of the 1970s, on a much wider basis than ever before, it is hoped that the book will prove timely and serviceable to all simuliidologists, so many of whom have collaborated herein in furnishing up-to-date synopses of all major aspects of the whole topic.†

Integrated control methodologies (whether designed to combat tropical blackfly disease vectors or boreal and other pest species) will certainly have a continuing place not only for appropriate synthetic chemical pesticides used as selectively as possible, but also for such "second generation" control agents as insect growth regulators (IGRs), whether chitin inhibitors or juvenile hormone mimics. The latter, specifically ALTOSID®, have recently shown promising results against larval blackflies in field tests with a minimum of harm to NTOs. There are, therefore, good grounds for optimism that still more intensive investigations (with particular attention to operational research to design complex integrated control methodologies against the background of a detailed understanding of already-existing natural factors limiting simuliid populations) will result in effective, enduring and environmentally acceptable integrated strategies for blackfly control in the foreseeable future.

March 1981   *Marshall Laird*

---

* Then, following the early-1980 suspected resistance to Temephos in the *S. sanctipauli/S. soubrense* species pair in the Forest Zone of Ivory Coast and its confirmation in May by susceptibility tests (Guillet *et al.* 1981) the intermittent use of chlorphoxim was begun, to control this population. This and other available alternative chemical larvicides pose more hazards to NTOs, including fish, than does Temephos. The early prospect of an environmentally acceptable biocontrol agent is thus most welcome; as is the information (Davies and Walsh, personal communication) that recent OCP field trials with *B.t.i.* have shown great promise.
† The latter's dynamic status is evident from not only the recent appearance of pesticide resistance mentioned above, but also the still more recent achievement of complete generation rearing of *S. damnosum s.l.* in a closed laboratory system of water circulation in the USA (Cupp *et al.* 1981).

Blackflies: The future for biological methods in integrated control

Acknowledgements

Support from the World Health Organization for investigations discussed in their chapters is acknowledged by B. V. Peterson and P. T. Dang, while J. F. Walsh, J. B. Davies and B. Cliff express their appreciation for help to all past and present OCP staff and administration and all members of the relevant Management Committees and National Onchocerciasis Committees. K. Ogata expresses gratitude for the assistance of the various entomologists from both countries participating in the Japan–Guatemala Onchocerciasis Control Pilot Project.

The contributions of M. H. Colbo, D. M. Davies, J. R. Finney, J. E. Mokry, R. A. Nolan, K. H. Rothfels and A. H. Undeen were supported by the Natural Sciences and Engineering Research Council Canada, and that of B. H. Thompson by the William and Flora Hewlett Foundation.

The International Development Research Centre assisted some of the studies discussed by M. H. Colbo, J. E. Mokry and A. H. Undeen, and sponsored the January 1972 scientific advisory group at the Memorial University of Newfoundland, from the proposals of which (IDRC-006e/006f, 1972, Ottawa) both RUVP and the present book evolved.

"The Origins of Blackfly Control Programmes" by H. Jamnback is published by permission of the Director, New York State Museum, State Education Department, and numbered 269 in this Institution's Journal Series.

Last, but far from least, the Editor acknowledges the forbearance and helpfulness of all who collaborated in this effort, with a special word of appreciation to his own Administrative Staff Specialist, Sadie Popovitch, for valuable assistance throughout all stages of the preparation of the book.

This book is dedicated to the late Professor Brian Hocking, FRSC (1914–1974) in recognition of his important research and teaching contributions to public health entomology, particularly as related to the physiology and control of biting flies, including Simuliidae.

I Systematics

Simuliid Taxonomy—The Contemporary Scene

R. W. Crosskey

Accurate identification of Simuliidae is becoming increasingly significant as more species become the target of pest and vector control operations, or the subject of epidemiological studies on the transmission of the blackfly-borne parasites of man (*Onchocerca volvulus* and *Mansonella ozzardi*). The characterization and differentiation of species are the most important practical functions of contemporary taxonomy.

Organization of species into higher categories and the definition of these categories (classification), plus the drawing of inferences about blackfly evolution (phylogeny), are also important parts of procedural taxonomy. Classification into taxa, based at least notionally on evolutionary affinity (there is no fossil record and our ideas of blackfly descent are entirely conjectural), is potentially helpful for inferring the likely biology of a little-known species when there are biological data available for its better-known relatives—morphological similarity tends to be accompanied by behavioural similarity.

The taxonomy of Simuliidae is now in a moderately satisfactory state although we still lack a universally agreed, stable classification for the world fauna. Despite its relatively small size (currently about 1270 species) the family is more difficult to classify satisfactorily than many larger dipterous families, on account of the remarkable morphological homogeneity that it possesses. The early stages, unlike those of many dipterous families with aquatic members, have essentially exploited only a single habitat: running water. Virtually the same morphotype occurs throughout the simuliids in association with this one kind of environment. So structurally uniform are the larvae and pupae that absence of larval head-fans or the occurrence of supernumerary hooks in the pupal onchotaxy, or of the absence of a pupal cocoon, in the few species concerned, constitute major departures from the norm. Such departures—which tend to occur in species living in unusual environments, such as phoretic associations with other arthropods, isolated islands or subarctic streams with little suspended food—are often inconvenient for tidy diagnosis of supraspecific taxa and their appropriate taxonomic ranking (it is interesting here to reflect how different the classification of simuliids would be if only the larval stage had ever been discovered). Being based on characters from the adult, pupa and larva simultaneously (there are no usable taxonomic characters on the eggs), classification tries to ensure that appropriate taxonomic weighting is given to such deviations.

All species or supraspecific taxa are not necessarily difficult to identify in all the taxonomically usable life-stages (larger larvae, pupae, males and females). Certain species in the better-known faunas can occasionally be quickly and easily identified on some distinctive feature (e.g. pupal gill shape, adult scutal pattern, male genitalia), but it is unusual for a taxon to be equally distinct in the larvae, the pupae and both sexes of the adults. Some African species of *Simulium* subgenus *Pomeroyellum*, for example, are easy to identify by their unmistakable pupal gills, but females are indistinguishable one from another. Conversely, some South American species with very obviously distinct adults have almost identical pupae. However reliable identification of simuliids, because of taxonomic difficulties, remains a task of the specialist, and one that cannot easily be undertaken by the field worker except on a local basis when dependable keys exist.

These problems have been compounded by the recent discovery that many supposedly single species, defined by morphological criteria, are actually complexes of biologically distinct entities showing few, if any, detectable structural differences (sibling species). This discovery was not unexpected by taxonomists, to whom the limitations of morphotaxonomy are only too obvious. However, the fact that blackflies (like many other Diptera) are prone to sibling speciation is profoundly changing the approach to simuliid taxonomy and imposing new constraints on how far the museum taxonomist (dependent largely upon the morphological attributes of preserved specimens) can go in identifying species. Naming a specimen from its morphology because no other criteria are available remains a legitimate part of the everyday "bread-and-butter" work of the "classical" or "museum" taxonomist (morphotaxonomist); but the identification has to be understood as carrying the proviso that the morphospecies for which the name stands may be a composite of two or more biological species.

The changing demands of simuliid systematics do not imply the obsolescence of morphotaxonomy, but require the incorporation into the taxonomic system of data derived from other sources. Currently, the two most important of these are taxonomically-orientated cytological studies on chromosomal micromorphology (cytotaxonomy) and electrophoretic studies on enzyme systems (zymotaxonomy), both of which are valuable for the characterization of species and probably for higher classification (though the value of enzymes at the supraspecific level has not yet been shown).

Descriptive Morphotaxonomy

In spite of recent developments in cytotaxonomy, and latterly in enzyme studies, relatively few simuliids have yet been characterized other than by conventional morphological means, largely because many areas of the world remain undercollected and there is no information available beyond that which can be derived from studying inadequate samples of preserved material (as most blackflies are harmless, there has seldom been direct stimulus for their taxonomic investigation). Today simuliids are being studied at the morphotaxonomic level in various parts of the world and description of new species, especially in lesser-known faunas, is commonplace.

At present, the overwhelming majority of blackflies are still recognized and defined on characters of their external ("hard-parts") morphology, even recently-described species; not because taxonomists are satisfied that morphology alone provides absolutely reliable criteria of species distinctness, but because data are seldom available (except in intensively studied species-complexes) from other sources. Species limits cannot be determined yet from experimental genetics because of the lack of success in cross-mating blackflies in the laboratory. Thus, most so-called species in the taxonomic literature are *morphospecies*, differentiated from other such taxa by morphological discontinuities deemed to be evidence of specificity in one or more life-history stage and/or (in the case of adults) in either or both sexes. Obviously, species status can only be inferred, not proved, from such evidence. Nonetheless, most of the morphospecies recognized in simuliid taxonomy are probably reproductively isolated from one another, although each of them may itself be a complex of sibling species.

Excessive taxonomic splitting may have occurred from time to time because of the difficulty of differentiating between morphological criteria that are indicative of intraspecific variation and those of true interspecific distinctness. There are some supposedly distinct morphospecies masquerading under different scientific names that are almost certainly conspecific, especially in little-worked faunas (e.g. South America). A continuing task of morphotaxonomy, in the course of revisionary work, is to uncover such synonymies and weed out the invalid names, just as it is one of its prime functions to describe newly-discovered species. Only when the tedious and unexciting taxonomic groundwork of species description (alpha-taxonomy) has been accomplished is it possible to advance taxonomy to the next stage of synthesis and to prepare comprehensive keys and monographs (beta-taxonomy). For some major parts of the world, again notably South America, this second stage has not yet been reached for the fauna as a whole, although there have been important advances on restricted groups.

Inequalities in morphological taxonomic knowledge, and in the quality and degree of fragmentation of the literature (see p. 13), vary considerably from one region to another. Of approximately 1270 currently recognized morphospecies, about half are from the Holarctic Region, a quarter from the Neotropical Region, and the remainder from the Afrotropical Region and the Oriento-Australasian Regions combined. Faunistic comparisons of this kind need to be made cautiously because apparent differences in species complements may have different explanations. Australia, the most arid continent, supports a relatively small simuliid fauna (morphologically determined) which is clearly due to natural factors.

On the other hand, the Palaearctic Region, currently with approximately 500 morphospecies, appears to be very species-rich compared to the Nearctic Region. The apparent difference is misleading because the Palaearctic fauna (whilst undoubtedly larger) has been more finely split morphotaxonomically than that of the Nearctic.

It is not practicable to give a list of the many morphological characters of black-flies that are used in conventional taxonomy: a useful summary exists (World Health Organization, 1977, Annex 2). Almost any structure of the external morphology (which includes the sclerotized parts of the female terminalia and the structures of the male hypopygium) is utilized for taxonomy if its variation (presence or absence, shape, morphometric proportions, etc.) is such as to suggest that it has discriminatory value for separating and diagnosing taxa, whether at the species level or above. The anatomical features that are used most regularly in taxonomic definitions and keys include:

Adult: eye features, antennae, female frons and mouth-parts, thoracic colour and vestiture, scutal patterning, leg colouring and form, shape of female claws, wing-vein conformation and vestiture, abdominal colouring and vestiture, development of abdominal sclerites, female terminalia (including spermatheca) and male genitalia.

Pupa: gill form and branching, cephalothoracic ornamentation and vestiture, abdominal sclerites, abdominal onchotaxy (arrangement of hooklets and spine-combs), cocoon shape and texture.

Larva: cephalic fans and antennae, cephalic apotome, hypostomium, postgenal cleft, mandibles, head pigmentation, body shape and cuticular ornamentation (notably presence or absence of tubercular swellings and scale vesti-ture) rectal gills and posterior circlet.

Species-Complexes

Speciation in blackflies has not always been accompanied by substantial, or even detectable, morphological differentiation; species-complexes of biologically distinct but morphologically inseparable (isomorphic) taxa exist in the Simuliidae, and probably exist throughout the family. The unmasking of such complexes has been the most influential taxonomic achievement of recent years, and owes much to intensive cytological studies on pest and vector species (e.g. *Prosimulium "hirtipes" s.l.* and *Simulium venustum s.l.* in North America, and *S. damnosum s.l.* in Africa), and explains certain behavioural, ecological and parasitological observations. These had seemed contradictory at the intraspecific level, but they made biological sense once it became known that a complex of sibling species was involved in entities formerly regarded as single morphospecies—each sibling species having its own interspecific bio-ecological and behavioural peculiarities.

The term *species-complex* should be rigidly defined, but is used haphazardly by taxonomists and others to mean different things. Here it is only used to mean an assemblage of obviously monophyletic sibling species that are isomorphic or virtually so, and that can be identified *reliably* solely on non-morphological taxonomic criteria. Thus restricted to true sibling species, the term can then be usefully contrasted with *species-group*, an assemblage of species sharing a similar morphotype but, in the existing state of knowledge, utterly dependent upon external morphological characters for their identification. Since members of a species-group are separable morphologically (even though sometimes with difficulty), they are not isomorphic. Whether an assemblage of species constitutes a complex or a group depends on the nature of the taxonomic criteria used at any time for its analysis. The distinction is a convenient one: e.g. *Simulium damnosum sensu*

lato at present constitutes the *damnosum*-complex because its member species are unequivocally identifiable only by cytological techniques. *Simulium amazonicum* and its allies, on the other hand, constitute the *amazonicum*-group as only morphological criteria are yet available for species recognition.

The taxonomic resolution of species-complexes currently depends mainly upon the techniques of the cytotaxonomist studying the polytene chromosomes of the larval simuliid salivary glands. In general, these are the only taxonomically usable chromosomes in blackfly tissues, although Bedo (1976) has found easily demonstrable polytene chromosomes in pupae and adults of some Australian Simuliidae. The larval salivary gland chromosomes possess a wealth of micromorphological attributes that can be used for inferring specificity (and therefore revealing the existence of sibling species complexes) and for determining probable evolutionary relationships by cytological means (cytophylogeny), but not necessarily for inferring the antecedence of one species in relation to another.

Cytotaxonomy is now a well established major arm of simuliid taxonomy, in which one of its special virtues is clarification of species-complexes. Rothfels (1979) has provided a masterly up-to-date exposition of its capabilities and its findings.

The work of the cytologist on species-complexes (in a sense morphological as chromosomes are structural) has been supplemented by studies on isoenzymes, thus bringing a truly non-morphological approach to bear on simuliid complexes. Although in its infancy, this work already shows taxonomic promise. Townson and Meredith (1979) have reported potentially very important advances in the taxonomy of the *Simulium damnosum* complex as a result of enzyme electrophoresis and comparable enzyme studies are now underway in Canada and Panama. Together, the studies of the cytologist and enzymologist are permitting us to recognize and diagnose species that, for want of morphological differentiation, cannot be discriminated in any other way. It is convenient to call these entities *cytospecies* or *zymospecies*, in contrast to morphospecies, at least until morphological differences are discovered.

The most advanced taxonomy in the Simuliidae is currently that of the *S. damnosum* complex of African onchocerciasis vectors. It has virtually reached the gamma-taxonomic stage, defined (after Mayr 1969) as the level of taxonomy dealing with biological aspects of taxa ranging from the study of infraspecific populations to those of speciation and evolutionary trends. The highly integrated taxonomy that is fast developing for this complex, utilizing data of several different kinds (cytomorphological, biochemical, zoogeographical, etc.) has already shown convincingly that *S. damnosum s.l.*, in the old morphological sense, is really an aggregate of perhaps as many as 25 sibling species (Dunbar 1976) of which about six are widespread in West Africa, the area from which this complex is best-known at present. Some of these occur sympatrically and some show considerable seasonal fluctuations in geographical range, thus complicating identification.

Dependence on chromosome study for reliable identification is inconvenient when rapid recognition is needed, as in control operations against the *Simulium damnosum* complex. The occurrence of morphological features enabling accurate routine identifications (especially of adult females), under field conditions, of course obviates the need for a cytotaxonomist in day-to-day work. Intensified morphological study including use of morphometrics and scanning electron microscopy (SEM) sometimes reveals usable morphological characters previously overlooked or thought to lack interspecific significance. Much recent progress has been made towards morphological discrimination of members of the *S. dam-*

nosum complex, both as adult females (see pp. 45–56) and as larvae. Fairly reliable identification of some species, or at least of species-pairs, is becoming possible using external morphological characters correlating with those cryptic ones dependent upon chromosomes and enzymes.

It cannot be assumed, however, that morphological interspecific differences will necessarily be found in simuliid species-complexes if we look hard enough. Many sibling species-complexes are known in insects that have resisted the most searching attempts at morphological discrimination (e.g. the well-known *Anopheles gambiae* complex)—genuine sibling species are by definition isomorphic. Neither should it be assumed that cytological and enzymatic studies provide immediate answers to all the taxonomic difficulties of simuliid species-complexes. As Townson and Meredith (1979) point out, speciation has not necessarily been accompanied by enzyme divergence (four of the six common West African siblings of the *damnosum* complex are enzymatically indistinguishable), and there are confusing polymorphisms in some enzymes such as esterases (Thomas 1977) in blackflies from different localities. Also, there may be interspecific enzyme differences that are not shown by enzyme electrophoresis. It has been shown that of 43 enzymes so far tested on the *S. damnosum* complex, only two (PGM and trehalase) have taxonomic value, as they permit the species-pair, *S. squamosum*/*S. yahense*, to be separated from the other four common West African species and from one another. Enzymatic proof of species-status for the individual members of this species-pair, when the chromosomal evidence for separate specificity was inconclusive (Townson and Meredith 1979), shows the potential of zymotaxonomic techniques.

Lastly, virtual identity in chromosomes is not a guarantee of conspecificity. We know of species with grossly different phenotypes that have similar chromosomes (e.g. in Tahiti, the sympatric *Simulium tahitiense* with normal larval head-fans and *S. oviceps* with greatly reduced head-fans and accompanying changes of head shape).

Supraspecific Classification

The homogeneity of blackflies makes supraspecific classification difficult, and there is some disunity among specialists as to the number of entities worthy of formal recognition as named taxa and their proper ranking in the taxonomic hierarchy. The difficulty arises largely from the almost unbroken evolutionary continuum characterizing the morphological attributes of simuliids and evident from any comprehensive study of the world fauna. The problem may be hardly noticeable if a more limited fauna (e.g. that of a country or zoogeographical region) is studied taxonomically, when seemingly "good" characters can be found for differentiating supraspecific taxa with some precision. Frequently, these criteria fail when applied on a global basis. In fact, the taxonomist commonly finds that characters apparently sound for the diagnosis and discrimination of subgenera, genera and higher categories, cease to be so with the discovery of simuliids possessing morphological attributes intermediate between those defining two taxa. Thus, the more we collect simuliids and discover previously unknown forms, or try to integrate a patchily-developed local taxonomy into a world system, the harder it becomes to construct a satisfactory classification (conforming to the old paradox that classification is only easy when one does not have much material). In recent years, only Rubtsov (1974c) has been bold enough to propose a world classification, although other taxonomists have made important contributions to this end, recognizing similar taxa (in many respects) to those of Rubtsov. Most taxonomists, however, are not completely in favour of Rubtsov's system because of the high degree of splitting and promoting of rather uniform groupings of simuliids that it entails, and its consequent inconvenience to non-taxonomists. Nevertheless, it represents an important attempt to deal with family classification comprehensively.

The fact that Rubtsov's classification is rather impracticable for routine use in simuliid studies is important, for taxonomy must serve the needs of all biologists. On the other hand, it is an overstatement to consider, as Rothfels (1979) states, that "assignment of supraspecific rank is entirely a matter of practical convenience", which would be the case if simuliid classification was *only* intended to serve practical ends. In reality, it attempts to serve two rather conflicting needs—those of reflecting the imagined phylogeny of the group and simultaneously providing a convenient system for identification. Combining these purposes can be difficult, particularly when insights derived from new data (e.g. cytology or the application of strict Hennigian phylogenetic principles) lead to the conclusion that the accepted classification is "wrong", though convenient, because it does not properly portray evolutionary relationships. An excellent instance was lately published by Wood (1978), who concludes that the aberrant genera *Gymnopais* and *Twinnia*, in which larval head-fans are absent, are evolutionary derivatives of *Prosimulium* with normal head-fans, and that these two taxa should, at most, be ranked as subgenera of *Prosimulium*, due to their phylogeny.

However, recognizing the great phenotypic difference between *Gymnopais/Twinnia* and all other blackflies (substantial enough for Rubtsov to place *Gymnopais* and *Twinnia* in their own subfamily), Wood retains these taxa as full genera in accordance with past practice, writing that he has "chosen to rank them as full genera because . . . blackfly workers at the present time will find such an arrangement most acceptable". It is much to be hoped that other workers will adopt an equally commonsense approach whenever such substantial conflicts between "convenience" and "evolutionary" classification arise (as they are likely to do if Hennigian phylogeny is too rigidly applied to blackfly systematics with all its implications for ranking and nomenclature). This is not to say, however, that simuliid taxonomy should not, in general, be influenced or amended in the light of phylogenetic considerations, providing these are widely accepted as sound and that it is borne in mind that (without a fossil record) phylogeny is, at most, informed guesswork as to the course of evolution.

Genera or Subgenera?

The rank given in conventional taxonomy to a recognized supraspecific taxon is subjective, as there are no fixed criteria by which subgenera, genera or suprageneric categories are defined. In simuliid taxonomy, two systems are widely used for ranking aggregates of species. In one, the subgeneric category is not employed and species are classified into a large number of relatively small genera; in the other, relatively few genera are recognized, but some of the larger ones are divided into subgenera. These differing approaches cause a schism in the taxonomic literature.

The subgeneric alternative, although not originating with him, was initially adopted by Rubtsov (1940) in his first comprehensive monograph of the simuliid fauna of the USSR. In his second (1956) edition of this work, he altogether rejected his earlier approach in favour of the full generic system. Since then, Rubtsov (1959–1964; 1974c) has consistently used the full generic classification. In this respect, he is currently followed by taxonomic specialists in eastern Europe, Italy, Germany and Scandinavia (an exception is Raastad 1979). The approach to classification that recognizes subgenera within large genera, adopted for instance by Crosskey (1969), is nowadays universally used by taxonomists in anglophone countries, France, Japan and Latin America, although not all are in complete agreement on the ranking of every taxon (Crosskey recognizes, for instance, some subgenera within *Prosimulium* that others treat as full genera).

The greatest practical effect of the "genera *versus* subgenera" taxonomic dichotomy is to cause inconvenient, and sometimes confusing, disparities in the binomial nomenclature of species belonging to the genus *Simulium s.l.* (which includes most of the world's important blackfly pest and vector species). In the full generic classification of Rubtsov's school, the large and worldwide *Simulium s.l.* is divided into about 45 restricted genera. A large number of generic names therefore appear in the binomina of important pest species, e.g. *Boophthora erythrocephala*, *Edwardsellum damnosum*, *Lewisellum neavei*, *Notolepria exigua*, *Odagmia ornata*, *Psilopelmia ochracea*, *Wilhelmia equina*. Under the subgeneric system which maintains a comprehensive single genus, *Simulium*, all these species, being congeneric, are nomenclaturally in *Simulium*, as *S. erythrocephalum*, *S. damnosum*, *S. neavei*, etc. (Terminal orthography of adjectival specific names must agree with generic gender. Hence, neuter endings of species names in *Simulium* often need to change to feminine or masculine endings when the full generic system is adopted for *Simulium s.l.* segregates.)

Although Rubtsov (1940) was an early proponent of the subgeneric system he now (Rubtsov 1974c) attacks it strongly on the dubious grounds that it complicates nomenclature by making it trinomial instead of binomial and that a genus has an ideal size of about 10–20 species. Davies (1975) has drawn attention to the weaknesses in Rubtsov's rationale over these

and other classificatory matters. The subgeneric system has many merits, one of
the strongest of them being that it does
not complicate nomenclature (as does the
splitting into a multiplicity of genera)
because use of the subgeneric name is
always *optional*. The subgenus is there as
a designated category for taxonomic purposes but its name can be disregarded for
everyday purposes by the non-
taxonomist, for whom it is frequently an
irrelevance (though *Edwardsellum* is often
cited in relation to the *S. damnosum* complex).

There are other virtues to preserving
Simulium as a large cosmopolitan genus
divided into subgenera, rather than shat-
tering it into many separate genera. For
example, the genus can still be easily
recognized and correctly named as such
by non-taxonomists, for whom the refinements of taxonomy have little or no
interest. Also, the concept of one large
genus known by the familiar name,
"*Simulium*", stresses the essential biological and structural similarity characterizing most blackflies. Recourse to restricted
Rubtsovian genera leads to much difficulty in distinguishing many of these from
one another, as not all are satisfactorily
characterized in all life-history stages or
both sexes. It also means that there are
few, if any, true *Simulium* spp. in the
tropics and certainly none at all in tropical Africa and Australasia. On this reason-
ing, *Simulium* s.s. becomes almost entirely an Holarctic genus, all human
onchocerciasis vectors (with the possible
exception of *S. metallicum*) having to be
placed in other genera.

Higher Classification: Tribes and Subfamilies

As with genera and subgenera, there is
no agreed classification at the suprageneric level. The taxonomic weight
that should be attached to morphological
discontinuities between genera or groups

of genera poses similar barriers to agreement over higher classification and,
therefore, over the formal ranking and
recognition of categories. A concomitant
problem of restricted-genera classification is the erection of additional higher
categories (tribes, subfamilies) for their
accommodation, some of them being,
obviously, much more closely allied to
one another than to more isolated
genera. Rubtsov (1974c) therefore uses a
rather complex classification into four
subfamilies (Parasimuliinae, Gymnopaidinae, Prosimuliinae, Simuliinae) of
which one (Simuliinae) is divided into
five tribes (Austrosimuliini, Cnephiini,
Eusimuliini, Wilhelmiini, Simuliini): a
total of eight suprageneric taxa. This system derives essentially from giving high
taxonomic weight to small morphological
differences.

Crosskey (1969), using subgenera at
lower classificatory levels, proposes a
much simpler higher classification, in
which all blackflies are accommodated in
two subfamilies (Parasimuliinae, Simuliinae) of which the latter is divided into
two tribes (Prosimuliini, Simuliini), a total of three suprageneric taxa. This system
gives less taxonomic weight to small morphological differences, being influenced
more by similarity than diversity and by
the need for some practical convenience
in classification and identification. Few
other, currently active, taxonomists deal
with higher classification in any way,
although Peterson (1977a) adopts the
two-subfamily system (that stems originally from Smart 1945), and Wygodzinsky and Coscarón (1973) provisionally
accept Crosskey's use of the tribes Prosimuliini and Simuliini. They emphasize
that strictly phyletic analysis will show
that a clear-cut tribal distinction between
Prosimuliini and Simuliini cannot be
maintained. In this, they are merely
reiterating the statement made by Crosskey (1969) when first proposing to recognize the two tribal taxa, i.e. that "no hard
and fast line can be drawn between the
two tribes, and no single character exists

that will hold for distinguishing every form in one tribe from every form in the other". All recent work only confirms that it is currently impossible to find clear-cut characters of absolute applicability for all life-history stages of blackflies globally, and unequivocal tribal assignment is sometimes difficult. It remains convenient, however, to recognize a distinction, which undoubtedly reflects some kind of phyletic cleavage, between the "prosimuliines" (typically forms without spiniform vestiture on the wing veins, often with forked radial sector, with feeble mesepisternal sulcus, without a hind-leg pedisulcus, often with strong pupal abdominal sclerites and tail-hooks, and usually with very ill-formed cocoon) and the "simuliines" (typically forms with spiniform as well as hair-like vein vestiture, with unforked radial sector, with well-formed mesepisternal sulcus, with undifferentiated pupal abdominal sclerites and no definite tail-hooks, and usually with a discretely-formed cocoon).

The genus *Parasimulium*, although known only from males, is so disjunct from other blackflies that it is placed in its own subfamily (Smart 1945; Crosskey 1969; Rubtsov 1974c, Peterson 1977a), a point on which all specialists agree. The two Holarctic genera *Gymnopais* and *Twinnia*, aberrant because of the absence of larval head-fans, are treated by Rubtsov (1974c) as forming his subfamily, Gymnopaidinae. Crosskey (1969), though, placed them in the Prosimuliini, which is supported by Wood's (1978) convincing demonstration that these genera are derived from *Prosimulium*, and therefore that the Gymnopaidinae is an untenable subfamily.

Although it is not sufficiently pertinent to this book to consider higher classification in more detail, an important disadvantage of Rubtsov's system should be mentioned, namely, its unbalancing effect on the classification of Simuliidae in relation to other Nematocera. In Culicidae, Chironomidae and Ceratopogoni-dae, for example, in which there are relatively stable and widely accepted higher classifications, the higher taxa are recognized and defined by substantial morphological character gaps such as do not exist in blackflies. Rubtsov's use of many subfamilies and tribes implies degrees of morphological diversity between higher taxa equivalent to those in other Diptera, when, in reality, blackflies show a most marked structural homogeneity.

Nomenclature and Types

Nomenclature remains fundamental for the communication of biological ideas, but in simuliids as in other organisms, is sometimes subject to change in the light of new or modified criteria affecting taxonomic concepts. Occasionally, a changed nomenclature is required for species or other taxa that are cited in the non-taxonomic, as well as the taxonomic, literature. Paradoxically, non-taxonomists widely recognize the fluid nature of taxonomy and the inevitability of changing concepts, yet, apparently, expect these not to be reflected in a changing nomenclature. This is unrealistic, and changes in blackfly scientific names must be accepted as an integral part of a continually improving simuliid taxonomy. The vast majority of names in Simuliidae are familiar only to a specialized fraternity. None is a household word that needs preservation at all costs, so that name changes required by the *International Code of Zoological Nomenclature* (which are mainly due to the uncovering of misidentifications or previously-unrecognized synonymy) should be made whenever appropriate. The occasional well-established and important name, if threatened with technical invalidity under the *Code*, can be validated, if necessary, by application to the International Commission on Zoological Nomenclature. There are, however, few blackfly names where such a course would be fully justified (preservation of *Simulium* itself and *damnosum* are examples).

Simuliid taxonomy is currently moving in many faunas, from the early stage of haphazard species description to a more organized (beta-taxonomic) stage of synthesis, in which old types are studied and correlated with early stages, keys prepared, and species re-described to modern standards. This is a transitional phase in which misapplications of names (sometimes of longstanding) become known or in which it is found that two or more names, that have been in use for supposedly distinct species, all apply to one species. Immediately preceding a new era of nomenclatural stability, based upon careful revisionary taxonomy rather than ignorance, it is not uncommon for there to be such a stage, characterized by major name changes. Some non-taxonomists are attracted to the idea of maintaining the names with which they have become familiar, despite their invalidity under the rules of zoological nomenclature, on the grounds that they are hallowed by "usage" and to change them creates instability. Though superficially attractive, this course is ultimately self-defeating. It is preferable to accept the occasional name changes as a living part of the taxonomic process, just as changes are accepted in the taxonomic concepts for which the names stand.

To persist, knowingly, with an erroneous nomenclature, because to change causes temporary inconvenience to a small coterie of blackfly initiates, only delays the development of a truly sound and stable nomenclature based on an international code of practice.

Application of names being determined largely by types, it is relevant to comment briefly on these in relation to simuliid taxonomy and the ways in which it is changing, particularly because of the common assumption that the existence of sibling species-complexes nullifies the value of morphological types. Clearly, the value of a preserved blackfly type-specimen (whether adult, pupa or larva, or part thereof) is diminished once it is known that the morphospecies for which it stood as name-bearer specimen is, in reality, a complex in which constituent species have to be differentiated non-morphologically. However, as refined morphological studies sometimes show that siblings, initially thought to be structurally indistinguishable, can be separated on preserved parts of their anatomy, the museum type on a pin or in an alcohol jar always retains its potential as the standard of reference. It may come back into use as the yardstick against which to judge conspecificity of specimens and to determine the rightful assignment of a name.

The *International Code of Zoological Nomenclature* does not rule that a type must be designated for each new species described, but, in practice, types are routinely designated for all newly-named blackfly morphospecies and occasionally for cytospecies. Most types are adult flies, but a few are pupae or larvae if associated adults were unknown at the time of original description. A type need not be an entire specimen, and Vajime and Dunbar (1975) legitimately designated larval salivary gland chromosome preparations as types of their cytologically-characterized new species in the *Simulium damnosum* complex.

Nearly all simuliid nominal species have existing type-specimens, preserved in accessible collections. These are important in routine morphotaxonomy as guarantors for correct identification and naming and other purposes. At present, the morphological type-concept is applicable both in conventional museum taxonomy and in cytotaxonomy (chromosomes are preservable items of morphology in miniature), but it obviously will not be directly applicable for any new-named taxa that can only be characterized by non-morphological criteria. In isoenzyme studies, for instance, a specimen has to be destroyed, or largely so, before it can be identified, so a new sibling species revealed by enzyme electrophoresis cannot have a type in the conventional sense of a permanently preservable specimen. However, work of this kind often involves specimens for which the conspecificity is certain (e.g. because they are from the same egg-batch) so that a morphological type could be designated from a specimen indubitably belonging to the same sibling species as the specimens used for enzyme analysis.

The value of a morphological type for a species non-morphologically characterized may be nil at the time of original description, but its potential value—because it came from the originally-studied population—cannot be foretold. There is a good case for designating a type in any newly-described and named simuliid species irrespective of the criteria employed for its characterization; and for the most precise recording possible of the type-locality, as inadequate data on type-localities of old nominal species cause practical problems in simuliid taxonomy (e.g. when a morphospecies is found to be a sibling complex but has an unknown or vague original provenance, and the former morphospecies name needs to be applied to the restricted sibling species coming as nearly as possible from the original locality).

Taxonomic Literature, Keys and Catalogues

The taxonomic literature on Simuliidae, spanning some 220 years since the description of the first species (by Linnaeus in *Culex*), is now very large and more than half of it has appeared in about the last 30 years. There is no comprehensive bibliography, but the most useful recent starting point for the world literature is in Rubtsov (1959–1964). Almost all the literature is morphotaxonomic (including faunistic), but the fast-growing field of

cytotaxonomy has already a substantial literature to which Rothfels (1979) has provided the *entrée* with an up-to-date bibliography. The new field of enzyme taxonomy of simuliids has almost no literature as yet: Townson and Meredith (1979) have published their own preliminary findings on the *S. damnosum* complex and reviewed pertinent literature.

Identification Keys

Simuliid identification remains difficult for little-worked areas, and even for better-known faunas often needs the services of specialist taxonomists with access to reference collections, because of the close similarities between many species. Even the best keys are quickly outdated by advances in morphotaxonomic knowledge and description of new species. Care is needed, on the part of the field worker, when identifying specimens that do not belong to some well-known species. Discovery of "good" key characters is not always easy, especially for separating larvae; and the hydrobiologist has to recognize that, although frustrating, it is not always possible to identify with certainty every larva or pupa collected from a stream, however good the taxonomy. Over-reliance on the infallibility of keys has led some simuliid non-taxonomists to report major extensions of known species ranges which *may* prove correct but equally could be quite erroneous—because they are based on identification of only one life-history stage in isolation when other associated stages are essential for confirmatory identification.

The following selected references contain starting-point keys for the areas or taxonomic groups indicated:

AFROTROPICAL REGION: Crosskey (1960, West African larvae; 1969, supraspecific taxa); Freeman and de Meillon (1953, adults and pupae of regional fauna); Lewis and Raybould (1974, *Simulium neavei* group).

AUSTRALASIAN REGION: Colbo (1976, Australian *Simulium s.l.*); Crosskey (1967, Australian and western Pacific segregates of *Simulium s.l.*); Dumbleton (1972, *Austrosimulium*); Smart and Clifford (1965, New Guinea fauna).

NEARCTIC REGION: Davies *et al.* (1962, Eastern Canadian adults and pupae); Merritt *et al.* (1978, Michigan larvae and pupae); Peterson (1970, Canadian and Alaskan *Prosimulium*; 1977a, *Parasimulium*; 1978, genera); Stone (1964, north eastern fauna); Stone and Snoddy (1969, Alabama fauna); Wood (1978, *Gymnopais* and *Twinnia*); Wood *et al.* (1963, eastern Canadian larvae).

NEOTROPICAL REGION: Coscarón and Wygodzinsky (1972, *Simulium* subgenus *Pternaspatha*); Dalmat (1955, Guatemalan fauna); Vargas and Díaz Nájera (1957, Mexican fauna); Wygodzinsky and Coscarón (1973, Prosimuliini excluding genus *Gigantodax*).

PALAEARCTIC REGION: Carlsson (1962, Scandinavian fauna); Crosskey (1967, Middle East fauna); Davies (1968, British fauna); Dinulescu (1966, Rumanian fauna); Grenier (1953, French fauna); Knoz (1965, Czechoslovakian fauna); Lewis (1973, Pakistani fauna); Peterson (1977b, Icelandic fauna); Rivosecchi (1978, Italian fauna); Rubtsov (1959–1964, regional fauna; 1962a, bloodsucking species of USSR).

There are no recent comprehensive key works available for the Oriental Region (including Japan). All the keys of the above-cited sources are morphotaxonomic. Few cytotaxonomic keys have yet been published, the most significant being the chromosomal key of Vajime and Dunbar (1975) to West African species of the *S. damnosum* complex. Townson (unpublished) has prepared a key to the same species discriminating them as far as possible on their isoenzymes. Keys to females of man-biting species in human onchocerciasis areas are given by Crosskey (1973b).

Taxonomic Catalogues

The world catalogue of Smart (1945), containing about 800 names, is useful for tracing taxonomic literature up to 1944, but is otherwise outdated. It is not yet superseded by a new world catalogue, which would contain about twice that number of names, but most of the world fauna is covered by recent regional catalogues, *viz* Stone (1965) for America north of Mexico, Vulcano (1967) for the Americas south of the United States, Crosskey (1973a) for the Oriental Region, and Crosskey (1980) for the Afrotropical Region. The last is up-to-date, but the others are not fully complete because of very recent description of additional species.

The Future of Simuliid Taxonomy

The interest shown in the Simuliidae almost everywhere in the world today, stimulated largely by the need to control them and to gain a better understanding of their relationship to pathogenic organisms, is making for a new awareness of the fundamental role of taxonomy in all blackfly studies and of the importance of accurate identification of pest and vector species involved in biological or chemical control operations—whether in the laboratory or the field.

The demand for accurate species diagnosis and identification has already been largely responsible for bringing about major developments in the field of cytotaxonomy as well as for more critical appraisal of morphotaxonomy. Undoubtedly, still greater use will be made of cytological characters in the immediate future as cytotaxonomic studies are made of those blackflies of which the chromosomes have not yet been investigated. The value of chromosome studies in taxonomy is now so widely appreciated, and the techniques so widely practiced, that cytotaxonomy is already conventional taxonomy in the sense that it is in routine use as a counterpart of morphotaxonomy. Today, an unparalleled degree of cooperation exists between the simuliid morphotaxonomist, who has much to learn from the revelations of the cytologist, and the cytotaxonomist who recognizes that "chromosome information needs to be hung on a conventional [i.e. morphological] taxonomic skeleton" (Rothfels 1979). In the next few years, this intermarriage of specialists is destined to produce rapid advances in simuliid taxonomy, not only in clarification of species-complexes, but through improved classification as the evolutionary relationships of blackflies become better understood. The merits of an integrated approach to taxonomic problems are exemplified by the North American simuliid fauna, in which about 14% of currently recognized species were first detected cytologically, and in which other presumed new species have been found by morphotaxonomists with new insight into the significance of small morphological discrepancies previously disregarded.

The future development of a highly integrated taxonomy is bound to require the participation of the enzyme specialist. Also, taxonomically-oriented enzyme studies of blackflies are likely to be a fast-expanding field, especially (as like chromosome studies) they lend themselves better to postgraduate theses than does morphotaxonomy. Although still at an early stage, zymotaxonomy is likely to make a major contribution to simuliid taxonomy, helping to offset the persisting lack of data from experimental genetics. It is much to be hoped that the problems of cross-mating and colonizing (pp. ix, 299–315) blackflies will soon be solved, for direct genetical data on species-limits would be taxonomically invaluable. It is less certain whether behavioural, physiological and other kinds of non-morphological characters can be discovered and usefully incorporated into the taxonomy in the near future.

However, there is no question that elucidation of isolating mechanisms in blackflies would be of great assistance in species-level taxonomy.

An enormous amount of routine morphotaxonomic work remains to be done, not only in under-collected parts of the world, but also in the best-known faunas. We can anticipate the appearance, in the next few years, of many descriptive taxonomic contributions enlarging our overall knowledge of blackflies and describing new species—providing there are the taxonomists willing and able to undertake the humdrum drudgery, over many years, that is always involved in major revisionary work. This is the main uncertainty for the future. Enzymes and chromosomes are more glamorous than "hard-parts" morphology, and few morphotaxonomists are on the scene to replace the dwindling band of present day taxonomists whose knowledge extends beyond localized faunas.

An Outline of Simuliid Classification

Because of the complexities of blackfly classification, the lack of an agreed system and the difficulty of expressing an uncertain phylogeny in tabular form, only the following simple outline of classification is given here as a means of ready reference for the non-taxonomist. Although specialists agree that certain genera (or subgenera) are evolutionarily much more closely allied than others, it is not practicable to place them in an order that expresses all the nuances of relationship. An alphabetical order is therefore used. The summarized distributions against each genus and subgenus serve to amplify the general account given in the "Geographical Distribution" chapter. Genera and subgenera treated as valid are in bold italics (latter indented) and synonyms in italics.

Subfamily PARASIMULIINAE
Parasimulium Malloch . . . . western USA
   Astoneomyia Peterson . . . . western USA
   Parasimulium s. str. . . . . western USA
Subfamily SIMULIINAE
Tribe PROSIMULIINI
Araucnephia Wygodzinsky & Coscarón . . . . central Chile
Araucnephioides Wygodzinsky & Coscarón. . . central Chile
Cnephia Enderlein. . . . . Holarctic Region (incl. North
                                             Africa)

Astega Enderlein
Cnesia Enderlein . . . . southern South America
Cnesiamima Wygodzinsky & Coscarón . . . . southern South America
Crozetia Davies . . . . Crozet Islands
Ectemnia Enderlein . . . . eastern North America
Gigantodax Enderlein . . . . Andean South America &
                                             Meso-America

Archicnesia Enderlein
Greniera Doby & David . . . . northern Holarctic Region
Gymnopais Stone . . . . northern Holarctic Region
Lutzsimulium d'Andretta & d'Andretta . . . . southern Brazil & northern
                                             Argentina
Mayacnephia Wygodzinsky & Coscarón. . . . western Nearctic Region to
                                             Guatemala
Metacnephia Crosskey[1] . . . . Holarctic Region (incl. North
                                             Africa)
Paraustrosimulium Wygodzinsky & Coscarón . . southern South America
Prosimulium Roubaud[2] . . . . Holarctic & Afrotropical Regions
   Distosimulium Peterson . . . . northern Nearctic Region
   Helodon Enderlein . . . . northern Holarctic Region
   Paracnephia Rubtsov . . . . southwestern Africa
   Parahelodon Peterson . . . . northern Nearctic Region
   Procnephia Crosskey . . . . eastern & southern Africa
   Prosimulium s. str. . . . . Holarctic Region (incl. North
                                             Africa)

   Ahaimophaga Chubareva & Rubtsov[3]
   Hellichia Enderlein
   Mallochella Enderlein (preocc.)
   Mallochianella Vargas & Díaz Nájera
   Taeniopterna Enderlein
   Urosimulium Contini . . . . western Mediterranean
                                             Subregion
Stegopterna Enderlein . . . . Nearctic & northern Palaearctic
                                             Regions
Sulcicnephia Rubtsov . . . . southwestern Asia
Tlalocomyia Wygodzinsky & Coscarón . . . . Mexico
Twinnia Stone & Jamnback . . . . northern Holarctic Region
Tribe SIMULIINI
Austrosimulium Tonnoir . . . . Australia & New Zealand
   Austrosimulium s. str. . . . . Australia & New Zealand
   Novaustrosimulium Dumbleton . . . . . Australia

Simulium Latreille . . . . nearly cosmopolitan (not New Zealand)

    Afrosimulium Crosskey[4] . . . . southern Africa
    Anasolen Enderlein . . . . Afrotropical Region
    Byssodon Enderlein . . . . Holarctic & Afrotropical Regions
    Echinosimulium Baranov
    Gibbinsiellum Rubtsov
    Psilocnetha Enderlein
    Titanopteryx Enderlein
    Chirostilbia Enderlein . . . . Neotropical Region
    Trichodagmia Enderlein
    Crosskeyellum Grenier & Bailly-Choumara . North Africa
    Dexomyia Crosskey . . . . Saint Helena Island
    Ectemnaspis Enderlein . . . . Neotropical & southern Nearctic Regions
    Edwardsellum Enderlein . . . . Afrotropical Region (excl. Malagasia)
    Eusimulium Roubaud . . . . Holarctic Region (incl. North Africa)
    Freemanellum Crosskey . . . . Afrotropical Region
    Gomphostilbia Enderlein . . . . Far East & Australasian Region
    Nipponosimulium Shogaki
    Grenierella Vargas & Díaz Nájera . . . . southern South America
    Hearlea Vargas, Martinez Palacios & Díaz Nájera . . . . Nearctic & northern Nectropical Regions
    Hebridosimulium Grenier & Rageau . . . Fiji & Vanuatu
    Hellichiella Rivosecchi & Cardinali . . . . northern Holarctic Region
    Hemicnetha Enderlein . . . . Nearctic & Neotropical Regions
    Dyarella Vargas, Martinez Palacios & Díaz Nájera
    Himalayum Lewis . . . . Himalaya Mountains
    Lewisellum Crosskey . . . . tropical Africa
    Meilloniellum Rubtsov . . . . Afrotropical Region
    Metomphalus Enderlein . . . . Afrotropical Region
    Montisimulium Rubtsov . . . . central Asia
    Morops Enderlein . . . . Australia & New Guinea Subregion

    Pselaphochir Enderlein
    Nevermannia Enderlein . . . . major regions (excl. Neotropical)
    Chelocnetha Enderlein[3]
    Cnetha Enderlein[3]
    Cryptectemnia Enderlein
    Inseliellum Rubtsov[3]
    Pseudonevermannia Baranov
    Stilboplax Enderlein
    Notolepria Enderlein . . . . Neotropical Region
    Obuchovia Rubtsov . . . . southern Eurasia
    Parabyssodon Rubtsov . . . . northern Holarctic Region
    Phoretomyia Crosskey . . . . tropical Africa
    Pomeroyellum Rubtsov . . . . Afrotropical Region

Psaroniocompsa Enderlein   .   .   .   .   . Neotropical Region
Psilopelmia Enderlein .   .   .   .   .   . Neotropical & western Nearctic
                                                        Regions
Lanea Vargas, Martinez Palacios & Díaz Nájera
Psilozia Enderlein .   .   .   . Nearctic Region (also Faeroe Is)
Neosimulium Vargas, Martinez Palacios & Díaz
     Nájera
Pternaspatha Enderlein   .   .   .   .   . Andean South America
Acropogon Enderlein
Dasypelmoza Enderlein
Schoenbaueria Enderlein   .   .   .   .   . northern Palaearctic Region
Miodasia Enderlein
Shewellomyia Peterson.   .   .   .   .   . eastern North America
Hagenomyia Shewell (preocc.)
Simulium s. str. .   .   .   . Holarctic, Oriental & northern
                                                        Neotropical Regions

Cleitosimulium Séguy & Dorier[3]
Danubiosimulium Baranov
Discosphyria Enderlein
Gnus Rubtsov[3]
Gynonychodon Enderlein
Odagmia Enderlein[3]
Phoretodagmia Rubtsov[3]
Phosterodoros Stone & Snoddy[3]
Pliodasina Enderlein
Pseudodagmia Baranov
Pseudosimulium Baranov (preocc.)
Tetisimulium Rubtsov .   .   .   .   .   . southern Palaearctic Region
Friesia Enderlein (preocc.)
Wilhelmia Enderlein   .   .   .   .   . Palaearctic Region (incl. North
                                                        Africa)
Xenosimulium Crosskey   .   .   .   .   . Madagascar & Comoro Islands

---

[1] This is a clearly characterized genus cytologically (Rothfels, 1979) that appears on recent evidence to be better placed in Prosimuliini than Simuliini.

[2] Subgenera shown in *Prosimulium* mainly follow Crosskey (1969) and Peterson (1970) but lack of taxonomic balance is recognized.

[3] Some specialists recognize valid genera or subgenera under these names.

[4] Here given subgeneric rank but hitherto treated as a genus.

Cytotaxonomy: Principles and their Application to some Northern Species-Complexes in *Simulium*

K. Rothfels

Sound taxonomy is fundamental to all other biological studies, for even the best observations are difficult to evaluate unless they can be referred to known and pure biological species. The chromosome studies of the past few decades beginning with those on the *Prosimulium hirtipes* complex (Rothfels 1956; Basrur 1959, 1962; Rothfels and Freeman 1977) and *Eusimulium aureum* (Dunbar 1959) have shown that, in the Simuliidae in general, the morphospecies of the taxonomist comprise a number of reproductively isolated, biologically distinct sibling species, sometimes referred to as *cytospecies*. They have also shown the need for the integration of the taxonomy of diverse faunas, particularly the Holarctic ones. A recent survey of the status of cytotaxonomic studies in blackflies is given in Rothfels (1979).

The present chapter proposes to delineate briefly the features and principles underlying the chromosome methodology, drawing on current studies of *Simulium vittatum* for examples. It will also update analyses for some other species-complexes in *Simulium*, both of economic importance (*S. venustum/verecundum*) and, as of yet, primarily theoretical significance (*S. tuberosum* and *S. argyreatum/decorum*).

Principles of Sibling Species Recognition by Cytology

Study of the giant polytene larval salivary gland chromosomes is uniquely able to reveal the existence of sibling species for the following reasons:

1 In their banding pattern, the polytene chromosomes possess enormous morphological detail with sequential specificity unaffected by convergence phenomena.

2 Sibling species generally differ by inversions, interchanges or other rearrangements in the banding sequence of one or more of their chromosomes. These are typically haploid $n = 3$, numbered I–III in decreasing order of size, with distinct long (L) and short (S) arms. The salivary gland chromosomes, as somatically paired double structures, reveal heterozygosity or hybridity for an inversion by local failure of pairing, if the inversion is small, and by "reversed loops", if it is large. When rearrangements are homozygous, i.e. present in both constituents, they can be discovered only by comparison with photographs or drawings of the standard or reference sequence.

3 The principal ways in which sibling species differ chromosomally are: (a) fixed sequence differences, commonly inversions or interchanges; (b) sex chromosome differences, for in many species the X and Y chromosomes are differentiated in their banding sequence (male heterogamety) and these differentiated sex chromosome systems are species-

specific; (c) inversion polymorphisms, i.e. the occurrence of alternative banding sequences that may occur heterozygously. Some such polymorphisms may be shared among siblings when they may differ in the frequency of the inverted sequence, yet others may be restricted to single species. Additionally, the occurrence of B chromosomes, band polymorphisms, and features of male meiosis, may also provide cytological criteria of value.

4 Cytological studies provide evidence for biological species-status of sibling species, directly in sympatric situations through the absence of hybrids, and inferentially in allopatric situations if cytological parameters of putative sibling populations have been determined over a sufficient geographical range to rule out the possibility that inversion clines or gradients exist, creating a semblance of species differences between widely separated populations.

5 Cytological studies can generate phylogenies based on sharing of sequences and subsequent divergence by successive unit rearrangement steps.

Simulium vittatum

Of all the species of blackflies that have been studied cytologically in any depth, *S. vittatum* has been cited as the one glaring exception to the universal existence of sibling species. This conclusion was based on the study of Pasternak (1964) who found *S. vittatum* to be a single species throughout its range in North America, though highly polymorphic both cytologically and morphologically. He noted one peculiar inversion interaction however: the Y chromosome inversion IIIL–1 showed mutual exclusion with the inversion designated as IS–7, almost always in individual larvae and frequently on a population basis. That is, virtually without exception, individuals that carried IIIL–1 did not carry

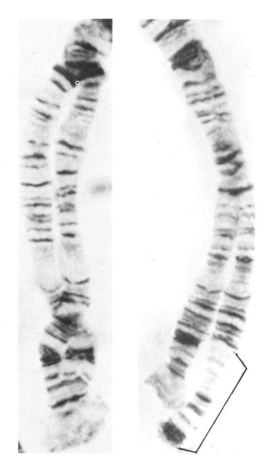

Photomicrograph of the sex chromosomes of the S. vittatum IIIL-1 sibling. On the left, the IIIL st/st sequence of the female (XX), on the right, the IIIL st/1 sequence of the male (XY). The bracket denotes the limits of the IIIL-1 inversion.

IS–7 heterozygously or homozygously. To explain this phenomenon, Pasternak invoked a lethal interaction between the two inverted sequences. Following Bedo's (1979) convincing demonstration of homosequential sibling species of *Simulium ornatipes* in Australia, differing, not in fixed rearrangements or sex chromosomes, but only in relative frequency of inverted sequences, R. Feraday (personal communication) suggested that a similar situation in *S. vittatum* would explain the mutual exclusion phenomenon, if IIIL–1 were restricted to one sibling and IS–7 to

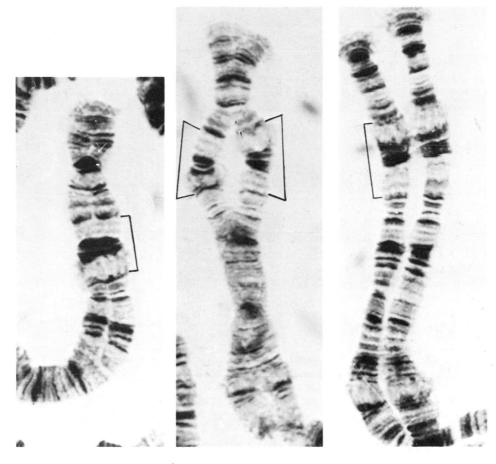

Photomicrograph illustrating the autosomal IIIS-2 polymorphism in S. vittatum. *Right, the standard homozygote; centre, the heterozygote; left, the inverted homozygote. Brackets show the limits of the inversion.*

the other. Accordingly, the situation was re-examined, with D. Featherston, in a population from Caledon, Ontario, in which Pasternak (1964) found both IS–7 and IIIL–1 with high frequencies. It became immediately clear that IS–7 not only showed exclusion with IIIL–1 but was, in fact, the differentiated sex (X) chromosome of a second sibling species. Grossly and provisionally, the two siblings may be characterized as: (1) *IIIL–1 sibling:* males IIIL–st/1, females IIIL st/st; IS–7 absent. (2) *IS–7 sibling:* males IS–st/7, females 7/7; IIIL–1 absent. Both siblings exhibit a very large number of poly-

morphisms most of which are shared. At least five of these polymorphisms are sufficiently common to be useful in analysis and all of these exhibit widely different frequencies in the two siblings.

The existence of separate sex chromosome systems and the maintenance of highly differential inversion polymorphism profiles in larvae which coexist in the same creek at the same time and in approximately equal numbers argues conclusively for the presence of two sibling species. Their reproductive isolation must be due to failure to interbreed and not to zygotic lethality. If it were the

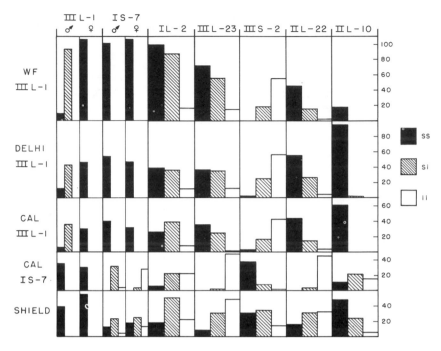

Histogram showing numbers of the standard homozygotes, heterozygotes and inverted homozygotes for the Y chromosome sequence IIIL-1, the X (Caledon) or autosome (Shield) sequence IS-7 and five autosomal inversions (designations from Pasternak, 1964). WF = Waterford, Ontario—IIIL-1 sibling; DELHI = Delhi, Ontario—IIIL-1 sibling; CAL = Caledon, Ontario—IIIL-1 and IS-7 sibling; SHIELD = pooled sample from Kahshe R., Muskoka and Costello Cr., Algonquin Pk, Ontario. (Norah Nolan, del.)

latter, and mating were essentially random, an intolerable genetic load would be imposed on the population. For instance, random mating, which would be 50% *inter*-sibling, would produce progenies at least one half of which would have to be eliminated in each generation because they combine IIIL–1 and IS–7 or because they are in other ways "improper" with regard to sex chromosome constitution.

What is the distribution of these two siblings within the range of the species (North America, southern Greenland and Iceland), and are there others? Pasternak (1964) recorded the IIIL–1 inversion from Ontario, New York State, and Wisconsin, and we can add records for Quebec and a number of additional sites in New York State (E. Gordon, coll.). None of Pasternak's collections from

Alaska, Saskatchewan and Manitoba, or ours from Alberta, had any IIIL–1 individuals. Thus, evidently, the IIIL–1 sibling is eastern. In contrast, IS–7, according to Pasternak (though occurring also in Ontario, Vermont and New Hampshire), characterizes all the western samples and we have found that a population from Prince George, B.C (B. McKague, coll.) is like the IS–7 sibling from Caledon, both in the sex chromosome system and the kinds of inverted sequences that are present. A population from Edmonton, Alberta (D. Craig, coll.) is also basically IS–7 sibling, though the Y chromosome carries an additional large basal inversion in IS which has not been found in the east.

Pasternak's (1964) records are difficult to interpret, for the data that relate IS–7 constitution to sex are not available. We have found that populations from the Canadian Shield in southern Ontario (Kahshe R., Costello Cr. and, probably, Kearney Lake outlet) are characterized by having IS–7 as an autosomal polymorphism, both males and females exhibiting all three classes of zygotes st/st, st/7, 7/7, with similar frequencies. Further, with respect to all other polymorphisms, these populations are intermediate between the southern Ontario IIIL–1 and IS–7 populations. Sex chromosomes are not known. Because his data are not broken down according to sex, we cannot say whether Pasternak's western IS–7 populations are the Caledon or the Shield type.

As to the interpretation of the populations from the Shield, two possibilities exist. Either these are forms of the IS–7 sibling in which the inversion has not yet become firmly associated with the X chromosome (such gradient situations are known both in blackflies and chironomids)—or they may belong to an altogether separate sibling. Shield populations thus illustrate, in miniature, the problems posed in allopatric comparisons. Our present feeling is that they represent a distinct sibling, precursor to the other two, that evolved following (1) the invention of the IIIL–1 Y chromosome and a co-adapted X, and (2) the restriction of IS–7 to X with a co-adapted standard sequence Y, and with shifts in the frequencies of the ancestral autosomal polymorphisms in opposite directions. A pleasing aspect of the salivary gland chromosome studies is that they allow observational discrimination between the alternative interpretations by intensive sampling in critical (intermediate) areas. Obviously, in view of the enormous range of *S. vittatum*, and the very limited sampling that has been done, other siblings may remain to be discovered. What is basically required for proper analysis of this or any other species complex in the Simuliidae is a cytological working over throughout the distributional range.

Simulium tuberosum

Simulium tuberosum, described by Lundstroem from Finland, has had a varied taxonomic history centering on the relationship between the Palaearctic and Nearctic forms. The latter, at one time, were given the specific epithet *perissum* (among others) but are now considered conspecific with Old World *S. tuberosum*. The situation is clearly more complicated, for Landau (1962) showed that four (or five) sibling species of *S. tuberosum* exist sympatrically in southern Ontario. Cytologically, all known siblings are set off from all related *Simulium* (e.g. *venustum/verecundum*), in that the nucleolar organizer is in the base of IIIS. Landau showed that, cytologically, the four (or five) siblings are extremely close. The basic autosomal sequences are identical for all siblings in all chromosome arms and some floating inversions are still shared among pairs of triplets of siblings. The distinction among the four common siblings rests solely on their sex chromosome systems based on inversions of IIS described as A, B, C, etc., each letter designating a single inversion step, differentiating, in general, X and Y of the sibling species as follows: (1) *CDE*: ♂♂ CDE/A, ♀♀ A/A. (2) *AB*: ♂♂ St/AB, ♀♀ AB/AB. (3) *FG*: ♂♂ St/FG, ♀♀ FG/FG. (4) *FGH*: ♂♂ and ♀♀ FGH.

Landau's findings raised a number of problems, namely: (1) which of the Nearctic siblings, if any, is the true *S. tuberosum* of Europe? (2) what is the geographic distribution of Landau's siblings and do others exist in North America? and (3) what is the situation in Europe? G. Mason (personal communication) obtained some additional data which are shown with Landau's records in Table 1. The only available European samples are from seven Norwegian populations (Raastad, coll.) and all of these were *AB* sibling with the typical sex chromosomes X = AB and Y = standard, and the basic sequence of the North

Table 1. Distribution of *S. tuberosum* sibling species. Data of Landau (1962) and G. Mason (unpublished).

Location	CDE	AB	FG	FGH	Collector
Ontario					
South (many sites)	+	+	+	+	R. Landau
					G. Mason
Central (many sites)		+	+	+	R. Landau
					G. Mason
Quebec, Laur. Pk				+	R. Landau
James Bay					
Sakami		+			C. Back
Eastmain	+	+			C. Back
Caniapiscau		+·			C. Back
Newfoundland		+			M. Colbo
Churchill, Manitoba		+			R. W. Dunbar
New Hampshire	+	+	+	+	J. F. Burger
(several sites)					
Arkansas (three sites)				+	W. S. Procunier
Norway (seven sites)		+			J. Raastad

American *AB*, though floating inversions appear distinctive. Until evidence to the contrary becomes available, we assume that *S. tuberosum*, from the Finnish type locality, is *AB*, like the Norwegian ones, and conclude that the *AB* sibling is the true *S. tuberosum*. Like all other truly Holarctic forms, it is northern in distribution, occurring at Fort Churchill, Manitoba, at three James Bay, Quebec, sites and in Newfoundland, though extending south at least as far as southern Ontario and New Hampshire. Mason's study so far failed to reveal new siblings of *S. tuberosum* in North America, but indicated that certainly the *FGH* sibling and, possibly, *FG* and *CDE* also have rather wide distributions.

Simulium argyreatum/decorum

Members of this species-complex occur both in the Old and New World and their relationship has been the source of considerable controversy. It is currently considered that the European species, *S. argyreatum* (*quondam noelleri*), is distinct from the North American *decorum, inter alia* because it lacks the unusually highly arched dorsum of female adult *S. decorum*, though, structurally and biologically, the two species are extremely close. In North America, Shewell (1958) suggested, on morphological grounds, that a distinction exists between the southern, typical *S. decorum* and a northern form "near *decorum*" (characterized by a flattened thorax of the adult female, as in *S. argyreatum*) reaching its southern limit at Churchill, Manitoba. In addition, there is a recurrent suggestion of ethological (mating behaviour) differences in *S. decorum* from southeastern Canada and certain New England states (J. Edman, personal communication).

In a cytological survey of the complex, R. Feraday (personal communication) could immediately distinguish three major entities: *S. argyreatum* from Finland (K. Kuusela, coll.) possibly conspecific with collections from central Europe (J. Grunewald, coll.); a separate sibling represented by several Fort Churchill, Manitoba, collections (R. Jefferies); and, the south central Ontario and northern USA *S. decorum*. His findings are summarized in Table 2. The basic sequence is identical in all populations of the complex, except that the European and

Table 2. Chromosome characteristics and distribution of taxa in the *S. argyreatum/decorum* complex (R. Feraday unpublished).

Location	IL heavy band	Sex chrom. arm	Autosomal inversion polymorphisms						
			IIIL–5	IIL–11	IIL–10	IIL–3	IIL–4,5	IIL–4,5,12	IIL–9
Finland	—	IIS	135/168	149/214	135/194	—	—	—	—
Norway	—	IIS	1/8	3/10	0/10	—	—	—	—
UK	—	?	?	?	?	—	—	—	—
Germany/Austria	—	None	—	0/34	1/44	—	—	—	—
Churchill, Manitoba	—	IL	—	42/62	40/62	—	—	—	—
James Bay, Quebec; Newfoundland	+	IIIL	—	0/20	13/16	—	—	—	—
Ontario (Central); NY and NH	+	IIIL	—	—	—	163/190	—	—	—
Ontario (South); Saskatchewan	+	IIIL	—	—	5/182	42/182	77/182	57/182	1/182

Churchill collections lack a very characteristic, heavy band in the base of IL, for which all North American collections (other than Churchill) are homozygous and which may have been transposed to its site by a small inversion. There is a very strong suggestion that, in the Finnish collections (three sites), the IIS arm is sex-determining and that two X chromosomes exist—X_0 with the basic sequence B and X_1 with the overlapping sequence K, whereas all Y chromosomes are B. It is known that K also occurs in Norway (Raastad, coll.). The central European (German, Austrian, Swiss) samples lack K or other recognized sex chromosome inversions. IIL–10, IIL–11 and IIIL–5 are characteristic floating inversions of the north European samples.

In contrast, in the Churchill populations IL functions as the sex arm, X is standard X_0, and the Y is always characterized by a complex basal rearrangement, i.e., the standard Y_0 has been displaced completely. The floating inversions, IIL–10 and IIL–11, characterize the Churchill populations as they do the north European ones, but IIIL–5 appears to be absent.

The remaining (southern) population of the complex in North America, already united by the heavy IL base band, further share a differentiated sex chromosome system based on IIIL. The X chromosomes are standard, or carry one additional inversion. The Y chromosomes are distinguished by two overlapping inversions, one of which is perinucleolar. Within this aggregate there are indications of three (geographical) subdivisions on the basis of floating inversions. Populations from James Bay (three sites) and Newfoundland (one) share IIL–10 and IIL–11 of the north European and Churchill populations. Populations from central Ontario, New York and New Hampshire, are characterized by the lack of IIL–10 and the prevalence of IIL–3. Finally, in populations from southern Ontario (and the few available larvae from Saskatchewan) the characteristic sequences are the overlapping IIL–4,5 and its near tandem IIL–4,5,12, which have not been found elsewhere. In 182 constituents, the standard IIL sequence did not occur, IIL–3 is common and IIL–10 very rare. In view of the paucity of the data it would be premature to speculate on the reality and status of the three sub-groups recognized in temperate North America. The clearly established proposition is that the north European, the Churchill and the temperate North American collections represent three different species, each with its own sex chromosome system based on a different

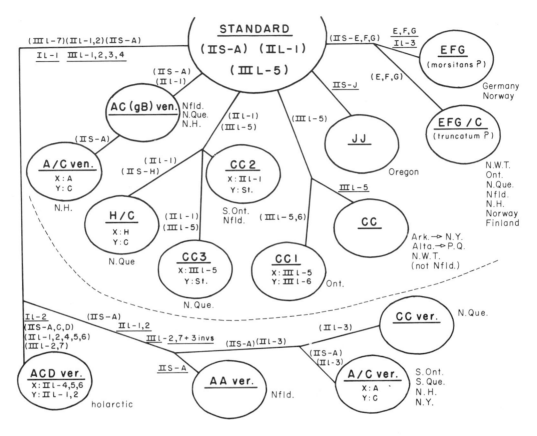

Cytophylogeny of the S. venustum/verecundum *complex. The phylogeny is to be read from Standard, which is a central arrangement in* Simulium. *Symbols in brackets denote inversion polymorphisms (floating inversions) coded for chromosomes (I-III) and arms (S-L). The fate of floating inversions is indicated along connecting lines following a branch point (carried floating—in brackets; fixed— underlined; lost—no longer shown). Designation of sibling species is according to their IIS sequence(s). X and Y chromosome sequences where known are indicated. Above the broken line, the sibling species related to* S. venustum; *below, those related to* S. verecundum. *(Norah Nolan, del.)*

member of the haploid complement, as in the three sibling species of the *S. pictipes* complex (Bedo 1975).

Simulium venustum/verecundum

This complex includes the main biter of man in North America and is, therefore, a prime target for control. For some aspects of its long and confused taxonomic history, see Rothfels *et al.* (1978).

That paper shows, in a first cytological survey, that in *verecundum* at least two sibling species exist: *ACD* and *AA–A/C* complex, which, in turn, might comprise two or even three siblings. In *S. venustum* at least seven sibling species were recognized; these were symbolized as CC, CC1, CC2, Ac(gB), A/C, EFG/C and EFG. Three new sibling species in *venustum* are described here, one from Oregon and two from James Bay, Quebec, and a revised phylogeny of the entire complex is presented.

JJ, The Oregon Sibling

This sibling species of *venustum* is represented in a substantial collection (L. Newman) from the Nehalem River. It differs from standard only in the single IIS inversion J which mimics F, but is larger, having both its breakpoints outside those of F. In regard to IIIL, the Oregon sibling is like *ACD* and *EFG* in not having IIIL–1,2,3,4 or *ACD*. Two rare subterminal floating inversions were found, one in IIIL and one in IIL. Sex chromosomes could not be distinguished. This sibling is closest to *CC*, from which it differs in only two inversions, IIS–J and lack of IIIL–5. Specific distinctness from *CC* is indicated however, since populations of typical *CC* have recently been obtained from a number of "quasisympatric" sites in Oregon (L. Newman, coll.) and from British Columbia (B. McKague, coll.).

The other two new siblings designated *H/C* and *CC3* [CC(gB)B] were obtained in 1977 and 1978 by C. Back (personal communication) in single large collections from Caniapiscau, James Bay, Quebec, remarkably without contamination by any other sibling. The two siblings may be differentiated from each other and from all other known *S. venustum* siblings as follows:

H/C Sibling

The standard sequence occurs in all arms; IIS is the sex arm. The standard sequence (C) is restricted to the Y chromosome, H being the universal X chromosome. Thus, males are always HC and females HH. The H sequence carries the Balbiani Ring very close to the IIS end (much closer than A). IIL carries the IIL–1, or grey Band (gBd, inversion as a common polymorphism. The inverted sequence predominates to the extent of approximately 73%, i.e., most individuals carry gB either heterozygously or homozygously. The IS sequence is always standard as are IL and IIIS. IIIL is basically standard, but carries a number of inversion polymorphisms that are limited to this species. B chromosomes were never found associated with the H sequence.

CC3 [CC(gB)B]

The standard sequence occurs in all arms of this sibling as well. IIIL is the sex-determining arm and the IIIL–5 inversion marks the X chromosome (with approximately 2% IIIL standard exceptions) and C remains as the Y chromosome. Thus, males are St/5 and females 5/5. In IS, the standard sequence occurs but a subterminal inversion, including sections 2–5 of Rothfels *et al.* (1978), predominates to the extent of 70% of the constituents. Thus, most individuals carry at least one inverted chromosome. In IIL, the gB polymorphism is common as in *H/C* with an inverted constituent frequency of about 65%. IIIL again carries a number of polymorphisms restricted to this species. The remaining arms, IL, IIS, IIIS, are standard. This species has the highest number of B chromosomes known in any blackfly with approximately 80% of the larvae having at least one B chromosome and with a mean B chromosome frequency per larva of about 1.7.

The *venustum/verecundum* phylogeny has been revised to include the three new siblings and to accommodate the findings that some *CC2* populations have retained the standard sequence in IIIL (as opposed to IIIL–5), and that *AC verecundum* populations seem to be of three types—*AA ver.*, *A/C ver.*, and *CC ver.* Though these are shown separately, it is still uncertain whether these forms are specifically distinct. The phylogenetic scheme is only one of several possible ones, basically because the occurrence of at least three sequences (IIS–A, IIL–1, and IIIL–5) in various combinations precludes an unequivocal interpretation of relationships. To simplify the scheme— and in contrast to Rothfels *et al.* (1978)— we have placed all three polymorphisms in Standard at the cost of appearing to imply that the *verecundum* line and *EFG/C* and *JJ* at one time carried IIIL–5, for which there is no evidence. From this Standard, the ordering of the various *venustum* lines is again somewhat arbitrary, depending upon which of IIS–A, IIL–1, IIIL–5 one chooses to believe was lost first (or what combination of these). Thus, the affinities shown in the diagram should not be read as presenting nearest neighbour relations unequivocally. These vagaries notwithstanding, integrating the changes from Standard will produce a correct chromosome description for each sibling. A major discontinuity exists between the *S. verecundum* and *S. venustum* lines involving some ten fixed inversions, far more than distinguish any of the siblings within *S. venustum*.

In relation to evolutionary processes, the sex chromosome constitution of the siblings is of interest. Of the ancient autosomal polymorphisms, IIS–A has become an X chromosome differential segment independently in *A/C ver.* and in *A/C ven.* It is retained as a polymorphism in *ACD ver.* and has become fixed in *AA ver.* from Newfoundland. IIL–1 (gB) characterizes the X chromosome of *CC2* and is also involved in sex chromosome differentiation of *ACD*. It has remained "floating" in *AC (gB) ven.*, *H/C*, *CC3* and (very rarely) in *EFG/C*. IIIL–5 has come to characterize the X of *CC1* and *CC3*, it has remained floating in *CC1*, and been fixed in *CC*. Certainly, this versatile use of ancient autosomal polymorphism in the elaboration of sex chromosomes, supports a concept of the primacy of sex chromosome co-adaptation as a means of speciation in the Simuliidae.

Conclusion

This chapter has attempted to give some insight into the cytological diagnosis of species in Simuliidae. The existence of sibling species is readily demonstrated in sympatric situations as in *S. vittatum IS–7* and *IIIL–1*, *S. tuberosum* and the several siblings of *S. venustum/verecundum*, including the three new ones described here as *JJ*, *C/H* and *CC3*. Allopatric situations pose problems—these can probably always be overcome where distributions are essentially continuous or parapatric as they must be for the *IS–7* type *S. vittatum* in Ontario or for *S. tuberosum* and *S. decorum* in North America. They may be insoluble where distributions are grossly discontinuous as they may be for East and West Coast *Prosimulium pleurale*. In these cases, individual judgment has to be used to "determine" whether isolated populations are mere geographical segregates with restricted or altered chromosomal polymorphism or are indeed separate species. This limitation applies equally to all other taxonomic criteria in blackflies.

In view of the difficulty in carrying out the critical laboratory crosses and the unlikelihood of exposure of the separate populations to each other in the field, it is more important to describe and specify cytological and other parameters than to enter into semantics of taxonomic rank and naming. The most extreme allopatric comparisons are those involving faunas of different continents. We would argue that where basic sequences, sex chromosome systems and prevailing chromosomal polymorphisms are identical, species identity should be assumed. In the *S. venustum* complex, this applies to the species named *truncatum* in Scandinavia and its unnamed and (except for cytology) unresolved counterpart *EFG/C* in North America. Conversely, we would argue that where populations differ in their sex chromosome systems, especially if non-homologous chromosomes are involved, these populations should be assigned to separate species, even though no differences may exist in the basic banding and some inversion polymorphisms are shared.

While disharmonious sex chromosomes (derived from two different species) may not necessarily be deleterious in F_1 hybrids, their reassortment in backcrosses and F_2 will lead to numerically unbalanced (single or triple) sex chromosome types and this, apart from all else, must constitute a strong reproductive barrier. This criterion applies to *S. argyreatum* from Finland (sex arm IIS) as compared to the form near *S. decorum* represented by the populations from Churchill, Manitoba (sex arm IL), as well as to northern v. southern *S. decorum* within North America. The decisive role assigned to sex chromosome differences is supported by the finding that in sympatric situations, e.g. in *Prosimulium* (Rothfels *et al.* 1978) and *Simulium pictipes* (Bedo 1975), sibling species generally differ in their sex chromosome constitution, and that the differentiated sex chromosome systems, i.e. X and Y taken together, are always species-specific. Indeed, these findings support a growing conviction that coadaptation of differentiated sex chromosomes in pairs may provide a primary mechanism in Simuliidae that permits sympatric speciation through progressively amplified disruptive selection.

Lest it be considered that, in a volume devoted to the prospects of integrated control, this chapter has been unduly concerned with minutiae of sibling species differentiation, let it be stressed once more that the concept of the sibling species is a provisional one, and that, generally, an intensive study will reveal morphological, distributional and biological differences between siblings. The biological differences may include life-cycle aspects, habitat preference, blood-meal host-specificity, vector potential, susceptibility to parasites and other features immediately pertinent to the purposes of this book. The various ways in which cytologists can help to provide pure material for taxonomic study have been listed repeatedly (Rothfels *et al.* 1978; Rothfels, 1979).

Cytotaxonomy of the *Simulium damnosum* Complex

R. W. Dunbar and Ch. G. Vajime

The cytotaxonomic study of the *Simulium damnosum* complex started with material sent from Uganda in 1963 by A. W. R. McCrae and M. A. Prentice. From the first three collections—including, significantly, one from the type locality of *Simulium damnosum* Theobald—four distinct categories were recognized (Dunbar 1966). Following up these observations, Dunbar (1969) reported nine discrete taxa in the *S. damnosum* complex. Since then, with the receipt of samples from more and more people, and from studies conducted in Africa, at least 26 distinct cytological categories have been recognized (Dunbar and Vajime 1972; Dunbar 1976).

Eight of these cytospecies have been formally described and named (Vajime and Dunbar 1975). Each of the remaining 18 siblings is identified by the locality from which it was first recognized. The phylogenetic chart is a summary of this information up to the beginning of 1978. Members of the *damnosum* complex from eastern and western Africa, with their country distribution, are summarized in Table 1. It is evident from this table that 18 and 10 siblings respectively have been recorded from East and West Africa. Only three siblings (*damnosum s.s.*, *sirbanum*, Menge) have been found on both sides of the continent. In West Africa, details of species distribution have been provided by Quillévéré and Pendriez (1975) for the Ivory Coast, Garms and Vajime (1975) for Liberia and Guinea, Vajime and Quillévéré (1978) for West Africa, but with particular emphasis on the OCP area.

When this project began, the status of *S. damnosum* as a taxon was a point of some controversy. From the start, the consideration that the cytological categories represented true species was acceptable to some blackfly taxonomists. Others, however, accepted it only with reservations.

The concept that a species is a population of individuals capable of interbreeding among themselves and producing fully fertile offspring, is the one considered here. Many of these 26 taxa have considerable ranges; overlapping, to varying extents, those of others. Also, two to four of these siblings may utilize the same breeding site, whether simultaneously or consecutively. Yet, with all this opportunity to hybridize naturally, there have been but a few examples of suspected hybridization noted from either side of the continent. This failure to find widespread hybrids highlights a corollary of the species concept, namely, that each population representing a species cannot outbreed with another such population to produce fully fertile offspring.

In distinguishing these 26 siblings cytotaxonomically, the basic consideration is the finding of populations differing by any or all of (1) a minimum of a single interspecific inversion; (2) a difference in the X and/or Y chromosomes; or (3) differences in the sets of intraspecific inversions. The several examples of different species-pairs within *Edwardsellum*, sharing the same chromosomal banding pattern, do not contradict the basic pre-

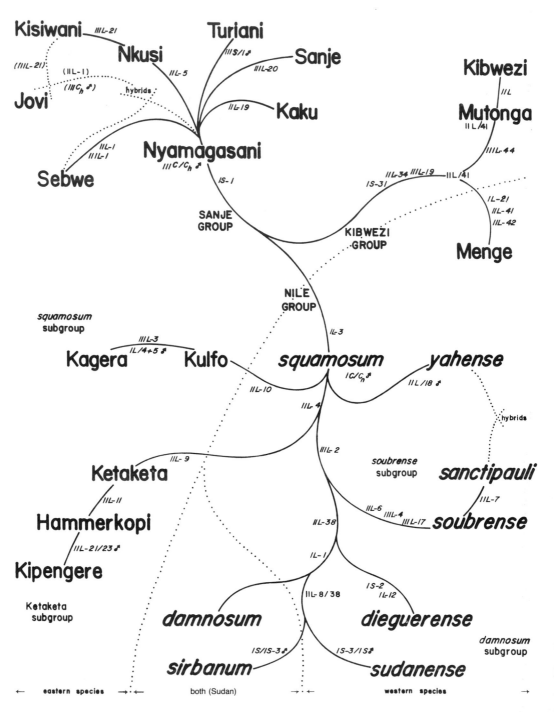

Phylogenetic relationships of members of the Simulium damnosum complex. Twenty-four siblings are included (another from Jimma, Ethiopia, has not been placed). Formal names for eight West African cytospecies are shown in italics; provisional local names for 16 East African siblings are capitalized. Chromosomal inversion differences between named members of the complex are coded beside lines between the siblings; sex-linked rearrangements are coded beside appropriate siblings. Group and subgroup names are provided for natural groupings of members of the complex.

Table 1. Members of the *Simulium damnosum* complex and their distribution in Africa.

Species and forms	Distribution	Habitat
SANJE GROUP		
Nyamagasani	Tanzania, Uganda	Upland river
Sebwe	Tanzania, Uganda	Upland river
Nkusi	Tanzania, Uganda	Sandy river
Kisiwani	Ethiopia, Kenya, Tanzania	Forest
Turiani	Tanzania	Upland river
Sanje	Tanzania	Upland river
Kaku	Uganda	Forest
Jovi	Tanzania	Upland river
NILE GROUP		
KIBWEZI GROUP		
Kibwezi	Kenya, Tanzania	Alkaline river
Mutonga	Kenya	Mountain river
Menge	Cameroon, Tanzania	Forest
squamosum subgroup		
Kagera	Tanzania	
Kulfo	Ethiopia	
squamosum Enderlein (1921)	Cameroon, Upper Volta,	Humid forest and riverine
(= Bille form)	Togo, Ghana, Ivory Coast	forests in Guinea
yahense Vajime and Dunbar (1975)	Ivory Coast, Liberia, Togo,	Humid forest and montane
(= Yah form)	Ghana, Guinea, Sierra	forest; extends into savanna
	Leone	in Guinea
Soubrense subgroup		
sanctipauli Vajime and Dunbar (1975)	Ivory Coast, Liberia, Ghana,	Humid forest
(= Bandama form)	Sierra Leone	
soubrense Vajime and Dunbar (1975)	Benin, Ivory Coast, Liberia,	Humid forest and riverine
(= Soubre form)	Upper Volta, Togo, Guinea	forest; extends into
		savanna in Guinea
damnosum subgroup		
damnosum Theobald (1903)	Nigeria, Sudan, Uganda	Nile River, Jos plateau
(= Nile form in part)		drainage
Volta	Cameroon, Ghana, Ivory	Guinea and Sudan
(= Nile form in part)	Coast, Liberia, Niger,	Savanna
	Nigeria, Senegal, Upper	
	Volta, Togo, Guinea, Mali	
sirbanum Vajime and Dunbar (1975)	Ghana, Ivory Coast, Mali,	Sudan Savanna; extends
(= Sirba form in part)	Niger, Nigeria, Sudan,	into Guinea Savanna
	Upper Volta, Togo,	and even forest.
	Guinea, Gambia	
dieguerense Vajime and Dunbar (1975)	Mali	Sudan Savanna
(= Dieguera form)		
sudanense Vajime and Dunbar (1975)	Ghana, Ivory Coast, Mali,	Sudan Savanna; extends
(= Sirba form in part)	Niger, Nigeria,	into Guinea Savanna in
	Upper Volta	Ivory Coast.
Ketaketa subgroup		
Ketaketa	Tanzania	
Hammerkopi	Tanzania	Ruaha river system
Kipengere	Tanzania	
UNASSIGNED		
Jimma	Ethiopia	

mise. However, by careful population analysis, they can be shown to represent two, or even more, species. For *Edwardsellum* analyses of three such pairs and one triplet have already been described (Vajime and Dunbar 1975). A similar complex situation is described herein for the Ketaketa subgroup.

Since there is frequent reference to the phylogenetic chart, a brief explanation for the subdivisions of the subgenus and the physical layout are presented here. Originally, there were only two groups, Nile and Sanje. These were designated on the strength of a claim by A. W. R. McCrea (1967) that he could distinguish adults between the two groups, but not within. This split coincided with a minimum chromosomal separation between the two closest species (Nyamagasani form and *S. (E.) squamosum*) of two interspecific inversions. Later, the Kibwezi form was distinguished. It is phylogenetically derived from between those two interspecific inversions, but has four additional ones. It was first placed in the Sanje group because of its geographical association with that group. With the discovery of two phylogenetically close forms sharing most of the same interspecific differences from the other groups, the Kibwezi group was established. The Sanje group has an internal coherence based on the radiating pattern from Nyamagasani of the different species. The other two groups have a deliquescent internal pattern. In the Nile group, the various branches, which are separated by at least two interspecific inversions within the main pattern, have been identified, for convenience, as subgroups.

As this work progressed it soon became apparent that all the taxa, occurring well within the eastern or western sectors of the continent, could be placed to the left or the right on the phylogenetic chart; while three found in both eastern and western localities, and following rivers flowing through the Sudan, could be placed centrally. The thin dotted line,

separating clearly-marked East and West African taxa, represents about two-thirds of Africa, from the Sahara to southern Africa—that part of the continent which has never been surveyed cytologically for species of *Edwardsellum*.

Groups, Subgroups and Species

Sanje Group

To date, this group is known only from East Africa. Of the eight presumed species, only one, the Nyamagasani form, is suspected to be a vector. Although the Kaku form comes from an area of southwest Uganda where onchocerciasis is found, too little is known about this taxon to comment further.

Nyamagasani Form

This is the arbitrarily chosen progenitor of the subgenus, being centrally positioned phylogenetically. On the slopes of the Ruwenzori Mountains of Uganda, and in several hilly areas of Tanzania, this taxon is found in areas where onchocerciasis is of high endemicity. It is present, and anthropophilic, in two other areas of Tanzania where onchocerciasis is absent—probably because the mean daily temperature, at the time of the year when transmission could occur, is too low for microfilarial maturation in the vector (Wegesa 1970a).

Sebwe Form

This is found in the same rivers, but at higher altitudes, as the Nyamagasani form, in both Uganda and Tanzania. However, the factor separating them is not altitude, but pH. Larvae of the Sebwe form occur where the hydrogen ion content is pH *c.*6.4, while the Nyamagasani one is found at pH *c.*7.0. Being known from an onchocerciasis-free area in the Ruwenzoris, it is not considered a vector.

Sebwe × Nyamagasani

These hybrids have been found in the Ruwenzoris and two places in central Tanzania. Backcrosses have also been found in Tanzania.

Nkusi Form

This is widespread throughout East Africa. Its occurrence in many onchocerciasis-free areas, particularly in Kenya, has led to its not being considered as a vector, despite the fact that elsewhere in Kenya, and in Tanzania too, it may also be present in onchocercal areas.

Kisiwani Form

It has been found locally in onchocerciasis-free areas in northern Tanzania, on Mount Kenya, and in southern Ethiopia.

Turiani Form

This form, from central Tanzania, differs from the Nyamagasani form only in a male character—the Y chromosome having a small inversion in the short arm of chromosome III. Chromosomally, the females are indistinguishable. Yet, this form is found in several villages, which are only a few kilometres away from similar settlements with fairly high onchocerciasis rates. That it does not seem to be a vector, emphasizes its distinctness from the Nyamagasani form.

Sanje Form

So far, this Tanzanian taxon is known only from the northeast, the south central highlands, and the far west of that country. It is the only *Edwardsellum* species in the Usambara Mountains. Experiments at Amani, on "*S. damnosum*", were usually concerned with this form.

Kaku Form

Only a few slides of this species, all from southwest Uganda, were available for study.

Jovi Form

This taxon is known from two sites in the central highlands of Tanzania. Cytologically, it exhibits characters present in three other presumed species, the Nyamagasani, Sebwe and Kisiwani forms. A double hybrid origin has been postulated, (Nyamagasani × Sebwe) × Kisiwani. Other origins are also possible, but have not been worked out in detail.

Kibwezi Group

The two eastern representatives are known from only a very few places.

Kibwezi Form

This occurs in the Kibwezi River in Kenya and the Pangani in Tanzania. It is suspected of biting man very occasionally, but not at a rate sufficiently high to warrant its being regarded as a possible vector. It favours decidedly alkaline water (pH *c*.9.0).

Mutonga Form

This is known only from the slopes of Mount Kenya.

Menge Form

This breeds in the River Menge in western Cameroon and has been recently found in Tanzania. In Cameroon, it was found sympatric with *S. squamosum*, a sibling considered to be anthropophilic. However, in areas surrounding larval habitats of this taxon, Menge breeding anthropophily is far less marked than in the vicinity of *S. squamosum* larval habitats (Thompson, personal communication). It is assumed that the Menge is not a vector of onchocerciasis.

Nile Group

The *Soubrense* subgroup is not found in the east of the continent.

Squamosum Subgroup

Its two eastern taxa, the Kagera and Kulfo forms, are known only from single sites, the rivers for which they are named; the western species, *yahense* and *squamosum* have been recorded from many sites.

Kagera Form

Possibly, this form is anthropophilic and therefore a potential vector of onchocerciasis in the Kagera valley.

Kulfo Form

This taxon has been collected on several occasions from the Kulfo River, in an onchocerciasis-free area of Ethiopia.

S. (E.) squamosum

This species was observed in small rivers of the Forest Zone. It extensively colonized heavily-shaded or forested areas within the Guinea Savanna, but there have been only two reports from the Sudan Savanna (near Bobo-Dioulasso, Upper Volta).

S. (E.) yahense

This blackfly is abundant along creeks and small rivers of the forest region. A few records within the Guinea Savanna concern the upper reaches of the Sassandra valley in Ivory Coast.

Soubrense Subgroup

S. (E.) sanctipauli was identified almost exclusively from the Forest Zone. A few savanna foci were found on the upper course of the Sassandra River in Ivory Coast, and on the Volta River, within the coastal savanna strip of Ghana.

S.(E.) soubrense was recorded from the Forest Zone through to the Guinea Savanna. Northern limits were established at the Leraba Bridge on the Upper Volta/Ivory Coast border.

Damnosum Subgroup

Two species [*S.(E.) damnosum* and *S. (E.) sirbanum*] of the subgroup found in the east share much of a common trans-African territory.

S. (E.) damnosum

The Nile River population (type locality) of this species is indistinguishable from the Nigerian population in which the sex determining system was discovered to be in chromosome I (Report, WHO/VBC-RU.1, July–Sept. 1979, Kaduna). From studies in Uganda this vector appears to migrate mainly along rivers rather than overland. Between this and the following species there is in Nigeria a sharp territorial boundary which coincides with the limit of rivers running off the Jos highlands.

Volta Form

This as yet unnamed species is the one incorrectly called *S. damnosum* throughout the OCP area. It is characterized by having its sex determining system in chromosome II. Probably this is the main species capable of migrating several hundred kilometres overland.

S.(E.) sirbanum

This species has recently been reported (Vajime, unpublished) from Sudan. The Sudan population is as monomorphic as that in West Africa.

S.(E.) dieguerense

This is a rare species. Six specimens were identified in a collection from the River

Bafing at Dieguera, Mali in 1971. Since then, it has been encountered only once.

S.(E.) sudanense

This is an uncommon species. It was described from a few examples from West Africa (Vajime and Dunbar 1975), but has been found since associated with dam outlets.

Ketaketa Subgroup

The three species in this subgroup have not been found outside Tanzania, where they are known only from the Great Ruaha River system of the south-central region. The Ketaketa subgroup is probably much more widespread than present data indicate, because of its association with onchocerciasis; but breeding sites in other river systems have yet to be located.

The internal phylogeny of the long arm of chromosome II for this subgroup is most confusing. At the present state of the study, no less than nine alternate arrangements and two crossover products have been recorded.

The phylogeny of the IIL arm in the Ketaketa subgroup showing the arrangements associated with each species. Heavy double bars depict the phylogeny from S(E.) squamosum; dotted lines show the origin of the two crossover products. Overlapping sequences are indicated thus, 4.9.11; and separate groups of such sequences thus, 4.9.11 + 22.23. Reference in the text for a sequence is by the heavy number only.

Ketaketa Form

This suspected vector has been found so far only in the Great Ruaha River just below the new dam at Kidatu, and three of its tributaries.

A sample received, in 1967, from Ketaketa on the Luhombero River, included a form which was recognized as new, and called after the village. But, because of the situation in the long arm of chromosome II (IIL), the complex of arrangements observed was not resolved until after good-quality material was obtained from a fresh location on the Luhombero River, just north of the new power station. From these first samples, eight alternate arrangements in IIL were recognized, seven being held in common and the Luhombero sample also having IIL–11. When the 50 specimens of the first Luhombero sample were checked against the Hardy-Weinberg equation, the probability of its being a single population was only 0.0001. When the two IIL–11 homozygotes (there were no heterozygotes with IIL–11) were excluded, the probability jumped to 0.99! Clearly, IIL–11, although central to the phylogeny of the Ketaketa IIL arrangements, had been lost to Ketaketa and had evolved into another taxon. Subsequent samples have confirmed these findings.

In contrast to the two taxa next considered, none of the IIL arrangements in the Ketaketa form could be associated with sex—just which are the sex chromosomes has not been determined.

With hundreds of slides to read, a novel technique for identifying a specimen was developed for this form. The seven known arrangements for this taxon, can pair in any of 28 different ways. For many of these complex heterozygotes, it was virtually impossible to follow the banding pattern in all but an "exceptionally good" preparation—of which hardly any were achieved. Instead of trying to read the banding pattern, the convolutions of the heterozygotes can be read. After the arrangements were labor-

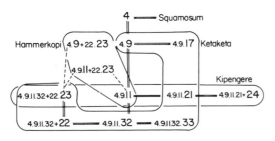

iously worked out from a few reasonable slides, Plasticine® model chromosomes were made to pair up, and sketches made of the resulting configurations. So far, three of the seven homozygotes and 12 of the 21 heterozygotes have been observed, the others being expected about once in a thousand or so times.

Ketaketa × Nkusi. A single slide from a Luhombero River sample did not fit any of the Ketaketa IIL heterozygote configurations. Eventually, it turned out to be IIL–5/IIL–4.9.11.32 + 22, a cross between species of two groups! Chromosome I did not have the two expected loops for IS–1 and IL–3, but was homozygous, indicating that the specimen was a backcross, apparently to the Ketaketa form. Chromosome III was in the long arm, almost completely unpaired. The short arm of both II and III appeared quite normal.

Hammerkopi Form

Only a very few examples of this taxon had been found at a time. They were associated with the Ketaketa form in some larger collections. In September 1976 the first sample of any size (17 specimens) was obtained. In this sample, two other arrangements were found which appear to be crossover products between IIL–9 and IIL–23, and IIL–11 and IIL–23. From this limited sample, it seems that IIL–11 represents the X chromosome and the two crossover products; IIL–9+23 and IIL–11+23 represent Y chromosomes; but IIL–11+23 may also be an alternate X chromosome. A much larger sample is needed to establish the true position.

Although the Ketaketa form has been implicated as a vector, the relative rarity of the Hammerkopi one has omitted it from such considerations. However, this recent finding of a sample with 31% Hammerkopi and 69% Ketaketa might mean that the former taxon, too, could be

a vector. The issue thus requires further careful study.

Kipengera Form

This form has been found in two tributaries of the Great Ruaha, flowing southward from the Kipengere Mountains in southwest Tanzania. At first, this was thought to be a distinctive population of the Ketaketa form, but is soon became evident that all males were heterozygous IIL–21/23 and most of the females were homozygous IIL–21/21. IIL–23 is the Y chromosome; IIL–21 the principal X chromosome; and IIL–11 and IIL–24 alternate X chromsomes.

This form originates in an area where there has been only a single cursory onchocerciasis survey, which failed to find evidence of the disease.

Unassigned

Jimma Form

Only two usable slides of this form were obtained. In both, the long arm of chromosome II could be read. It proved unlike any of the others. Unfortunately, chromosome I could not be read, so its assignation to a group was impossible.

Detailed Study of Ketaketa

In central Tanzania, onchocerciasis is found in a variety of biotopes, on well-watered plateaus in mountainous areas, and the slopes of these mountains, as well as in the country below, which has a very pronounced dry season. It is this latter type of country with which onchocerciasis is associated and in which the Ketaketa form may be the vector. In the low hills to the southwest of Morogoro the rivers are temporary. While there is a very high incidence of onchocerciasis, no larval site has yet been located for any *Edwardsellum* species. In the rainy season, biting by an *Edwardsellum* species has been recorded, the villagers complaining of severe biting as they cultivate their plots. North and south of Kilosa, and around the Uluguru and Mahenge Mountains, similar situations exist, but there are permanent rivers present. Except in the Luhombero, the Nkusi form is present everywhere. The Turiani form, the only other representative of *Edwardsellum* present, occurs only occasionally.

At one site, north of Kilosa, studies are being conducted on the transmission of onchocerciasis by the vector. Although biting rates are extremely high, the river by the site is the breeding place for Nkusi only, with Turiani appearing about 10 km upstream. Which cytological form is the vector, has not yet been determined. Everywhere, the behaviour of the vector and its elusive nature are of such constant pattern, that it seems probable that only a single form is responsible for transmission. Yet, only the Nkusi and Turiani have been found to breed regularly in these areas. Whichever is the vector, it seems to have considerable ability to adapt. Its reappearance and biting behaviour are highly seasonal. It appears able to invade areas annually, not through mass migration over vast distances, but by a relatively few individuals establishing new populations by stages; either in temporary rivers, in full spate, or in permanent ones, as they change their characteristics.

The Ketaketa form, because of the IIL chromosome polymorphism, has the potential to store intact several alternate genetic systems which could allow it to adapt rapidly to different environmental conditions, or change host preferences. A few observations lend credence to this proposition and are illustrated in Table 2.

First, an examination of the proportions of IIL–17 and IIL–21 (August 1975) from the Great Ruaha and its nearby tributary, the Luhembe River, shows a marked difference (Table 2). The site on the Ruaha is only 20 km from that on the

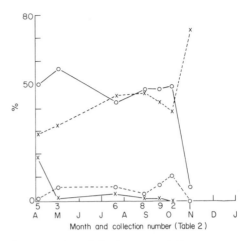

The proportion of the four most frequent arrangements in the Ketaketa form, from the Luhembe River in different months. This is a composite picture from four years' data to shown trends only.

Luhembe, and about 10 km from the confluence. The site on the Ruaha has been examined a number of times in the past, without any *Edwardsellum* being found. The site between the new dam and the power house tailrace is, at times (such as when this sample was taken), only a tiny trickle because of the diversion of much water for hydroelectric purposes. Was

The proportion of Ketaketa and Nkusi forms, from the Luhembe River in different months. This is a composite picture from four years' data to show trends only.

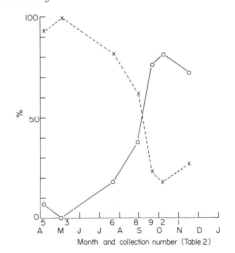

Table 2. Samples of Ketaketa from three rivers.

River and date	Ketaketa sample size	Probability of single population			Percentage of each frequency							Ketaketa: Nkusi: Hammerkopi frequency	Ketaketa: Nkusi %
		X^2	N	P	9	17	21	22	23	32	33		
Luhembe													
8, 17 November 1973	8	0.9	3	0.8		6	75		19			8:22	27:73
5 October 1974	9	0.4	3	0.9		50	39	11				9:42	18:82
1 May 1975	55	7.7	10	0.6	1	58	33	6	2			55:	100:
15 September 1975	74	2.2	10	0.9+	3	49	42	3		3		74:11	87:13
5 April 1976	40	5.0	6	0.5	19	51	29	1				40:3:1	93:7
20 July 1976	80	11.1	15	0.7	3	43	46	6		1	1	80:18:1	82:18
30 July 1976	76	12.2	10	0.3	3	52	41	3		1		76:20:3	79:21
29 August 1976	101	0.7	6	0.9+	1	49	47	3				101:62:2	62:38
20 September 1976[a]	35	19.5	6	0.006	1	49	43	7				35:120	23:77
Great Ruaha													
19 October 1975	79	10.3	10	0.5	1	78	19		1			79:0:1	100:
Luhombero													
6 June 1966	7	1.1	1	0.3		71	29					7:20	26:74
29 August 1967	48	2.4	15	0.9+	4	41	49	4		1	1	48:0:2	100:
20 July 1971	2					50	50					2:6	25:75
17 October 1975	6	3.9	6	0.7	8	26	58	8				6:	100:
25 August 1976	38	6.2	10	0.8	3	48	41	4	4			38:3:17	93:7

[a] In this sample, 62% of the chi square comes from an unexpected but not improbable heterozygote, III–9/22.

this a rapid adaptation made possible by a particular set of characters harboured by IIL–17 to a different set of ecological conditions created by man?

Another point is the shift in proportions of particularly IIL–9, IIL–17, IIL–21, and IIL–22. The figures below are based on data from several years from the Luhembe and should only be interpreted as showing a trend, i.e. the shift in proportions of the Nkusi and Ketaketa forms at the site throughout the year. This could perhaps explain why elsewhere Nkusi is always found when Ketaketa is hoped for—collections not being made at the right time of year. Although both graphs are composites of data from four years, a two-year longitudinal study was started in July 1976, on the Luhembe River (Table 2) to check this trend; the initial findings reveal similar changes.

Morphological Identification of Species and Vectors

The most vexing blackfly problem throughout Africa is the inability, in terms of the cytologically-determined populations, to identify, morphologically, the adults of any *Edwardsellum* species. Only an indirect method is available for determining which taxa are vectors. This consists of noting the breeding sites for a species which is consistently, or exclusively, associated with endemic onchocerciasis or anthropophily, the assumption being that such species are vectors. By this method, in East Africa, *S.(E) damnosum* and the Nyamagasani form have been implicated to a high degree of certainty by their association with the Nile and Ruwenzori foci, respectively, in Uganda; and, in the Sudan, *S.(E) sirbanum* has been implicated by similar observations (Gassouma, personal communication). Parallel correlations for Ketaketa are scanty, but, except for Hammerkopi, there appears to be no other contender. Positive morphological

identification of females caught biting, or identification through the examination of their polytene chromosomes, would remove any remaining doubt. Thus, one project to facilitate the search for adult morphological characters, and two to seek positive identification, through polytene chromosomes, have been started at Amani.

Reared Adults

Collections were made of all stages. These include chromosome preparations from sites where cytological studies have shown the presence of either (a) populations of a single taxon, or (b) mixtures of two taxa; one predominating and the other also known to be the only one present at another site. The target for these taxa was a minimum of 100 pinned adults, with their pupal cases. This was achieved for five forms—Sanje, Kisiwani, Nkusi, Kibwezi and Turiani. Furthermore, the process was started for Jovi, and is finished for Ketaketa. After appropriate labelling, these specimens were put in the hands of a leading blackfly taxonomist. Cursory examination of male mesonotal patterns have revealed quite marked differences, so perhaps there is yet hope of finding female morphological distinguishing characters.

Adult Polytene Chromosomes

Usable polytene chromosomes from adult Malphigian tubules have been observed in North American and Australian species. Freshly-reared adults of two *Edwardsellum* species, Sanje and Kibwezi forms, have been so examined. In both, the polytene nuclei were too small to be useful. Similar results have been observed for species of West Africa. However, other species of *Simulium* from the same sites do have useful Malphigian tubule polytene nuclei.

Larvae from Blood-fed Adults

The rearing of larvae from eggs of females caught biting to a stage where their giant chromosomes can be read has so far been unsuccessful. First-instar larvae have been achieved, but they failed to develop further. The probable reason is that the water in which those larvae were maintained, in the laboratory, was unsuited to them, having been taken from a river which was only assumed to be the breeding site. The dying off of first-instar larvae is typical if the water lacks the trace chemicals appropriate to them (Grunewald 1973).

The Whole Complex

The splitting of the familiar "*Simulium damnosum*" into numerous species, although deprecated by some[5], goes far towards explaining the bewildering diversity of behaviour recorded for this "species"; a diversity which, incidentally, occasioned the present cytotaxonomic appraisal. It is submitted that, instead of just one species, many occur in this complex—each with its own distinct biological and ecological requirements. Studies of the biology of the various entities within this complex have only just begun, particularly in West Africa, under the impetus of the research engendered by OCP. Much of the earlier information, garnered from populations referred to "*S. damnosum*", may eventually prove to be attributable to a particular taxon, defined as a result of this splitting. Even so, there seem to be fundamental differences between the groups of new taxa found in East and West Africa. Anthropophily (and therefore vector

---

[5] The "splitters" are, however, deriving support not only from chromosomal evidence but also through the use of cytotaxonomic clues in discovering cryptic morphological features of systematic value (see pp. 45–56) . . . Ed.

capacity) is associated with all but perhaps two, West African species, the Menge form and *S.(E.) dieguerense*. The "best" western vectors are *S.(E.) damnosum*, *S.(E.) sirbanum* and *S.(E.) squamosum*. Conflicting reports of zoophily come from areas where other western members of the *damnosum* complex are known to be present. On the other hand, apart from the Nile Valley, with *S.(E.) sirbanum* and *S.(E.) damnosum*, the ratio of vectors to non-vectors is reversed. East Africa has almost twice the number of species found in the west, yet only two probable, and two possible, ones have been implicated from this part of the *damnosum* complex.

Western species prefer long rivers, running through gently rolling country, while those of East Africa usually prefer mountain streams. The trace hydrochemical composition of the streams concerned is apparently more important than the size of rivers. Kisiwani, for example, was identified from a tiny semipermanent stream (a small dam outflow at the foot of the Usambara Mountains). Yet, the Sigi River (a large river flowing through the coastal plain near Tanga), has a large permanent population of this taxon. Both rivers, though, have approximately the same trace chemical composition (Grunewald, personal communication). River size then should not be considered the critical controlling factor, although, of course, stream velocity, length, etc., may well influence hydrochemical conditions.

In East Africa, several species of the Sanje group have been observed breeding in the same river, but at different locations. The controlling factor was earlier thought to be altitude. It has now been shown to be hydrochemistry, with pH the best single indicator. Thus, the Sebwe form occurs in the higher reaches at pH 6.4; and with increasing hydrogen ion content, it is replaced by the Turiani, Sanje and finally the Nkusi forms, the last-mentioned occurring at pH 7.8 (Dunbar and Grunewald 1974). In both East and West Africa, exceptions exist as regards this general statement on blackfly larval sites. *S.(E.) squamosum* and *S.(E.) yahense* prefer rivers coming from upland areas; as do the eastern Kagera and Kulfo forms. The Kibwezi group species also seem to prefer such rivers, but not necessarily those flowing through really mountainous areas. On the eastern side of the continent, the Ketaketa subgroup is found in rivers running through lowland country, like the Nile group species in the west. Also, while these three taxa show a preference for water of high pH values, western ones are found under more acidic conditions.

Morphological Means of Separating Siblings of the *Simulium damnosum* Complex (Diptera: Simuliidae)

B. V. Peterson and P. T. Dang

Prior to 1966, *Simulium damnosum* Theobald was considered to be a single, but morphologically and biologically variable, species. A cytotaxonomic study of Ugandan larvae, identified as *Simulium damnosum*, revealed the presence of four distinct cytological categories, representing four distinct but closely-related species (Dunbar 1966). Since that time, at least 25 cytotypes in this complex have been discovered as a result of further chromosomal studies (Dunbar 1969; Dunbar and Vajime 1971, 1972; Vajime and Dunbar 1975). These cytotypes and their distributions were recently summarized in tabular form in a report of a WHO Informal Consultation (1977) and by WHO (1978). The West African members of this complex include *S. damnosum* Theobald *s.s.*, *S. sirbanum* Vajime and Dunbar, *S. squamosum* Enderlein, *S. sanctipauli* V. and D., *S. soubrense* V. and D., *S. yahense* V. and D., *S. dieguerense* V. and D., and *S. sudanense* V. and D. The latter

two species are rare, or of uncertain status.

Several species of the *S. damnosum* complex are serious pests of man and important vectors of *Onchocerca volvulus* (Leuckart), the filarial worm causing the human disease known as onchocerciasis or river blindness. Reliable means of identification of these vector species are therefore prerequisite to the necessary biological studies leading to a practical control program. To complement the preceding chapter we will discuss the morphological features useful in separating the adults of the six common siblings of the *S. damnosum* complex in West Africa.

The family Simuliidae is small in comparison with many other families of Diptera. It is also one of the most homogeneous and easily recognized. Generic and, especially, specific determinations are, however, often difficult to make because of this homogeneity. The probable recent radiation of the group, as well as the highly specialized and restricted habitat to which the immature stages have become adapted, might account for the limited differentiation of their morphology. Whatever the cause, the resulting homogeneity severely decreases the number of distinctive morphological characters available for use in delimiting species.

As our knowledge of blackflies increases, the value of species characters, and even supraspecific group characters, tends to decrease as intermediate forms are discovered. This results either in a marked increase or a notable decrease in the numbers of specific-group names due to individual authorities defining the species with varying degrees of latitude.

The taxonomic separation of sibling, cryptic or aphanic species, whichever term is to be used (Steyskal 1972), has never been easy. Usually, such species have been studied by cytogenetical, ecological or behavioural techniques before the investigation of their morphological characters. Although most species of blackflies have been segregated on the basis of their morphological characters, the importance of the degree of difference between these characters is often extremely difficult to decide. Aside from the works of D. M. Davies (1949), L. Davies (1957), Neveu (1973), Jedlička (1978a, b), and Rubtzov (1962, 1974), there are very few studies of the variation of morphological characters within any given species or group of species to aid the taxonomist in his judgment of the importance of the variation, or lack of it, that he sees. This is the current situation with members of the *S. damnosum* complex.

A number of workers have recognized and commented on the morphological variation that has been seen in specimens of the complex collected from various African countries and from various types of habitats. Because of space limitations only a few of those papers mentioning variation in the adult stage are referred to below.

Gibbins (1933), in light of material described by Pomeroy (1920) and Bequaert (1930), alluded to the possibility, ". . . that Pomeroy's specimens are at least racially distinct from *damnosum*". Therefore, he described the larval, pupal and adult stages of *S. damnosum*, ". . . in order to avoid further confusion," over the true identity of this species. Nevertheless, after re-examining Pomeroy's material, Freeman and de Meillon (1953), considered it to be true *S. damnosum*. Regarding the latter as very constant in its structural features, these authors mentioned only the variability of the dorsal thoracic pattern of the male. On the other hand, they discussed the variation reported in the biting habits of the species. Grenier *et al.* (1960) noted a seasonal change in size of the adults, while Lewis (1960) drew attention to some regional differences in size and in some aspects of behaviour. Marr and Lewis (1963, 1964) pointed out some variation in colour of the antennae of females from Ghana, and Lewis (1965), comparing the specimens from West Cameroon with specimens from Ghana, noted differences in the setal tuft of the fore tarsus, in colour variations and in wing size.

After studying variations in colour, thickness of the setal tuft of the fore tarsus, and wing length, Lewis and Duke (1966) concluded that *S. damnosum* from northern savanna areas differed in various ways from specimens from forest areas, and that these differences may have been partly of a clinal nature and due partly to factors affecting individual blackflies during development. Working with specimens of the *damnosum* complex from Uganda, McCrae (1967) stated:

There is considerable evidence, other than chromosomal differences, that these different forms are distinct. Morphological and morphometric studies, still in progress, provisionally allow two groups to be distinguished among the forms from Uganda on the basis of differences in male scutal pattern and distribution of spicules on the larvae.

McCrae (1968, 1969) subsequently reported morphologically recognizable varieties of adult males from East Africa, and also some correlation between his morphological varieties and the cytotypes distinguished by Dunbar (1966, 1969). Gouteux (1975) briefly mentioned and figured some differences in the male genitalia of two undescribed members of the *damnosum* complex from Zaïre.

Recently, Quillévéré *et al.* (1977b) presented the results of a study on the females of the six common members of the complex known to occur in the Ivory Coast. Their results were summarized in the form of a key to the females based on differences in the antennae and the number of teeth on the blade of the maxilla. However, their key fails to separate *S. squamosum* and *S. yahense* from one another. Soon afterwards, though, by a revision and expansion of this key to include the number of spinules on the radial vein of the wing (Soponis and Peterson 1976), Quillévéré and Sechan (1978) were able to separate the females of all six West African species of the group. How useful and reliable this key will be remains to be determined. The most recent study of various morphological characters for separating the West African females of the complex is that of Garms (1978). He confirmed that the colour of the basal wing tuft (setae on the stem vein) was useful in separating the savanna species from the forest species; that the length, shape and colour of the antennae separated the former from two of the latter; and that different populations of the *damnosum* complex are distinguishable by morphometric means.

A detailed morphological study of the adults of the six common sibling species of the *S. damnosum* complex occurring in West Africa has been in progress since 1976. Its purpose was to try to find previously overlooked external characters, that could be relatively easily observed and used with a high degree of reliability to separate these species. According to current opinion *Simulium squamosum*, a member of this complex described by Enderlein in 1921, is a distinct species both on the basis of chromosomal and external morphological features. Even though there is great similarity between *squamosum* and *damnosum* s.s., there are specific external characters that can be used to separate the majority of adult specimens of these two species. Presumably, it should also be possible to separate most, if not all, of the other known siblings in the complex.

The preliminary morphological study by Soponis and Peterson (1976), was based on various measurements and/or setal counts of 55 different characters of a series of female specimens from Togo. This study suggested that, on the basis of the frequency distributions of these characters, some differences could be recognized between two populations of these blackflies. However, species separation by this method proved to be impracticable. Following the same approach Quillévéré *et al.* (1977b), Quillévéré and Sechan

(1978), and Garms (1978) used various measurements of the female antenna, the number of serrations on the blade (lacinia) of the maxilla, the number of spines on the radial vein of the wing, the colour of the basal wing tuft, and morphometric features to separate the six common siblings occurring in West Africa.

In a current study by the present authors, a number of readily visible external characters have been found and are being evaluated. It is anticipated that this will provide an accurate, but simpler and more practical, method of identification for the adults of most of the siblings involved. The most important of these male and female criteria appear in character distribution Tables 1 and 2. These features were evaluated for their usefulness in differentiating known adults of the complex, including the six common siblings of West Africa. Also, at least four new species are now recognized by the authors from East Africa.

Results so far achieved in this study have shown that the siblings from the savanna areas (*S. damnosum* s.s. and *sirbanum*) are collectively separable from those of the forested areas (*squamosum*, *sanctipauli*, *soubrense*, and *yahense*) by the following respective combination of characters: pale setae (versus dark setae) on the stem vein and basicosta of the wing; and all pale setae (versus at least some dark setae mixed with pale) on the vertex and postcranium. The characters separating the individual species of these two groups, plus those separating the few females of *squamosum* that apparently overlap into both groups, will be published separately. Furthermore, it has been possible to correlate the morphological variation of these adults with the cytotypes as distinguished by Vajime and Dunbar (1975). The present study, based mainly on large numbers of adults reared from cytotyped larval populations, has also demonstrated that *squamosum* is probably further from *yahense* than was thought by Vajime and Dunbar (1975). In fact, *squamosum* probably has a closer

Table 1. Distribution of selected characters in females of six West African species of the *Simulium damnosum* complex. Key: +, character present; −, character absent; (), character occasionally present or absent; L. flagellum longer than distance between apices of ocular triangles; S, flagellum shorter or equal to distance between apices of ocular triangles; D, scales on anterior surface of hind femur entirely dark except for a few silvery scales at base, and anterior surface of hind tibia with silvery scales on no more than basal three-fifths; P, anterior surface of hind femur and tibia almost entirely covered with silvery scales. Numbers 1–6, indicate approximate ratios between pale and dark scales or setae when vesture is of mixed colours.

Region	Character (SPECIES ♀)	State	damnosum	sirbanum	squamosum	sanctipauli	soubrense	yahense
HEAD	VERTEX (SETAE)	PALE	+	+	+	2+	2+	− 1(+)
		DARK	−	−	−	1+	1+	+ 4(+)
	POSTCRANIUM (SETAE)	PALE	+	+	2 , 4 +	1, (2)+	1, (2)+	1+ (−)
		DARK	−	−	1 , 1 +	1, (1)+	1, (1)+	4+ (+)
	CLYPEUS (SETAE)	PALE	+	+	+	+	+	+ −
		DARK	−	−	−	−	−	+ +
	FLAGELLUM OF ANTENNA	L	−	−	−	+	+	+ (−)
		S	+	+	+	−	−	− (+)
THORAX — SCUTUM	SUPRA-ALAR RIDGE (SCALES)	PALE	+	+	+	+ 2+	+ 2+	1+
		DARK	−	−	−	− 1+	− 1+	1+
	SIDE (SCALES)	PALE	+	+	+	+ 2+	+ 2+	1, 1 +
		DARK	−	−	−	1+	1+	1, 2 +
	PRESCUTELLAR AREA (SCALES)	PALE	+	+	+	1+	1+	−
		DARK	−	−	−	2+	2+	+
	SCUTELLUM (SETAE)	PALE	+	+	+	+ 1(+)	+ 1(+)	−
		DARK	−	−	−	− 2(+)	− 2(+)	+
THORAX — WING	BASICOSTA (SETAE)	PALE	+	+	+ + −	4+	4+	−
		DARK	−	−	− + +	1+	1+	+
	STEM VEIN (SETAE)	PALE	+	+	+ + −	−	−	−
		DARK	−	−	− + +	+	+	+
	ARCULUS	PALE	+	+	+ −	−	−	−
		DARK	−	−	− +	+	+	+
	CALYPTER MARGIN	PALE	+	+	+ −	−	−	−
		DARK	−	−	− +	+	+	+
THORAX — LEG	COXA I (SETAE)	PALE	+	+	+	1+	1+	1+
		DARK	−	−	−	1+	1+	2+
	COXA III (SETAE)	PALE	+	+	+	−	−	−
		DARK	−	−	−	+	+	+
	TROCHANTER II (SCALES)	PALE	+	+	+	1+ (+)	1+	1+
		DARK	−	−	−	1+ (−)	1+	2+
	TROCHANTER III (DORSAL SETAE)	PALE	+	+	+	+ 4(+)	+ 4(+)	−
		DARK	−	−	−	− 1(+)	− 1(+)	+
	HIND FEMUR & TIBIA	D	−	−	+	+	+	+
		P	+	+	+	−	−	−
ABDOMEN	ABDOMINAL TERGITES V-IX (SETAE)	PALE	+	+	+	+ 4(+)	+ 4(+)	−
		DARK	−	−	−	− 1(+)	− 1(+)	+

Table 2. Distribution of selected characters in males of six West African species of the *Simulium damnosum* complex. Key: as for Table 1.

CHARACTER				damnosum		sirbanum		squamosum		sanctipauli	soubrense	yahense
HEAD	POSTCRANIUM (SETAE)		PALE	+		+		−	1+	−	−	−
			DARK	−		−		+	2+	+	+	+
	CLYPEUS (SETAE)		PALE	+		+		+		−	−	−
			DARK	−		−		−		+	+	+
THORAX	SCUTUM	SUPRA-ALAR RIDGE (SCALES)	PALE	+		+		+		1+	1+	1+
			DARK	−		−		−		4+	4+	4+
		SIDE (SCALES)	PALE	+		+		+		1+	1+	1+
			DARK	−		−		−		2+	2+	2+
		SCUTAL DARK PATTERN				?		AS DAMNOSUM		AS DAMNOSUM		AS DAMNOSUM
		SCUTELLUM (SETAE)	PALE	+		+		−		−	−	−
			DARK	+		+		+		+	+	+
	WING	BASICOSTA (SETAE)	PALE	+	6+	+	6+	−	+ +	−	−	−
			DARK	−	1+	−	1+	+	+ −	+	+	+
		STEM VEIN (SETAE)	PALE	+	6+	+	6+	−	+ +	−	−	−
			DARK	−	1+	−	1+	+	+ −	+	+	+
		HALTER		+		+		+		−	−	−
				−		−		−		+	+	+
	LEG	COXA I (SETAE)	PALE	+		+		+		1+	1+	1+
			DARK	−		−		−		1+	1+	1+
		TROCHANTER II (SCALES)	PALE	+		+		+		−	−	−
			DARK	−		−		−		+	+	+
		TROCHANTER III (DORSAL SETAE)	PALE	+		+		+		−	−	−
			DARK	−		−		−		+	+	+
ABDOMEN	ABDOMINAL TERGITE VII (SPOT)		VII	+		+		−		−	−	−
			VII	−		−		+		+	+	+

affinity to *damnosum s.s.* and *sirbanum* than to any of the other species.

Some females of *squamosum* studied have entirely pale setae on the stem vein and basicosta, giving them a pale hue similar to the savanna species. Nevertheless, it is still believed that *squamosum*, as now recognized, is intermediate between the savanna and the forest groups of species. However, recent findings suggest that *squamosum* might even be a complex of two species. This study has also confirmed, contrary to previous opinion, that the scutal patterns of the males of these species are important, easy to see, and reliable for distinguishing some taxa. Finally it has been demonstrated that the setal and scale patterns of the male are very similar to those of their corresponding female. In the past, the variations in all these patterns were regarded simply as intraspecific variations.

Short descriptions utilizing the visible external characters mentioned above and a key for separating the males and females of the six important West African species of the *damnosum* complex will be published elsewhere (Dang and Peterson, in preparation). The following general description of the complex is applicable to the species known to the authors from West and East Africa. It has been prepared to meet the immediate needs of current research concerning these blackflies in the OCP area.

General Description of the Adults of the *Simulium damnosum* Complex

Female: body generally dark, with banded legs. Size varying from about 3.0 to 4.0 mm. Head slightly wider than long. Eye large, rather deeply emarginate at ocular triangle. Antenna with nine flagellomeres, black except for scape, pedicel and first flagellomere which are pale and usually reddish brown, densely covered with fine silvery pubescence; first flagellomere largest, flagellomeres IV–X somewhat compressed and usually slightly wider than long, terminal flagellomere conical, longer than wide and usually bearing 2–3 terminal setae. Frons grayish pruinose, narrowest just above antennae, gradually broadening dorsally. Clypeus subquadrate, narrowing and somewhat pointed dorsally, more broadly rounded ventrally, convex in profile and slightly longer than wide. Vertex rather densely covered with fine, appressed, lateroproclinate setae forming a pair of triangular patches; postcranium with

rather dense, appressed setae radiating from occipital foramen; postocular setae not strongly differentiated. Frontal setae usually situated in two irregular rows running along inner margins of eyes to near bases of antennae; clypeus rather densely covered with proclinate setae, those along ventral margin longer. Setal colour patterns on vertex, postcranium, frons and clypeus are of some taxonomic importance. Labrum subtriangular, as long as clypeus, with two strongly sclerotized apical sclerites each armed with 2–6 hook-like teeth. Mandible broad, with a broadly pointed apex and with about 21–29 fine serrations distally along inner margin. Blade of maxilla (lacinia) with 11–18 serrations along apical one-third of outer margin and 7–12 serrations along apical two-fifths of inner margin; palpus distinctly longer than antenna, segments III and IV subequal in length and each as long as first two segments combined; segment V slender, at least twice as long as segment IV; sensory vesicle of segment III relatively large, subspherical, its greatest diameter more than half as long as maximum width of segment. Labellum, in ventral view, somewhat reniform, densely covered with minute setae, basal thecal plates each with a single posteriorly directed seta.

Thorax black, lightly grayish pruinose. Antepronotum much reduced, greatly narrowed medially, pear-shaped laterally with a tuft of dorsally directed silvery scales; postpronotum paler than scutum. Scutum broad, slightly longer than wide, widest near anterior end of supra-alar ridge. Colour and distribution pattern of scales, particularly dark scales along lateral side, supra-alar ridge and depression is of great taxonomic importance. Scutellum dark brown to black, broadly pointed posteriorly, densely covered with scales and with long setae along lateral margins; colour and distribution of scales and setae is of some taxonomic value. Postnotum dark brown, lightly grayish pruinose, bare. Precoxal bridge of proepisternum complete, bearing on each side a tuft of appressed post-eroventrally directed silvery scales; pleural region of thorax grayish pruinose, bare except for mesepimeral tuft of dorsally directed silvery scales; ventral portion of katepisternum smoothly rounded. Fore and hind legs robust, mid leg more slender and slightly shorter, all legs densely covered with appressed silvery and brown or black scales imparting a banded appearance; integument dark brown to black except hind basitarsus which has a broad white band (in newly emerged females legs are almost entirely white except tarsus (excluding hind basitarsus), small areas at joints and dorsal surface of tibiae which are dark). Fore coxa densely covered with silvery scales on anterior surface and silvery or brown setae along ventral margin; mid coxa with dense silvery scales on dorsal surface and with silvery or brown setae along ventral margin; hind coxa with at most a few brown scales intermixed with brown setae on dorsal surface and along ventral margin. Fore and mid trochanters usually with dense silvery scales on dorsal surface of each, hind trochanter without scales. Pale and dark colour of scales and setae on coxae and trochanters is of some taxonomic value. Femora predominantly covered with dark brown scales mixed with a few scattered silvery scales; silvery scales more abundant on dorsal surface of basal third of segment. Fore tibia deeply concave anterodorsally and densely covered with shiny silvery scales; mid tibia slender, with mixed silvery and dark brown scales; hind tibia moderately broad with a moderately large patch of shiny silvery scales on basal one-third to one-half of anterodorsal surface. Fore tarsus greatly dilated with a dense crest of setae dorsally, this crest especially pronounced on basal tarsomere, degree of dilation (maximum width) in relation to length of basal tarsomere is of some taxonomic value; mid tarsus of normal shape, basal three tarsomeres armed with two rows of strong spine-like setae along ventral margins; basitarsus as long as or slightly longer than remaining segments combined; hind basitarsus almost entirely white, covered dorsally and anteriorly with scales and posteriorly with dense even setae forming a brush-like surface, also with a sharp ventral carina bearing an even row of fine and stout and somewhat flattened setae providing a finely serrated appearance, ventral side of this segment also with a row of dark bristle-like setae running parallel and adjacent to anterior side of carina. Claws of all legs similar, sickle-shaped and sharply pointed, basal tooth large and obtuse. Wing with anterior veins pale whitish to yellowish brown; basicosta, second axillary sclerite and stem vein each with a dense tuft of setae whose colour is often of taxonomic value especially in separating savanna from forest species of complex; radius with 9–27 setae dorsally, number of these setae varies somewhat between species and is of some taxonomic value. Colour of arculus (=MA) and base of A_2 (second anal vein), as well as colour of fringe and margin of calypter, varying from creamy white to smoky brown to dark brown and is of some taxonomic importance.

Abdominal tergites well developed, strongly sclerotized; basal scale (tergite I) with silvery scales and sometimes dark scales in addition to fringe of long setae; tergite II densely clothed with silvery scales but without fine setae; tergites III and IV narrower than tergites II and V, dull black, densely clothed with dark brown to black scales but without fine setae; tergites V–IX shiny, dark brown to black, sparsely setose and without scales, these tergites appearing as a half shell-like covering on distal half of abdomen; colour of setae on these tergites is of some taxonomic importance. Anal lobe (=paraproct) expanded ventrally forming a broadly rounded ventral lobe. Cercus half moon-shaped or nearly so. Abdomen membranous laterally and largely so ventrally; pleural region rather densely covered with leaf-like tufts of fine silvery scales. Sternite I small; sternite VII vestigial; sternite VIII well developed, with a pair of attenuate and usually curled or hook-like ovipositor lobes (gonapophyses, hypogynial valves); stem of genital fork (sternite IX) in lateral view, blade-like and usually narrowing at middle.

Male: body generally darker than female and with sharply contrasting spots or pattern; slightly smaller in size than female. Head with shape of eye and angle of line of demarcation separating dorsal and ventral portions of eye of taxonomic value especially in East African (Tanzania) species of this complex. Antenna similar to that of female but usually more slender. Facial area noticeably hourglass-shaped due to small clypeus; clypeus convex, varying from pear to fig-shaped, silvery pruinose, all setae directed dorsally except those along ventral margin which are decumbent; colour of clypeal setae is of taxonomic value especially in separating savannah (light) from forest (dark) species. Postcranium with rather dense, long, appressed setae radiating from occipital foramen, dorsal portion and postgenal areas bare; pale or dark colour of postcranial setae is of some taxonomic value. Mouth parts greatly reduced, blade of mandible absent or vestigial. Sensory vesicle of third palpal segment small, its diameter much less than half width of its segment, spherical in shape.

Thorax similar to that of female except more strongly convex giving a humpedback appearance; grayish or silvery pruinose and with contrasting velvety black markings usually consisting of a median V-shaped vitta and two large sublateral oval spots; size and shape of these markings and colour of scale pattern on scutum of some taxonomic value, the latter usually comparable to that of female. Scutellum dark brown to black, lateral margin paler and with long dark brown to black setae. Legs more slender than those of female, especially fore basitarsus which is not as greatly dilated. Claws small, similar in all legs, but those

of hind leg usually somewhat stronger and slightly thicker; each claw with a thumb-like basal tooth and shielded mesally at basal third by a large, finely striated quadrate plate. Wing similar to that of female except anterior veins sometimes slightly darker and number of setae on vein R usually fewer than in female of same species. Halter pale yellowish to smoky brown; extent of pigmentation of halter is useful in separating savannah species from forest species.

Abdomen slender, with velvety black and silvery pruinose markings dorsally, usually grayish black laterally and ventrally; pleural membrane of all segments with fewer scales than in female. Tergite I black, densely fringed with long dark brown to black setae; tergite II almost entirely silvery pruinose and bare except for a narrow subquadrate, median black spot, with black setae; tergites III and IV entirely black and rather densely covered with long dark brown to black setae laterally and shorter and fewer setae dorsally; tergites V to VII largely silvery pruinose laterally, tergites V and VI each with a small median subquadrate black spot whereas black spot on tergite VII is larger and much wider than long; tergite VIII entirely black or sometimes with a narrow silvery pruinose band along each lateral margin; tergite IX black, subshining, lightly grayish pruinose. Sternites moderately sclerotized except sternite II membranous. Terminalia structurally similar in most species. Gonocoxite subquadrate with a prominent conical lobe distally on outer lateral margin bearing long setae. Gonostylus somewhat triangular in shape, rather flattened dorsoventrally, sharply curving dorsally, finely setose with a single stout apical spine. Median sclerite Y-shaped, with long stem and short arms. Ventral plate of aedeagus, in dorsal view, sub-triangular, with a prominent median notch apically due to a deep V-shaped dorsal groove, covered dorsally with short laterally directed spinules; apex in lateral view, smoothly rounded and strongly curved ventrally, finely setose; degree of curvature and depth of dorsal groove variable and of some taxonomic value.

Female and male characters from this description, having salient taxonomic importance for species-discrimination, are summarized in Tables 1 and 2.

Appendix[6]

Morphological characters that are usually described or discussed, in one way or another, in most blackfly descriptions, are listed hereunder. They include most of the basic features having some value for distinguishing and relating simuliid species.

MORPHOLOGICAL FEATURES OF SIMULIIDAE
Basic list of morphological features to be examined in conventional blackfly taxonomy

I. EGG
   Dimensions: greatest length and width
   Chorion sculpture
II. LARVA
   Total length of mature specimens in alcohol
   Comparative examination of (1) smallest available larva (preferably 1st instar); (2) half-grown larva; and (3) full-grown larva with dark histoblasts.
   (A) Head capsule
      General shape, as seen in dorsal and lateral outline
      Shape of cephalic aptome
      Distribution of setae
      Colour pattern
      Hypostomal cleft: shape
                        width
                        depth
      Antenna: length
                        length of individual segments
                        colour pattern
      Labrum: size
                        shape
                        setal patterns
      Labral fan: number of rays in main fan
                        width of specified rays and length of the end-spine
                        pectination
                        width of expanded fan
      Mandible: shape
                        apical teeth
                        preapical teeth
                        comb teeth of preapical margin:
                           number
                           comparative form and size

---

[6] This list is expanded and modified from a similar one by the senior author that appeared as Annex 2 in "Species Complexes in Insect Vectors of Disease (Blackflies, mosquitoes, tsetse flies)." Report of a WHO Informal Consultation, Geneva, 15–19 November 1976. WHO DOCUMENT WHO/VBC/77.656; WHO/ONCHO/77.131, 1977; 56 pp.

Maxillary palpus: relationship of length to basal width
                  setal patterns
Hypopharynx
Hypostoma: length
                  tooth pattern: relative heights, widths and spacings
                  serration of lateral margins to head capsule
                  hypostomal bristles, including their position in relation to hyposto-
                  mal margin
Mandibular
phragma: length of ventral prolongation
(b) Larval body
Shape
Colour pattern
Shape and presence or absence of tubercles
Cuticular spines, setae or scales
Abdominal sclerotizations
Proleg circlet: number of rows of hooks
                  number of hooks per row
Proleg: shape of lateral plate and extent of sclerotization
Anal papilla: number
                  branching
                  length
Anal sclerite: relative lengths of arms
                  presence of internal struts
Scales around anal sclerite
Shape and presence or absence of ventral tubercles of segment eight
Posterior circlet: number of rows of hooks
                  number of hooks per row

III. PUPA
Total length of specimens in alcohol
Cocoon: shape
        texture and density
        incorporation of foreign matter
Position and size of pupa relative to cocoon
Respiratory organ: number of filaments
                   branching pattern
                   shape
                   length and any variation in filaments
                   ornamentation of filaments
Chaetotaxy: head and thoracic trichomes
                   abdominal spines and hooks
                   size and shape of terminal spines, setae

IV. ADULT FEMALE
Total length of specimens in alcohol or freeze-dried
(A) Head
Shape and size in relation to thorax
Outline in side view (specimens in alcohol or freeze-dried)
Area of compound eye: mean dimension of corneal facets
                      width of interfacetal tissue
                      interfacetal setae
                      size and shape of ocular triangle
Frons and clypeus: length
                     colour and pruinosity
                     width and form
                     setal patterns

Antenna: colour
               number of flagellomeres
               relative sizes and shapes of segments
Scale and setal vestiture of various head regions
Maxillary palpus: length
                          size and form of segments
                          size and shape of sensory organ and its opening
Maxillary and mandibular blades: degree of sclerotization
                                 number of teeth
Shape of median proximal space of cibarium and cornua, and presence of teeth
(B) Thorax
Degree of convexity of scutum in side view
Colour of scutum: integumental markings
                    degree of shine or pruinosity
Colour of scutellum, pleuron, anepisternal (pleural) membrane
Setal colour and length of vestiture of scutum, scutellum and mesepimeron
Presence of vestiture on: anepisternal (pleural) membrane
                          katepisternum
                          postnotum
Wing: Venation and colour
       Membrane colour
       Shape, length and width
       Length of basal section of radius in relation to total wing length
       Basal cell, presence and features
       Colour and distribution of setae on main wing veins
Legs: Colour patterns
       Setal and spine patterns and colour
       Calcipala and pedisulcus
       Length and width of basitarsus in relation to succeeding tarsomeres and tibiae
       Compression of fore tarsus, particularly depth of all tarsomeres and proportions
       of basitarsus
Claws: size, curvature and presence of basal tooth and size, in controlled planes of view
(C) Abdomen
Colour, sclerotization, shine and pruinosity of various tergites
Distribution of vestiture and its colour
Number and size of sternites particularly shape of 8th sternite
Abdominal "scale" fringe length and colour
Terminalia: size
               cercus, anal lobe and hypogynial valve: shape
                                         sclerotization
                                       setal distribution
               shape of genital rod (sternite 9)
               spermatheca: size
                          shape
                          sclerotization
                          pattern
                        presence of internal hairs
V. MALE
Total length of specimens in alcohol or freeze-dried
(A) Head
Relative and absolute sizes of typical upper and lower eye facets
Angle of line of demarcation between upper and lower eye facets
Other features as for females
(B) Thorax
All features as listed for female

(C) Abdomen
    All features as listed for female
    Terminalia: size of genitalic complex in relation to body size
               shape and vestiture of gonocoxite and gonostylus in ventral, lateral views
               shape of ventral plate of aedeagus, median sclerite and paramere in ventral,
                   lateral and terminal views
               vestiture of ventral plate and membrane of aedeagus
               shape and size of dorsal sclerite when present

Geographical Distribution of Simuliidae

R. W. Crosskey

Nothing is known and little can be inferred about the historical biogeography of blackflies. It is conjectured that the family Simuliidae is an ancient one that was differentiated from other early Diptera in Mesozoic times, but evidence for this is lacking. No Cretaceous fragments are known, and the two Mesozoic fossils from the Jurassic rocks of England that have been supposedly simuliids have recently been assigned elsewhere: Craig (1977a) has convincingly shown that *Pseudosimulium humidum* (Westwood) is non-simuliid (though possibly ceratopogonid), and Rohdendorf (1964) has assigned *Simulidium priscum* Westwood to his extinct family, Protopleciidae. It is not until the Tertiary era that genuine blackflies make their first (and only) appearance in the palaeontological record, occurring as rare inclusions in Baltic amber of Oligocene age—their rarity in amber reflecting the fact that simuliids are unlikely insects to be caught very often in the forest resins from which fossiliferous amber is formed. Although several names have been applied to amber simuliids, only one species, *Simulium oligocenicum*, has been adequately de-scribed and figured, and this is unmistakably a blackfly closely allied to modern forms (Rubtsov 1936).

Since the contemporary distribution of blackflies cannot be interpreted in the light of evidence from the past, and we really know nothing of the geographical pathways by which simuliids have dispersed around the globe, this chapter is confined to a factual account of faunal distribution as it is known at present. It should be remarked, however, that Rubtsov (1974c) has speculated interestingly, if not always convincingly, about the significance of continental drift and the rôle of birds in simuliid dispersal. Rubtsov asserts that carriage of blackflies on birds has played a large part in dispersal, at least in some groups, but this supposition should be treated cautiously. Because many blackflies feed on birds, it is tempting to assume that what Rubtsov calls "bird phoresy" over large distances actually occurs, whereas there is no evidence in support of this hypothesis. To judge from their claw structure, simuliids found on remote oceanic islands are usually ornithophilic, or presumably so. However, this is not proof that they were originally carried to the islands on their avian hosts. As pointed out elsewhere (Crosskey 1977) in relation to colonization of St Helena, it is only blackflies pre-adapted to feed on birds that would be likely to survive—most oceanic islands having no indigenous land mammals on which arriving haematophagous simuliids could feed. Thus, birds might account for the survival, but not necessarily the arrival, of isolated insular blackfly faunas. Recent evidence for the occurrence of large meteorologically-aided migratory movements in the *Simulium damnosum* complex suggests other explanations for trans-oceanic colonization, but data on aerial dispersal of simuliids remain extremely meagre. At least we can be sure, though, that all the blackfly distributions we see today are the result of natural occurrences, for the Simuliidae are a group that cannot be inadvertently transported by the activities of man—as has happened with many other Diptera.

General Picture of World Distribution

The Simuliidae are cosmopolitan and occur in all major land-masses, other than Antarctica, and in many archipelagos and isolated islands of both continental and oceanic origin. Broadly speaking, there is a blackfly fauna present in every part of the world where there is running water of sufficient permanence to provide the biological requirements of the developmental stages. Some species are able to maintain themselves in areas where the streams are of quite temporary duration. In general, blackflies are only absent from regions that are wholly hostile to them because of the total absence of flowing surface water—polar regions,

barren deserts, coral islands, or islands of volcanic lavas devoid of streams. Even in hot desert areas, blackflies can occur in very localized sites where natural watercourses or irrigation channels exist, e.g. the central Saharan massifs and the Sinai oases. Family distribution extends to high latitudes, including Baffin Island in arctic Canada, Greenland and Novaya Zemlya. Some species survive in the melt-water below glaciers, but at very high latitudes the number of species is small and bloodsucking species few or none. The most northerly locality for any simuliid is Bear I. (Björnöya) in the Arctic Ocean at 74° 30' N, where a non-biting and parthenogenetic *Prosimulium* (*P. ursinum* Edwards) occurs. In the southern hemisphere, the extremes of range are in Tierra del Fuego, with several species, and in the subantarctic islands of Crozet and Campbell, each with one species. The absence of Simuliidae from the Falkland Islands is notable, as this represents one of the few instances where suitable

The world distribution of the family Simuliidae *(heavy outlines). Heavy circles indicate isolated islands or island groups where at least one species occurs; light circles indicate islands without a simuliid fauna. Some presumed routes of dispersal are shown by arrowed broken lines. Dotted lines indicate the approximate boundaries between the zoogeographical regions named.*

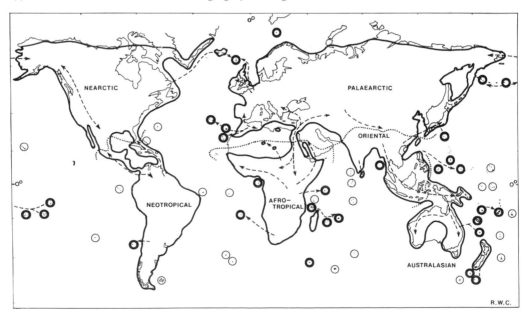

habitats exist but there is no blackfly fauna.

The genus *Simulium s.l.* is so widespread that its range is almost fully coincident with that of the Simuliidae as a whole. The only significant area that lacks a *Simulium* fauna is New Zealand, where other blackflies (*Austrosimulium*) occur. Other than this, the family range differs from the *Simulium s.l.* generic range only in so far as a few primitive prosimuliine forms occur on a few isolated islands where *Simulium* is unrepresented.

The Regional Faunas

The world fauna of blackflies consists at present of about 1270 recognized species, of which about 1000 belong in *Simulium s.l.* The representation of the fauna in the different continents is by no means uniform. The relatively small and arid continent of Australia has, as would be expected, a much smaller fauna in terms of species than the large and extensively tropical continents of South America and Africa. Table 1 summarizes the numbers of species and supraspecific taxa for the six main zoogeographical regions, to enable comparisons to be made. However, care should be exercised in interpreting the figures, because taxonomic knowledge is not equally advanced for all regions, and the Palaearctic Region (which *prima facie* is richest in species) has been subject to a more finely-split species taxonomy than has been applied elsewhere.

Blackfly distribution and faunal composition are discussed here under the usual regional zoogeographical terms, but only some of the main features are touched upon for each region. It is impossible, in a short chapter, to detail the differential distributions of the generic, subgeneric and species-group components in each region, but it should be noted that most regional faunas comprise a mixture of supraspecific taxa, some with wide and general distributions, others with more restricted ones (e.g. some Palaearctic taxa are confined to the hot, dry southern parts of the region, and some Neotropical taxa are confined to the Andean mountain areas).

Palaearctic and Nearctic Regions

These, which jointly form the Holarctic Region, can conveniently be treated together. Their faunas are similar in character, both composed of some 22 named generic or subgeneric taxa, both with a high proportion of more primitive prosimuliine forms as well as a rich *Simulium* fauna, and both including the genera *Gymnopais* and *Twinnia* that occur nowhere else in the world and are remarkable for the absence of larval headfans. In addition, these regions share many other supraspecific taxa. Cytological studies are confirming that certain individual species have genuine Holarctic distributions, stretching, to a greater or lesser degree, from Canada to Siberia. The North American *Simulium vittatum* Zetterstedt is a species (or possibly species-complex) of remarkable range from the borders of Mexico to Greenland, Iceland and the Faeroe Islands (but not yet Shetland).

Together, the Nearctic and Palaearctic Regions have a described fauna of over 600 species, or about half the world simuliid species; but only about 150 of these are North American. The disparity in size of fauna between these regions is, however, largely artificial. It is partially accounted for by past differences in taxonomic approach to the significance of minuscule morphological differences, which Russian workers have made the basis for many species in the Palaearctic Region (this region thereby *seeming* to be preponderant over all others in number of species). Although the North American fauna has still not been comprehensively studied, Peterson (1970) has revised the regional *Prosimulium* and has stated (Peterson 1978) that there are about 75 species of *Simulium s.l.* in North America. On this continent, the most remarkable blackfly taxon is *Parasimulium*, a genus of four species (Peterson 1977a) apparently confined to the western USA. It has so far defeated all attempts to find its breeding sites and early stages, and has no counterpart in any other region.

Blackflies occur throughout the Palaearctic Region, the distribution including Iceland, Faeroes, Azores, Madeira, Canaries, the Mediterranean islands and Sakhalin. The British fauna consists of about 34 species, and there is a well-developed fauna in Japan. The fauna of Europe (except Iberia and the southern Balkans) and the USSR has been well studied, but that of drier southern areas from North Africa through Turkey to Iran, Afghanistan and China is poorly known. Omitting city-states, the only countries of the region from which simuliids are unrecorded are all virtually waterless—Western Sahara and the states of the southern Persian Gulf. Blackflies of Pakistan (Lewis 1973) are predominantly Palaearctic, with few if any Oriental elements. The Palaearctic fauna has been monographed by Rubtsov (1959–1964), but this work is already outdated by the description of another 130-odd species in the past fifteen years. Zwick (1978) has tabulated the European species by faunistic area.

Oriental Region

No monograph exists for this region, which has been only haphazardly studied. Distribution range includes the islands of the East Indian archipelago, Sri Lanka, Andaman Is, Hong Kong and Taiwan. Simuliids remain unknown from many of the Indonesian Is, including the large island of Celebes, and are virtually unknown from Burma, Thailand, southern China, and the Indochinese nations.

The region is remarkable for the complete absence, on present evidence at least, of prosimuliine forms; all the *c.* 110 known species belonging to *Simulium s.l.* But Himalayan India, where prosimuliines can be anticipated as intruders from the Palaearctic, has been poorly prospected. Even so, it appears certain that the Oriental Region has the least diversified fauna of any of the major zoogeographical regions. There are few specially distinctive elements in the *Simulium* fauna except for the man-biting pest of the Himalayas, *S. indicum*, which is assigned to its own subgenus (*Himalayum* Lewis). The uniformity of the fauna is reflected in the small number of supraspecific taxa (Table 1).

Table 1. Approximate numbers of species and genera/subgenera of blackflies currently recognized in the world fauna, arranged alphabetically by zoogeographical region.

Region	Total number of species	Species of *Simulium* s.l.	Number of genera/subgenera[a]
Afrotropical	166	155	15
Australasian	102	78	7
Nearctic	147	75	22
Neotropical	300	257	21
Oriental	110	110	6
Palaearctic	485	366	22
World fauna[b]	1270	1000	80

[a] It is necessary to use the term genera/subgenera because taxonomists differ in the rank that they accord to named aggregates of species.
[b] Figures given for the world fauna are lower than the totals for the six regions because a few species and genera/subgenera occur in more than one region.

Australasian Region

With little more than 100 species, this region has the smallest regional fauna. Only 13 *Simulium s.l.* species are known in Australia, and the genus is absent from New Zealand. However, about 65 species are found in Melanesia (mainly

New Guinea), Micronesia and Polynesia. The Polynesian fauna includes *S. oviceps* Edwards from Tahiti, in which the larval head-fans are remarkably reduced (Craig 1977b), and which Rubtsov (1974c) has placed in its own genus, *Inseliellum*. The genus *Austrosimulium* is restricted to the region, a dozen species occurring in Australia and another dozen or so in New Zealand (where *Austrosimulium* is the only genus present in the simuliid fauna). A few primitive prosimuliine forms occur in Australia and Tasmania, and current cytological studies seem likely to reveal the presence of more species in southwestern Australia. The inhospitable centre of Australia is largely devoid of a blackfly fauna, and all but two of the 13 *Simulium* species occur only on the eastern side of the continent. Although simuliids occur on many Pacific island groups they are absent from Hawaii and unknown from Tonga and Samoa, though recently discovered in the Cook Islands (Crosby, personal communication).

Afrotropical Region

Simuliids are widely distributed throughout this region and have been recorded from every mainland African state south of the Sahara except for the waterless Djibouti. They are also known from Yemen and South Yemen and from many of the islands falling within the regional limits (Madagascar and the Mascarenes, Seychelles, Fernando Póo). Though not exceptionally rich, still numbering well under 200 species, the fauna is varied. It consists almost entirely of endemic taxa confined to the region, such as the *Simulium* subgenera *Edwardsellum* and *Lewisellum* that contain, respectively, the *S. damnosum* complex and *S. neavei* group of disease vectors. A small relict prosimuliine fauna, localized mainly to southern Africa and isolated inselbergs in east-central Africa, accounts for about 7% of the species, the others belonging to *Simu-*

lium s.l. The regional fauna is remarkable for the occurrence of many phoretic species in which the larval and pupal stages live on river-crabs, prawns and mayfly-nymphs. Such a phenomenon is unknown in other regions, except for the very rare occurrence of mayfly-phoretic forms in Central Asia. Freeman and de Meillon (1953) monographed the fauna, but the work is now outdated by subsequent descriptive work and the recognition of many of their "pupal-forms" as "good species". Although a comprehensive review of the fauna was provided by Crosskey (1969), a new regional monograph is required.

Neotropical Region

The blackfly fauna of this region is extremely rich, and will probably prove to outnumber, in species, any other region when it has been adequately worked out. Already, some 300 species have been described of which 43 belong in the Prosimuliini. The other species all belong to *Simulium s.l.*, about 170 species being recorded in South America and 85 in Central America (Mexico to Panama) and the Caribbean. Regrettably, there is no monograph as yet for the South American fauna. Its preparation is one of the outstanding needs in contemporary taxonomy. Clearly, however, the Neotropical fauna consists almost entirely of endemic groups, scarcely any of which extend their range into North America (i.e. regarding Mexico as entirely Neotropical and not North American). In broad character, therefore, the fauna resembles that of the Afrotropical region is so far as *Simulium s.l.* is concerned. The prosimuliine fauna (Wygodzinsky and Coscarón 1973) consists of several small and uniquely South American taxa, and the moderately large endemic genus *Gigantodax*. The latter ranges along the whole Andean mountain chain from Colombia to Tierra del Fuego (also Juan Fernandez Island) and probably it has more species awaiting description. In addition to mainland distribution, the range of *Simulium s.l.* embraces several islands of the Greater and Lesser Antilles (including Cuba, Jamaica, Puerto Rico, Guadeloupe, Dominica, Trinidad and Tobago), but there are no blackflies in the Galapagos or the Falklands.

Insular Faunas

A remarkable facet of simuliid biology is the success that the group has had in colonizing so many of the world's islands. Some of these are remote from the continental land-masses or stepping-stone island chains—or at least remote today, whatever their proximity might have been in geomorphological history. Dispersal to islands cannot occur in the larval or pupal stages, and is extremely unlikely at the egg stage. Thus, the propagule for establishment of an insular fauna must be the adult (mated female or male-and-female). It is uncertain whether transmarine movements of adults occur entirely by airborne dispersal, or partially through the agency of birds. It is pertinent, though, that all contemporary blackflies have broad, fully developed wings in both sexes (though the weakness of their posterior veins is aerodynamically puzzling). Even the aberrant *Crozetia crozetensis* (Womersley) on the isolated, bleak and windswept, subantarctic Possession I. (Crozet group), shows no tendency to brachyptery.

Many islands are known to have a blackfly fauna of at least one species, but some island groups, where simuliids probably occur, remain unprospected (e.g. Nicobar Is., Cape Verde Is.) and others remain very poorly investigated (e.g. Solomons chain and the smaller islands of Indonesia and the Caribbean). At present, it is therefore difficult to ascertain how far the Simuliidae conform to the basic theory of island biogeography (MacArthur and Wilson 1967). Of

course, many islands have to be discounted because they totally lack fluvial habitats that could be colonized by immigrant blackflies. European islands, however, appear to conform roughly to expectation. Thus, in number of species, Sicily exceeds the smaller island of Crete, which, in turn, exceeds the much smaller islands of Majorca and Rhodes. In the British islands of the continental shelf, repopulated in postglacial times, the species components of the fauna show a similar diminution, Great Britain having more species than Ireland, and each having very many more species than occur in the smaller or more distant islands such as the Isle of Man and the Shetlands. Comparisons of this kind cannot be carried too far, though, for blackfly occurrences can be profoundly influenced by such factors as geology and the recent activities of man in altering the environment. Rhodes, though only half as large as Majorca, has twice as many simuliid species; but it has many running streams, whereas much of the water in the porous limestone of Majorca is underground. In the Canary Islands, water control for agriculture has left few sites at which blackflies can survive.

Oceanic islands show some unexpected situations in view of their small size and the relative uniformity of ecological niche that each has to offer. Surprisingly, three blackfly species occur in Tahiti (Craig 1975) and three in St Helena (Crosskey 1977). The latter island of 120 km^2 and 1900 km from the West African coast is comparable to Mahé (Seychelles) with an area of 140 km^2 and 1600 km from the East African coast, but although both islands are rugged and well-watered, only one simuliid species occurs in the Seychelles. Such comparisons can be misleading, as it cannot be assumed that a stream-watered island without a blackfly fauna today has never had one, or that an island with very few species has never had more. Periodic establishment and extinction of insular populations has probably been a regular phenomenon in the biogeographical history of Simuliidae. In modern times, some extinctions by man have probably occurred (at least two blackfly species used to occur on San Miguel in the Azores, but there now appears to be only one).

Information on insular distributions is very scattered, and the following list, arranged alphabetically by ocean, is given to summarize knowledge of the presence or absence of blackflies on islands as far as practicable (large islands such as those of Japan, New Zealand, Indonesia–New Guinea, the UK and Mediterranean, Iceland, Sri Lanka, Madagascar, Taiwan—all of which have blackfly faunas—are omitted for this purpose). Islands that probably support a blackfly fauna of at least one species, but for which this is not yet established, are listed as "unprospected".

Arctic Ocean

Present, Bear I.; absent, Jan Mayen, Spitzbergen.

Atlantic Ocean

Present, Azores, Canary Is, Faeroes, Fernando Póo, Greater Antilles, Lesser Antilles (some still unprospected), Madeira, St Helena; absent, Ascension, Bahamas, Bermuda, Cayman Is, Fernando da Noronha, Tristan da Cunha group; unprospected, Annobon, Cape Verde Is, Principe, São Tomé.

Indian Ocean

Present, Andaman Is, Comoro Is, Mauritius, Réunion, Rodriguez, Seychelles, Zanzibar; absent, Aldabra and Cosmoledos, Amirante Is, Amsterdam and St Paul, Chagos Is, Cocos Is, Laccadive and Maldive Is, Tromelin I.; unprospected, Nicobar Is, Socotra.

Pacific Ocean

Present, Aleutian Is, Bonin Is, Caroline Is (Palaus and Truk), Commander Is, Cook Is, Fiji, Juan Fernandez Is, Kurile Is, Mariana Is (including Guam), Marquesas Is, New Caledonia, Norfolk I., Ryukyu Is, Solomon Is, Tahiti, Vanuatu; absent, coral islands and atolls (Tuvalu, Kiribati, Line, Marshall, Phoenix Is, etc.), Easter I., Galapagos Is, Hawaii, Kermadec Is, Lord Howe I.; unprospected, Samoan Is, Tonga.

Southern Ocean

Present, Auckland I., Campbell I., Crozet Is; absent, Falkland Is, Heard I., Kerguelen, Macquarie I.; other islands presumed absent.

Distribution of Pest Species

It is impossible to attach a precise meaning to what constitutes a pest species of simuliid, either directly in relation to man and onchocerciasis or in relation to the effects of blackfly biting on livestock and poultry. Nevertheless, it can be said that species that can reasonably be classified as pests, either because of the harmful effects of their mass biting or their involvement in disease transmission, occur in all zoogeographical regions. An important pest, however, may be geographically very localized, and the same species may be relatively harmless in part of its range, but a savage pest in another. Furthermore, certain species of larger rivers may occur periodically in phenomenally abundant outbreaks, and species that have not previously been known as pests may become such if their populations become excessive. *Simulium austeni*, for instance, had not been known as a man-biter in Britain prior to the 1960s, but for several years during the early 1970s, it became a localized man-biting

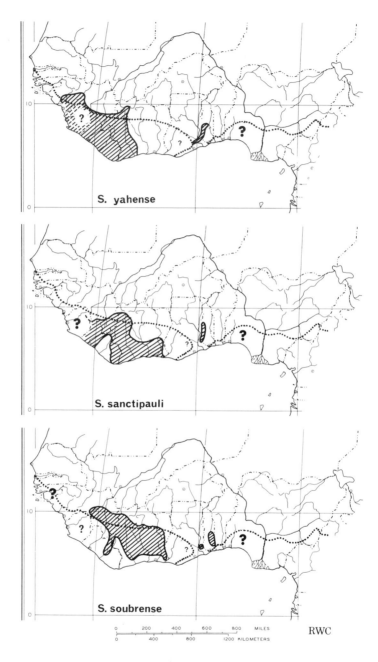

Geographical distribution of members of the Simulium damnosum *complex in West Africa. (The dotted line represents the boundary between forest/forest–savanna mosaic areas to the south and savanna areas to the north.)*

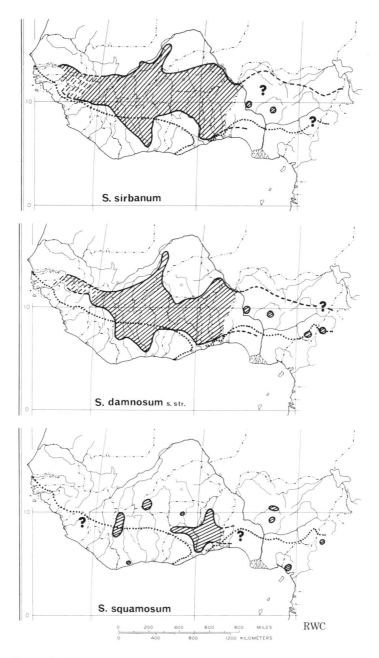

S. sirbanum

S. damnosum s. str.

S. squamosum

RWC

Geographical distribution of members of the Simulium damnosum *complex in West Africa. (The distribution shown for the savanna cytospecies* S. sirbanum *and* S. damnosum s. str. *is the natural breeding range prior to control measures in the OCP. For explanation of the dotted line see legend on p. 64.)*

Table 2. Summary of the more important simuliid pest species, showing their geographical distribution and some of their principal effects. Species are listed alphabetically and names indented in square brackets are those that are widely used in the literature when a generically split taxonomy is adopted for *Simulium s.l.* Subgeneric names are omitted for brevity.

Pest species	Area of occurrence as a pest	Effects of the pest
Austrosimulium australense Schiner	New Zealand	Biting of man
Austrosimulium bancrofti Taylor	Southwestern Australia	Biting of man
Austrosimulium pestilens Mackerras and Mackerras	Queensland	Mass biting of man and livestock, transmission of bovine onchocerciasis
Austrosimulium ungulatum Tonnoir	New Zealand	Mass biting of man
Cnephia ornithophilia Davies,[a] Peterson and Wood	Mid-western USA	Transmission of *Leucocytozoon simondi* to waterfowl
Prosimulium mixtum Syme and Davies [formerly "*hirtipes*" in North American literature in error]	Northeastern USA and eastern Canada	Mass biting of man (disruption of lumbering and tourism)
Simulium amazonicum Goeldi	Northern Brazil	Mass biting of man, transmission of *Mansonella ozzardi* in man
Simulium anatinum Wood	Canada (Ontario)	Transmission of *Leucocytozoon simondi* to waterfowl
Simulium antillarum Jennings	Caribbean Is	Biting of man
Simulium arcticum Malloch	Canada (Alberta and Saskatchewan)	Mass biting and death of livestock
Simulium buissoni Edwards	Marquesas Is	Mass biting of man
Simulium callidum Dyar and Shannon	Mexico and Guatemala	Probable transmission of human onchocerciasis
Simulium cholodkovskii Rubtsov [*Gnus cholodkovskii*]	USSR (Baikal area)	Mass biting of man and livestock
Simulium congareenarum Dyar and Shannon	Eastern USA	Transmission of *Leucocytozoon smithi* to turkeys
Simulium damnosum Theobald species-complex (several closely allied species)	Tropical Africa	Transmission of human onchocerciasis (most important African vectors)
Simulium decimatum Dorogostajskij, Rubtsov and Vlasenko [*Gnus decimatum*]	USSR (Urals)	Mass biting of man and livestock
Simulium equinum Linnaeus [*Wilhelmia equina*]	Northern Eurasia	Biting of man and livestock
Simulium erythrocephalum [*Boophthora erythrocephala*]	Northern Eurasia	Mass biting of man and livestock
Simulium exiguum Roubaud	Northern South America	Probable transmission of human onchocerciasis in Venezuela and Colombia, also man-biting in northern Brazil
Simulium guianense Wise s.l.	Brazil (northern Amazonas)	Transmission of human onchocerciasis
Simulium haematopotum Malloch	Northern South America	Biting of man
Simulium "incrustatum"-like sp. (prob. not true *incrustatum* Lutz)	Northern Brazil and Guyana	Biting of man and livestock
Simulium indicum Becher	Southern Himalayas, Pakistan to Assam	Biting of man
Simulium jenningsi Malloch	Eastern USA	Biting of man and livestock, death to poultry
Simulium jolyi Roubaud	Vanuatu	Biting of man
Simulium maculatum Meigen [*Titanopteryx maculata*]	Across USSR	Mass biting of man and livestock
Simulium meridionale Riley	Mid-western USA	Mass biting of poultry
Simulium metallicum Bellardi	Mexico to northern South America	Transmission of human onchocerciasis in Venezuela, mass biting of man
Simulium neavei Roubaud	Eastern tropical Africa	Transmission of human onchocerciasis
Simulium nigroparvum Twinn	Eastern USA	Transmission of *Leucocytozoon smithi* to turkeys
Simulium ochraceum Walker	Southern Mexico and Guatemala	Transmission of human onchocerciasis, mass biting of man
Simulium opalinifrons Enderlein	Paraguay	Mass biting of livestock

Table 2. (contd)

Pest species	Area of occurrence as a pest	Effects of the pest
Simulium ornatum Meigen [Odagmia ornata]	Northern Eurasia	Biting of man and livestock, transmission of bovine onchocerciasis
Simulium pusillum Fries [Schoenbaueria pusilla]	Northern USSR	Mass biting of man and livestock
Simulium quadrivittatum Loew	Central America, especially Belize	Mass biting of man
Simulium rugglesi Nicholson and Mickel	Canada and northern USA	Biting of man and livestock, death of poultry, transmission of Leucocytozoon simondi to waterfowl
Simulium slossonae Dyar and Shannon	Eastern USA	Transmission of Leucocytozoon smithi to turkeys
Simulium sanguineum Knab s.l.	Northern South America	Mass biting of man and livestock, transmission of human onchocerciasis in northern Amazonas, probable transmission of Mansonella ozzardi in Colombia
Simulium transiens Rubtsov [Parabyssodon transiens]	Across USSR	Mass biting of man
Simulium truncatum Lundström	Scandinavia	Biting of man
Simulium venustum Say	Canada and north-eastern USA	Mass biting of man and livestock
Simulium vittatum Zetterstedt	North America	Biting of livestock
Simulium woodi de Meillon	North-east Tanzania	Local transmission of human onchocerciasis

[a] While vectors of two of the economically important species of Leucocytozoon (L. simondi, L. smithi) are listed, readers are referred to Fallis et al. (1974) for a more complete tabulation.

nuisance on one small river system in southern England. Over one hundred simuliid species have been reported to bite man, but relatively few are of major importance. In the tropical African fauna, some dozen species (counting the Simulium damnosum complex as one, for this purpose) attack man, but only the major onchocerciasis vector species do so regularly and persistently.

Major pest species mostly occur in continental land areas with large rivers or in well-watered and well-vegetated areas with abundant streams, such as the Amazon basin, the African tropics, and the middle and higher latitudes of North America and Eurasia. Intensely annoying man-biting pests also occur in much smaller land areas (e.g. Austrosimulium ungulatum in the South Island, New Zealand, and Simulium buissoni in the Marquesas Islands). Drier areas such as the southern Mediterranean basin and the Middle East are free from serious pest species, but Austrosimulium pestilens is an important veterinary problem in semi-arid areas of Queensland.

Table 2 summarizes the distributions and the effects of the major pest species.

Africa showing the geographical distribution of the Simulium damnosum *complex (solid line and broken line extensions together). The solid line delimits the area in which man-biting occurs, and the broken line shows additional range for the complex without man-biting.*

1000 km

The Golubaz fly (*Simulium colombaschense* Fabricius) of the Danube region of Europe, and the true Buffalo-gnat (*Cnephia pecuarum* Riley), of the lower Mississippi plains are omitted from the list. Neither of these once-notorious simuliids is any longer of pest significance because of environmental changes.

The *Simulium damnosum* complex (including most important vectors of human onchocerciasis in Africa) is confined to mainland Africa, except for minor extensions of range into Yemen (southwest Arabian peninsula) and the island of Fernando Póo. Man-biting and pest status for the complex are confined to the tropical area delimited by the solid line (man-biting appears to be still unreported from Yemen, but probably occurs because of the presence there of human onchocerciasis). (Geographical distribution of the six common members of the *Simulium damnosum* complex in West Africa is shown by the maps on pp. 64–65.)

II Control Methods

The Origins of Blackfly Control Programmes

H. Jamnback

The major thrust of blackfly control campaigns has been directed against the larvae which are confined to a highly-restricted, easily treated habitat, running water. Adult blackfly control, although effective in limited areas and for special purposes, is too costly for general use and is not considered further here (see pp. 79, 81, 82, Ed.).

Until the appearance of DDT, the methods proposed and tested for blackfly control were notably ineffective. They were uneconomical (sweeping larvae from stones or removing vegetation in rivers), ineffective, and/or involved the application of materials highly injurious to non-target fauna (oil emulsion, xylol, kerosene, arsenicals). The earliest record in our files of the use of DDT against blackflies is an unpublished report by F. C. Bishopp dealing with work carried out in 1943. As Assistant to the Chief of the United States Bureau of Entomology and Plant Quarantine, he had early access to a new "wonder" insecticide—DDT. A paper entitled "Black Flies and Related Pests as Economic Problems, and Research on Control" described this work as follows:

> In 1943 I carried out a crude experiment which convinced me of the high susceptibility of *Simulium* larvae to DDT. I dripped an emulsion consisting of kerosene solution of DDT, soap, and water into a small stream, Sligo Creek, in Montgomery County, Md. Accurate measurements of the rate of application were not made, but it is thought to be about 10 p.p.m. applied during a period of about 15 minutes. Within a few hours the well-developed larvae of *Simulium venustum* on rocks in the stream disappeared for a distance of at least several hundred feet downstream.

By far the most significant reports testifying to DDT's effectiveness against blackflies were published by Fairchild and Barreda in 1945 and by Garnham and McMahon in 1947. Some of the background is given below.

Dr Fairchild, now a Professor Emeritus at the University of Florida at Gainesville, was kind enough to supply the following information:

> During W.W. II I was a Captain in the Sanitary Corps, A.U.S., stationed at Gorgas Memorial Laboratory in Panama, and working on testing insecticides and repellents under contract with O.S.R.D. In the spring of 1943 our group was augmented by the arrival of Capt. W. C. MacDuffie, bringing with him a few pounds of the then miraculous and mysterious D.D.T. The main push was then towards malaria control, and at $12.00 per lb. none could be spared for other work. After seeing the astounding effects of small quantities of the stuff on mosquito larvae and other aquatic insects in Panama, it occurred to me that D.D.T. might prove useful for the control of the aquatic stages of black-flies, a group on which I had recently worked. Suitable numbers of *Simulium* larvae were not readily available in lowland areas of Panama, so I took advantage of a visit to the highlands of Chiriqui Province in August 1944 in another connection to make some crude preliminary observations. These consisted in pouring some stock D.D.T.-xylene–Triton X-100 into a concrete sluice connected with an abandoned irrigation project. Larvae were quite

abundant on the bottom and sides of the sluice, and the speed and completeness with which they disappeared indicated to me, at least, the D.D.T. would kill *Simulium*.

The idea of using D.D.T. for black-fly control was so obvious that it no doubt occurred to many people simultaneously. The catch was that the total supply was then controlled by the military. So when Dr. Tony Donovan of the Pan-American Sanitary Bureau was assigned to Guatemala, I was enlisted as being the nearest available entomologist with both a knowledge of black-flies and access to D.D.T. In those days D.D.T. was so critical that it was not allowed out of the hands of the military, so to try D.D.T. against black-flies, P.A.S.B. was obligated to include me in the deal.

On arrival in Guatemala in the fall of 1944, I was teamed with Eloy Barreda, a Sanitary Engineer from U.S.P.H.S. assigned to work with P.A.S.B. on control of Onchocerciasis. Barreda furnished the essential know-how connected with estimating stream flow, and his knowledge of the area around Yepocapa, where most of the testing was done, was invaluable. I shudder now to think what we did to the streams of the area. Fortunately the only stream inhabitants utilized by the local population were crayfish, but they were decimated.

As an aside, Professor Fairchild indicated that this was ". . . the only piece of control work I ever published . . . something of a record in this day when 'Entomologist' has become synonomous with 'bug killer'."

Professor P.C.C. Garnham, FRS, CMG, now at the Imperial College, and Mr J. McMahon, kindly provided the following commentary on their early work in East Africa.

In 1946, during the course of the first antimalarial campaign with DDT on a large scale in rural Africa, the emulsion had to be carried on donkeys across rivers and streams. On one occasion a donkey stubbled and the emulsion spilt into the river. We seized the opportunity to discover if the breeding places of other *Simulium* species were eliminated—they were completely destroyed. Accordingly, a small plot scheme was instituted in the Kodera district, which entailed dosing the rivers Sanda and Kitare, with DDT at the rate of 2.5 ppm for periods of 30 minutes, for a total of 13 applications, between January and June 1946. The last adult fly was captured on 23 March, and none has been caught since. Eradication was thus achieved in this focus.

This result made it certain that *S. neavei* bred in rivers and with this knowledge we decided to treat one of the larger foci and the Kakamega/Kaimosi area in North Nyanza was chosen. The first attempt was made in 1947 when all rivers and streams in an area of 1500 sq. miles were treated with 11 applications of DDT at the rate of 1–2.6 ppm/30 minutes. However *S. neavei* reappeared within a few months and the scheme was temporarily abandoned. This was before the phoretic association between *S. neavei* and crabs had been discovered, and at that time it was impossible to determine the full extent of the breeding sites.

The discovery (van Someren and McMahon, 1950) that the earlier stages of *S. neavei* bred attached to the crab, *Potamonautes niloticus*, enabled us to survey the Kakamega/Kaimosi area, this time paying particular attention to the breeding places in the rivers, and not, as in the first survey, in relation to the distribution of the adult fly.

In 1954, with the new knowledge regarding the more extensive distribution of breeding, the total Kakamega/Kaimosi focus was treated, including rivers outside the forest zone. Seven treatments of DDT at 0.5–2.0 ppm for 30 minutes were applied at 10 daily intervals and achieved almost complete success. A small stretch of the Yala river was found to be still infested with *S. neavei* in December 1955, and retreatment of this river during 1956 resulted in the complete disappearance of *S. neavei* from the district. No flies have been captured since that year (McMahon *et al.*, 1958).

These demonstrations of the great efficacy of DDT in controlling blackflies led to a number of field trials elsewhere in the world, especially Alaska, Canada and the USA with subsequent trials in Uganda, Ghana and Nigeria. At this time, in the late 1940s and early 1950s, almost no one anticipated that within two decades DDT would be banned for blackfly control in most parts of the world because of its adverse long-term consequences to NTOs.

Research, beginning in the late 1950s, concentrated on identification of substitutes for DDT which were both non-persistent and effective for blackfly control. Two of the most effective substitutes discovered were temephos (Abate ®) and methoxychlor. Both have been widely used and studied in detail, both as to their effectiveness against blackfly larvae and their impact on NTOs (e.g. Dejoux 1978, Escaffre et al. 1976 for temephos, and Fredeen 1975 and Wallace et al. 1976 for methoxychlor). Both have now seen

wide use for a number of years, and the developing fear of blackflies becoming resistant to them has now been realized as regards temephos (Guillet et al. 1981, and see p. ix). This has encouraged research on a number of possible alternatives including other larvicides and formulations, insect growth regulators and biological control agents, some of which are promising (e.g. Lacey and Mulla 1977, 1978; Thompson and Adams 1979; Undeen and Nagel 1978). It should be emphasized, however, that these control agents require at least as much carefully controlled testing both in the laboratory and the field as earlier larvicides.

Blackfly Control Occasioned by Major Hydroelectric Projects in the USSR from 1955–1965

A. M. Dubitskii

Blackflies are prominent among the blood-feeding insects of the wooded, forest-steppe and mountainous regions of the USSR (Dubitskii 1978), but methods for simuliid control were poorly developed until the early 1950s. The stimulus for relevant scientific investigations, and for the implementation of practical control measures, was provided by the construction of the Bratsk, Krasnoyarsk, Zeisk, Sayansk and Viluyisk hydroelectric schemes, and consequent economic development in these vast and potentially wealthy regions of Siberia. Less important blackfly problems were already associated with European Russia (e.g. in the valleys of the Volga, Don, and Dnieper), as well as many smaller streams.

Intensive hydroelectric construction led to the new populations, directly and indirectly, experiencing massive blackfly attack. This clearly necessitated the urgent development of appropriate control methodologies which, at that time, meant the massive application of environmentally persistent chlorinated hydrocarbon pesticides. The cessation of regular and heavy blackfly control measures (coinciding with growing appreciation of the adverse environmental consequences of the massive use of such insecticides) reflected the sharp reduction in numbers of biting Simuliidae, due both to successful insecticidal treatments and to far-reaching hydrological changes.

The contribution reports the chemical methods actually used during the phase of intensive hydroelectric construction. Although these data thus have important historical significance, they should only be interpreted today as a methodological, tactical and strategic basis for future integrated control measures posing much less prospect of harm to the environment. The information presented is summarized from Rubtsov (1957–1963), Nabokov (1959), Usova (1961), Timofeefa *et al.* (1962), Nefedov (1964), Dubitskii (1964), Zvyagintsev (1965), and particularly Grebel'skii (1958–1963).[7]

Information concerning blackfly control in the smaller types of streams is deliberately left out of consideration. Important though these are as simuliid production sites, the relevant control needs are analogous to those used elsewhere. The measures to be concentrated on are those that were implemented on a very large scale at a time when vast, new territories were being opened up, particularly in the Ob-Lenski river system. The work in Srednii (Middle) Priangar will be emphasized, for it became a blueprint for anti-simuliid measures (via both prophylaxis and control) in other parts of Siberia with only rather minor practical modifications necessitated by local considerations.

---

[7] The author has chosen not to acknowledge individual references. . . . Ed.

Distributional and Biological Background of the Major Pest Blackflies

About 50 simuliid species were known from Priangar in the mid-1950s, but only six of them proved to be significant pests of humans (occurring throughout the area)—*Gnus cholodkovskii, G. jacuticum, Simulium morsitans, S. reptans, S. (Schoenbaueria) pusillum* and *S. (Sch.) brachyarthra*, and differ considerably in pest importance. *Gnus cholodkovskii* proved, by far, the most important of them, not only in Priangar, but also in large areas of the Yenesei-Angara and Angara-Lenski river systems. While the proportion of adult females of this species in the overall pest blackfly population fluctuated markedly, it never comprised less than 50–60% of the total. In the immediate vicinity of the Angara and Oka, *G. cholodkovskii* was the only blackfly attacking man at certain times, while, over the whole biting fly season, it accounted for 97–99% of the pest problem.

Blackfly eggs were collected from shallow water in the Angara and Oka, both near the banks, and in mid-stream, at 30–50 cm. Usually, they were found at the surface, characteristically on large rocks, logs and smaller fragments of timber; and, though far less frequently, on aquatic vegetation. Eggs were discovered with equal frequency, densely agglomerated and diffusely disseminated. The fact that huge numbers of eggs were often concentrated together reflects the rather tranquil river bed conditions. The larvae became scattered across the complete transect of streams. From the 2nd-instar onwards, it was scarcely possible to link any particular physio-chemical factors of the habitat with choice of site. In the Angara, larvae exhibited greater preference for logs and smaller pieces of timber, or true aquatic plants, colonization of either rock surfaces or terrestrial vegetation dangling into the water being rarely seen. In addition to the usual, rather evenly dispersed population, massive accumulations of blackfly larvae were encountered annually in comparatively restricted sectors of streams, notably, in rapids or at river bends and narrows; this was especially the case for *G. cholodkovskii* and *S. reptans*.

An important trait of blackfly biology is the ability of the larvae to migrate downstream, the reasons for such behaviour varying considerably. Besides larval displacement due to alterations in their immediate environment, age-induced dispersal (i.e. deliberate move to another attachment site) was characteristic in some Angara species. Thus, such displacement was typical of *G. cholodkovskii*, the larvae of which migrate to deeper sections of the river bed to pupate. This accounts for the fact, so striking to field collectors, that insignificant numbers of *G. cholodkovskii* pupae are found in general surveys. Obviously, individual long-distance displacements do not occur simultaneously under natural conditions. They reflect the operation of various factors, including the direction of the current, in relation to individual larvae, on the chances of the latter encountering new substrates quickly enough, etc. Nevertheless long distance individual displacements may occur under appropriate conditions: the total distance a larva covers during the course of development has much significance from both ecological and control standpoints.

A major proportion of the blackfly species of the Angara fauna, including those of greatest importance, exhibit two generations annually. The duration of their larval and pupal stages is unusually long; larval migration being significant in this connection. The earlier instars are commonest in the headwaters of the Angara, and downstream drift was found responsible for the location of later instars in the Oka. The latter situation contrasts with that in the Angara where, generally speaking, larvae of different ages are mixed together below the headwaters. There is, therefore, little expectation of finding a clear sequence of increasingly older larval instars with increasing distance downstream in such rivers. Wide diffusion of adult blackflies throughout extensive areas takes place before the urge to feed on blood. Thus, the first reports of *G. cholodkovskii* bites were made almost simultaneously at field stations remote from one another. Moreover, the extent of dispersal is different from species to species. Thus it was established that *G. cholodkovskii* spread within a radius of as much as 200 km from the larval sites, while *S. reptans* and *S. pusillum* only scattered to a radius of 20–30 km from the breeding streams. These data explain not only why *G. cholodkovskii* often predominates far from blackfly streams, but also why associations of species account for localized problems elsewhere.

The extent of adult *G. cholodkovskii* dispersal varies from year to year. It reflects *inter alia* differences in aggressiveness between populations and strains. In years when the females were biting heavily, their numbers declined with increasing distance from the Angara. Passive dispersal of simuliids is well-known, winds carrying them for long distances. Such flights of *G. cholodkovskii* from untreated zones to places as much as 50 km distant in only 24 h, were recorded during the period of intensive blackfly control efforts. The search for appropriate blood sources and the ability to seek them over long distances leads to the concentration of female blackflies around congregations of humans or domestic animals. However, the nature of the biotope has bearing on the numbers of such blackfly concentrations, which can vary appreciably even in closely adjacent locations.

Near Bratsk, the highest numbers of blackfly adults were detected from late June to early July, and in the last third of August. Lower population peaks were observed in the final third of July, and late in the first and second thirds of August. Again, purely seasonal fluctuations in the numbers of pest blackflies were recorded from the valley of the River Chuna. There, three peaks were observed, at the end of the first third of June, in the first third of July, and in the second third of August. The first rise was usually scarcely perceptible. In some years, only two population peaks were registered. The total number of blackflies biting at population peaks proved variable from zone to zone of Priangar. The operational control data from the middle reaches of the Angara showed that in barely a minute, as many as 1046 female blackflies could be captured in the immediate vicinity of one person. In the valley, the corresponding figure was 255 and, in the Chuna valley, 400.

The Basis and Choice of the Methodology for Controlling Blackflies in Central Siberia

In this part of the USSR, effective control measures were developed against vectors of diseases of both man and domestic animals. Of course, relevant practical experience did not prove transferable from one group of blood-feeding arthropods to another. Usually, individual pest species are best dealt with by specific measures, following investigation of their particular habitats. For example, at the moment of natural, simultaneous and rapid declines in the numbers of adults, the application of control measures could hardly be expected to yield the best results. Major differences in methodologies may also be occasioned by peculiarities in the specific composition and ecology of the complex of pest blackfly species from one zone to another.

The unique composition of the simuliid fauna of Priangar is of much interest. This was not properly realized at the start, when the biology of other vectors there was appreciably better-known. Moreover, it must be remembered that prior to 1955, satisfactory methods for blackfly control had not been developed in the USSR. Nevertheless, the seriousness of the pest problem, posed to humans by these insects, was evident enough. Working capacity was known to be seriously reduced through not only actual biting, but also the persistent and extreme annoyance caused by these insects forming a cloud about the head and entering the mouth, nose, ears, and eyes. Sometimes, too, the toxic effects of blackfly bites occasions feelings of nausea, besides swellings—the latter being especially troublesome around the eyes and sometimes being associated with generalized lethargy and fever. During periods of massive blackfly attack, human productivity was shown to decline by 20–30%. Also, the number of occupational injuries increased significantly. At the major construction sites of Priangar, any productivity decline brought about by such pest problems entailed decided economic loss. In areas where biting fly control was not undertaken, there were recorded instances of consequent economic losses ranging from tens of thousands to hundreds of thousands of roubles daily.

In the dairy industry, the average daily milk yield was shown to be reduced by up to 45% through blackfly attack; and egg-laying of domestic birds was severely reduced as well. The calculation of losses in meat production is more difficult, but fattening of stock was found to be substantially retarded.

In the light of expanded construction in the northern parts of the USSR, blackfly control thus became a major and urgent priority. Acceptable working conditions had to be established at and near construction sites; the areas concerned being so vast, that the remedial measures needed to be planned on a very large scale. Blackfly control procedures varied in detail with the characteristics of the zone, relevant factors being: the overall dominance of *G. cholodkovskii* in adult pest populations; the capability of this species to disperse far from its breeding sites; the concentration of the females around blood sources; the confinement of the immature stages to a limited number of streams, also the widespread distribution of the larvae along these streams; and their ability to migrate far downstream, thus eventually colonizing the entire water-course.

Factors of at least equal importance, reflecting both natural, environmental conditions, and the artificial ones associated with agriculture, influenced the designing of the necessary control measures. They included the complex navigating channels of the Oka and Angara; the unsuitability of many stretches of these rivers for navigation; the lack of a developed road network and, hence, of convenient access; the fact that the control areas were widely scattered; the rugged nature of the country; the uneven density of forests; and the requirement of a developing network of power transmission lines.

In the overall complex of control measures adopted, aerial spraying of chemical pesticides occupied an insignificant place. However, at times it became the leading (and sometimes the only) approach at, for example, isolated construction sites. The purposes of the developing industrial enterprises throughout the control zones naturally influenced the types of treatment procedures adopted, too. Clearly, the requirements of control at a large hydroelectric project (with the inevitable separation of building activities by the river itself and widespread scattering of production subdivisions) are different from those posed by other types of industrial construction or, for that matter, town-building—these latter types of projects involve more concentrated activities and, of course, the circumstances posed by forestry development and large collective farms are different again. Although in these two cases, the total size of the control area may be relatively small, the actual sites requiring treatment can change quite frequently in a single season. For all these reasons, the application of control was under constant review.

Some hold that the best general black-fly control procedure is by means of precisely applied larvicides. When chemical adulticiding is practised, the treatment zones reflect not only the nature of the overall area, but also the direction of prevailing winds, location of larval sites, etc. In other words, the application of control measures is not so much to specific territories infested with blackflies, but to mobile concentrations of these insects. Therefore, repetitive treatment of breeding sites and adult blackfly ranges is complicated by airborne dispersal problems negating the effectiveness of control, perhaps immediately after its application. Thus, the methodologies developed in the USSR in the 1950s were aimed at human settlements, whether temporarily or permanently, so as to achieve maximum pest reduction from a control effort that otherwise would have had to be applied to extremely large areas.

Larviciding was not concentrated on particular sections of major streams, because the widespread dispersal of the immature blackflies necessitated treatment of as much as 3–500 km of river bed in order to reduce the population at a given point downstream; something hardly feasible for such a large and meandering river as the Angara, with its many tributaries in country of rugged relief. For full control of blackflies under such circumstances, moreover, the adults would have to be attacked over distances from the larval sites—to at least the minimum radius of their dispersal. Around such a project as the Bratsk Hydroelectrical Station, such treatment of approximately 500 km^2 would have been required. Instead, it was decided to undertake frequent, repetitive treatments aimed at destroying both adults and larvae. The former goal was, by far, the easiest to realize, for increases in adult numbers and the severity of their attack is easily demonstrable. To establish the

peaks of larval population fluctuation, in a large and remote river, demands far greater effort on the part of those making larval counts and, for example, the species- and age-compositions of larval populations together with analysis of the ratios of one species to another. Under the most favourable circumstances, the results from such efforts are not always decisive enough to serve as a basis for effective larviciding.

Again, even the near-elimination of immature blackflies from a stream system is not followed by an immediate reduction in the magnitude of the biting pest problem; but carefully pinpointed adulticiding will normalize working conditions very quickly. It was thus decided that at least pending the regulation of flow in the Priangar river, larval control could not be the only measure adopted; nor, for that matter, even the main measure—for this, adulticiding was felt to offer the best prospect of practical success.

Insecticidal aerosols generated on the ground, supported by aerial spraying, was agreed upon as the method of choice for such construction sites as the Bratsk Hydroelectric Station. Aerosol fogging was used for immediate amelioration of the problem facing workers; a longer-term benefit being expected from the residual action of DDT formulations sprayed by aircraft onto vegetation. Larviciding was reserved for those parts of rivers where especially large concentrations of the immature stages were recorded, in order to avert what would otherwise have been massive biting by the resultant adults. Such integration of chemical adulticides and larvicides proved highly beneficial to the working environment in individual remote working sites, for example, those where transmission lines were under construction, and the beds of future reservoirs were being cleared. The workers in such sites also wore protective clothing.

The fact that such measures proved necessary year after year on a very large scale reveals neither that the actual procedures employed had inherent short-comings, nor that the overall control plan was unsatisfactory. It was simply due to the biological inevitability of reinvasion of control zones from outside. Even immediately after an almost complete suppression of blackflies in a particular locality, further adults were reinvading in large numbers. For example, the spring generation of blackflies, throughout the whole of Priangar, was suppressed in 1955 to the extent that blackflies rarely caused trouble to humans or domestic animals in June, July, and the first half of August of that year. Nevertheless, by the final third of August, the incidence of adult blackflies had risen to the usual level for that period. Such a rapid resurgence during a single generation testified to the exceptional difficulties often posed by blackfly control. The need for radical destruction of these pests during such large construction projects, obviously demands the intermittent application of control measures, even where the areas in question are of vast extent. It is thus hardly surprising that technical setbacks were encountered.

Application of Blackfly Control Measures

Chemical larvicides were applied against blackfly larvae, both from aircraft in the form of sprays, dusts and caked formulations, and from the ground, using fire equipment. It should be emphasized that even small boats could not be navigated in extensive sections of the channels of such rivers as the Angara and Oka. Concentrated emulsions of DDT and oil (3.5% dilution) were used for aerial spraying, dosage being estimated on the basis of water velocity, and modified as required, for particular conditions of treatment (including the local topography and hydrology, and the number of flights).

Under the conditions prevailing in the Angara, $0.1–0.2\,g/sec/m^3$ water was the desired dosage. Some 500 kg of formulation were applied per treatment where the stream velocity approximated $300\,m^3/sec$ (droplet size being 50–100 μm). The full dosage for any particular reach of a river was only achieved after several flights, which took place consecutively. While this is lower than some dosages recommended elsewhere, such applications led to a sharp reduction in the number of simuliid larvae of treated areas (e.g. 1845 to 5, as sampled at one treatment site over the 24 h following spraying). Sometimes, the treatment zone became completely free from larvae. Operational impediments responsible for reduced effectiveness of some applications, included failure of spray aircraft to maintain course so as to achieve the necessary even dispersal of the larvicide. This was due to topographic factors, the steep and high banks of the Angara and Oka rivers making it hazardous for low flying aircraft to maintain optimum flight patterns, and also precluding transverse treatment of the streams (which would have been preferred by the programme organizers). A further difficulty posed by the rugged terrain was that of ensuring that the larvicide actually contacted the water some 30–50 m upstream from major concentrations of immature blackflies. Under such circumstances, an increase in the amount of formulation to 800 kg/flight failed to improve the situation. More than 12–18 km of stream bed were never cleared of blackfly larvae by a single aerial application of insecticidal spray. Always, too, migration of fresh larvae from upstream habitats caused the reappearance of a substantial simuliid population within 3–6 days of treatment. Also, the Bratsk rapids—above which larviciding was not undertaken—had to be treated twice or three times as frequently as downstream reaches.

From 1960 onwards, aircraft were used not only for spraying as described above, but also for the application of caked formulations. Blocks of locally-available inert materials (sand, clay) containing DDT paste with emulsifiers, and weighing 0.5–1.0 kg (40–60% DDT concentration) were dropped over the stream from fixed-wing aircraft or, for preference, from helicopters. The usual load for such formulations was less than that for spray emulsions. The resultant increased manoeuverability and greatly enhanced toxic effect of the insecticide released for some time from the blocks once they were soaked, resulted in much improved control efficacy.[8]

A third method of aerial application sometimes used, was dusting.[9]

The work volume for the aerial spray programme varied considerably from year to year, despite the fact that the chief blackfly larval sites treated remained the same. The reasons for this include changes in hydrological conditions affecting downstream dispersals from untreated areas into the control zone, and overall and unforeseen fluctuations in the blackfly population.[10]

Turning to ground application procedures, DDT emulsions were sometimes applied as high-pressure jets from fire hoses. The specially adapted equipment was either installed on bridges so that the stream could be sprayed from side to side or, when river conditions allowed, from barges or launches in mid-stream. In the latter case, high-pressure hoses were occasionally operated from such craft brought as close as possible to the downstream side of rapids or swiftly-flowing narrows.

To summarize the overall effectiveness of larviciding by these various procedures, after three multiple treatments at a dosage of $0.1 \, g/m^3$ water (30 minutes exp-

osure time), total mortality of the simuliid larvae was achieved for as much as 170 km downstream. Reduction of the dosage to 0.08 g technical DDT/m^3 water achieved a total kill for 70 km, while only 35 km were fully controlled at a dosage of 0.06 g technical DDT/m^3 water. Also, when the exposure time was reduced from 30 to 20 minutes at a dosage of 0.1 g technical DDT/m^3 water, 100% mortality was extended for only 50 km.

Pesticidal aerosols were the chief means of controlling adult Simuliidae. A "green oil" base was used; additional materials including "black oil", oil waste and dichlorethane, in various combinations, increased viscosity, specific gravity, and sometimes the solubility coefficient of DDT. Aerosol generators of various sizes were used. Sometimes, applications were made from launches operating at speeds of up to 25 km/h with 6–12 small generators installed on the bow-hatch and operating simultaneously. This led to a uniform distribution of aerosols across the water surface and proved valuable in adulticiding locations difficult of access (e.g. islands and actual power lines). It led to amelioration of the pest problem over approximately 47 ha/h. By contrast, the same number of small generators, operated from stationary positions ashore, could treat only 12 ha/h. The frequency of aerosol treatment varied with the locality. In the case of relatively small and scattered sites, it was necessary on an almost daily basis. When heavier equipment was used for area fogging, as much as 8000–12 000 ha/h could be treated. In such instances, no more than ten treatments were required over the entire season.

Barrier treatments to prevent adult

---

[8] Correspondence with the author indicated that adverse consequences of these DDT dosages to NTOs were not reported; and that data are not available on the significance of natural predation on blackflies in large Siberian rivers . . . Ed.

[9] No further details were supplied by the author . . . Ed.

---

[10] By analogy with 1944 applications of DDT to Costello Creek, Ontario, by R. R. Langford (Davies 1950), DDT-induced destruction of relatively long-lived aquatic predators may have facilitated post-spraying resurgence of blackfly populations at least temporarily freed from regulation by natural enemies . . . Ed.

A powerful vehicle-mounted jet engine aerosol generator used for Siberian blackfly control.

blackflies dispersing from untreated areas to treated ones, were undertaken from low-flying aircraft, applying 3–5% DDT emulsions in 2–6 km strips of 40–50 m width; the strips ranging from 100–200 m apart. Drift posed problems on occasion. Such aerial treatments against adult blackflies, replaced the use of aerosols in sparsely vegetated areas of flat relief within easy access of airstrips. However, such treatments proved generally uneconomical, except for decelerating the reinvasion of scattered control zones.

Returning to the methodology of aerosol applications, the greater the blackfly-infested area simultaneously fogged by a number of generators, the fewer treatments were necessary. Of course, the capacity for dispersal of adult blackflies meant that more frequent aerosol fogging (sometimes on a daily basis) was necessary for restricted areas, such as construction camps. When significant tracts around such sites could be treated at the same time, treatments could be spaced 3–5 days, and sometimes as much as 7–10 days apart. Once again, as in the case of larviciding, effectiveness varied annually depending upon blackfly population fluctuations, besides other factors includ-

ing meteorological ones, as well as major topographical alterations due to construction activities. In the latter connection, the thinning out of previously heavily-wooded areas was associated with differences in the dispersal of the aerosol wave range, from some 300 m to as much as 3000 m. Reduction of blackfly biting was evident immediately after the passage of an aerosol wave. In the affected area, the activity of these insects ceased altogether within 30–40 minutes. However, early re-emergence took place unless barrier treatment was undertaken.

As Grabel'skii (1958–1963) reported, the decline in overall simuliid biting populations at the Bratsk hydroelectric dam construction site averaged 71–86%, thanks to control measures, as outlined above. After completion of the installations, when control was relaxed, the pest problem still remained much less than before, barely reaching 51% of pre-control levels and that for short periods only. On average, the post-control blackfly population remained at about one-quarter to one-sixth the pre-control level. However, it must be understood that this is a rather rough estimate, because the absolute size of the original estimate was subject to fluctuation for purely natural reasons. Explanations for the drastically lowered post-control incidence are offered in the following section.

The Influence of Hydrological changes upon Blackfly Incidence

During the construction of hydroelectric facilities, environmental changes are clearly inevitable. In Siberia, it proved that their effect on the blackfly populations was so adverse that there was no need for continuing chemical control after construction was completed. First of all, any sharp reduction in river levels prejudices survival of blackfly eggs, where the latter are left well above the surface. Lowered such levels, and even actual exposure of the river bed, where it had previously been at a depth of 2 m, took place along the entire course of the Angara below the dam, as the reservoir began to fill in the autumns of 1961 and 1962. Such low water levels had never been observed before. Flow rate changed immediately after massive blackfly ovipositing and thus the great majority of that generation of G. *cholodkovskii* eggs were desiccated. This sharply reduced the numbers of larvae from the spring of 1962 onwards.

Secondly, significant water temperature alterations prejudiced the immature stages of blackflies. Frequent and sharp fluctuations in water level (averaging about 1.7 m) proved to cause changes in the quantity of suspended matter present. At the same time, cooling water entering the stream from the dam delayed warming to 10°C for as long as 81 days in the near vicinity of the installation. Satisfactory completion of blackfly larval development was thus delayed until the early autumn (September) so that larvae could not be found in the first 100 km below the dam. This postponement of adequate warming of the water to the level normal for larval development of course became less and less noticeable with increasing distance from the dam. However, for at least 350 km downstream from Bratsk, varying degrees of delay in larval development still remained noticeable. Also, larval numbers remained low by comparison with pre-control ones, while the representation of G. *cholodkovskii*, within the developing complex of blackfly species, ceased to be characteristic of the normal Angara situation.

Again, the development of heavy concentrations of various species of unicellular algae resulted from changing hydrological circumstances associated with the formation of the new artificial lakes, e.g. intense silting-up of former larval attachment sites was observed along the whole course of the Angara. The destructive effect of this to G. *cholodkovskii* populations is evident enough from the few findings of significant numbers of larvae and pupae in the summers of 1962 and 1963. Such discoveries as were made, concerned reaches where tributaries flowed into rapids and narrow river bends, i.e. sections of the stream still remaining deficient in silt or freshwater algae.

Damming a large river of course excludes the possibility of any blackfly development in the new static water impoundments. Therefore, since flooding of the Bratsk reservoir began in the autumn of 1961, the following year's spring generation was prevented from completing development down the whole courses of the Angara and Oka below the dam. It was not until August, 1962, that the flood conditions approached those suitable for the initiation of emigration flights of G. *cholodkovskii* to the middle reaches of the Angara. Until then, dispersal of adult blackflies downstream from above the impoundment of course remained possible.

Comparisons of the efficacy of various hydrological monitoring techniques was complicated by the post-construction exposure of shoals, water level fluctuations and silting. It proved indeed that the deliberate exposure of shoals in autumn offers good prospects for the environmental management of blackfly populations. Even sharp rises in water levels, other conditions being equal, proved useful control measures.

Hydrological changes, as a whole, induced by the filling of the Bratsk artificial lake, led to stabilization of blackfly populations throughout Priangar at the new low levels already mentioned. From then onwards, only one relatively low blackfly population peak was reached in place of the two pre-control ones. Following the subsequent decline in numbers, no significant blackfly attacks were registered at all. Parallel results, involving hydrological changes leading to a dramatic lowering of blackfly populations, were achieved in large rivers of European Russia as well as along the Volga and Kama.

To conclude, the end result of major hydroelectric construction on large rivers of Siberia was the unexpected realization that long-term blackfly control had been brought about by ecological changes, the streams thus being spared future major pollution by blackfly larvicides. It is thus believed that the deliberate incorporation of environmental procedures of blackfly control, into the construction of dams and reservoirs for hydroelectric installations, could maximize work efficiency while minimizing adverse environmental impact of the control measures upon NTOs.

World Health Organization Onchocerciasis Control Programme in the Volta River Basin

J. F. Walsh, J. B. Davies and B. Cliff

In West Africa, onchocerciasis is a widespread and socio-economically important disease, the only known vectors of which are members of the *Simulium damnosum* complex (pp. 272–274). *Simulium damnosum s.l.* can be found in most parts of the savanna, at least in the wet season, and the focal distribution of disease and vector typically found in Central America and in parts of Central and East Africa is absent. Most large streams and rivers scattered across the savanna between 8°N and 12°N provide suitable breeding places, especially where the waters flow across the Pre-Cambrian Basement Complex (Crosskey 1956). Areas of sedimentary rock are generally less suited to the requirements of *S. damnosum* and the ancient dune systems, deriving from the Sahara in more arid periods, which now cover parts of the Sudan Savanna Zone and all of the Sahelian Zone, are generally inimical to *S. damnosum s.l.* breeding. These Sahara sands determine the northern limits of both vector and disease, save where man-made structures have produced rapids.

Several attempts at control of *S. damnosum s.l.* in West Africa were made in the 1950s and 1960s. Among the more notable were the Mayo Kebbi scheme in 1955 (Taufflieb 1955), the Abuja control scheme of 1956 (Crosskey 1958; Davies *et al.* 1962), the Kainji Dam Scheme in 1961 (Walsh 1970a), and the FED campaign in the Farako area of Western Mali in 1966 (Le Berre 1968; Prost 1977; Thylefors and Rolland 1977). All gave some degree of success as long as the control activities were maintained. More importantly, they pinpointed many difficulties, in particular that posed by the regular influx of blackflies from unstreated surrounding areas (Hitchen and Goiny, 1966).

A meeting of OCCGE, USAID and WHO in 1968 concluded that at that time the control of the disease depended on control of the vector *S. damnosum s.l.* by means of larvicides (WHO 1969). It was felt that this method of control would be feasible if carried out on a large scale in the West African savanna and that "such control would stop the further occurrence of onchocercal blindness which under present conditions is such a frequent sequel to the infection, having dire socio-economic repercussions on the afflicted communities and obliging them to abandon much fertile land in river valleys". It was agreed that the Volta River Basin as originally proposed by Waddy (1963) should form the nucleus of the area to be brought under control.

Subsequently, several West African governments expressed the desire to undertake a large scale campaign against the vector of onchocerciasis in collaboration with their neighbours, the relevant Specialized Agencies of the UN and with the cooperation of interested donor nations. WHO, which was requested to act as Executing Agency in association with FAO, placed a mission in the field in 1970 to gather the necessary information, to plan and cost an operation, and to obtain objective estimates of the likely economic benefits. No particular emphasis was to be placed on the humanitarian advantages of such a Programme though these can be expected to be considerable (Laird 1977). The Preparatory Assistance to Governments Mission (PAG Mission) produced its detailed Report in August 1973, with two objectives: (1) to combat a disease that is widespread and severe in the area; and (2) to remove a major obstacle to economic development.

At a meeting of interested parties in November 1973 the findings of the PAG Mission were endorsed. In January 1974 the Onchoceriasis Control Programme (OCP) was brought into being; it has an expected duration of 20 years, a time based on the postulated life-span of the parasite *Onchocerca volvulus* in man.

Management Structures

Supervision of the policies to be adopted in the planning and execution of OCP is in the hands of the participating countries (the seven West African countries in whose territories control is being undertaken), the donor countries and the four responsible International Agencies— FAO, IBRD, UNDP and WHO, meeting at least annually as the Joint Coordinating Committee (JCC). WHO acts as the Executing Agency and meets about four times a year with the other responsible Agencies in a Steering Committee. In order to provide WHO and JCC with the necessary intellectual input and support of the international scientific community, three statutory groups were set up, each expected to meet once or twice a year. These are the Ecological Panel (EP), which is comprised of five experts, representing the four responsible agencies and UNEP; the Scientific and Technical Advisory Committee (STAC), which consists of 12 eminent members from the fields of tropical medicine and the biology and control of vectors; and the Econo-

mic Development Advisory Panel (EDAP), composed of 12 members eminent in the field of tropical development economics, agronomy and sociology.

Each of these bodies reports annually through the Steering Committee to the JCC, giving its views on the aims, methods, achievements and shortcomings of the Programme.

OCP Area

The original area chosen for the OCP consisted of 654 000 km^2 (in earlier documents estimated as 700 000 km^2) of Guinea and Sudan Savanna in seven countries of West Africa, Mali, Ivory Coast, Upper Volta, Ghana, Togo, Benin and Niger.

The whole of the savanna portion of the Volta Basin was included. This considerable area was flanked by the Komoe-Leraba and the White Bandama valleys to the west and by the northward flowing tributaries of the Niger River to the northwest and the northeast. It was hoped to reduce the problem of reinvasion of the treated area by taking in this large area, utilizing the natural northern and northeastern limits of vector distribution, and incorporating the Volta and Kossou man-made lakes. The southern boundary of the control area skirted along the northern edge of the Forest Zone and it was thought unlikely that forest cytospecies of the *S. damnosum* complex would reinvade the savanna in substantial numbers.

In order to cover this large area, control was introduced in three phases during the period from February 1975 to July 1977. An extension of control to a further 110 000 km^2 in southern Ivory Coast (Phase IV) was approved by the JCC in December 1977. Control began in June 1978 in approximately half of this additional area. The remainder was brought under control during the wet season of 1979.

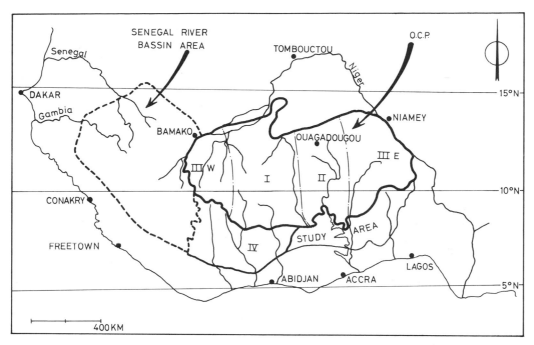

Map showing the location of the OCP area in West Africa, the operational Phases, and possible extension areas.

Besides the area presently under control, entomological and epidemiological studies are underway in a further 111 000 km² in the southern parts of Ghana, Togo and Benin. These studies are designed to show the medical and socio-economic importance of onchocerciasis, to determine what influence, if any, is exerted on the existing OCP area by adult blackflies emanating from those uncontrolled zones and to assess the likely cost of any successful vector control operation.

A further scheme to cover the Senegal River Basin area is under consideration.

Activities of the Programme

In order to undertake this vast project, it was necessary to choose methods which were appropriate to attaining its objectives, to ensure that the results of its activities were properly assessed and to

plan, as accurately as possible, OCP's likely life-span.

At the planning stage, no medical solution to the onchocerciasis problem in West Africa was available, a situation which has not materially improved subsequently. The only feasible method of controlling the disease remains the reduction of populations of the vectors, by chemical or other means, to levels below which transmission of *Onchocerca volvulus* is insufficient to maintain onchocerciasis in its clinically pathological forms.

Before control could be envisaged, it was necessary to establish networks of entomological and epidemiological evaluation as well as to develop an aquatic monitoring protocol and to support teams of independent workers in their studies of invertebrate and fish populations in the rivers scheduled to be treated. OCP also acted as coordinating body in the fields of economic development. In the research sphere, it initiated and sustains work particularly in the fields of cytotaxonomy, classical taxonomy, laboratory colonization, methods of sampling and biogeographical studies of the vector. Intensive studies have been made yearly on the invading populations of vectors which appear at the onset of each monsoon season. Considerable efforts have gone into devising small-scale methods of testing insecticides, which will give satisfactory information on the likely effects of the insecticide when applied in the field. A wide variety of compounds and formulations have been tested in the field, with particular emphasis upon the study of micro-encapsulated formulations. A full-scale river trial of an alternative candidate larvicide has been completed.

Studies are underway on distinguishing between the microfilariae and other developmental forms of the various *Onchocerca* species which may occur in West Africa, and in elucidating the vector relationships. Medical studies have been heavily concentrated on the search for and testing of chemotherapeutic agents. In addition, OCP staff have amassed a great deal of invaluable data concerning the impact of onchocerciasis at village level, and the natural history of the ocular lesions caused by the disease.

Entomological Evaluation

In order to operate the control programme, it is necessary to obtain regular information concerning the suitability of rivers for breeding *S. damnosum s.l.* and their rates of discharge. This information, together with details of the presence of eggs, larvae and pupae and the number of biting flies, forms the basis on which the aerial operations are planned. To carry out this work, the OCP area is currently divided into 24 Sub-Sectors, each in charge of a technician. Each Sub-Sector consists of a small office and laboratory with its own radio-communications links. Each has three to five vector-collecting teams, consisting of two collectors and a driver, usually with an all-terrain vehicle. Each week, these ground-operating teams visit over 200 breeding sites where searches are made for larvae and pupae of *S. damnosum s.l.* Notes are made on the presence of other species of *Simulium*, and of blackfly eggs; also the general state of the breeding site. No attempt is made to rigidly quantify this work. Searches usually last between one-half and one hour and the occurrence of *S. damnosum s.l.* larvae and pupae is recorded using a four-point frequency scale, from absent to abundant. At certain times, breeding sites are innumerable. At high-water levels, it is often difficult, or impossible, to make thorough searches of the rivers. Thus, owing partly to this lack of quantifiability and the difficulty of access, most reliance is placed on the vector-collector catch.

In OCP, two-man teams are employed, each man working alternate hours, between 07.00–18.00 h local time each day. The catch is expressed in flies/man/day, and from this work an annual index—the Annual Biting Rate (ABR)—is calculated (Walsh *et al.* 1978). Ideally, the catches should be carried out from dawn to dusk. However, this is extremely difficult to achieve in an operational situation involving many widely dispersed teams, working very long hours. Since the start of OCP, adult blackflies have been caught at over 600 points, though only about 320 of these are now regularly visited. To mount this effort and the ground back-up to the aerial operations, the Vector Control Unit employs 563 men full-time. They include 14 graduate entomologists, one cytotaxonomist, and seven operations managers and technical officers.

The scale of the evaluation effort can be judged from Table 1, which gives a summary of the catches made during the last completed 12-month period, November 1977 to October 1978. In Table 2, details of the entomological evaluation for each February, from 1975 to 1979, are given.

Table 1. Entomological evaluation—adult blackfly catches.
Annual summary (November 1977–October 1978).

Month	No. of capture points	No. of days worked	No. of adults caught	Average daily catch	% Dissected	% Parous	No. of infective: adults	larvae
November	294	1293	23 832	18	66	83	380	722
December	256	1126	7 881	7	85	81	87	169
January	248	1019	5 171	5	78	78	43	91
February	217	858	4 000	5	50	86	10	24
March	213	1202	5 009	4	60	78	56	115
April	228	1293	58 669	45	23	94	305	601
May	265	1505	85 857	57	29	84	915	1648
June	290	1641	98 585	60	34	75	889	2336
July	310	1627	85 278	52	38	72	950	2174
August	304	1772	73 761	42	44	73	790	1695
September	324	1238	41 679	34	47	64	354	903
October	321	1315	42 808	33	45	61	379	952
Total		15889	532 620	34	39	75	5158	11 466

Too much should not be read into the figures concerning the number of adult blackflies caught, as variable numbers of catches were made outside the control zone at different seasons and in different years. Control started in mid-February 1975, so that the figures for that particular month represent virtually the pre-control dry season situation. In February 1978 62 days' catching was undertaken at 21 points outside the control zone; whereas, in February 1979 there were 210 days' catching at 66 such points. Thus, 7% of the evaluation work was carried out outside the control area in February 1978 and in February 1979 this had risen to 18%.

The adults caught are either preserved directly in alcohol for possible taxonomic or morphometric study (Garms 1978) or kept alive and dissected as soon as possible to determine their physiological age and the presence and stage of any *Onchocerca volvulus*. In addition, the presence of other filarial worms, and also entomopathogens (e.g. Mermithidae and Fungi) is recorded. Details are entered in triplicate on a work sheet. One copy is kept in the Sub-Sector, one is sent to the Parent Sector, and the third goes to Geneva for keypunching and storage in the WHO Headquarters Computer.

Control of *Simulium damnosum*

During the PAG Mission, careful consideration was given to the choice of an insecticide for the OCP. Field trials were

Table 2. Entomological evaluation effort. Summary of adult blackfly catches (February 1975–1979).

Year	No. of capture points	No. of days worked	No. of adults caught	Average daily catch	% Dissected	% Parous	No. of infective: adults	larvae
1975	98	395	13 760	35	88	60	143	—
1976	154	782	6 713	9	96	78	—	119
1977	256	1236	12 215	10	70	75	—	141
1978	217	858	4 000	5	50	86	10	24
1979	287	1154	16 671	14	49	68	127	560

carried out on several compounds and formulations using a variety of application techniques. It was eventually decided that an emulsifiable concentrate of temephos (Abate ®), formulated by Procida, could efficiently eliminate *S. damnosum s.l.* larvae while, at the same time, being very safe to handle (Laws *et al.* 1968), of low toxicity to fish, and relatively innocuous to the non-target invertebrate fauna and of fairly rapid biodegradability.

In West African conditions, larval habitat temperatures normally vary between 22 °C and 32 °C and may rise to about 35°C. Development of the aquatic stages is thus very rapid. The larva which alone is killed by the insecticide can develop from egg to pupa in as little as 8 days.

It was assumed from the start that to operate satisfactorily on the requisite scale, with the necessity of a weekly treatment cycle, it would be vital to use aircraft for the application of the insecticide. Operational research undertaken at the time of the PAG Mission indicated that not only was this the case but that only by the use of helicopers could the insecticide be satisfactorily delivered to the smaller, heavily-wooded streams and to the small, isolated breeding sites which are a conspicuous feature of the dry season. Consideration was given to the purchase of aircraft by WHO but eventually it was decided that the employment of an experienced aerial contractor with an international reputation would prove most efficient. During the life of the project there have been two contractors, Evergreen Helicopters of Oregon, USA, who used Bell Jet Ranger 206B helicopters and Pilatus Porter PC6 fixed-wing aircraft, and Viking Helicopters of Ottawa, Canada. Viking use the same type of fixed-wing aircraft but a different helicopter, the Hughes 500C.

For a specialized aerial operation, West African conditions pose many problems. Communication systems are poor and unreliable, excessive dust causes rapid engine wear, Harmattan winds from the NE give poor visibility and make navigation hazardous, and high temperatures result in decreased aircraft performance. For an aerial contractor, all these factors have to be evaluated and overcome to meet the terms of the aerial contract.

On its initiation, the aerial operation had to be carefully geared to the biology of *S. damnosum* and it was imperative that breeding sites received treatment regularly, once a week. This involved the choice of an aerial contractor capable of performing without fail, and the negotiation of a strict contract which ensured this. The contract drawn up between WHO and the selected contractor was hence lengthy, very specific in detail, and with a severe penalty clause in the event of default.

The aerial operations have now been in progress for almost five years. It has been possible during this period to improve the techniques of larviciding and to work directly with the contractor in the development of more efficient release equipment for this unique programme. At the start of the campaign, only one type of specially developed treatment equipment was requested, the *"vide-vite"* system, which allows the rapid discharge of carefully measured quantities of insecticide (Lee 1973; Lee *et al* 1973; Parker 1975). As OCP progressed, it became clear that treatment in the dry season posed a much greater problem than in the wet.

During, or shortly after the wet season (May–October), when the rivers are in full spate, it has been found that treatment is effective if made by spot application of the larvicide as a mass, rather than employing distribution methods. At this time, the flow of water is sufficient to mix and distribute the insecticide and to carry it an effective distance downstream. Consequently, application points are relatively few, one application often being sufficient on larger rivers for stretches of up to 40 km.

In the dry season (November–April), when the rivers are low, the flow of water is generally not strong enough to mix and distribute the insecticide. Different application equipment and methods have then to be used to achieve a satisfactory kill of the *S. damnosum s.l.* larvae. A band of insecticide, stretching from bank to bank, has been found to be effective and ensures that all parts of the breeding site downstream receive treatment. Two or three such bands may be required at some particularly difficult breeding sites in order to obtain an adequate contact time between larvae and insecticide. In the dry season, therefore, treatment involves careful placement and distribution of insecticide. Sometimes there is virtually no flow between suitable breeding places and each one may have to be treated individually. Thus, several treatments may have to be made on one stretch of river only 1 km long.

Treatment rates vary, with dosage usually between 0.03 and 0.10 ppm, measured over a 10-minute discharge period. On larger rivers during the wet season the lower dosage rates pertain; during the band treatment of the dry season, the higher rates of treatment may be used. In extreme cases, the insecticide may be diluted two or three times in the aircraft's insecticide tank at the start of the treatment circuit, in order to provide a dose of sufficient volume of liquid to enable the pilot to make an adequate band treatment reaching from bank to bank.

The aerial operation involves a huge logistic exercise. Both the helicopters and fixed-wing aircraft have an endurance of about 2.5 h. This means that a refuelling stop, often for insecticide also, must be made every two hours. Over 100 fuel

Extent of rivers treated in the wet season in the last week of August 1978 (454 bush landings made in August).

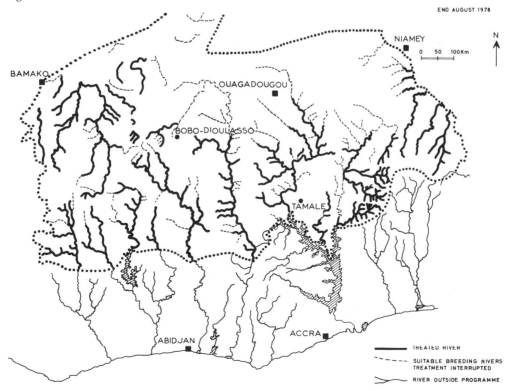

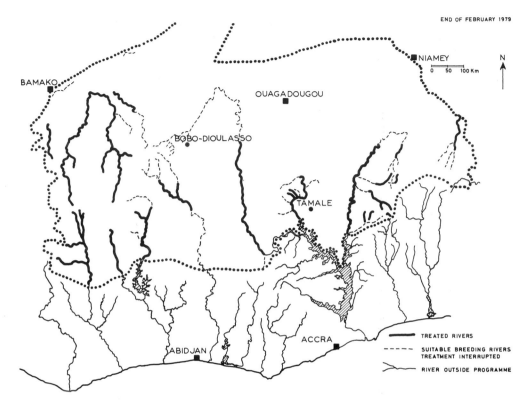

Extent of rivers treated in the dry season in the last week of February 1979 (191 bush landings made in February).

caches are hence scattered over the OCP area, where stocks of fuel and insecticide must always be present. The stocking of these depots is, in itself, a carefully planned exercise, for many roads are impassable during the wet season and supplies have to be distributed before the rains start. In 1978, over 800 000 litres of aviation fuel (JET A1) and 250 000 litres of insecticide had to be so distributed. The work is markedly seasonal; in the wet season up to 18 000 km of river are treated and surveyed weekly, and only about 6000 km of river in the dry season.

The first aerial contractor was installed at the base in Bobo-Dioulasso in October 1974 and final survey flights by both fixed-wing aircraft and helicopters were made before treatment flights (using one Pilatus Porter and one Jet Ranger) commenced in December. Operations recom-

menced in February 1975 using one Jet Ranger, with a back-up machine available. It has continued uninterrupted in Phase I ever since. On the White Bandama River between Ferkéssédougou and the Kossou Barrage in central Ivory Coast, over 220 consecutive weekly treatments have been made.

Activities extended to cover most of northern Ghana and central Upper Volta (Phase II) in March 1976. During wet-season operations, the size of the aerial fleet operating from the Bobo-Dioulasso base, increased to five helicopters and one fixed-wing aircraft.

At the end of 1976, the aerial contractor changed and his successor was installed with half his fleet at Bobo-Dioulasso and the other half at Tamale in Northern Ghana. In March, 1977, the control activities were extended to cover Mali as far west as the Niger at Bamako (Phase III West) and later into northern Togo and Benin (Phase III East). The present wet-season aircraft fleet operating in this area and the Ivory Coast extension, consists of eight Hughes 500C helicopters and two Pilatus Porter fixed-wing aircraft.

In order to mount the aerial operation, the various factors determining the scale and pattern of the control operations have to be reassessed each week. To achieve this objective, a resumé of the findings of the entomological evaluation teams and, in particular, the vital hydro-logical gauge readings are radioed each Saturday morning to their respective Aerial Operations bases. Details of the occurrence of rainfall, line squalls, etc., and the position of the ITCZ are also transmitted. The data from the eastern half of the OCP area are recorded by the radio operator at the OCP Headquarters in Ouagadougou.

The daily biting rate for each point visited by the vector collecting teams is plotted on 1:1 000 000 scale maps, the presence of catches containing nullipar-ous females being separately indicated. Sites of larval searches are plotted, in red where *S. damnosum s.l.* breeding is occur-ring successfully (presence of pupae or late instar larvae which could not have developed in the period since the pre-vious dose if that had been successful), and green where no evidence of breeding *S. damnosum s.l.* has been found.

Entomologists and technical officers of the Unit based in Ouagadougou hold a brief meeting late each Saturday morning to consider the significance of the week's entomological results and of the overall hydrological and meterological situation. After reflection, over the remainder of the weekend, discussion with the Chief, Aerial Operations, are held first thing on Monday morning. Decisions relating to the intensification or reduction of treat-ment for each river basin and the necessi-ty of any prospection flights are made. Details of these decisions are recorded in the Operations Room Log Book. The

Aerial Operations Manager at the Tamale base is brought into the discussions by radio and the necessary advice and in-structions given. Sector Chiefs are con-tacted and advised of any modifications to their surveillance programmes. In a limited number of cases, the Sector Chiefs may be asked to carry out ground application of insecticide. Such treat-ments are made routinely when survey teams find evidence of treatment failures.

In the better-known western half of the OCP area, the Aerial Operations Mana-ger and the entomologists based in Bobo-Dioulasso make their own decisions, only contacting Unit Headquarters when major changes in operations are contem-plated. Nevertheless, the weekly data from the western areas are passed to the Headquarters as soon as more urgent radio traffic ceases. This information thus usually reaches Ouagadougou late on Monday or early on Tuesday.

The Aerial Operations Officers work throughout Monday, calculating river discharge rates and insecticide quantities required. Treatment maps are prepared for each aircraft, indicating the treatment circuit, refuelling and night stopping points. The rivers to be treated are marked and the dosage rates on each stretch of river shown. In ideal condi-tions, the exact location of each insecti-cide application is given. However, it is often impossible to be precise about this last factor. In such cases, more general instructions are given, indicating the maximum number of dropping points on particular stretches of river.

A pilots' briefing is held by the Aerial Operations Officers late on Monday afternoon when any problems relating to the treatment circuits are discussed. At first light on Tuesday, the treatment air-craft leaves on what is usually a 3-day circuit. On return to base, each pilot is de-briefed. Treatment maps are re-turned, with the pilot's annotations showing what treatments were actually made. Reasons for any divergence from the original plan are discussed and the more experienced pilots provide the Aerial Operations Manager with an invaluable source of up-to-date information concerning hydrological and weather conditions. In order to facilitate this feedback, pilots are usually kept on the same treatment circuit for several consecutive weeks and are thus able to appreciate the changes occurring to individual rivers and even specific breeding sites of importance. The work is hazardous and arduous, cockpit temperatures regularly exceed 40 °C in the hot season and the skill and enthusiasm of the pilots is a key factor in the success of the whole operation.

Effects on the Larval Populations of the Vector

Within two or three weeks of the start of larviciding operations, S. damnosum s.l. larvae had disappeared from most sites. This situation has continued since then. However, in no treatment cycle are all breeding sites effectively treated, and sporadic outbreaks of breeding occur from time to time. Certain sites have proved exceptionally difficult to treat and various combinations of ground and aerial application of insecticides have been tried. Treatment problems are frequently caused by man-made breeding sites formed by causeways, broken bridges and small and large dams. Many of these seem to have been designed specially to plague the Simulium controller. Several of the more troublesome breeding sites could probably be reduced by a small-scale engineering effort, as could many of those formed by redundant man-made features. Programme staff will be looking into these problems in 1979–1980.

Larval searches following successful applications often reveal a surprising abundance of healthy non-target organisms, These frequently include the blackflies Simulium adersi, S. griseicolle and S.

schoutedeni. Viable well-grown larvae of these species may be present 24 hours after larvicidal treatments which have eliminated S. damnosum s.l. Thus, it is not just a question of rapid growth, allowing development from egg to pupae between successive treatments, though this aspect undoubtedly contributes to the success of S. adersi, in particular. The survival of S. griseicolle, at Kainji, during successful anti S. damnosum s.l. operations using DDT, has been noted previously (Walsh 1970b), differences in feeding behaviour being suggested as offering a possible explanation (Walsh 1970b).

Routine susceptibility tests have indicated no development of resistance to Abate® so far. This is not surprising, given the relatively short period during which control has been carried out and the regular and large scale influx of adult blackflies from untreated areas.

Effect on Biting Densities and Transmission

The overall effect of OCP is best considered by reference to two criteria; the Annual Biting Rate (ABR) and the Annual Transmission Potential (ATP) (Walsh et al. 1978). These are estimates of the total number of bites that would be received by a person sitting at the capture-point all day for one year, and the number of infective larvae that he would receive from these simuliids.

Pre-control data (see Table 5) have been compiled from many sources since, for example, in Phase I, OCP was not able to collect a full year's pre-control data before larviciding began. Much of this information has been obtained from the OCCGE/IRO at Bouaké, Ivory Coast. The two pre-control maps are remarkably similar. They indicate that over most of the OCP area, roughly one infective larva indistinguishable from O. volvulus was found for every 10 S. damnosum s.l. examined.

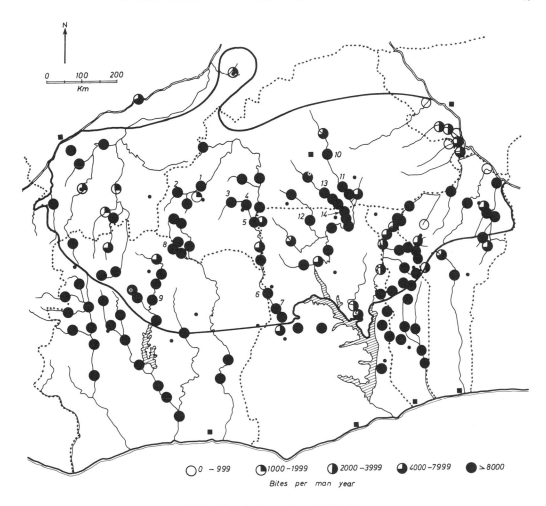

Annual Biting Rates—pre-control. Numbered sites refer to Table 5.

At a WHO Scientific Working Group, which met in Geneva (June 1977), to assess the Biomedical Criteria for Resettlement in the OCP area, it was concluded that vector biting rates and transmission potentials, based on dissection results, should be taken in conjunction, when assessing results of control or making predictions. The concept of a Tolerable Level of *Onchocerca volvulus* Transmission in resettlement areas was proposed. This was basically intended to indicate those levels of infection at which there was no likelihood of dangerous ocular lesions developing.

From the pre-control evidence available, it was agreed that an ABR of less than 1000 and an ATP of less than 100 should be satisfactory in this respect. It was felt necessary to introduce the concept of ABR to provide a yardstick for uninhabited areas where ATPs might be low, but which might later be resettled by human populations already infected with onchocerciasis.

Data suitable for calculating ABRs exist amongst old records for many areas of West Africa, whereas adequate information concerning transmission potentials is rarely available.

In only five sites, mainly in the extreme NE, were pre-control ABRs less than 1000 and at only 10 of the points was the ATP less than 100. Between November 1977 and October 1978 however, the majority of points in the original programme area showed acceptable ABRs and ATPs. In most cases, the levels of ABR and ATP are in agreement. The most interesting exception lies at the confluence of the White and Red Voltas, a densely inhabited area of hyperendemic onchocerciasis. There, in spite of largely successful control, infection rates in the simuliids are exceptionally high so that the level of transmission is still unacceptable. In this particular instance, there was incomplete control during May 1978 when adult blackfly densities reached about 10/day at Nangodi Bridge (Site 14) before a solution could be found. Although at this site the ABR remained below 1000 at 587, the ATP rose to 102 (Table 3), indicating the importance of using both criteria together. In Table 3, more detailed annual data on ABRs and ATPs are given for 14 of the best-known and often most dangerous sites within the Phase I and II zones from which at least adequate pre-control ABR data were available. The positions of the sites are given by the reference numbers on the map. As can be seen, with the exception of those sites located within the invasion area, ABRs and ATPs have generally been reduced to less than 10% of the pre-control figures, even at these difficult sites. In the heart of the Volta River Basin

in the upper Red and White Volta Valleys transmission seems to have been completely interrupted and, in many cases (e.g. Wayen Bridge) no simuliids have been caught for several years.

However, as a result of the occasional occurrence of adult blackflies at distances of over 400 km from their probable sources, and of the very rapid development of the aquatic stages, it has not been possible to terminate the larviciding campaign over large areas except in the upper reaches of the Black, Red and White Volta Rivers. As can be seen many suitable tributary streams were left untreated in August 1978 but most of the main rivers were treated.

The Invasion Problem

Following the commencement of larviciding in Phase I (February 1975), a rapid fall in adult blackfly densities was observed over the whole of the treated area. All indications were that the helicopter/insecticide application techniques were successful in controlling *S. damnosum s.l.* breeding on a large scale. This initial success was short-lived, for in March 1975 biting rates on the Bandama and parts of the Leraba Rivers began to rise. Strenuous efforts were thus made to find unsuccessfully treated breeding sites which could account for the numbers of adults observed. These careful searches failed to reveal a breakdown in control.

Table 3. Growth size of OCP area.

	Original OCP area				Ivory Coast extension	Extension study area
	Phase I	Phase II	Phase III W	Phase III E		
Area in 1000 km^2	247	134	77	196	100	111
Length of river (km) treated in wet season	4500	2352	3000	4500	3500	4135
Month and year of starting control	II/1975	III/1976	III/1977	III to VII/1977	V/1978– III/1979	—

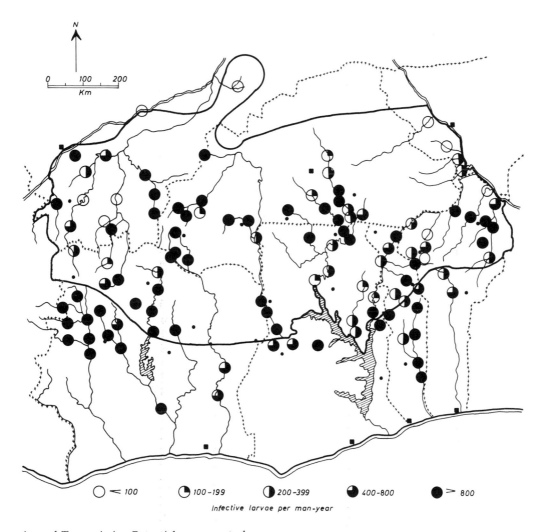

○ < 100   ◑ 100-199   ◑ 200-399   ● 400-800   ● > 800

Infective larvae per man-year

Annual Transmission Potentials—pre-control.

Dissections indicated that the adult blackflies were predominantly old, many of them carrying infective stages of *O. volvulus*. It was thus assumed that the OCP area was being reinvaded by substantial numbers of infective vectors. This phenomenon has proved to occur annually at the end of the dry season, as the Intertropical Convergence Zone (ITCZ) moves north of the OCP area.

Investigations made between 1975 and 1978 are reported in detail by Garms *et al.* (1979). They have indicated that there are extensive movements of adult black-flies from southwest to northeast, the direction of the prevailing winds south of the ITCZ. Vast numbers of adults must be involved, many of them penetrating 250–300 km into the control area. Occasional examples travel at least 400 km. Most such long-distance migrants appear to be old. They are almost invariably parous when caught by the vector collectors. Presumably, they arrive some time after taking a blood meal to oviposit before feeding again nearby. Many appear to be very old and a high proportion (up to 15%) may be carrying infec-

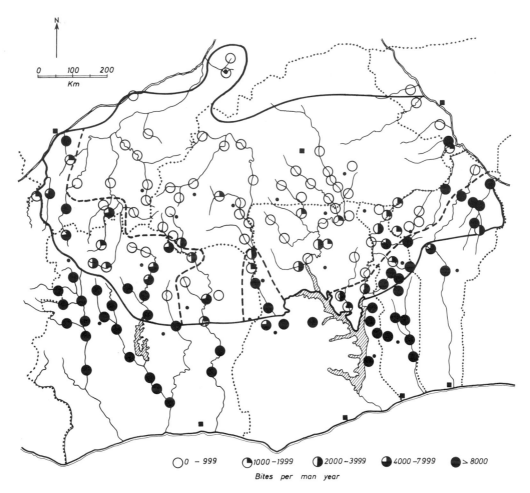

○ 0 – 999   ◔ 1000 – 1999   ◑ 2000 – 3999   ◕ 4000 – 7999   ● ≥ 8000

Bites per man year

Annual Biting Rates: 1977–1978. The area considered to be subject to significant numbers of invading blackflies is indicated by the broken line.

tive third-stage larvae in their heads. Thus, it is likely that they are capable of making more than one long-distance migratory flight. The pattern of arrival at different sites and the coincidence of peaks at sites such as Chaussée Niakar-amandougou (No. 9) and Leraba Bridge (No. 8) suggest that these movements are wind-borne.

The presence of infective parasites being extremely adverse to the Programme, a considerable effort was made to determine the source of such adults. Rearing, cytotaxonomic and mor-phometric studies have show that vir-

tually all migrating flies are either *S. damnosum s.s.* or *S. sirbanum*, and that *S. sanctipauli*, *S. soubrense* and *S. yahense* do not make such movements. The role of *S. squamosum* is not yet clear. Energy-dispersive X-ray spectroscopy has so far failed to identify any particular source area, but has indicated that adult black-flies arriving at Leraba Bridge are not all of the same origin (Bennett, personal communication), a finding confirmed by Garms (1978) using morphometric methods.

Following extensive surveys, it became clear that the two savanna cytospecies (*S.*

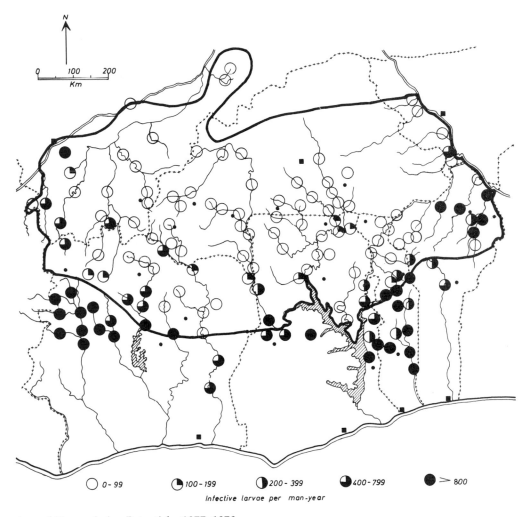

O 0-99   ◑ 100-199   ◐ 200-399   ◕ 400-799   ● > 800

Infective larvae per man-year

Annual Transmission Potentials: 1977–1978.

damnosum s.s. and *S. sirbanum*) were much more widely distributed in the forest and transitional vegetation zones than had orginally been thought (compare maps in Vajime and Dunbar, 1975, with those in Vajime and Quillévéré, 1978). This has resulted in the extension of control into the valleys of the Sassandra, Marahoué and southern Bandama rivers in Ivory Coast. Preliminary results suggest that this area was largely responsible for the simuliids which invaded the upper Bandama and Leraba rivers.

About 25% of the original OCP area was subjected to a fairly substantial influx of *O. volvulus*-carrying blackflies. However, within this zone, and under control conditions, the adult simuliids are fairly strictly confined to the close vicinity of the rivers in marked contrast to the situation in the pre-control period. A good example of this is the case of the village of Dangouadougou, 7 km from the Leraba River on the frontier between Upper Volta and Ivory Coast. In 1972, between April and August, *S. damnosum* s.l. could be caught in the village at a

Table 4. Annual totals of flight hours and insecticide dispensed.

Year	Hours flown H/C[a]	Hours flown H/W[b]	Insecticide dispensed (10001.) H/C	Insecticide dispensed (10001.) F/W
1974	144	104	0.56	3.1
1975	2783	541	46	30.0
1976	4265	614	80	50.0
1977	5358	1026	84	71.0
1978	5356	1204	95	121.0

[a] Helicopter.
[b] Fixed-wing aircraft.

mean density of 13/day, compared with 136/day at the river by the Leraba road bridge. In 1975, 1976 and 1977, mean biting rates were 0.008, 0.12 and 0.57/day in the village compared with 103, 95 and 45/day at the bridge. So, while reinvasion had been responsible for little post-control change at the river, the effect at the village was considerable. It was with great surprise, therefore, that we learned from the epidemiologists that, in their 3-year follow-up survey of Dangouadougou, they had found that amongst the villagers there was no change in the ocular lesions of onchocerciasis victims, compared to changes in other areas within the control zone (Rolland and Thylefors, 1979). In effect, transmission was continuing at pre-control levels, contrary to the entomological evidence. The villagers, when questioned, denied that they ever visited the river where infection could be expected to take place.

In an attempt to resolve this disturbing anomaly, a detailed helicopter survey was made of the area surrounding the village. This revealed the existence of a footpath which left the village on the opposite side from the river, but, after avoiding a swamp, returned to the Leraba River some 10 km upstream from the bridge. A large area of wet-season farms was located here, with huts where whole families stayed during the cultivation season (which coincided with the peak reinvasion period). Some farms were as close as 0.5 km from the river bank, and it is inevitable that the river formed the main human water source. Adult catches confirmed the presence of infected S. damnosum s.l. It is thus clear that the infection was taking place at the river after all.

Pathogens and Parasites of Blackflies

During the routine evaluation dissections, pathogens and parasites of Simulium damnosum s.l. have been recorded on numerous occasions. There are several instances of populations with a relatively high prevalence of ovarian fungal infections. One such population was found on the River Bou, a tributary of the Bandama River in Ivory Coast, in August, 1975 (G. Zerbo, personal communication), where 64% of nulliparous S. damnosum s.l. were found harbouring a fungus superficially similar to that described by Garms (1975).

By far the commonest parasites, apart from Onchocerca spp., in adult blackflies are mermithid worms. These have been found widely scattered across the OCP area. Several collections have been identified as Isomermis lairdi and, on occasions, this species may be quite common. At Nakong Bridge (Site 12), a notorious focus of onchocerciasis transmission on the River Sissili in Ghana, during the late wet season in 1974, over 50% of S. damnosum s.l. caught biting humans were carrying mermithid worms (I. lairdi was identified from several samples on different dates). Almost invariably, the infected flies were sterile, but because there were also several parous blackflies infected with O. volvulus, control commenced with Phase II. With hindsight, this control may not have been necessary, as it is possible that the infected parous blackflies were not of local origin.

During the mermithid season (October–November), it might be possible to interrupt chemical control. However,

Table 5. Comparison of Annual Biting Rates and Annual Transmission Potentials
at 14 sites within Phases I and II.

Ref. no.	Place name	Phase	Annual Biting Rate					Annual Transmission Potential				
			1975	1975	Pre-1976	1977	1978	1975	1975	Pre-1976	1977	1978
1	Samandeni Bridge	I	11 223	2 163	232	120	51	>260	4	7	13	11
2	Lanviera	I	52 245	3 713	1 511	538	318	>920	19	55	65	29
3	Nabere	I	32 984	6 273	3 026	935	1 576	929	162	105	64	96
4	Diebougou Bridge	I	62 955	3 357	2 144	613	360	—	42	42	29	0
5	Mombello	I	9 230	1 015	600	135	165	—	45	3	0	0
6	Chache[a]	I	34 957	14 841	2 724	642	2120	—	833	101	14	266
7	Agbolekame[a]	I	55 644	19 033	15 381	10 223	24 757	>1 072	1 300	1 119	700	2 686
8	Leraba Bridge[a]	I	158 800	26 289	16 827	8 632	6 833	15 800	1 660	1 324	1 020	488
9	Ch. Niaka[a]	I	25 333	22 689	5 286	11 570	9 057	—	1 959	403	504	454
10	Wayen Bridge	II	9 479	54	0	0	0	249	0	0	0	0
11	Yakala	II	22 419	4 246	219	226	110	1 080	172	5	25	0
12	Nakong Bridge	II	19 378	3 987	722	1 882	1 794	—	43	27	95	95
13	Po Bridge	II	16 782	7 779	300	62	217	982	77	0	0	0
14	Nangodi Bridge	II	66 903	7 854	602	795	587	2 695	310	29	67	102

[a] Sites within the invaded area.

since substantial breeding takes place on this river from June to November, the controller would need a great deal of convincing biological information before he could consider suspending regular conventional insecticidal control in favour of integrated methods. Garris and Noblet (1975) report that parasites of Simuliidae (including mermithids) in the streams of South Carolina were not affected by Abate®. Therefore, provided that *I. lairdi* is not *S. damnosum s.l.*-specific, such treated rivers could retain this capability of exerting biological control at certain seasons, by the parasite surviving in other *Simulium* species less affected by the treatments.

Aquatic Monitoring

From the outset, WHO treated the environmental aspects of the programme very seriously, as is indicated by the fact that the first of the advisory committees to be established was the Ecological Panel which met as early as January 1974. An *ad hoc* Ecological Group met in May 1974 to prepare an aquatic monitoring protocol. Unfortunately, owing to the paucity of biological information on tropical rivers in general, and on those of West Africa in particular; the absence of indigenous hydrobiologists in six of the seven countries in the OCP area; and the political pressure to implement control at the earliest possible moment, the aquatic monitoring activity in Phase I zone was preceded by the control operations. However, monitoring activities were eventually developed in Ivory Coast, Ghana and Upper Volta. At present, the invertebrate and fish faunas are both monitored by standard procedures (Lévêque et al., 1977) at nine sites. Two additional sites are monitored for fish only, and two for invertebrates only. Of these sites, four have been studied both before and after the implementation of *Simulium* control operations, and two are located outside the OCP boundary. Numerous other sites have been visited by consultants from Canada, France and the UK, in order to collect control data.

All this field activity has been carried out by independent workers but WHO has, of necessity, had to provide financial support.

So far, it seems as though the larviciding activities of OCP have resulted in no serious overall diminution of the faunal diversity, that the numbers of invertebrate organisms have been reduced by roughly 30% (Dejoux and Elouard 1977), and that no significant changes have occurred in the fish populations.

A detailed statistical analysis of the data from all sites was prepared, and the results considered by the field workers involved, together with a small international group of experienced hydrobiologists, in late September 1979. The findings of this *ad hoc* working group will be placed before the Ecological Panel, which numbers, among its members, eminent hydrobiologists and toxicologists. Their conclusions are awaited with considerable interest.

In addition to these specialist studies, the OCP staff maintain general surveillance of the health of the rivers. Several reports of fish mortalities have been investigated, but none have so far proved well-founded. The Programme was once accused of causing serious reductions in fish populations, but the 1974 decline in fish catches in Northern Ghana, which occasioned these charges, could well have resulted from several consecutive years of drought. Moreover, OCP had not then begun its insecticide campaign in the area, though its aircraft had made several prospection flights at different seasons. More amusingly, it has even been suggested that OCP was responsible for a decline in the wild rabbit[11] [*sic*] populations of northern Togo. By precisely what mechanism this striking performance was supposed to have been produced was never elucidated! (Walsh 1979).

Conclusion

It has been the object of this chapter to

describe the vector control activities in the Volta River Basin, and to give a brief outline of the progress to date. Although insecticidal treatment of the rivers included in Phase I has continued without interruption every week since February 1975 the rivers of the extreme east were not included in the scheme until July 1977. Therefore, the whole of the original OCP area has been operational for only 20 months.

In many ways, the project has been an enormous field trial. Although the instigators of the scheme were confident of success, one wonders whether they would have attempted it had they known the number, variety and complexity of the entomological, logistic and administrative problems they were due to face. The invasion problem, although anticipated, was only revealed completely by the efficient elimination of all local breeding which had previously masked the presence of the invaders. It turned out to be greater than at first envisaged. For the entomologists, field staff and many others associated with OCP, it generated an enormous interest and the studies which were undertaken to try to solve the mystery of the origin of the invading blackflies involved large tracts of Ivory Coast, Upper Volta and Mali. Currently, similar studies, aimed at clarifying the situation in the eastern sector, range over Ghana, Togo and Benin.

As a result of this activity, knowledge of the morphology, cytotaxonomy, and bionomics of *S. damnosum s.l.* in West Africa has advanced rapidly, and the empty spaces in the puzzle of its life are slowly being filled. However, many still remain. In particular, collection methods are still limited to catching hungry or ovipositing females. Also, virtually nothing is known for certain about what happens between emergence and the first blood meal, or between blood meals. There is an urgent need for a trapping device to replace, or at least augment, the Vector Collector, so that the low blackfly densities that it is hoped will continue to

---

[11] Actually the animal in question was the Togo Hare, *Lepus capensis zechi* (Brer Rabbit of Noel Chandler Harris). True rabbits, *Oryctolagus* spp., do not occur south of the Sahara.

exist over most of the OCP area can be monitored more efficiently and cheaply.

Operationally, there have been advances; such as the development, by aerial operations staff in collaboration with the contractors, of the "dribble" release system for dry season applications. So far, the helicopter has proved to be indispensable not only as an application tool, but for survey and research work. It is expensive, but when the work that can be achieved in one day is compared with the cost of a similar ground operation in terms of time, manpower, vehicles and effectiveness, it proves a very attractive investment. The use of aerial application methods also has a very important aspect which is often not fully realized. In OCP, the application of the insecticide to the 17 800 km of river, is the responsibility of only 10 men—the helicopter and aircraft pilots (compared to the hundreds of knapsack-sprayer operatives in malaria and yellow-fever schemes), so that they can be trained and supervised to a much higher level, and any failures can be quickly brought to the notice of the pilot responsible.

As to the future, we are optimistic. Given the present level of cooperation between the peoples and governments of the seven participating countries, the continued support of the donor countries, and the very encouraging results of the first three-year follow-up epidemiological and opthalmological surveys which are now beginning to come in, we feel that the aim of OCP—to reduce the disease to a level at which it is no longer of economic importance—is almost within our grasp. This optimism would appear to be shared by others, judging from the requests that are being received for extensions or similar schemes in Togo, Benin, Nigeria and the Senegal River Basin.

Preliminary Report of Japan-Guatemala Onchocerciasis Control Pilot Project

K. Ogata

In 1915 Dr R. Robles became the first to record American onchocerciasis from Guatemala, where it is accordingly called "Robles' Disease". He assumed that the vectors would prove to be blackflies and recommended that nodulectomy would be the effective treatment. Since his discovery, both original research and experimental trials concerning treatment and control measures have been carried out in this country. Indeed, Guatemala was the first country from which results of the practical use of DDT, as a larvicide against a blackfly vector of onchocerciasis, were published [Fairchild and Barreda 1945; see pp. 71–72 . . . Ed.]. Denodulization has continued to be practised there as a control measure for six and a half decades since this was first urged by Robles. The difficulty of vector control under local conditions was a contributory reason for this. Meanwhile, after detailed and careful preparation, OCP was launched in West Africa, late in 1974, by the Executing Agency, WHO.

According to progress reports issued by WHO, this programme continues to show real promise [see preceding chapter and p. ix for progress to date . . . Ed.].

Against this background, the government of Guatemala requested that of Japan to furnish technical cooperation with respect to onchocerciasis control. In March 1975 Japan sent a mission to Guatemala to study the feasibility of such cooperation. Its aims were to investigate the epidemiology of onchocerciasis, and existing disease control measures; and to recommend how best to organize the proposed venture. Accepting the report, Japan despatched a second mission in June 1975 to negotiate a protocol for a collaborative project. The activities discussed in this chapter, duly commenced in May 1976. By then, Japanese experts were on the spot and the necessary supplies and equipment had been received.

Entomological aspects of the project and progress over the first two years are outlined below. Much of the data so far obtained [until December 1978 . . . Ed.] remaining unpublished, credit for this work should be assigned to the entomologists concerned with the project.

Object of the Onchocerciasis Control Pilot Project in Guatemala

Measures for the reduction of onchocerciasis include vector control, nodulectomy and chemotherapy. In the last-mentioned connection, diethylcarbamazine and suramin both have a rôle. However, because of the possibility of massive side effects, the greatest care is needed for the administration of these drugs. While they may be applicable to the treatment of individual cases, we do not yet have a basis for mass treatment, as ORSTOM and WHO experience in West Africa has shown.

While nodulectomy has been found beneficial in achieving some decrease in

actual eye damage due to *Onchocerca vol-*
vulus, it seems only of supportive (largely
cosmetic) value, in a major control effort.
It is submitted that the only really effective approach is the actual interruption of
onchocerciasis transmission through vector control. Historically, the most conspicuous success in the latter connection
was achieved against *Simulium neavei* in
Kenya, a quarter of a century ago (McMahon 1958). Currently, WHO's OCP, commenced in the Volta River Basin in late
1974, is achieving very promising results
[pp. 85–103 . . . Ed.] in suppressing the
adult density of *Simulium damnosum s.l.*
by the aerial spraying of Abate® formulations.

In Guatemala, large-scale vector control trials took place in the early 1950s
(Lea *et al.* 1955). Approximately 1500
streams, in an area of 205 km², were then
treated with DDT—without achieving
more than a slight decrease in adult
blackfly population density.

In 1971 WHO sent Dr J. P. McMahon to
Guatemala to investigate the feasibility of controlling onchocerciasis vectors
(McMahon 1971). He recommended larviciding, by ground or aerial application,
emphasizing, however, the difficulty of
implementation. He also outlined an
appropriate research and training plan.
In Mexico, the small-scale control of
onchocerciasis vectors was initiated at
the national level in 1930. After 1965,
though, organizational changes reduced
effectiveness (Mallén 1974). When the
present project commenced, no organized onchocerciasis control activities
existed in Latin America.

The aim of this venture is to determine
the feasibility of blackfly control in
Guatemala. A pilot-project area has been
established in the endemic zone, and
intensive entomological, parasitological
and epidemiological investigations have
been initiated. This, though, is purely
the preparatory phase for a control programme directed towards interrupting
onchocerciasis transmission, by larviciding operations against *Simulium*

ochraceum. From the results of this pilot
project, it is anticipated that a control
programme suited to the special difficulties faced in Guatemala will be
developed. At the same time, local personnel will gain experience enabling
them to staff a future, national-level campaign.

Onchocerciasis in Guatemala

The Central American Republic of Guatemala has an area of 108 780 km² and a
population of 6 500 000.[12] Of the 22 Departments, under which the country is
administered, at least seven are subject to
onchocerciasis. There are four foci of the
disease. Zones I and II are small, and
located in Huehuetenango (northwest
Guatemala, adjacent to Mexico). Zone III
is an area of high endemicity, where
workers on some coffee plantations have
an infection rate of almost 100%, with

Onchocerciasis foci in Guatemala.

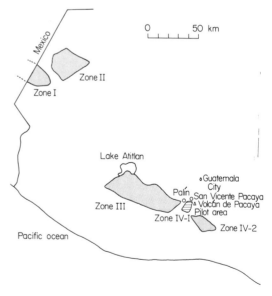

12 1978 estimate from *Background Notes*, US
Department of State Publications, Office of
Media Services, Bureau of Public Affairs,
Washington DC, USA . . . Ed.

>10% ocular and skin lesions and 1% blindness (Figueroa 1974). Zone IV is subdivided into Zone IV–1 and IV–2. It is somewhat isolated, being located at the boundary of the Departments of Escuintla and Santa Rosa. In Zone IV, endemicity is low and ocular lesions and blindness are seldom seen. Zones III and IV total approximately 125 × 35 km, with an area of between 4000 and 5000 km^2. They extend from 600–1500 m above sea level on the Pacific slopes of the volcanic region. Scattered through them, are numerous coffee plantations.

According to the 1970 report of the Section of Onchocerciasis, Ministry of Public Health and Social Assistance, 56 820 inhabitants in the endemic area of 5232 km^2 (total population: 385 000) were examined by the nodulectomy teams. Altogether, 5189 were found to exhibit nodules. On this basis, it was estimated that there might well be approximately 31 145 such cases throughout the country. However, the actual number is now assumed to be several times that—obviously, this must be a multiple of the figure, for those showing nodules. Before 1976, there had not been any comprehensive surveys for endemicity (based upon microfilaria rate, as revealed by skin snips). According to data obtained during the period from 1946 to 1964 (Garcia 1965), Acatenango has the highest nodule rate (32.7%). Cuilco, Yepocapa, San Lucas Tolimán and Santiago Atitlán show 30.7, 22.2, 21.5 and 21.6% respectively. At Recreo and Panajabal (in Yepocapa), microfilarial rates of 97% and 96%, respectively, were obtained.

The chief manifestations of American-type onchocerciasis are: onchodermata, nodules, elosion and blindness. The most severe symptoms relate to eye damage. Of 663 onchocerciasis cases examined (by Estrada *et al.* 1963) on seven plantations, keratitis, iritis and blindness were present in 50, 10 and 2.1%, respectively.

Until 1976 the Onchocerciasis Section of the Health Service Bureau, Ministry of Public Health and Social Assistance, was administratively responsible for measures against this disease. For more than 40 years, the only such measure had been nodulectomy. This work had been performed by brigades, each including six teams, consisting of one operator and one assistant. These brigades travelled through the endemic area twice or three times annually. Although some authorities considered that their efforts reduced blindness to some extent, there was no observable alteration in the annual incidence of nodules. One interesting finding, however, was that onchocerciasis endemicity is closely related to the distribution of coffee plantations. A distinct socio-economic factor must therefore be recognized as regards the onchocerciasis problem in Guatemala.

Description of the Pilot-project Area

In selecting this area, the following criteria were given special weight. The area must: be representative of the endemic zones of Guatemala; be isolated, to avoid vector infiltration from surrounding localities not under blackfly control; have topographic barriers, whether mountains, valleys, lakes, etc.; be at least 50–100 km^2; exhibit moderate onchocerciasis endemicity; include several coffee plantations; have permanent inhabitants; be easily accessible; and be as free as possible from flood hazards, etc.

Zone IV–1, located in San Vincente Pacaya, Department of Escuintla, proved to satisfy most of these relevant criteria, and it was decided to establish the pilot project there.

San Vincente Pacaya lies approximately 30 km south of Guatemala City. Volcán de Pacaya (2500 m) is northeast of the project area, which extends to the southern Pacific slopes at 500–1500 m. Municipio San Vincente Pacaya covers 236 km², and has approximately 6300 inhabitants. The selected area comprises some 168 km² of the Municipio, measuring about 12 km from west to east and 14 km from north to south. It is on a slope in which some 1000 m of altitude are lost in only 10 km. The equivalent of a natural barrier on the western border of the area is a highway, running through the valley from Guatemala City to Puerto San José. The other borders are with non-endemic areas to the north, east and south. The chief village of the project area is San Vincente Pacaya, near the northern border. A number of coffee plantations are scattered through the southern localities, their populations being rather stable. Northward there are few permanent settlers, but some workers are employed outside the pilot area, in Palín, to and from which they travel daily.

Endemicity of Onchocerciasis in the Pilot-project Area

The initial microfilarial survey in this area, revealed positive skin snips for 513 (33.2%) of the 1540 inhabitants examined (Tada *et al.* 1977), with higher rates from villages at 700–1200 m, than in those still higher (on the plateau above 1700 m) or in the lowlands below 500 m. Of 1318 people locally examined, a few years earlier (Garcia 1971), 127 (9.6%) exhibited nodules. In the present survey, which concerned 13 villages, the ratio of the microfilarial rate (based on a single skin snip) to the nodule rate (by palpation) showed close correlation. In males, the microfilarial rate rose with age (highest value, 75% in the 40–49 year age group). In females, though, it was about half that for males in each of the age groups. It was thus suggested that the heavier parasitaemias of males may result from their being more frequently exposed than females to infective bites, for occupa-

tional reasons; as is well-known from other filariases.

Among 500 persons found to have nodules, 56.5% exhibited these on the head (25.2% from accipital and 22.1% from parietal regions). The remaining 43.5% were on the trunk (16.4% from the iliac regions and 6.2% from the scapular ones). A slit lamp was used to examine the eyes of 1037 subjects, of whom 988 were also examined by skin snip. Sixty-one (6.2%) of these people had microfilariae in the anterior chamber, the incidence being 8.7% in males, and 2.5% in females. There was clear correlation between the anterior chamber positive rate and microfilarial density rate in the skin. As already mentioned, the only earlier onchocerciasis control measure practised in Guatemala was nodulectomy. It is thus interesting to report that of 1513 inhabitants examined, from San Vincete Pacaya, 540 (35.7%) had formerly experienced this operation.

Time Scale for Entomological Aspects of the Pilot Project

The onchocerciasis control pilot project in San Vincete Pacaya was scheduled as a five-year plan; with preparatory, attack and evaluation phases (Table 1). The entomological activities in each phase were foreseen as set out below.

Preparatory Phase

First Year (1976)

The first priority was to determine the vector. Several anthropophilic blackfly species had already been recorded from the project area. It was thus crucial to success, to ascertain which of these was responsible for onchocerciasis transmission, so that future control could be focused accordingly. Towards this end, and in the interests of gathering adequate

Table 1. Time schedule of five year plan of the onchocerciasis control pilot project in Guatemala.

Phase	Entomological activities
Preparatory Phase	
First Year 1976	Vector determination
Second Year 1977	Vector biology studies
Third Year 1978	Control measure studies
Attack Phase 1979	Vector control operation
Evaluation Phase 1980	Evaluation of the project

data on local Simuliidae and developing necessary methods for control evaluation, investigations were programmed as follows in the pilot area: survey of the blackfly fauna; investigations of the Simuliidae found to be anthropophilic; studies of the natural incidence of *Onchocerca volvulus* in adult blackflies; establishment of suitable methods for laboratory rearing of adult blackflies; observation of the development of *O. volvulus* in adult blackflies, under laboratory conditions; mapping of blackfly-breeding watercourses; investigation of larval habitats of the presumed chief vector, *Simulium ochraceum*; investigation of the biting behaviour of Simuliidae; seasonal prevalence of blackflies; and studies of methods for evaluating vector control.

Second year (1977)

Biological studies of the chief vector (probably *S. ochraceum*, as it proved from the results of the first year's work . . . Ed.) continued, with particular respect to aspects relevant to control decisions, and particularly towards elucidation of the vector, with the mapping of the streams utilized as its larval habitats.

(1) Vector elucidation: further studies of the natural incidence of *O. volvulus* in adult blackflies; and investigations of experimental infections with *O. volvulus* in Simuliidae in the laboratory, with further studies of parasite development.

(2) Vector biology: mapping of larval habitats of *S. ochraceum*; studies of the streams utilized by *S. ochraceum*; observation of the seasonal prevalence of adults and early stages of *S. ochraceum*; determination of the larval development period of *S. ochraceum*; investigation of the gonotrophic cycle of *S. ochraceum*; flight-range survey of adult *S. ochraceum*; and studies of methods for evaluating vector control.

Third Year (1978)

Decisions on control measures to be made in preparation for the Attack Phase in the following year, with particular attention to:

(1) Larviciding procedures: selection and formulation of insecticide; determination of dosage; decisions on the frequency of larvicide applications; determination of the optimum season for larviciding; and selection of treatment points.

(2) Operational planning: organization; and training of personnel and obtaining of supplies and equipment.

(3) *Onchocerca volvulus* transmission analyses: determination of the vector density level, at and below which transmission is interrupted; and designing of a transmission model.

(4) Pretreatment evaluation: studies of adult blackfly density; investigation of larval blackfly density; and estimation of annual transmission potential.

(5) Investigation of the environmental impact of insecticiding.

(6) Survey of predators, parasites and pathogens of blackflies.

Attack Phase (1979)

This phase of the five-year plan envisaged intensive larviciding against the vector throughout the pilot area. It was anticipated that ground parties would apply larvicides to the headwaters of streams on a weekly or fortnightly basis, for three months.

Evaluation Phase (1980)

After the first four years, there would be a comprehensive evaluation of procedures used and results achieved. Changes in vector density and the Annual Transmission Potential in the project area would be analysed from entomological data collected before and after treatment. Any alterations in actual disease incidence would be determined by parallel epidemiological evaluation. The objectives of the pilot project would, hopefully, be attained by the end of this fifth year. Thereafter, the resultant standard vector control methodology would be available for country-wide implementation.

Structure and Management of the Pilot Project

The enterprise is conducted by the Guatemalan government, with the technical and financial cooperation of the Japan International Cooperation Agency (JICA). The project team operates within the Servicio Nacional de Erradicación de Malaria (SNEM), Dirección General de Servicio de Salud, Ministerio de Salud Pública y Asistencia Social. SNEM comprises five Departments. One of these, Departamento de Programas Adscrito, is responsible for vector-borne diseases other than malaria, such as onchocerciasis, yellow fever and Chagas disease. The onchocerciasis control project team is composed of four groups—epidemiological, parasitological, entomo-

logical and administrative. Japanese and Guatemalan scientists, technicians and administrators are assigned to these. Operational tasks, such as mapping and insecticide application, are assisted and reinforced by the Departamento de Operaciónes del Campo.

Some 10 Japanese specialists have long-term assignments to the team at all times. They include five entomologists, two parasitologists, two epidemiologists and a coordinator. In addition, at least six other experts are sent annually, for short-term involvement as occasion demands. The attack phase involved the temporary allocation, to the team, of from 10 to 20 field personnel from the Departamento de Operaciónes del Campo, for actual control operations.

The Vector

As indicated, Robles (1919) suggested that blackflies might be responsible for the transmission of onchocerciasis in Guatemala. Subsequently, experiments in Mexico (Hoffman 1930, 1931) demonstrated that *O. volvulus* will develop in *Simulium callidum* and *S. metallicum*. Strong (1931, 1934) narrowed the choice of possible Guatemalan vectors to *S. ochraceum*, *S. callidum* and *S. metallicum*. De León (1947) and Gibson (1951), after natural infection surveys, reported the levels of *O. volvulus* in *S. ochraceum*, *S. callidum* and *S. metallicum*. On the other hand, Gibson and Dalmat (1952), on the basis of their successful laboratory experiments, made the further suggestion that *S. exiguum*, *S. haematopotum* and *S. veracruzanum* might be potential vectors too. Dalmat (1955) claimed that *S. ochraceum* might well prove to be the chief vector in Guatemala, for the following reasons: its distribution coincides closely with the area where onchocerciasis is endemic; although the distribution of *S. ochraceum* does not always exactly coincide with that of the Guatemalan onchocerciasis, its population density is substantially higher in the endemic zone than in unaffected areas; while *S. ochraceum* is anthropophilic, the other blackfly species present are strongly zoophilic; and *S. ochraceum* feeds longer from humans than do the other species, and its behaviour is sluggish afterwards.

Hamon *et al.* (1974) have lately pointed out, in a PAHO review document, that of three possible vectors in the Western Hemisphere, *S. ochraceum* might be the most important.

From the start, the project team accorded top research priority to pinpointing the chief vector. This was to avoid an otherwise unselective control campaign, costing greatly more than an operation aimed at a single target species. The conclusion was duly reached that the control campaign should be conducted against *S. ochraceum*, for the following reasons:

(1) No adults of any other blackfly species were every found with infective larvae of *O. volvulus* in the head, during the team's natural infection survey. The parous rate, and *O. volvulus* infection level of blackflies from human bait, were recorded for a full year (1977–1978) in the pilot area (Ochoa, personal communication). Of the approximately 7300 adult simuliids collected, 60.7% were *S. ochraceum*, 33.8%, *S. metallicum*, and 5.5%, *S. callidum*. Only 0.2% of the *S. ochraceum* proved to harbour infective larvae of *O. volvulus*. Considering the adults of all three species, some 1.6% exhibited larvae of this worm in the thorax and abdomen. Notably, although *S. metallicum* (thought to have been a possible secondary vector) *never* had *O. volvulus* in the head, 1.0% of the adults of this species dissected, had worms in other parts of the body. With respect to the third suspected vector, only one adult of *S. callidum* (out of approximately 400 examined) contained *O. volvulus* larvae; and this involved the abdomen. These data indicate the likelihood that only *S. ochraceum* is of importance to onchocerciasis in the pilot area.

(2) As mentioned earlier, there is good agreement between the range of *S. ochraceum* and the endemicity of onchocerciasis in the pilot area. Larval blackfly surveys revealed that the Chilar range (northwestern part of the pilot area) includes the main breeding sites of *S. ochraceum*. Although the rugged Chilar area (altitude 700–1300 m) has few inhabitants, people from nearby villages maintain cultivations there. This region includes the headwaters of three significant streams, flowing southward through the pilot area. The distribution of human-biting *S. ochraceum* adults coincides with that of the larvae, such attacks being registered at altitudes of from 900–1300 m above sea level. According to Tanaka *et al.* (personal communication), the relationship between biting adult density and altitude is self-evident (Table 2).

Table 2. The relation between the adult density of *S. ochraceum* attacking human bait and the altitude of the project area.

Locality	Altitude (m)	No. flies collected
Los Jazmines	1300	284
Los Lavaderos	1300	357
Rio Guachipilín	1200	136
El Rodeo (Palín)	1200	65
Finca Las Chilcas	900	30
Finca Las Esperanza	600	3
Finca Berlín	600	6
Finca Peña Blanca	400	5
Finca Agua Blanca	300	0

Tanaka *et al.* (personal communication).

The parasite index of the human population in the Chilar region is significantly high, too. Tada *et al.* (1978), using the skin snip method, obtained positive microfilarial rates of 7.7% in El Camarón, and 16% in Los Chagüites, at an altitude of about 500 m. Higher such rates were recorded from El Patrocinio (38.4%, 1600 m), Los Ríos (49.5%, 1500 m), Caña Vieja (62.3%, 700 m) and Berlín (77.3%, 600 m). Sato *et al.* (1978) demonstrated the applicability of antigen, prepared from

Table 3. The relation between the skin test and skin-snip
positive rates of onchocerciasis, and the altitude of the
project area.

Altitude (m)	No. tested	Positive rate %	
		Skin test	Skin test + skin-snip test
1800–1700	126	23.1	24.0
1300–1200	114	42.1	48.0
1000– 800	88	54.7	69.2
800– 600	42	53.3	77.8
400	25	35.7	37.1

Sato et al. (1978).

adults of the dog heartworm, *Dirofilaria immitis*, to onchocerciasis diagnosis via skin tests. They applied the technique to determining the distribution of the disease in the pilot area. Table 3 shows the relationship of altitude to onchocerciasis rates, as estimated antigenetically and by skin snips. Clearly, the highlands (600–1300 m), where *S. ochraceum* attacks humans most vigorously, are the regions of highest endemicity.

(3) To explore the feasibility of altogether dismissing *S. metallicum* as an object of control, Matsuo et al. (1978) investigated anthropophilism in the pilot area. Adults collected at Guachipilin proved to comprise: *S. metallicum* (57.8%), *S. ochraceum* (21.9%), *S. callidum* (20.0%), and *S. downsi* and other infrequently encountered species (totalling 0.2%). Because *S. metallicum* is held to be the onchocerciasis vector in Venezuela (Convit 1974), it obviously merited searching investigation in the pilot area. Ito and collaborators (personal communication) investigated the potential of *S. metallicum* as a vector, through experimental infection in the laboratory. They concluded that *O. volvulus* microfilariae were indeed able to develop to the infective larval stage, and to reach the head of this blackfly. It may therefore be asked *why* infective larvae have not yet been found in the head of *S. metallicum* in nature? Possible reasons are that this blackfly is so strongly zoophilic that an individual is unlikely to bite man more

than two or three times in its entire life; and that it is physically damaged when (through too many microfilariae being ingested) the worms invade its various organs (Omar et al. 1977). Ito (personal communication) discovered that most *S. metallicum* females die within a day of feeding from experimental subjects with high microfilarial density in the skin, because of such physical damage. Few *S. ochraceum* were found to die even after ingesting large numbers of microfilariae, because many of the latter were cut through or otherwise injured by the buccopharyngeal armature. Thirdly, *S. metallicum* shows a strong preference for biting the lower part of the body, where microfilarial density is lower than in the upper parts. Under natural conditions, their chances of actually ingesting microfilariae is thus reduced accordingly. For the three reasons suggested, it is considered that *S. metallicum* is unlikely to be of practical significance as a vector.

Therefore, while recognizing that *S. metallicum* could certainly act as a major vector under circumstances where few animals exist as sources of blood-meals, or where human onchocerciasis is of low endemicity, it is concluded that the species can be safely ignored in the target area—at least during the earlier phase of the campaign. However, it is submitted that once onchocerciasis endemicity falls appreciably, following successful *S. ochraceum* control, *S. metallicum* might well become an alternative vector.

Vector Biology

Larval Habitats of *S. ochraceum*

Dalmat (1955) proposed a classification for Guatemalan streams having blackfly breeding sites, on the basis of relative age. He graded them as "infant", "young", "adolescent", "mature" and "old", in accordance with their size, form, current-flow, substrate and vegetation. He further pointed out that the principal anthropophilic blackflies, then assumed to transmit onchocerciasis, prefer "infant" or "young" streams, or parts thereof. The preparatory phase of the pilot project confirmed that *S. ochraceum* principally used "infant" streams and only occasionally "young" ones.

The characteristics of *S. ochraceum* larval habitats are: convergence of several minute trickles from rocks; no definite bank trough; stream bed hardly distinguishable from the dry surroundings; tributaries of "young" or "adolescent" streams; and "miniature waterfalls", perhaps no more than a moving film of water over rock.

Physical parameters for typical *S. ochraceum* larval habitats have been summarized by Tanaka *et al.* (personal communication). They range from 3–100 cm in width, 1–10 cm in depth, 10–500 m in length, 0.1–10.0 l/sec in volume, 15–50° in incline, and have a water temperature of 18–22 °C. In the pilot area, such habitats are typical of the rugged Chilar terrain (700–1300 m above sea level), whence most of the rivers originate.

Seasonal Prevalence of *S. ochraceum*

In Yepocapa, larvae, pupae and adults of *S. ochraceum* may be found throughout the year, with adult population peaks in January and August (Dalmat 1955). This same authority stated that the ensuing larval populations peak in April and October. However, not all these findings

were confirmed by our preparatory phase data. While larvae, pupae and adults were indeed found throughout the year, there was only a single population peak for each life-history stage. The larval peak was especially noteworthy, being reached during the dry season, between October and April. During the rainy season, from May to October, the density of *S. ochraceum* larvae remained rather low. Adults failed to show so evident a population peak. Their numbers were lowest between July and November. From the control standpoint, the best season for larviciding thus seemed to be during the dry months— from October onwards.

Feeding Behaviour of *S. ochraceum*

S. ochraceum is a day-biter, which has not been found to exhibit any noticeable diurnal rhythm. Its attack is usually confined to upper parts of the body. This behaviour is in sharp contrast to the other characteristic blackflies of the area (*S. metallicum*, *S. callidum*, *S. gonzalezi* and *S. downsi*)—these, as already mentioned prefer to bite lower down (Dalmat 1955; Matsuo *et al.* 1978). It is submitted that these contrasting behaviour patterns are of great importance, epidemiologically; for as microfilarial density is highest in the skin, e.g. the face and trunk, it is much reduced in the lower extremities (De León and Duke, 1966). *Simulium ochraceum* clearly has better opportunities than the other species for feeding from parasitized humans.

Gonotrophic Cycle of *S. ochraceum*

Laboratory and field observations (Watanabe 1978) suggest the following gonotrophic cycle for this species: females mate one day after emergence, taking a blood meal within a further day; maturity of their ovaries takes place about four days later; the period from one blood meal to the next is approximately five days at 22 °C; and the extrinsic period for *O. volvulus* microfilariae to develop into infective larvae and reach the head of the vector is at least eight days at 25 °C (JICA, 1978) or four days at 30 °C (Collins *et al.*, 1977).

These data suggest that microfilariae ingested at the first feeding develop into infective larvae at the time of the third or fourth blood-meal.

Vector Control Measures

Simulium ochraceum has been found breeding in some 50 trickles and streams of four of the main rivers in the pilot area. This species is also suspected of utilizing several times more such larval habitats; from which, however, it has yet to be recorded.

The results of laboratory and field ex-

Map of the pilot area for the onchocerciasis control project.

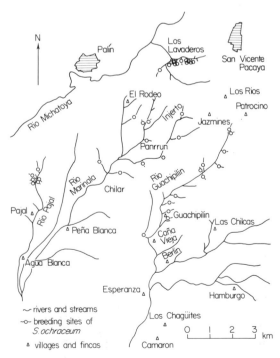

periments in the preparatory phase (Tabaru and Shimada, personal communication), have shown that temephos 10% slow-release briquettes are appropriate to the project. Not only are they larvicidally effective, but they are convenient to use, and pose minimum adverse effects to NTOs.

Field experiments using these briquettes so that the dosage of temephos was 10 ppm/minute, led to the continuing absence of *S. ochraceum* larvae for two weeks following treatment. It was concluded that successive insecticide applications should probably be undertaken in just under this period, treatment of each stream or trickle taking place at only two sites.

On this basis, and assuming the need for 400 application sites (i.e. on 200 trickles or streams in the pilot area) a single spraying team, composed of two field operators, could treat eight sites daily. Further assuming bi-weekly intervals of application and a five-day working week, only 10–15 field operators, including the supervisor, would suffice for the entire operation.

POSTSCRIPT: In the Lavaderos River basin located in the northern part of the pilot area, the larviciding operation was launched in March 1979 applying 10% temephos briquettes at two-weekly intervals. Twenty-four tributaries having been mapped by then for the Lavaderos River, insecticide treatments were implemented at 47 sites of the watershed; by September 1979 coverage had increased to 88 sites. Two kinds of entomological evaluation were adopted. One involved the periodical estimation of adult blackfly density by means of collections from human bait; the other called for the periodical assessment of larval and pupal density, one such reading always being obtained nine days after each application. It proved that larval density decreased rapidly soon after each temephos application, subsequent searching revealing neither old larvae nor pupae in most tributaries. Collections from a bait-boy showed that the number of adult simuliids attempting to bite at first fluctuated between 26 and 524 during three-hour observation periods before pesticide treatments, but then decreased remarkably, so that from June to December, adult density remained between 0 and 21 (Nakamura, personal communication).

In the Barretal and Zapote River basin, adjacent to the Lavaderos River, two-weekly larviciding was commenced in June 1979 at 83 sites on 44 tributaries. Although it has been difficult to find larvae there, during the post-treatment assessments, adult blackfly population decreases have proved much less marked in the Barretal/Zapote zone than in the Lavaderos River basin. The reason for this is suspected to be invasion from the southern area not under blackfly control (Takaoka, personal communication).

The larviciding of the Guachipilin River basin will commence in March 1980. The mapping of this watershed has shown that there are approximately 140 tributaries, totalling some 33 km in length. This entire operation (Lavaderos, Barretal, Zapote, and Guachipilin River basins) requires four application teams and one evaluation team, each composed of two or three field personnel (Takahashi, personal communication).

Control of *Simulium arcticum* Malloch in Northern Alberta, Canada

W. A. Charnetski and W. O. Haufe

Simulium arcticum Malloch has been a long-standing problem for livestock producers in some areas adjacent to the Athabasca and Saskatchewan Rivers in northern Alberta and north-central Saskatchewan. It breeds generally in all the headwaters of rivers rising on the eastern slopes of the Rocky Mountain range in Alberta. For unknown reasons high potential for breeding in certain reaches of the Athabasca and Saskatchewan Rivers leads to massive outbreaks of adult flies that severely attack cattle, horses, and swine at distances exceeding 150 km from the river courses.

The problem has assumed increasing importance, both to man and animals, with the need to expand the livestock industry into northern regions of Alberta. Although blackflies were always particularly troublesome to man, they were accepted as a natural hazard in the early development of northern agriculture. At that stage, agricultural enterprises were primarily based on cropping practices. Under cultivation, however, most north-

erly soils were found to be low in organic matter so that rotations were necessary to include up to 60% of cropping in the form of forage production. This requirement, in addition to the uncertain maturation of cereal grain crops during short frost-free periods, became a major incentive for the development and expansion of livestock production as the basic agricultural enterprise for farmers already established in the region. Many areas have natural economic potential for cattle raising in utilizing native roughages on river slopes, ridges, muskegs, lake shores, and meadows between mixed conifers and deciduous bluffs as pasture resources, in addition to cultivated grasses, legumes, and other livestock fodder.

Development and expansion of the livestock industry rapidly emphasized the economic importance of blackflies in northern Alberta as a serious problem of both man and animals. A damaging outbreak in 1962 caused farmers and residents to petition agricultural administrators for relief from the infestations. Since the scope, severity, and persistence of the problem were not fully appreciated for several years, early work was limited to surveys and small-scale studies to identify the pest species, define their distribution in rivers and streams, and start tests of pesticides to control their breeding in small streams. As outbreaks increased in severity and frequency, in a five-year period that threatened continued livestock expansion, owners organized and developed an initial economic assessment of damages and losses from infestations in 1971. In a survey area of 53 245 ha, containing 13 008 cows, losses in 1971 included 973 dead animals and 38 sterile bulls. The most extensive loss, the reduced gain in weight of animals on pasture, estimated to be about 45 kg/animal, was then valued at $390 000. Loss of production in unbred cows on pasture was more than $90 000. These losses, in addition to inefficiency of operations with interrupted calving schedules, sterile bulls, and dead animals, amounted to an immediate annual monetary loss of about $600 000 or about $46/cow–calf unit for all cattle enumerated in the survey area. These direct monetary losses still excluded labour expended and time lost by livestock owners in providing emergency relief to cattle in housing, movement of cattle from pastures, preparation of smudges, and application of spray treatments during periods of heavy blackfly attack.

During early agricultural development in northern Alberta, it was thought that infestations of biting flies on farms were localized and originated in neighbouring swamps. Surveys and preliminary studies, between 1962 and 1972, showed that attacks on cattle and horses during the most damaging outbreaks were comprised of about 92% *Simulium arcticum*, 6% *S. venustum*, and 2% *S. vittatum*. Major persistent outbreaks were caused by massive breeding of *S. arcticum* confined to the Athabasca River from which emerging adults, under favourable weather conditions, were capable of dispersing for distances of at least 150 km to invade farms. For reasons of economics and practical requirements of effectiveness in control operations, it was necessary to focus on abatement of *S. arcticum* at its breeding sites in the Athabasca River.

The Athabasca is one of the largest rivers in Canada and drains an extensive watershed representing about one-quarter of the area of Alberta or an area of about 155 00 km². As such, it is the habitat for an extensive and highly diversified aquatic life-web. The ecological implications of the impact of blackfly larvicides on so large an environment necessitated extensive multidisciplinary studies, not only to evaluate the effectiveness of pesticides as a practical economic control for *S. arcticum*, but also to determine environmental acceptability in an economic compromise for evaluation of alternative approaches to the problem. These considerations justified the development of the Athabasca River Black Fly Research

Program [ARBFRP] (1973–1977) as a multidisciplinary study of a large river system.

Review of *Simulium arcticum* Control in Canada

Early studies on control of *S. arcticum* were stimulated by outbreaks of the insect before 1948 that prevented farmers from fully using livestock in their enterprises in an area of 52 000 km² adjacent to the Saskatchewan River (Rempel and Arnason 1947). Tests with DDT as a blackfly larvicide were initiated in the Saskatchewan River system in 1948 (Arnason *et al.* 1949; Fredeen *et al.* 1953b). These early studies recognized the role of adsorption of DDT on suspended solids in effective larvicidal control of blackflies in river water (Fredeen *et al.* 1953a). DDT was applied to the Saskatchewan River in most of the years between 1948 and 1967 at rates of 0.1–0.3 ppm for 15 minutes (Freedeen *et al.* 1971). Single injections in these treatments eliminated most blackfly larvae from the river for distances of 185 km with adsorption of the pesticide to silt particles contributing to long-distance transport downstream. Invertebrates and fish remained relatively abundant in the South Saskatchewan during the period of these treatments but were periodically reduced in the North Saskatchewan by severely polluted water entering from the west (Reed 1962; Fredeen 1977b). Analyses of pesticide residues in fish from both branches of the Saskatchewan River in 1968 indicated levels of DDT, DDD, or DDE in pooled samples of muscle tissue to be less than 0.01 ppm. The highest concentration, 0.05–0.06 ppm, was found in one sample of Goldeye [*Hiodon alosides*]. Concentrations of chlorinated hydrocarbons were similar to those in fish from the untreated part of the South Saskatchewan above the Gardner Dam (Fredeen 1977b).

As a result of restrictions on the use of

DDT, extensive tests on the use of methoxychlor as a blackfly larvicide were carried out in the Saskatchewan River (Fredeen 1974, 1977b). By the end of 1972, 14 treatments with dosages of 0.143–0.443 ppm for 15 minutes were evaluated. Single 15-minute injections of 0.18–0.24 ppm methoxychlor as an emulsifiable concentrate removed 75–99% of *S. arcticum* larvae from 23-km to 34-km sections of the river. The higher concentrations eliminated 98% of larvae from rapids 64 km downstream and an estimated 46% at 139 km. A single 7.5-minute injection of 0.6 ppm eliminated 100% of larval instars three to six from sites 40–80 km downstream, 91% at 121 km, and 66% at 161 km (Fredeen 1975). The treatment removed 96% of larvae less than 1 mm long at the 161-km site. Some of the tests were performed in exceptionally low river volumes while a reservoir on the South Saskatchewan was being filled (Fredeen 1977b) and this may have accounted for some variation in larvicidal effectiveness.

In tests not exceeding a concentration of 0.3 ppm, fish including caged rainbow trout [*Salmo gairdneri*] yearlings were not visibly affected. Larvae of Simuliidae, Chironomidae, Plecoptera, and Ephemeroptera collected from the annually treated areas on artificial substrates were more abundant in 1972 than in 1969 (Fredeen 1977b). Larvae of Trichoptera were less abundant in one treated area in 1972 but more abundant in another (Fredeen 1974). Populations of these orders were also observed to decline suddenly for undetermined reasons on one occasion in an untreated section of river. Although immatures of Plecoptera were affected similarly to *S. arcitcum* by treatment at the higher concentration of 0.6 ppm in 1973, those of Chironomidae, Ephemeroptera, and Trichoptera were less affected. Populations of nontarget organisms (NTOs) recovered in two to four weeks. No recovery of *S. arcticum* was observed in a 10-week post-treatment study because it has one major generation each summer (Fredeen 1977b).

Fredeen (1977b) suggested that rapid recolonization reduces the impact of methoxychlor treatments in the Saskatchewan River. He estimated that chironomids surpassed pre-treatment densities within one to three weeks, ephemeropterans within one to four weeks, trichopterans within 1–7 weeks, and plecopterans within 4–5 weeks. Rapid recolonization is to be expected when a significant upstream section of the river is left untreated.

On the basis of extensive treatments in 1968–1973, methoxychlor has been considered as an effective replacement for DDT in the Saskatchewan River (Fredeen 1977). Neither it nor its known metabolites have posed any lasting threat to the environment (Kapoor et al. 1970; Fredeen et al. 1975). Fredeen et al. (1975) concluded from a study of residues in 1972 that a single 15-minute injection of methoxychlor at 0.309 ppm contaminated water and biota for a short time at least as far as 22 km downstream. However, no detectable residues of methoxychlor remained from similar treatments of previous years. Methoxychlor was indicated to have some selective action against filter-feeding species, especially blackfly larvae. No residues were detected in insect larvae or mussels that inhabited the river bed. Of all fish species examined, only Goldeye showed tendencies to accumulate methoxychlor. Of the residue accumulated in edible flesh of Goldeye, 80% were lower than the concentration of methoxychlor injected into the water 8–9 days earlier. Absence of methoxychlor residues in other species have been attributed to relatively low oil content of muscle tissues as well as ability of most species to eliminate it rapidly from their systems. No methoxychlor residues were detectable in any fish 17 weeks after treatment.

Available knowledge on larvicides for *S. arcticum* has been insufficient to model operations in physical and hydraulic terms for environmental accountability and to justify transfer of methodology between large river systems. As compared with the Saskatchewan, the Athabasca is a deep turbulent river with higher average rates of discharge. It poses significantly different hydraulic dimensions relating to injection, mixing and fate of the larvicide and, consequently, to acceptable criteria for impact on the aquatic environment. For this reason, an integration of interdisciplinary studies is required to justify larviciding methods in large rivers.

The Athabasca River Blackfly Research Programme

The Programme

The ARBFRP was developed from a feasibility study in 1973 to develop and evaluate chemical control of *S. arcticum* in the Athabasca River (Haufe and Croome 1980). The use of a larvicide appeared to be a most immediately achievable and economically practical approach to prevent severe pest outbreaks and to reduce farm losses of livestock production.

The programme was designed not only to examine the more extensive problems of biting flies in northern agriculture, but also to provide information necessary to establish the environmental impact of a larvicide on a large, deep river, and to manage blackfly problems as they relate to agricultural, industrial, and recreational resource development.

The required interdisciplinary and interagency research programme was initiated in 1974 and completed with a post-treatment baseline study in 1977. Total participation was outlined within six objectives:

(1) To identify and characterize the breeding sources of the blackfly *S. arcticum* in the Athabasca River and to develop methods of treating the river with a pesticide to reduce the production of the pest in a selected area.

(2) To determine the level and extent of reduction in breeding sources required in abatement operations to provide economic reductions of infestations in contiguous agricultural areas.

(3) To estimate infiltration rates for pest populations reinfesting agricultural areas from sources outside the area of the abatement operation.

(4) To develop methods of monitoring an abatement operation for deleterious effects of treatments on aquatic NTOs and the river environment.

(5) To develop criteria for acceptable impact on the river environment in conjunction with specifications for pesticidal treatment of river systems.

(6) To assess the impact of infestations of *S. arcticum* and other related biting flies and the effect of abatement procedures on the productivity of livestock and development of livestock enterprises, and to evaluate the benefits of animal protection in the area.

The Study Area

The study area consisted of a reach of the Athabasca River from 40 km upstream of the Athabasca townsite to 320 km downstream together with sampling locations immediately upstream of Fort McMurray, in the Athabasca River Delta, and in Lake Athabasca.

The river discharge at the Athabasca townsite dramatically fluctuates in the spring and summer, influenced by rain (mean annual precipitation is 560 mm) and snow melt in the upper watershed. Except for a few short reaches in which the river bed widens, the flow is relatively rapid and turbulent even at low rates of discharge.

The river valley is stream-cut with moderately forested walls and is about 75 m deep. The top and bottom widths are generally about 2.5 and 0.3 km, respectively. The channel is entrenched in the valley and exhibits an irregular pattern of considerable sinuosity (ratio of

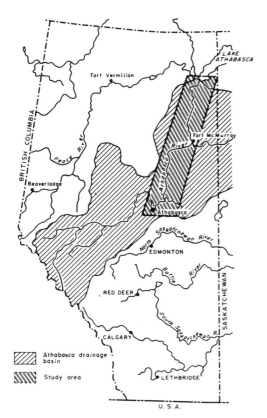

The Province of Alberta showing the Athabasca River drainage basin and the total study area of ARBFRP.

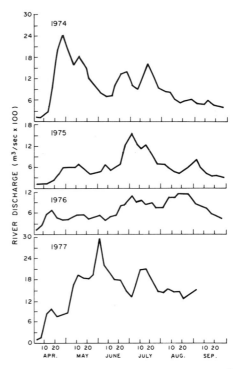

Daily discharge of the Athabasca River at Athabasca Townsite (Environment Canada gauge station 07BE001) for the periods 1 April–30 September, 1974–1977.

channel length to corresponding distance along the valley axis is 1:2) with occasional islands. The channel is laterally stable, although the banks are subject to occasional slumping. The channel bed consists of a shallow layer of gravel with local sand over soft cohesive clay. Beginning at Athabasca, the river steepens progressively until Fort McMurray and flattens drastically thereafter. The frequency of extensive rapids continually increases from about 140 km downstream of the Athabasca townsite until Fort McMurray. The reach of the river of interest in this programme can be divided into five sections (Table 1) with more or less constant slopes (Beltaos 1980).

High river velocity, frequent flooding over extended periods, and the confined channel bed, combine to maintain an intensively scoured river bed throughout most of the river course above Fort McMurray. These, plus a high concentration of suspended organic matter (7–26 mg/l) and an ideal substrate, provide a

Table 1. Estimated slope of five sections of the Athabasca River downstream of the Athabasca Bridge.

Distance downstream from bridge (km)	Slope (m/km)
0–95	0.26
95–140	0.44
140–208	0.71
208–258	0.61
258–396	0.98

highly favourable environment for the rapid and prolific development of *S. arcticum*.

Hydrometric Surveys

The concept of a river treatment for blackfly control requires an understanding and measuring of hydraulic processes to describe and interpret the dispersion of the chemical and to account for the detection and control of the target species, and the reduction of its effects on NTOs and environmental quality.

Therefore, numerous river cross-sections were surveyed (Beltaos 1980) within a reach from 60 km upstream of Athabasca to about 400 km downstream at Fort McMurray. In addition, a longitudinal depth profile of the river near the thalweg was obtained from 60 km above Athabasca to 260 km downstream. Echosounders and Raytheon recorders were used to measure depths while transverse locations in the stream were estimated using optical range finders. Within the study area, the Athabasca River varied in width between 150 and 420 m and commonly reached depths of 8 m. Cross-sectional velocity distributions were measured using Price current meters at selected cross-sections.

Determination of Treatment Time

Monitoring of *S. arcticum* larval numbers and development began as soon after spring breakup of ice as river travel was safe; usually the first week in May. Larvae were sampled using artificial substrates (Depner *et al.* 1980a). These substrates consisted of white plastic cones 17.5 cm high with a basal diameter 10 cm and surface area 353.6 cm². Cones were attached by a 24-cm leader with swivel snaps to rings tied into a 20-m polypropylene line at points one and 10 m from the top, with an anchor at the bottom and a float at the top. When this

"set" was established in three metres' depth of water, the cones were suspended about 0.3 and 1.5 m below the water surface. Three such sets were established by boat at each sampling station. A sampling period was 7 days, after which the cones were replaced with clean ones and the set replaced. Retrieved cones were carefully washed with alcohol to remove the attached larvae and the instar distribution and larval numbers determined.

From the feasibility study, it was established that an average of 500 larvae/cone beyond 100 km downstream of the Athabasca consituted a possible economic outbreak and that larviciding was required. The actual treatment was scheduled to coincide with the period in which a significant number of larvae had reached the seventh instar, but before any significant number had pupated.

Larvicidal Applications

The Athabasca River was treated at one location (the Athabasca Bridge) in both 1974 and 1975 and at two separate locations in 1976; the first treatment, 160 km downstream of Athabasca and the second, immediately downstream of the Athabasca Bridge. The current was metered at the designated injection sites to determine velocity and flow distributions

Hydraulic characteristics of section 30 m upstream of injection site, 4 June 1974.

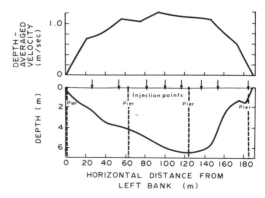

(Beltaos 1980). This information, together with the depth profile and water level on the day of treatment, was essential to establish the quantity of insecticide required and its distribution along the specific river cross-section.

On 4 June 1974, using the previous day's hydrometric determination, the river was treated with 796 litres of 25% emulsifiable concentrate (EC) methoxychlor (Stan Chem 25 EC, PCP Reg. No. 11617) to achieve a source concentration of 300 ppb (μg/l) for 15 minutes. The water temperature was 14 °C and the river discharge was 759 m^3/s at an average velocity of 1.07 m/s. Treatment was carried out from seven point sources on the downstream side of the Athabasca Bridge. Each of the seven injection sources consisted of a steel drum whereby the rate of flow of larvicide was controlled by a calibrated orifice at the end of a 3-m long drop-tube designed to dispense the required amount of insecticide over a 15-minute period (Depner *et al.* 1980a).

This application technique lacked the flexibility necessary for large river larviciding because of its dependence on man-made structures. Therefore, an improved procedure was devised (Depner *et al.* 1980a) and used in 1975 and 1976. With this technique, application could be made from several boats, thus permitting the introduction of the chemical at any location along a river without contamination of the terrestrial habitat. The rate of application from the drums was regulated by a calibrated orifice at the end of a short drop-tube and the use of a constant flow apparatus (modified Mariott flask) within the drum.

On 4 June 1975 the Athabasca River was treated 100 m downstream of the Athabasca Bridge with 291 litres of methoxychlor (Stan Chem 25 EC from 1974, analysed as 21%) applied by boat. The water temperature was 15 °C and the river discharge was 477 m^3/s at an average velocity of 0.79 m/s. This application was calculated to give a source concen-tration of methoxychlor of 300 ppb for 7.5 minutes at the point of injection. Methoxychlor movement was monitored through residue analysis of river water samples taken at various time and space intervals within a 240-km downstream area as well as at 396 (Fort McMurray), 410, and 580 km downstream.

On 20 May 1976 the Athabasca River was treated about 160 km downstream of the Athabasca townsite with 213 litres of methoxychlor (Methoxol 25% EC, PCP Reg. No. 6763) at a river discharge rate of 566 m^3/s. On 25 May, 1976, the Athabasca River was treated a second time at the Athabasca Bridge with 227 litres of methoxychlor (25% EC) at a river discharge rate of 416 m^3/s at an average velocity of 0.72 m/s. Both treatments were also applied from three boats and were calculated to give a mean source concentration of 300 ppb for 7.5 minutes.

Larvicidal Control of Blackflies

To assess the degree of larval control achieved in the abatement operation (Depner *et al.* 1980a), two groups of three cone-sets were established about one week before the estimated treatment date at each sampling site. Knowing the hydrometric parameters of the river, the time of methoxychlor arrival and departure was calculated for each sampling site and, allowing for a 10% safety factor, pre-treatment and post-treatment times for cone retrieval were established. Three cone-sets were retrieved for each pre-treatment and post-treatment sampling.

In 1974, river treatment achieved larval control for the entire 160-km evaluation distance. In 1975, however, treatment gave adequate control for only 80 km. It is hypothesized that this was due to the lower discharge and thus velocity (1.07 m/s in 1974 v. 0.79 m/s in 1975) and the resultant decrease in the capacity of the river to transport silt (wash-load).

In 1976, a double treatment was initi-ated so that larval control could be achieved in the lower reaches of the river. The first treatment essentially removed larvae from the 160-km reach investigated. However, the second treatment achieved better than 90% control for only 100 km.

Larvicidal treatments were effective. However, the range of efficacy appears to vary with the overall river velocity. It was concluded that the distance of effective control was about 120 km at a discharge of less than 560 m³/s, while at higher rates the effective range exceeded 160 km. Modifications to the formulation could increase the effective range at lower velocities. In addition, the method of injection is also critical in the effectiveness of the larvicide in large deep rivers. The use of powered boats, as in 1975 and 1976, rather than bridge-based application, as employed in 1974, improved dispersion and insecticide mixing, thereby reducing the distance required for complete mixing (discussed later under modelling) as well as decreasing the concentration of insecticidal "hot spots". Further treatment improvements could reduce these still further.

It is possible, therefore, using the physico-chemical and hydrometric parameters of the insecticide formulation and the river, to describe a river abatement operation for most large, deep rivers for any required distance.

Methoxychlor Residues in Water

The distribution and persistence of methoxychlor in the Athabasca River was determined (Charnetski et al. 1980b). Time–concentration curves were established for 12, 14, and 12 cross-sections for the three treatment years, respectively, based on about 2800 residue analyses by gas–liquid chromatography utilizing electron capture detection.

Through the use of separators (Charnetski, unpublished), it was shown that methoxychlor remains in emulsion close to the treatment but, as dispersion pro-

cesses reduce the concentration of the formulation, adsorption takes place and a greater amount of the pesticide is carried by the suspended particulate material known as wash-load. With increasing adsorption, more material is lost to the moving and the static bed of the river, thus reducing the methoxychlor concentration in the water and its associated wash-load. In addition, using a point sampler, a cross-sectional profile of methoxychlor distribution showed, not unexpectedly, that pesticide-laden washload particles are distributed according to well-known hydraulic phenomena; resulting in significant variations in methoxychlor concentration from one bank to another.

Time–concentration curves were established for each cross-section. There were as many as five sampling sites across a cross-section, and 25 samples were taken at each site. These samples were taken at intervals equally spaced over a previously calculated period for the passage of the methoxychlor. From these curves, the total methoxychlor dose (expressed as ppb-h) was calculated. Clearly, in 1974 the amount of methoxychlor in the water was excessive at locations up to 80 km downstream of application. After this point, loss of methoxychlor was more gradual. The curves for the 1975 and 1976 second treatment are very similar, and show a rapid decrease in dosage to about 80 km downstream with subsequent levelling-out and a substantially reduced

Typical concentration curve for instantaneous injection where t_a is the time of methoxychlor arrival at the location and t_d is the time of departure.

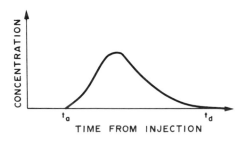

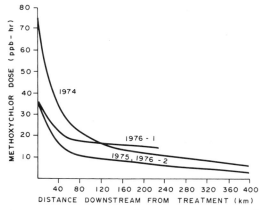

Relationship of methoxychlor dosage in water and associated wash-load to distance downstream from treatments in the Athabasca River, 1974–1976.

However, the steeper slope of the river reach over the initial 1976 treatment allowed for a much better recovery of methoxychlor. This was undoubtedly due to the increased wash-load capacity of the water and, consequently, a reduction in the loss to the bed-load and static bed. In addition, complete mixing of the methoxychlor injection undoubtedly occurred within a short distance.

Relationship of Methoxychlor Concentration in Water and Larval Control

loss of methoxychlor. The dose curve for the initial 1976 treatment is generally higher. This relatively high dosage is attributable to the fact that this treatment was made in a reach of river with a significantly higher velocity than the other three treatments.

Calculation of the "recovery ratio" (ratio of the total estimated methoxychlor recovered at a location divided by the total amount injected into the river) provides the best description possible of the overall loss of the compound from the water and its associated wash-load. The recovery ratios for 1974, 1975, and the second 1976 treatment clearly showed that, for the same reach of river, varying discharge and velocity had no effect.

It is well-known that methoxychlor kills blackfly larvae, but that sub-lethal concentrations only cause larvae to detach and drift. This detachment was important because the reach of river below Fort McMurray offers no substrate for larval development and, therefore, drift by the larvae results in certain "ecological death".

Three effects of methoxychlor on blackfly larvae are clearly evident by relating pesticide dosage (ppb-h) to larval mortality (Charnetski *et al.* 1980b). In the 1975 and 1976 second treatments, mortality was less at the lower methoxychlor dosage (7–12 ppb-h). This first effect undoubtedly resulted in some mortality, but the reduced population was probably due largely to detachment resulting from

Relationship of larval blackfly control to estimated methoxychlor dosage based on data from four treatments of the Athabasca River, 1974–1976.

Relationship of measured methoxychlor recovery ratio in water and associated wash-load to distance downstream from treatments in the Athabasca River, 1974–1976.

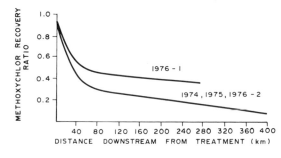

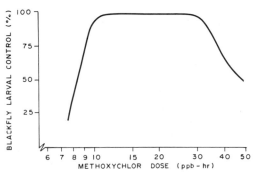

irritation by low methoxychlor concentrations. The two curves independently established an LD$_{90}$ (lethal methoxychlor dosage to kill 90% of the population) of 10.7 and 9.6 ppb-h, respectively. This similarity is not surprising in view of their concentration profiles. The second effect, mortality-release, of methoxychlor is described by the dosage-mortality curve for the initial 1976 treatment (15–30 ppb-h). This curve conforms to the upper portion of the curve established by the 1976 and 1976 second treatment curves. The third effect of methoxychlor, at still higher dosages, is mortality without release (25–50 ppb/h). The 1974 treatment resulted in excessive amounts of methoxychlor immediately downstream of treatment. This, plus the probability of incomplete adsorption (high concentrations in the water), lead to the hypothesis that the larvae are killed but do not immediately release from the sampling cones. This unidentified mortality would account for the apparently poor control (< 90%) at methoxychlor dosages above 33 ppb-h in 1974.

Distribution and Persistence of Methoxychlor in Athabasca River Mud and Bed-load

The second major component of the ecosystem for harbouring residues is the bed-load. Bed-load can be defined as

Modified Bogardi bed-load sampler.

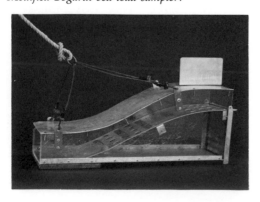

material moving along the static bed of the river and may, in fact, be composed of particle sizes found in appreciable quantities in the shifting portions of the bed. Bed-load samples were collected using a modified Bogardi sampler. Particle size analysis established that 85% of the particles retained by the sampler were within the range of 0–425 μm.

Mud samples were collected from areas of significantly reduced water velocities, such as shorelines, sand bars, and backwaters. Although these areas account for an extremely small proportion of the river bed, they may be sinks for methoxychlor residues.

The major reduction in the concentration of methoxychlor in the water and associated wash-load is the result of loss to the moving and static bed as shown by the high residue methoxychlor concentrations ranging to 5000 ppb immediately downstream of the 1974 application (Charnetski and Depner 1980). These levels dropped to 885 ppb within 48 h in 1974 and to less than 185 ppb in 8 h in 1976 and significantly lower levels thereafter. Concentration decreased rapidly to about 10 ppb within 60 days. Overall, the greatest loss to the bed occurs within 40 km of application subject to a continuing displacement of the methoxychlor "slug" downstream. Generally, the levels were reduced from the peak to about 25 ppb or less at all sites within 10 days. This post-treatment level in the bed-load corresponds with the residue levels found in the mud from 10–75 days after injection.

Nevertheless, it must be kept in mind that not all of the methoxychlor-laden particles would be biologically available; some may be too large and others too small. Therefore, the values reported here express overall environmental contamination.

Temporary sinks of methoxychlor-laden particles occur because of variable hydraulic conditions in the river, resulting in subsequent intermittent movement and redistribution during increases in river discharge. This intermittent movement and re-entry of methoxychlor accounts for some infrequent exceptions to the general loss curves.

The overall loss of methoxychlor from the measured system has been attributed to loss to the static bed, degradation (biological or chemical), or desorption at non-detectable levels from particulate material into the water. Conclusive evidence has been shown (Charnetski and Depner 1980; Charnetski *et al.* 1980a) that methoxychlor residues do not persist in river bed material. Neither bed-load nor mud contained determinable residues (above 0.1 ppb) 17 months after the three successive years of treatment in either the study reach or Lake Athabasca and the associated river delta.

Analytical Modelling of Mixing and Effects of Insecticides

The effectiveness, predictability, and environmental control of a larvicide introduced into a flowing aquatic ecosystem is dependent on the dispersion, uniformity of concentration, and reduction in availability of the chemical as influenced by hydraulic action, together with the physico-chemical properties of the chemical.

Beltaos (1980) formulated an analytical model that accounted for processes associated with physical and chemical properties of methoxychlor that can influence mixing patterns, in addition to mixing processes imposed by the river flow itself. The quantitative expression of these processes is based mainly on a study of observed downstream variations of insecticide recoveries in water (Charnetski *et al.* 1980b) and in bed-load and mud (Charnetski *et al.* 1980a). Two hypotheses, adsorption by suspended particles and losses to the channel boundaries, were shown to provide fair predictions of the downstream decline of the recovery ratio.

The degree of mixing for the first methoxychlor treatment (1974), assuming seven point sources of a neutral tracer, was calculated as 90, 95, and 98% at 1.5, 19, and 48 km downstream, respectively. Although methoxychlor is subject to substantial losses (it is not neutral), it was shown that these estimates were satisfactory. Subsequent injections (1975 and 1976) approximate a line-source application across the river and were believed to result in more efficient mixing.

The present model has been partially verified. However, because of the many assumptions involved in its derivation, additional field or laboratory data, or both, are necessary to provide satisfactory confirmation. This verification of the model is prerequisite to adoption for use on other rivers.

Because quantitative prediction in the previous model depended on the determination of three empirical coefficients to fit the observed variations of the recovery ratio and because some further research was required, a second, more practical, model was developed (Beltaos and Charnetski 1980). This model involved a semi-empirical approach to the analysis of the time-concentration data (Charnetski *et al.* 1980b) within the framework of one-dimensional theories, normally applicable to neutral substances (tracers).

The simplified model provides rough predictions of concentration–time curves. Much of the discrepancy found among the observed concentration–time curves at the same cross-section may be due to non-uniformities in the spatial distribution of the capacity of the channel boundaries to retain methoxychlor. To account for such non-uniformities empirically may be impractical, if not impossible. Beltaos and Charnetski concluded that if this were true, then numerical solutions would result in little improvement over the predictive method of this second model. However, the model as published would require some modifications for application to rivers other than the study area of the Athabasca River.

The application of hydraulic principles

in interpreting the effectiveness of treatment and in accounting for pesticide behaviour in explaining variations in results, is unprecedented for blackfly larviciding operations. Although still incomplete for the extension of river treatments to large rivers other than the Athabasca, the preliminary modelling in hydraulic terms is a major step towards the objective of achieving full accountability in future development and use of pesticides in rivers and streams.

Impact of Methoxychlor on Invertebrates

The lack of adequate criteria from studies of aquatic ecosystems in rivers and conflicting intepretations óf the relationship between drift of invertebrates and standing crop, as expressed in various sampling methods, are major obstacles to quantitative assessment of impacts of blackfly larvicides on NTOs, particularly in large rivers. The ARBFRP included several studies relating to larvicidal treatments and employed various methods of sampling for drift and standing crop (Depner et al. 1980b; Flannagan et al. 1980a, b; Haufe et al. 1980a, b). Of these studies, those of Depner et al. and Haufe et al. were coordinated with studies of residues in water, wash-load, and bed-load, and of their distribution in relation to hydraulic parameters of the river (Beltaos and Charnetski 1980b; Charnetski et al. 1980b; Charnetski and Depner 1980). Studies by Flannagan et al. were related to some of the larvicidal treatments but were independent of hydraulic parameters in the coordinated framework for other integrated sampling procedures.

All related studies of drift showed that some of the non-target invertebrate taxa are chemo-receptively sensitive to very low concentrations of the methoxychlor formulation. A treated pulse of water stimulates activity, particularly at the leading edge, and temporarily increases the density of organisms in the drift as it passes downstream (Flannagan et al. 1980a, b; Haufe et al. 1980a). Of 47 genera encountered in the drift, eight (represented by Cheumatopsyche spp. and Hydropsyche spp. in the Trichoptera, Baetis spp., Heptagenia spp., and Ephemerella spp. in the Ephemeroptera, and Isogenus spp., Isoperla spp. and Hastaperla spp. in the Plecoptera) were identified with dramatic increases in drift densities associated with passage of the pulse of larvicide (Haufe et al. 1980a). Consequently, these genera have been considered as biological indicators to monitor impacts of methoxychlor in blackfly larviciding operations (Haufe 1980). Casualty rates measured as a proportion of the drift and including all organisms showing abnormal behaviour as well as those moribund and dead were generally related to the level of drift sensitivity (Haufe et al. 1980a). They did not exceed 23% of the drift as an average or 40% as individual samples in a replicated sampling of the river cross-section.

Flannagan et al. (1980a, b) have interpreted invertebrate samples in qualitative terms and on the basis of a definition of "catastrophic drift" (Waters 1965, 1972) to indicate massive displacement of taxa downstream during blackfly larviciding. They concluded that the increase in drift density is of sufficient magnitude to seriously interfere with the distribution of the standing crop. This conclusion, however, has not been supported by other studies in the programme. Comparable analyses of the drift during treatments, indicated that drift density of sensitive indicator species returned to, and was maintained at, about the same level after passage of the larvicidal pulse as before, except possibly, for plecopterans (Haufe et al. 1980a). In the case of plecopterans, general emergence of maturing nymphs coincided phenologically with optimum timing of blackfly larviciding, and had to be considered as a factor additional to methoxychlor in reducing drift density after treatment in late May and early June. Furthermore, if massive drift interfered with the distribution of standing crop, it was not reflected significantly in regular time-series sampling of the resident benthic population at a large number of sites covering the effective range of treatments (Depner *et al*. 1980b). Drift samples in regular time-series during a summer with no treatment disclosed various seasonal cycles in drift density for different genera including some abundant taxa that decline in numbers after late May (Haufe *et al*. 1980b).

Risk to NTOs is thus evident for deep rivers during the mixing phase of the pesticide immediately after injection. Nevertheless, there was no substantive evidence, from any of the studies, that increased drift densities caused displacement over distances sufficient for serious disruption of balance in the standing crop (Depner *et al*. 1980b; Flannagan *et al*. 1980a, b). The quantities of certain benthic taxa were observed, at some sampling sites downstream in the treated reach of the river, to be randomly lower in June than in other months. They coincided with the first few weeks after treatment and, without reference to other sampling data, would be consistent with a minor disturbing effect of the pesticide. However, lower numbers of the same relative order of variation were also observed in the same month at control sites upstream of treatment and also for a year with no treatment (Depner *et al*. 1980b). These variations would also be consistent with effects of the first general emergence of some genera in late May and early June (Haufe *et al*. 1980a).

Invertebrate studies have identified a serious need to resolve the relevance of various sampling methods for adequate interpretation of perturbations in complex aquatic ecosystems. There was no clear evidence, however, that methoxychlor, at a concentration of 300 ppb for 7.5 minutes, had any serious effect on standing crop in the benthos.

Impact of Methoxychlor on Fish

Concern over the effects of methoxychlor on NTOs, especially fish, was crucial to the design of the ARBFRP. Major concerns included direct toxicity during exposure to the larvicidal pulse moving downstream, trophic interference in food chains supporting the fish population, and objectionable residues of the pesticide in fish tissues including muscle, fat, liver, and gonadal tissue. Studies were documented with reference to background insecticide and hydrocarbon residues (Charnetski and Currie 1980), methoxychlor residues in wild and caged fish (Lockhart *et al*. 1977; Charnetski and Currie 1980; Lockhart 1980) as well as toxicity and trophic relations (Lockhart 1980). These studies involved many species of fish (both native and exotic) and a large number of variables. Unfortunately, space in this chapter permits only a summary and discussion of the more salient points.

There is no certainty that environmental conditions for measurement of residues in toxicological studies are completely representative (whether conducted in field cages or in the laboratory) of those pertaining to free exposure to the larvicide in the natural river ecosystem. Nevertheless, all the studies conducted provide reasonable estimates of the risks to fish populations in larviciding operations with methoxychlor.

There is no evidence that concentrations of methoxychlor comparable with river treatments were either seriously toxic or lethal to fish. Laboratory tests indicated that tissue residues in fish exposed to the 1974 river treatment approached to within a factor of 10 of those causing gross morbidity and death. This reduced the risk through effective larvicidal treatments in 1975 and 1976 at a rate of 300 ppb for 7.5 minutes in which the potential exposure was half that in 1974. In laboratory experiments, exposure dosages (ppb-h) with high methoxychlor concentrations over a short period are likely to produce higher tissue residues than low concentrations over a long period. Effectively, this relationship reduces the concern for fish exposure as the treatment pulse moves downstream.

Examination of the gut contents of fish suggests that they suffered no deprivation of food. However, low values for blood proteins, suggestive of starvation, were identified with one sample of four White Suckers (*Catostomus commersoni*) captured in July 1974 about 77 km downstream of the treatment site. Low protein levels may be induced in fish by a number of factors, one of which was demonstrated in the laboratory as starvation for 8–16 weeks. This neither clearly indicates that river fish were starving, nor that it was the result of treatment. All other samples of the same species of fish caught had normal levels of blood protein.

There was no evidence from the accumulated laboratory analyses and field observations that river treatments pose any serious risk for the trophic relations essential to the food chain for fish.

Alternatives to Larvicides in Blackfly Control

The options in protecting livestock from infestations of *S. arcticum* are limited to reduction or elimination of larvae in breeding sites, control of adult dispersal into agricultural areas, individual protection of animals on farms, and change in livestock management. Of these, larvicidal control of breeding sources is by far the most practical and economical because larval development is confined to one major river (Haufe 1980).

Although changes in livestock management have been seriously considered, they are not compatible with a competitive and viable industry. Selective area control of adult dispersal with adulticides has doubtful potential effectiveness until more is known of the behaviour of adult

S. arcticum and of the factors and patterns that control the displacement of adults from the breeding source. Protective measures for animals on farms are the only promising alternative to larvicides at the present time.

Repellents

Some progress has been made in the development and evaluation of repellents for use on animals. Compounds developed for human use have been found in field tests to provide generally longer periods of protection on livestock than on man. For example, diethyl toluamide (DEET), hexandiol (6–12), dimethyl phthalate (DMP), and 3-acetyl-2-(2,6-dimethyl-5-heptenyl)-oxazaladine (R69) prevents blood-feeding by blackflies for periods up to 2.5 days after application on cattle as compared to several hours on man (Shemanchuk 1980). The new pesticide permethrin shows promise in providing an even longer-lasting barrier to blood-feeding of blackflies although it may act more as a protectant against contact than as a repellent. Repellents with effectiveness in the order of 10 days are considered necessary for practical acceptance in economic livestock operations.

Shelters

Animal shelters used alone and in combination with pesticidal sprays and automatic treatment devices have been evaluated in field tests in the Athabasca area (Khan 1980). Shelters provided significant relief for cattle from blackfly invasions on farms. However, the expense of construction and limited flexibility of their use in normal grazing operations discourage their acceptance. Shelters are not compatible with utilization of pastures and roughages for maximum rate of growth during the summer grazing period, nor with the economic use of cattle as a natural harvester of native forages in marginally agricultural areas.

Evaluation of Pesticides in an Integrated Approach to Control of *Simulium arcticum*

In the context of economically viable options at present, control of *S. arcticum* for the protection of new and expanding livestock production in the drainage basins of major rivers such as the Athabasca and Peace Rivers in northern Alberta is highly dependent on continued use of environmentally acceptable larvicides. The results of the ARBFRP, although unprecedented in terms of multidisciplinary studies to evaluate methoxychlor in an extensive highly complex aquatic ecosystem, emphasize not only the potential for selective use of chemicals, but also the need for a more thorough understanding of the dynamics of drifting and colonizing behaviour in invertebrate populations of large rivers (Haufe 1980). The physical dimensions of velocity, turbulence, and volume of discharge in large rivers such as the Athabasca offer little scope for the effectiveness, not to mention manipulation and management, of biological controls as economic or practical alternatives to larvicides.[13] It is obvious, with the high potential for massive dispersal of blackflies from relatively well-defined breeding sources in such rivers, that livestock protection depends heavily on the development of more sophisticated larviciding procedures as an essential part of an approach to integrated pest management.

---

[13] See pp. ix, 181–196 for views on the possible application of microbial agents (as distinct from, e.g. such conventional biological controls as predators) to such situations to supplement chemical approaches in future integrated methodologies . . . .Ed.

Accurate evaluation of larvicides in large river systems is hindered by a critical need for information on the trophic status of aquatic invertebrate populations and their importance to the food chain and to the fish community. The relationship of standing crop and drift patterns of invertebrates to such components as larval survival, growth, fecundity, and body condition in the life-cycle of the river fauna is insufficiently understood for knowledgeable management of river systems in the long term. These studies are essential to the development of adequate criteria for environmental acceptability and intelligent use and formulation of pesticides in rivers and streams. In combination with greater adherence to hydraulic parameters, this kind of information affords yet undeveloped approaches in ecological chemistry to exploit degradable pesticides to advantage in environmental compatability.

New integrated forms of quantitative methodology and correspondingly appropriate sample systems are needed to resolve conflicting interpretations from too-meagre analysis of aquatic systems. Concepts of a quantitative ecological approach to balance in patterns of natural systems (Williams 1964), in conjunction with more comprehensive approaches to the graphic and mathematical analyses of biotic communities in rivers and streams (Wilhm 1972), should encourage more extensive data bases for the evaluation of larvicides in integrated pest management of blackflies in large rivers. In this respect, the ARBFRP has produced an unprecedented bank of data and samples for more detailed analyses of environmental impact along these lines of study. Computer analyses of about 200 000 generically collated samples of invertebrates from regular time-series sampling systems of the drift and the benthos in the Athabasca have been partially completed (Haufe, unpublished). These analyses, with a mathematical model to represent Williams' concept, have already shown

clearly that invertebrates in the Athabasca River have maintained the taxonomic balance demonstrated, in concept, by
Williams (1964). Further analyses are required to evaluate similar relationships
for numbers of individuals distributed
among the taxa as a potential criterion for
larvicidal impact on ecosystems with reference to baseline studies with no treatment. Preliminary results from these unpublished findings, in addition to the
consolidated general analyses of all published studies in the ARBFRP, support
the conclusion that methoxychlor can be
managed in large rivers as an environmentally acceptable larvicide (Haufe
1980).

A major requirement still remaining
from the ARBFRP is a clear resolution in
methodology of the relation between
drift and standing crop invertebrates.
With a satisfactory understanding of this
relation, drift measurements may become an extremely convenient economic
means of monitoring larviciding programmes for public, administrative, and
operational accountability.

A Small-scale Environmental Approach To Blackfly Control in the USA

P. A. La Scala and J. F. Burger

Whenever parks, campgrounds, resort hotels, and other recreational facilities are developed in woodlands with abundant running water sources in the northeastern USA, blackflies may cause varying degrees of annoyance to visitors. The importance of Simuliidae as pests of humans and other animals in North America has been documented by many authors (Webster 1902; Strickland 1913; Bradley 1935; Remple and Arnason 1947; Fallis 1964; Fredeen 1969; Craig 1972; Jamnback 1973; Fredeen 1977).

Large numbers of tourists visit New Hampshire in spring and summer and vacation at the many resorts in the northern half of the state. The biting and annoyance of female blackflies in certain areas create a nuisance for residents and vacationers during the tourist season. This has resulted in suspected economic losses in resort areas where blackfly populations are high. One resort where such losses are suspected is the Balsams Hotel in Dixville Notch, New Hampshire.

Dixville Notch is located at 44° 52'N, 71° 17'W, in the Northern Upland Region of New Hampshire, in predominantly mixed Spruce-Fir, northern hardwood (birch, maple) forest. This area receives about 430 mm precipitation annually and the mean annual temperature 14.0 °C. Mountains 610–760 m high surround the hotel complex, creating a bowl-shaped valley. The mountainous topography, with an extensive drainage basin, supplies the requirements of clean, well-oxygenated, fast-flowing water that can support large blackfly populations. The presence of ideal natural breeding habitats plus the lake and pond impoundments, and their outlets, constructed by the hotel since the late 1800s for water conservation and flood abatement have resulted in ideal habitats for outlet breeding *Simulium* species.

A severe blackfly season in 1975 prompted the Balsams Hotel management to seek assistance from the Department of Entomology at the University of New Hampshire. A survey of simuliid breeding sites was initiated during the summer of 1975. The survey showed that large populations of *Simulium decorum* Walker were breeding in streams near the hotel. A project was then initiated to study the life-cycle, abundance, and seasonal distribution of *S. decorum* in the study area leading to the development of a practical population management programme.

Major Breeding Habitats

Two major types of outlet breeding habitats were studied by the senior author in Dixville Notch. The first type is a subterranean overflow pipe (7.9 m × 0.6 m/dia.) originating from a spring-fed pond. A first-order stream meanders through a small wooded area of spruce, fir, and northern hardwoods. Blackfly breeding was most concentrated in the pipes, and continued into the woodland for 3–5 m.

Breeding was reduced further downstream due to algal growth on stream
substrates. The second type consisted of
reinforced concrete dams, with subterranean drainage conduits 3.9–30.0 m
long × 1.4–2.4 m/dia.

Only *S. decorum* and *S. venustum/vere-
cundum* complex larvae were abundant at
outlets. Other species occurring in
streams below outlets included *Simulium
corbis* Twinn, *S. vittatum* Zetterstedt, *S.
croxtoni* Nicholson and Mickel, *S. tuberosum* Lundstrom complex, *Stegopterna
mutata* (Malloch) and *Cnephia dacotensis*
(Dyar and Shannon). Since *S. decorum*
was the dominant species throughout the
study area it was selected for studying
the effectiveness of control procedures.

Target Species

Simulium decorum is a multivoltine species
that over-winters in the egg stage. It is
widely distributed in North America
from Alaska to Newfoundland, south to
Oregon, Colorado, Arkansas, and Florida. Immature stages are most abundant
at the outflows of artifically or naturally
impounded waters, on the sticks of beav-
er dams, and on dam faces (Stone 1964).
Other blackflies exhibiting similar habitat
distribution patterns include *Cnephia ere-
mites* Shewell, in Alaska, *C. dacotensis* and
in the Palearctic Region *S. sublacustre* L.
Davies in Britain and the Palearctic sister
species of *S. decorum*, *S. argyreatum*
Meigen (Hynes 1970).

Mature larvae and pupae of *S. decorum*
were collected at water temperatures
from 10 °C to 24 °C. Immature stages
were found in water with current velocities of 0.2–1.0 m/s, and at depths from
2 cm to 25 cm.

The eggs of *S. decorum* are nearly mature when the female emerges from the
cocoon. Females emerge with ovarian
follicles in Stage IV in mid-May, producing the first batch of eggs autogenously.
However, subsequent generations have

follicles in Stage II, II–III, and probably
require a blood meal for maturation of
eggs. Females were observed ovipositing
in overflow pipes, on rocks, and the
splash zone of dam weirs. Dead females
were found attached to egg masses, on
the wooden weir boards, and inside the
overflow pipes. The feeding of *S. decorum*
on mammals and avian hosts in Canada
and North America is well documented;
(Nicholson and Mickel 1950; Hocking
and Richards 1952; Sailer 1953; Davies
and Peterson 1956; Anderson and De-
Foliart 1961).

Early Control Attempts at Dixville Notch

Weed (1904) experimented with a larvicide known as Phinotas oil (soluble and
insoluble) against blackflies. This material which had been used previously for
control of mosquito larvae, duly proved
equally effective against blackfly larvae.
However, concern was expressed about
the effects of chemicals on fish populations.

At the time of these investigations,
Weed was experimenting with mechanical removal of concentrated blackfly larvae
at pond outlets on the University of New
Hampshire campus (then called New

*Mechanical removal of immature stages of S. de-
corum at Lake Gloriette subterranean conduit.*

Hampshire College). Weed suggested using stable-brooms at Dixville Notch to dislodge masses of larvae, and collecting them on wire netting downstream. He found that the sweeping method was entirely practical; offering, in some breeding grounds, a simple means of destroying larvae.

Non-chemical methods of blackfly control have been documented by authors in the USA (Weed 1904; O'Kane 1926), Germany (Wilhelmi 1920) and Africa (Quélennec et al. 1968). The work mentioned by Weed and O'Kane has been mentioned above.

Wilhelmi (1920) suggested that the removal of breeding substrates early in the season would be the most productive method for reducing blackfly populations. His methods included transferring heavily infested rocks onto land, and cutting and removal of submerged vegetation in rivers and streams to which the larvae attach.

Quélennec et al. (1968) compared the various forms of dam spillways in Upper Volta as blackfly breeding sites. They found that stepped spillways, frequently employed in rural regions, provided ideal breeding habitats for the S. damnosum (Theobald) complex. Vertical spillways, being found less favourable for blackfly breeding, were recommended for use in future dam construction projects.

Assistant demonstrating mechanical removal, modified pole brush, with metal diversion chute in place.

migrate downstream from outlet breeding sites; females are known to be pests of mammals, birds, and occasionally humans; and females tend not to migrate far from breeding sites.

An experimental programme, using mechanical removal and habitat manipulation, was designed to eliminate the immature stages of S. decorum from pond and lake outlets within a 1.6 km radius of the Balsams Hotel complex. Four major breeding sites were sampled by the senior author for the immature stages of S. decorum, beginning in May 1978.

A system for rapid determination of the larval instar was employed as a guide to the timing of control measures. This method consisted of grouping the larvae

Closeup of wire bristle broom with removed eggs, larvae, and pupae.

Population Management of *Simulium decorum* Larvae

The breeding characteristics of S. decorum, in addition to the desire to minimize the environmental impact of blackfly control measures, lead to a re-examination of the techniques employed by Weed in the early 1900s. The decision to employ these techniques was influenced by the fact that: S. decorum larvae are concentrated at outlets of artificial impoundments; the larvae tend not to

according to respiratory filament histoblast development. Development was
divided into five stages. Stage I comprises first-star larvae with egg-burster present and larvae with no visible respiratory histoblast (instars 1–4). Stage II was
used for larvae with a round undifferentiated histoblast (instar 5). Stage III was
identified by the histoblast being differentiated into rudimentary filaments (instar 6). Stage IV was determined by the
histoblast being fully developed but un-
pigmented (instar 7). Stage V was designated for larvae with fully developed
black histoblast (later instar 7; prepupa).

*Subterranean drainage conduit with metal diver-
sion chute in place.*

On June 22 1976 mechanical removal
was initiated by the senior author, at all
major sites with *S. decorum* in Stage IV.
The measures consisted of physically
scraping the eggs, larvae, and pupae
from their attachment sites with wire
bristle brooms and modified pole
brushes. Most larvae and pupae were
removed by this method but some where
observed reattaching downstream after
scraping was completed. To prevent re-
attachment downstream, metal roofing
sheets (3.6 × 0.9 m) were constructed as
stream diversion chutes. These chutes
were attached at each site having an
overflow pipe or subterranean drainage
conduit. During the scraping procedure,
the chute directed detached eggs, larvae,
and pupae out of the stream channel and
onto land.

Habitat manipulation included removal of heavily-infested rocks, vegetation,
and other breeding substrates from control sites. Lake discharge was reduced
for two days by adding boards to the
dam weirs. Diminishing the discharge
through manipulation lessened oviposition and reduced the area of habitat available to larvae, thereby so concentrating
the latter that residual populations in
zones of reduced flow were eliminated
rapidly. Adverse effects on NTOs were
not observed during the experiments.

Initial manipulation techniques at all
sites were completed in two days requiring about 10–20 man hours. Each site was
checked for re-infestation on the 2nd,
4th, and 7th day after treatment. Mechanical removal or habitat manipulation
was used when larval sampling showed
10–20% of larvae in Stage IV. Since re-
infestation occurred at all sites every
seven days following manipulation, each
site was monitored weekly to prevent
accumulation of mature larvae and
pupae. To eliminate breeding, control
procedures were applied to each major
breeding site every 10 days. Control procedures were required from late June
until the end August.

We estimate that at least 90% of the
larval populations of *S. decorum* could be
eliminated by the measures discussed
above. Where blackflies breed in restricted areas, the control procedures described above are feasible, practical, and
environmentally sound. These methods
provide an alternative to more traditional
control measures when species to be controlled are restricted to outlet sites and do
not migrate and re-attach downstream
following removal.

III Predators, Parasites and Pathogens

Predators upon Blackflies

D. M. Davies

Predators were neatly contrasted with parasites by Elton (1927) who said "The difference between . . . a carnivore and a parasite is simply the difference between living upon capital and income." They doubtless exert an important natural control on simuliid populations under certain conditions, especially on their immature aquatic stages. However, most records of such predation are random observations (Jenkins 1964), and there are few studies that provide much understanding of the impact such predation has on blackfly populations. The observations of some workers led them to believe that the effect of predation is secondary to parasitism (Abdelnur 1968; Ezenwa 1974). By contrast, Speir (1976) determined that in western Oregon streams (USA) predation accounted for 82.6% of simuliid larval mortality, with parasitism contributing only 0.5% (and drift 6.2%). Hynes (1960) stated that the normal enemies of *Simulium* are fishes, stoneflies and caddisworms, the last two being the most affected by soluble DDT (Formerly widely used to control simuliids). In another study, caddisworms and mayflies were the most reduced by

DDT (Graham and Scott 1959). Heavy reduction of these predators did not occur when particulate insecticides were used (Jenkins *et al.* 1968). Ide (1967) remarked that simuliid populations rebound in years subsequent to insecticidal treatment of streams. He found that predator populations, such as those of Trichoptera and Plecoptera, were very susceptible to insecticides and recover less quickly. Laird (1978) warns that measures for suppression of immature blackfly levels must not destroy control elements in the ecosystem—particularly simuliid predators, of which there are many.

Whether simuliids are eaten by a certain predator depends on the species, age or stage and size of both predators and prey; on the predator's habitat preference in relation to that of the prey; on the relative availability and abundance of the simuliids in relation to non-simuliid prey; and on the time of day or season. Thus, before attempting to use a predator as a control agent, an intense ecological assessment should be undertaken to determine which of these factors most significantly affects simuliid predation. Yet another factor which may be involved is movement by the larvae. Predators (e.g. caddisworms), may be more attracted to undulating or crawling simuliids (Kurtak 1978). Many larvae escape caddisworm predation by forming a "horseshoe" pose with head and posterior to the substrate (Kurtak 1978), or "inch-worming" or detaching and drifting on a silk thread (Hart 1978). When a larvae is touched, its violent twisting may discourage or dislodge smaller predators. There are reports of salivary silk entangling and ultimately killing predators such as gammarids (Grenier 1945a, 1945b) and ephemerid nymphs (Bradt 1932).

As increasing focus is directed toward methods of bio-control of immature and adult blackflies, it is important that the potential of various vertebrate and invertebrate predators be evaluated. A review of the known simuliid predators follows.

Vertebrate Predators

Mammalia

One might expect that bats would feed on adult blackflies during the evening crepuscular period when simuliid flying activity often remains near its maximum, especially if prolonged by the stimulus of the presence of vertebrate hosts. Male blackflies form relatively tight pre-mating swarms over a swarm-marker; females congregate above larger prey animals. Such swarming behaviour also facilitates predation. M. B. Fenton (personal communication, 1978) says that bats do, in fact, spend much time exploiting insect swarms and are quite opportunistic. I have found no reference to such predation in the literature for *Myotis* or *Pipistrellus*. This has been confirmed by personal communication with J. R. Tamsitt (University of Toronto), J. O. Whitaker (Indiana State University), and M. B. Fenton (Carlton University).

The insectivorous North American water shrew (*Sorex palustris*) may feed on blackfly larvae and pupae, although these shrews are less frequent in rapid currents preferred by Simuliidae. Only one larval blackfly was found in 88 stomachs examined (Conaway 1952). In Australia, 14.8% of 61 platypus (*Ornithorhynchus anatinus*) had a dominance of dipteran larvae in their diet, including simuliids, as gauged by cheek-pouch contents (Faragher *et al.* 1979).

Aves

Some birds are opportunistic feeders, and species in several avian families are reported to feed on adult and immature simuliids (Table 1). Appreciable numbers of blackflies are ingested by only a few of these (Table 1). Dalmat (1955) stated that seven of 25 species of Guatemalan birds had adult remains in their stomachs. However, such blackfly fragments were

Table 1. References to specific birds feeding on Simuliidae.[a]

1. Dalmat (1955) in Guatemala (A)
    Amazibis candida pacifica (Griscom): Trochilidae
    Basileuterus c. culicivorus (Lichtenstein): Parulidae
    Cissilopha m. melanocyanea (Hartlaub): Corvidae
    Myiodynastes l. luteiventris Sclater: Tyrannidae
    Onychorhynchus m. mexicanus (Sclater): Tyrannidae
    Vermivora peregrina (Wilson): Parulidae
    Vireo s. solitarius (Wilson): Vereonidae
2. Davies, D. M. (1955 unpublished notes) in Ontario, Canada (A)
    Anas p. platyrhynchos L.: Anatidae
3. Dinulescu (1966) in Roumania
    Cinclus sp.: Cinclidae
4. Hocking and Pickering (1954) in N. Manitoba, Canada (A)
    Zonotrichia leucophrys (Forster): Fringillidae
5. Ladle, M. (personal communication 1978) in England (L)
    Gallinula chloropus L.: Rallidae
6. McAtee (1932) in USA (L,P,A)
    Archilochus colubris L.: Trochilidae
    Chaetura vauxi Townsend: Micropodidae
    Cinclus mexicanus unicolor Bonaparte: Cinclidae
    Nettion carolinense (Gmelin): Anatidae
    Querquedula discors (L.): Anatidae
    Vireo philadelphicus (Cassin): Vireonidae
7. Millais (1902 in Kortright 1942) in Iceland (A)
    Anas penelope L. (as *Mareca penelope*): Anatidae
8. Riley (1887), Johannsen (1903) in USA (A)
    Mimus p. polyglottos L.: Mimidae
    Troglodytes t. hiemalis Viellot: Troglodytidae
    Gallus g. domesticus L.: Gallidae
9. Rivosecchi (1978) in Italy (A)
    Swallows: Hirudinidae
    Acrocephalus palustris: Syliviidae

[a] A = adult, P = pupa, L = larva.

so few as to indicate that these insects are probably not a normal dietary constituent.

Dippers (*Cinclus* spp.) frequent rapid streams in western North America and Europe, and may capture many larvae and pupae under water (McAtee 1932; Carlsson 1962; Dinulescu 1966; Jost 1975). The scarcity of this bird at any particular site makes it of minor importance in simuliid control.

Given shallow rapids, abundant larvae, and suitable perches just above the water surface, Common Grackles (*Quiscalus quiscula*) fed on simuliid larvae (Snoddy 1967).

Table 2. Bird predation on Simuliidae.

	Predators		Prey			
Species	No. examined	No. with prey	Species	Av. no./predator	Country	Authority
Bulbulcus ibis (L.)	>1	>1	*Cnephia pecuarum* (Riley)	few A	USA	Snoddy (1978)
Buphagus erythrocephala (Stanley)	58	4	*Simulium* sp.	c. 0.4A	Kenya	Moreau (1933)
Cinclus mexicanus unicolor Bonaparte		3	Simuliid	125L, 13P	USA	McAtee (1934)
Nettion carolinense (Gmelin)	80 000	1	Simuliid	900L	USA	McAtee (1934)
Querquedula discors (L.)		1	Simuliid	L	USA	McAtee (1934)
Clangula hyemalis (L.)	2	1	*S. vittatum* Zett.	0.5A	Greenland	Longstaff (1932)
Histrionicus histrionicus (L.)	31	31	Simuliidae	>96% L[a]	Iceland	Bengtson (1972)
Gallinula chloropus L.	5A, 5J	1		0.1L	England	Hartley (1948)
Quiscalus quiscula quiscula L.	<6+/time	<90%	*Simulium pictipes* Hagen		USA	Snoddy (1967)

L = larva, P = pupa, A = adult, J = juvenile.
[a] Percentage of simuliid larvae in relation to other animal and plant food.

Young and mature Harlequin Ducks (*Histrionicus histrionicus*) fed almost exclusively on larval simuliids in Iceland and rarely took adults (Bengtson 1972). Stomach contents of 20 adults averaged 12.7 g (wet weight) of larval simuliids per duck.

A male European Widgeon was seen to eat hundreds of a sort of "stinging house fly" from stones at the edge of a small Icelandic lake (Millais 1902 *in* Kortright 1942). B. V. Peterson (personal communication, 1978) considers that these "flies" were probably simuliids.

Riley (1887) reported that chickens have, on occasion, ingested such large numbers of blood-filled simuliids from each other that they have become helpless.

In southern USA, mockingbirds (*Mimus polyglottos*), Winter Wrens (*Troglodytes troglodytes*) and domestic hens have been noted to feed on blood-engorged blackflies (Osborn 1896 *in* Johannsen 1903).

Carlsson (1962) and Rivosecchi (1978) observed swallows with open beaks flying through simuliid swarms, as did R. Glatthaar (personal communication, 1980).

Amphibia

Martof and Scott (1957), Martof (1962) reported that, in Georgia, USA, 7.5% of 150 shovel-nosed salamanders (*Leurognathus marmoratus*) dissected contained simuliid larvae. Hocking (1950) stated that the northern wood frog (*Rana sylvatica cantabrigensis*) feeds on simuliid larvae.

Pisces[14]

Most trophic studies on fish (Table 3) only consider consumption, but there are some workers who have considered prey availability by determining the abundance of benthic organisms (Allen 1941;

---

[14] Also see p. 157 for additional records provided since this chapter was submitted.

Hynes 1950; Maitland 1965; Straskraba *et al.* 1966).

Salmonid species, because of their economic importance, have been studied more than other fish in regard to their feeding habits (Tables 3 and 4). Fry of Atlantic salmon (*Salmo salar*), when their yolk supply is expended, swim from under stones into the open and begin devouring chironomid and simuliid larvae (Badcock 1949). White (1936) observed that the diet of fry contained only 0–3.5% simuliid larvae, although no information on abundance of any aquatic insects was given. Mills (1964) reported much heavier predation of simuliid larvae by salmon fry; heavier, in fact, than that seen by Ladle (1978) for salmon parr. The consistently high utilization of simuliid larvae, when abundant, may be due to their conspicuousness on surfaces exposed to water currents (Allen 1941). Maitland's (1965) findings differ, however.

Trout seem to feed more from the surface than salmon and thus simuliid adults are often more numerous in their diet than larvae (Emery 1913; Edwards 1920; Levander 1923 *in* Kuusela 1978). According to Gibson and Galbraith (1975), however, trout and juvenile salmon feed mainly on benthic Trichoptera, Ephemeroptera and Simuliidae. Simuliids, mainly larvae, form one of the major foods of cut-throat trout (*Salmo clarki*) except for the largest fish (Dimick and Mote 1934). In a western Canadian stream in winter, 10–20% of rainbow and cut-throat trout fed on simuliid larvae (Idyll 1942). Eidt (1978) suggests that feeding on drifting larvae, whether caused by insecticiding or not, might be considered as scavenging rather than predation.

Predation may be selective, not reflecting general prey availability. The work by Elliot (1967) disclosed that brown trout (*Salmo trutta*) less than one year old, selected simuliid larvae at a rate greater than 25 times their availability. This predation occurred mostly at night. Fish

Table 3. References of fish predation on Simuliidae.

Acipenseridae
Acipenser ruthensis L.: Bening 1924; Baranov 1937; Dinulescu 1966; Rubtsov 1967
Scaphirhynchus platorhynchus (Raf.): Modde and Schmulback 1978
Salmonidae
Oncorhynchus kisutch (Walbaum): Mundie 1969
Oncorhynchus rhodurus Jordan and McGregor: Tanabe 1970
Salmo clarki Richardson: Dimick and Mote 1934; Schultz 1939; Idyll 1942; Hynes 1950
Salmo gairdneri Richardson: Dimick and Mote 1934; Idyll 1942; Hynes 1950; Peterson 1960; Jenkins *et al.* 1970; Elliott 1973; Tippets and Moyle 1978
Salmo salar L.: White 1936; Allen 1941; Badcock 1949; Frost 1950; Hynes 1950; Thomas 1962; Mills 1964; Maitland 1965; Maitland and Penney 1967; Rubtsov 1967; Mann and Orr 1969; Gibson and Galbraith 1975; Back 1978; Eidt 1978; Ladle 1978
Salmo trutta L.: Edwards 1920; Stankovitch 1924; Levander 1923; Pentelow 1932; Idyll 1942; Frost 1945, 1950; Hartley 1948; Hynes 1950; Nilsson 1954; Smyly 1955; Carlsson 1962; McCormack 1962; Thomas 1962; Mills 1964; Maitland 1965; Zivkovic and Kacanski 1965; Dinulescu 1966; Straskaraba *et al.* 1966; Maitland and Penney 1967; Rubtsov 1967; Elliott 1967; 1973; Chaston 1968; Mann and Orr 1969; Kuusela 1978; Ladle 1978
Salvelinus fontinalis (Mitchell): Twinn 1939; Ide 1942; Wolfe and Peterson 1959; Miller 1974; Gibson and Galbraith 1975; Back 1978; Eidt 1978
Brachymystax lenok (Pallas): Rubtsov 1967
Coregonus clupeaformis (Mitchell): Hocking and Pickering 1954
Coregonus (*?laveretus* L.): Rubtsov 1962, 1967; Kuusela 1978
Coregonus peled (Gmelin): Rubtsov 1962
Coregonus sardinella Valenciennes: Rubtsov 1962, 1967
Thymallus spp.: Carlsson 1962; Rubtsov 1962, 1967; Zivkovic and Kacanski 1965; Dinulescu 1966; Bobrova 1971; Kuusela 1978; Ladle 1978
Galaxiidae
Galaxias vulgaris Stockell: Cadwallader 1974
Hiodontidae
Hiodon alosoides (Rafinesque) (as *Hyodon chrysopsis*): Cameron 1922
Esocidae
Esox lucius L.: Hocking 1950
Characidae
Hydrocynus ?forskahlii: Service and Lyle 1975
Cyprinidae: Riley 1887
Barbus meridionalis: Zivkovic and Kacanski 1965
Couesius plumbeus (Agassiz): Burger 1978
Cyprinus carpio L.: Eder and Carlsson 1977
Gobio gobio (L.): Hartley 1948; Ladle 1978
Leuciscus idus (L.): Rubtsov 1962, 1967

Leuciscus leuciscus (L.): Hartley 1948; Rubtsov 1967; Dinulescu 1966; Ladle 1978
Moroco jouyi: Tanabe 1970
Notropis ardens (Cope): Bettoli 1978
Notropis hudsonius (Clinton): Price 1959
Phoxinus laevis Ag.: Tack 1940
Phoxinus phoxinus L.: Frost 1943; Hartley 1948; Maitland 1965; Straskraba *et al.* 1966; Maitland and Penney 1967; Rubtsov 1967; Mann and Orr 1969; Ladle 1978
Platygobio gracilis (Richardson): Cameron 1922
Rutilus rutilus (L.): Hartley 1948; Rubtsov 1962, 1967
Rutilus pigus virgo: Zivkovic and Kacanski 1965
Semotilus atromaculatus (Mitchell): Baranov 1937; Twinn 1939; Keast 1966
Squalius cephalus (L.): Hellawell 1971; Mann 1976
Varicorhynchus tanganica Boulenger and other spp.: Marlier 1952; Dinulescu 1966
minnows: Emery 1913; Pomeroy 1916
Cobitidae
Cobitis biwae: Tanabe 1970
Nemacheilus barbatula (L.): Smyly 1955; Maitland 1965; Dinulescu 1966; Maitland and Penney 1967; Ladle 1978
Castostomidae
Catostomus catostomus (Forster): Hocking 1950
Catostomus commersoni (Lacépède): Cameron 1922; James 1951; Eder and Carlsson 1977
Hypentelium nigricans (Lesueur): Bettoli 1978
Amblycipitidae
Liobagrus reini: Tanabe 1970
Amphiliidae
Amphilius jacksoni: Corbet 1961
Phractura ansorgei Boulenger: Crisp 1956a
Ictaluridae
Ictalurus nebulosus (Lesueur): Keast 1966
Ictalurus punctatus (Rafinesque): Price 1959; Stauffer *et al.* 1976
Pylodictus olivaris (Rafinesque): Stauffer *et al.* 1976
Anguillidae
Anguilla anguilla L.: Frost 1946; Hartley 1948; Thomas 1962
Gadidae
Lota lota L.: Nilsson 1954
Gasterosteidae
Culaea (as *Eucalia*) *inconstans* (Kirtland): James 1951; Keast 1966
Gasterosteus aculeatus L.: Idyll 1942; Hynes 1950; Maitland and Penney 1967; Mann and Orr 1969
Pungitius (as *Pygosteus*) *pungitius* (L.): Hocking 1950; Hynes 1950; James 1951
Percidae
Etheostoma blennioides Rafinesque: Bettoli 1978
Etheostoma caeruleum Storer: Bettoli 1978
Gymnocephalus (= *Acerina*) *cernua* (L.): Rubtsov 1962, 1967
Cichlidae
Tilapia aurea (Steindachner): Service and Lyle 1975

Table 3 (contd)
Eleotridae
  Gobiomorphus huttoni Ogilby: Crosby 1975
Gobiidae
  Gobius similis: Tanabe 1970
Cottidae
  Cottus asper Richardson: Idyll 1942
  Cottus bairdii Girard: Coffman 1967
  Cottus gobio L.: Straskraba et al. 1966; Zivkovic and
    Kacanski 1965; Dinulescu 1966; Rubtsov 1967;
    Mann and Orr 1969; Kuusela 1978; Ladle 1978;
    Adamicka 1979
  Cottus poecilopus Heckel: Straskraba et al. 1966;
    Kuusela 1978
  Cottus pollux: Tanabe 1970
Cyprinodontidae
  Profundulus (?punctatus Günther): Dalmat 1955,
1958
Clariidae
  Clarillabes petricola Greenwood: Corbet 1958a, b
Pleuronectidae
  Platichthys (as Pleuronectes) flesus L.: Ladle 1978
Unknown families
  Fish: (Rubtsov 1962)

older than two years, however, selected these larvae below availability.[15] This trend was shown also by Nilsson (1937), McCormack (1962) and Mann and Orr (1969). A greater nocturnal feeding on drifting simuliid larvae was noted also for rainbow trout (Salmo gairdneri) (Elliot 1973), which could see them even against the night sky (Jenkins et al. 1970). These fish selected blackfly larvae above their availability (Jenkins et al. 1970). On the other hand, juvenile coho salmon (Oncorhynchus kisutch) selected below the availability of drifting larvae, despite the fact that benthic simuliid larvae were their most important diet item in summer and autumn (Mundie 1969). Brook trout (Salvelinus fontinalis), the diet of which is composed of 16% simuliid larvae (Ide 1942), has peak feeding from 17.00–20.00 h and at 08.00–11.00 h (Hore 1940 in Ide 1942).

---

[15] And, when observed under January ice on Broad Cove River from Memorial University's Benthobservatory (Laird et al. 1974), one was seen to face upstream from abundant larval blackflies covering rock-surfaces, while—at lengthy intervals—ingesting fragments of vegetation drifting downstream. [Original observation, Ed.]

In the USSR, a study of 238 graylings (Thymallus spp.) showed that over 50% of the diet was composed of simuliid larvae and over 25% of pupae (Bobrova 1971), which is similar to that reported by Rubtsov (1962). Carlsson (1962) and Ladle (1978) also noted that these fish fed on larval simuliids, while Kuusela (1978) found one grayling with more adults than larvae in its stomach.

Baranov (1937) stated that the little sturgeon (Acipenser ruthensis) fed principally on larval Simulium columbaschense in Yugoslavia and that one fish contained 200 larvae. In contrast, Modde and Schmulbach (1978) found that simuliids were never important in the diet of the shovelnose sturgeon (Scaphirhynchus platorhynchus) in the Missouri river, South Dakota—even in March they constituted only 0.6% (dry weight) of the diet.

Whitefish (Coregonus spp.) at times may make simuliid larvae a major part of their diet in northern Ontario (Hocking and Pickering 1954), in the USSR (Rubtsov 1962, 1967) and Finland (Kuusela 1978).

Various genera of Cyprinidae are reported to prey on simuliid larvae. However, they formed only 1.0–1.4% of the diet of carp (Cyprinus carpio) in Colorado (Eder and Carlson 1977). Shiners (Notropis spp.) take occasional larvae and adults (Price 1959; Bettoli 1978). European minnows (Phoxinus spp.) have been more closely studied, but, again, simuliid larvae (and occasional adults) comprise a minor dietary component (1.0–1.4%) (see references in Table 3). Stomachs of creek chub (Semotilus atromaculatus) sometimes contain 20% simuliid larvae by volume (Keast 1966), and may occasionally be gorged with them (Twinn 1939). In New Hampshire, USA, lake chub (Couesius plumbeus) actively fed on larval Simulium decorum Walker, massed on the downstream lip of a dam [Burger 1978]. In England, another chub (Squalius cephalus) was shown to eat 1–10% simuliid larvae (Hellawell 1971; Mann 1976) and the roach (Rutilus rutilus) to eat 6%

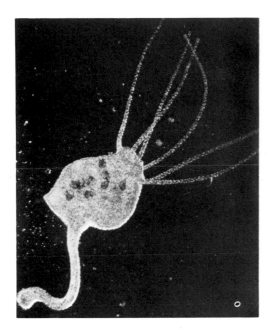

Hydra *containing head capsules of first-instar simuliid larvae in Germany (J. Grunewald 1978).*

(Hartley 1948). In Africa, out of 120 examples of *Varicorhynchus tanganicae*, 51% was found to have fed on simuliid larvae (Marlier 1952). The stone loach (*Nemacheilus barbatula*) ate up to 5% simuliid larvae (Smyly 1955), although Maitland (1965) reported <1%.

Cameron (1922) reported that white suckers (*Catostomus commersoni*) were the most important predators of simuliid larvae in the Saskatchewan river in the Canadian prairies. In Colorado, however, simuliid larvae formed only 0.1–1.4% of this fish's diet (Eder and Carlsson 1977). During the month of May, Keast (1966) found, in Ontario streams, that medium-sized (70–120 mm) brown bullheads (*Ictalurus nebulosus*) had 19% simuliid larvae (by volume) in their diet, while smaller and larger fish had 0% and 2% respectively. In England and Wales, the common or yellow eel (*Anguilla anguilla*) may, at times, include 4.0–12.5% simuliid larvae in its diet (Frost 1946; Hartley 1948; Thomas 1962). Keast

(1966) found that larger (36–50 mm) brook sticklebacks (*Culaea inconstans*) had a diet of 35% simuliid larvae in May and 15% in October, but that smaller fish did not feed on them. The diet of the gudgeon (*Gobio gobio*), in England, averaged 9% simuliid larvae (Hartley 1948). Sculpins show varied proportions of simuliid larvae in their diet: 2% in USA (Coffman 1967); 5% in Czechoslovakia (Straskraba *et al.* 1966); and 74–79% in a few Finnish fish (Kuusela 1978).

Dalmat (1955, 1958), from his Guatemalan blackfly studies, found simuliid remains in all but 10 small examples of 200 *Profundulus* spp. sampled. He reported that fish of this genus were the most effective predators of simuliid larvae and of emerging and ovipositing adult blackflies; suggesting that, in certain restricted areas, they might be useful biocontrol agents, but noting that they prefer deeper water than is typical of simuliid larval habitats.

In Ghana, according to Crisp (1956a) *Phractura ansorgei* is an important predator of simuliid larvae, individuals often being crammed with larvae. In Australia, *Gobiomorpha* only feeds to a minor extent (1.4%) on larval simuliids (Crosby 1975).

In general, it appears that, under conditions where blackfly larvae are abundant and concentrated or where many adults are emerging or ovipositing, fish such as trout and salmon may eat large numbers and exercise considerable natural control. Graylings, chub and even sculpins and sticklebacks may make simuliid larvae a considerable portion (one-third to three-quarters) of their diet in temperature climes. In the tropics, fish of the following genera might prove worthy of further study as biological control agents: *Profundulus* in Guatemala, *Phractura* in Ghana, and *Variocorhynchus* in Tanzania.

When applying insecticides, care should be taken to avoid debilitating or killing fishes which are providing a measure of natural blackfly control. On the other hand, there is a recorded case

Table 4. Salmonid fish predation on Simuliidae.

Species	No. examined	Size	No. with prey	Species	% in stomach	% in fauna	Country	Date	Authority
Salvelinus fontinalis	—	5–14 cm	31	*Simulium* sp.	1.9L[a]		New Brunswick, Canada	2–6/18/72	Eidt (1978)
Salmo salar	—	8.5–9 cm	6		1.6L				
Anguilla rostrata	—	21.5 cm	1		1.0L				
Salmo salar	4	1.5 lb	2	*Cnephia saileri* +3 other spp.	830L[a] 330A 10P		Quebec, Canada	12/7/77	Back (1978)
Salvelinus fontinalis	1	1 lb	1						
Salvelinus fontinalis	127	—	64.6%L 38.6%P	*Simulium* sp.	18.1L 2.8P		Ontario, Canada		Ide (1942)
	?	—	some A						
Salmo trutta	40	0–20 cm	0–15%	Simuliid L	0–2		British Columbia, Canada	1937–1939	Idyll (1942)
Salmo gairdneri	70	20–51 cm	0–50%	Simuliid L	0–63				
	118	0–20 cm	8–33%	Simuliid L	1–49				
	102	20–51 cm	25–67%	Simuliid L	66–97				
Salmo clarki	70	0–20 cm	0–22%	Simuliid L	0–48				
	50	20–51 cm	0–41%	Simuliid L	0–85				
Salmo clarki	326	1–19 in		Simuliid	5.1L 0.8P 1.1A		Oregon, USA		Dimick and Mote (1934)

Fish	No. of fish	Age/size	No. Simuliidae	Prey	No. per fish	No./% in stomach	Availability	Location	Date	Reference
Salmo salar	?	1°–yr	?	*Simulium* sp.	0.6P 5.5L 2.2P	0.8L	4.3[b]	Scotland and England		Allen (1941)
Salmo salar	?	2°–yr	?	*Simulium* sp.	0.46	2.6L	2.4[b]			
Salmo trutta	30		1	*Simulium* sp.		3.5L		Cambridgeshire, England	1939–1941	Hartley (1948)
Salmo trutta	24 16 16	2°+yr 2°+yr 2°+yr	1 1 0			0.0018° 0.0024 0		England	15/04/67 16/08/67 26/09/67	Chaston (1968)
Salmo trutta	178	1°-yr 2°-yr 3°-yr		Simuliid		2.5L° 0.5L° 0.5L°		England	10/63– 10/64	Elliott (1967)
Salmo trutta	62			Simuliid		32L	41	Pyrenees, France	June– August 1972	Elliott (1973)
S. gairdneri	174			Simuliid		22L	41			
Trout	1			Simuliid		c.300	A>>L	Finland	31/05/21	Levander (1923)
S. trutta var. *fario*	28 1 2	1°–4°-yr 2°+yr 64–67 mm	24 1 2	Simuliid Simuliid Simuliid		145L>>P 443L 29P 39L 3P	14.7	Finland	15–19/06/67 16/08/78 17/07/78	Kuusela (1978)

E = egg, L = larva, P = pupa, A = adult. Percentage in stomach, °by weight or +by number.
a Average number per fish.
b Availability (% in stomach/%in fauna by numbers).

where the use of insecticides (on the
Victoria Nile, Uganda) initiated fish predation on simuliid pupae by fish seeking
a food substitute (Corbet 1958a).

Insecta

Trichoptera (Caddisflies)

Larval caddisflies are in many instances
the most important insect predators of
immature blackflies (Miall 1895, 1912).
The latter are often scarce or absent in
habitats (vegetation or rocks) where caddisworms   are   abundant   (Ontario—
Twinn 1939; Davies 1950; Utah, USA—
Peterson 1960; Italy—D. M. Davies 1958;
USSR—Usova 1964). Of all the trichopteran families, the hydropsychids, which
share rocky rapids with simuliids, exert
the greatest natural control on their larvae and pupae; they are reported as
blackfly predators from many countries
(Table 5; and Burton and McRae 1972).
However, all hydropsychid species may
not exert equal pressures on immature
simuliids. For example, caddisworms
appeared to be the most important predators of simuliid larvae in New Zealand
streams, yet the predation rate proved
low; only 1.4% of 135 larval *Hydrobiosis
parumbipennis* (4th to 5th instar stage) ate
simuliid larvae (Crosby 1975). In the

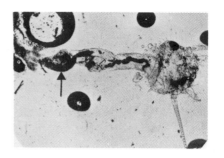

A stonefly nymph, Neoperla spio, *dissected to
show a simuliid larva in its midgut in Ghana (G.
Crisp 1956b; courtesy of the British Empire Society
of the Blind).*

USA, Coffman (1967) found *Hydropsyche*
spp. included only 0.24–1.16% simuliid
larvae in their diet and *Cheumatopsyche*
sp. only 1.32%. In Central Africa, 6% of
Hydropsyche larvae and 2% of *Cheumatopsyche* larvae contained simuliid larvae
(Hynes and Williams 1962). In Ghana,
however, Burton and McRae (1972)
found that 10% of 50 larval *Hydropsyche*
sp. and 12% of 25 *Cheumatopsyche* sp. fed
on larval *Simulium damnosum.* In Ivory

*A chironomid larva feeding on a larval blackfly in
Swedish Lapland (A. Kureck, personal communication, 1978).*

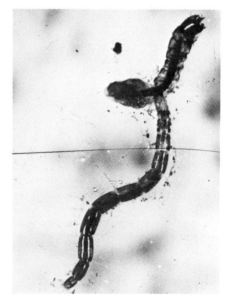

A lake chub (Couesius plumbeus) *with larvae of*
Simulium decorum *in its mouth. These larvae
were torn from the downstream face of a spillway
over a small dam in New Hampshire, USA (J. F.
Burger, personal communication, 1978).*

Table 5. References to caddisflies (Trichoptera)[a] predatory on Simuliidae.

Trichoptera: Baranov 1936; Twinn 1939; Hocking 1950; Peterson 1960; Giudicelli 1961; Usova 1964; Dinulescu 1966; Abdelnur 1968

Philopotamidae:
Chimarra spp.: Crisp 1956a; Peterson 1960; Peterson and Davies 1960 (E, L, P); Service and Lyle 1975; Gorayeb and Mok 1978?; Gorayeb and Pinger 1978

Polycentropodidae:
Plectrocnemia sp.: Pacaud 1942
Plectrocnemia conspersa Curtis: Jones 1949; Hildrew and Townsend 1976
Polycentropus? sp.: Peterson 1960 (L?)
Polycentropus flavomaculatus (Pictet): Jones 1950

Hydropsychidae: Miall 1895, 1912; Noyes 1914; Pomeroy 1916; Davies 1949, 1950; Grenier 1953; Chutter 1971; Service and Lyle 1975; Hart 1978
Arctopsyche sp.: Peterson 1960 (L?)
Arctopsyche grandis (Banks): Mecom 1972
Diplectrona ?modesta Banks: Peterson and Davies 1960 (L?)
Cheumatopsyche sp.: Coffman 1967; Hynes and Williams 1962; Burton and McRae 1972
Cheumatopsyche ?analis (Banks): Peterson and Davies 1960 (E, L, P)
Hydropsyche spp.: Howard in Riley 1887; Howard 1888, 1901; Pomeroy 1916; Davis 1934; Wolfe and Peterson 1959; Peterson and Davies 1960; Carlsson 1962; Hynes and Williams 1962; Chutter 1968; Burton and McRae 1972; Mecom 1972; Rhame and Stewart 1976; Strnad and Gibbs 1978
Hydropsyche angustipennis (Curtis): Pavlichenko 1977a, b
Hydropsyche colonica McLachlan: Crosby 1975
Hydropsyche dorsalis (Curtis): Scott 1958
Hydropsyche instabilis (Curtis): Jones 1949, 1950
Hydropsyche ornatula McLachlan: Zivkovic 1955 (L, P); Pavlichenko 1977a

Hydropsyche pellucidula (Curtis): Pavlichenko 1977a; Kuusela 1978 (L,P)
Macronema spp.: Gorayeb and Pinger 1978; Gorayeb and Mok 1978
Smicridea sp. (or near): Dalmat 1955

Rhyacophilidae: Miall 1912; Kurtak 1978
Rhyacophila spp.: Grenier 1943, 1953; Peterson 1960; Carlsson 1962; Speir 1976; Hart 1978; P. Zwick 1978
Rhyacophila dorsalis (Curtis): Slack 1936; Jones 1950; Ladle 1978 (L?)
Rhyacophila fuscula Banks: Peterson and Davies 1960; Ezenwa 1974
Rhyacophila nubila Zett.: Kuusela 1978 (L, P, A)
Rhyacophila obliterata McLachlan: Jones 1949
Hydrobiosus parumbripennis McFarlane: Crosby 1975

Glossosomatidae
Agapetus sp.: Grenier 1953

Hydroptilidae: Burton and McRae 1972
Orthotrichia spp.: Burton and McRae 1972 (P); Disney 1973 (E, P); Tsacas and Disney 1974
?Phryganeidae (*Phryganea* sp.): Howard 1888; Petersen 1924; Grenier 1953

Brachycentridae
Brachycentrus sp.: Peterson 1960
Brachycentrus americanus (Banks): Mecom 1972

Limnophilidae: Friederichs 1919; Peterson 1960
Anabolia sp.: Enderlein 1931; Rubtsov 1940
Halesus ruficollis Pictet: Pacaud 1942 (L, P)
Neophylax sp.: Peterson and Davies 1960

Leptoceridae
Athripsodes sp.: Kuusela 1978

Odontoceridae: Gorayeb and Pinger 1978 (?L); Gorayeb and Mok 1978 (?L)

Helicopsychidae: Gorayeb and Pinger 1978 (?L); Gorayeb and Mok 1978 (?L)

Microcaddis larvae: Burton and McRae 1972

[a] Simuliid larvae unless otherwise indicated: E = egg, L = larva, P = pupa, A = adult. (These abbreviations apply to following Tables also.)

Coast, Service and Lyle (1975) reported that 31.5% hydropsychid larvae fed on *S. damnosum*. In the USSR, Pavlichenko (1977a) determined that 42% of 255 *Hydropsyche* larvae, mainly *H. angustipennis*, fed on larval blackflies. Considering only those attached to vegetation (i.e. excluding those on stones), a small majority (52% of 117) proved to have so fed. He also found that, on average, a larval *H. angustipennis* (on plants), ingested 2.6 simuliid larvae (but fewer eggs and pupae). Some hydropsychids contained up to 75 small larvae. He considered this species to be a significant population-limiting factor for blackflies (Pavlichenko 1977b).

The relative extent to which larval hydropsychids, and those of other trichopteran groups, control simuliids by consuming them and by dislodging them to drift into unsuitable habitats, is not well understood (Peterson and Davies 1960; Hart 1978; Kurtak 1978). However, Speir (1976) calculated such predator-caused drift mortality in western USA to be no more than 6% in relation to actual consumption (83%). *Rhyacophila* spp., whose non-case-building larvae actively search for prey, accounted for 34.9% of simuliid larval mortality and *Hydropsyche* spp. for 2.1% (Speir 1976). On average, a *Hydropsyche* sp. larva fed on 1.0 simuliid larva per day and a *Rhyacophila* sp. on 2.6 per day (Speir 1976).

In Britain, Jones (1949) also found that caddisworms fed well on simuliid larvae, i.e. 25.6% of *Rhyacophila*, 16% of *Hydropsyche* and 20.7% of *Petrocnemia*. Scott (1958) discovered that more larvae of *H. fulvipes* (28%) than of *R. dorsalis* (13%) fed on *Simulium* larvae. Again, 11% of *Ryacophila* larvae were shown by Slack (1936) to contain blackfly larvae. Younger hydropsychid larvae tend to be herbivorous, becoming increasingly carnivorous as they grow (Muttkowski and Smith 1929; Burton and McRae 1972; Crosby 1975). Pavlichenko (1977a), however, reported that younger larvae consumed more simuliids than did older ones.

Observations indicate that downstream drifting is a characteristic of simuliid larvae and that drifting examples are readily eaten if they pass into hydropsychid larval nets (Howard *in* Riley 1887; Chutter 1971; Burton and McRae 1972). Hydropsychid larvae also leave their retreats to hunt larval blackflies

(Hart 1978). Burton and McRae (1972) and Disney (1973) recorded, from Africa, that a larval hydroptilid (*Orthotrichia* sp.) feeds on simuliid pupae and exuviae. The former authors postulated, from its voracious feeding habits, that, when abundant, it might well kill many simuliids.

Odonata (Dragonflies and Damselflies)

Various authors (Table 6) mention nymphal dragonflies and damselflies consuming larval simuliids on various continents. As many as 30% of nymphal odonates contain simuliid remains in Ivory Coast and Brazil respectively (Lyle and Service 1975; Gorayeb and Pinger 1978). Australian naiad dragonflies were introduced into New Zealand during 1931–1932 to control larval blackflies (Anon. 1932). Adult odonates, mainly dragonflies, have been reported to devour simuliid adults swarming around vertebrate hosts (Twinn 1939; Dalmat 1955; Pritchard 1964, 1978; Service and Lyle 1975). Newly-emerged and ovipositing adult blackflies are also subject to odonate predation (Peterson and Davies 1960). Crisp (1956–00) spoke of adult 1960). Crisp (1956a) spoke of adult odonates as probably the most important

Table 6. References to Odonata predatory on Simuliidae.

Odonata: Riley 1887 (A); Lugger 1896; Cameron 1922 (L); Anon. 1932 (L); Bradt 1932 (L); Drummond 1933 (L); Twinn 1939 (A); Davies 1949 (A); Grenier 1953 (L); Peterson and Davies 1960 (A)
Gomphidae:
Paragomphus hageni Selys: Crisp 1956a (A)
Aeschnidae
Aeschna canadensis Walker: Pritchard 1978 (A)
Aeschna eremita Scudder: Pritchard 1978 (A)
Aeschna umbrosa Walker: Peterson and Davies 1960 (L)
Corduliidae
Cordulia shurtleffi Scudder: Pritchard 1978 (A)
Epitheca (as *Tetragoneuria*) *spinigera:* Pritchard 1964 (A), 1978 (A)
Libellulidae: Johannsen 1903 (A); Emery 1913 (A); Gorayeb and Mok 1978 (L); Gorayeb and Pinger 1978 (L)
Brachythemis lacustris Kirby: Crisp 1956a (L, A)

Libellula sp.: Dalmat 1955
Libellula quadrimaculata L.: Sharleman 1915 (A)
Orthetrum chrysostigma Burmeister: Crisp 1956a (A)
Sympetrum internum Montgomery: Pritchard 1964 (A)
Trithemis donaldsoni Burmeister: Crisp 1956a (A)
Zygonyx flavicosta: Service and Lyle 1975 (L)
Platycnemididae
Metacnemis singularis: Service and Lyle 1975 (A)
Unplaced
Phaon iridipennis Burmeister: Crisp 1956a (A)
Calopterygidae
Calopteryx (as *Agrion*) *maculata* (Beauvois): Peterson and Davies 1960 (L)
Hetaerina sp.: Dalmat 1955
Agrionidae?: Gorayeb and Mok 1978 (*L?); Gorayeb and Pinger 1978 (L?)

predators of adult *Simulium* in Ghana, but only 7/100 adult *Brachythemis lacustris* Kirby, taken from swarms during two weeks, exhibited simuliid remains. Pritchard (1964) found, by dissection, that only 1% of a sample of dragonflies flying around humans in western Canada had fed on simuliids. It seems that even at the best of times nymphal and adult odonates exert only a minor influence on simuliid populations.[16]

Ephemeroptera (Mayfly Nymphs) [Table 10]

Under certain circumstances baetid and heptageniid nymphs (especially the larger ones—Agnew 1962) may become important predators of larval blackflies (Cameron 1922; Crisp 1956a). Other authors have reported no dissection evidence of such feeding. Bradt (1932), in fact, observed that, when attacked by

---

[16] *Zygonyx* spp. nymphs might be exceptions to this generalization (R. N. Gambles, personal communication, 1973 and see Service and Lyle 1975) . . . Ed.

mayfly nymphs in aquaria, *Simulium* larvae lashed around secreting salivary silk, completely enveloping and ultimately killing these predators.

Plecoptera (Stonefly Nymphs)

It would appear that Perlidae and Perlodidae are the families most predatory on larval blackflies. Sheldon and Osborn (1977) hypothesized that stonefly predation was responsible for the seasonally-low population of simuliid larvae at a lake outlet in California.

Crisp (1956a, b) considered that *Neoperla spio* was the most formidable enemy of immature blackflies in the Red Volta River, Ghana; Some specimens containing up to seven *Simulium* spp. larvae. The latter were reported from 14% of the same predator in Uganda (Hynes and Williams 1962). In the USA, another member of the genus, *Neoperla clymene*, may devour large numbers of blackfly larvae and pupae from March to May (Vaught and Stewart 1974). Of Britain's 14 stonefly species, Hynes (1941) listed six that prey on larval simuliids. He

---

Table 7. Reference to stoneflies (Plecoptera) predatory on Simuliidae.

Plecoptera: Cameron 1922; Dinulescu 1966; Gibbs 1978
Perliidae: Muttowski and Smith 1929; Grenier 1953; Gorayeb and Pinger 1978
  Acroneuria californica Banks: Sheldon 1969; Speir 1976
  Claassenia sabulosa (Banks): Fuller and Stewart 1977
  Hesperoperla pacifica (Banks): Fuller and Stewart 1977
  Neoperla sp.: Hynes and Williams 1962
  Neoperla clymene (Newman): Vaught and Stewart 1974
  Neoperla spio Newman: Crisp 1956a, b
  Paragnetina (as *Perla*) *immarginata* (Say): Wolfe and Peterson 1959
  Perla bipunctata Pictet (as *carlukiana* Klapálek): Hynes 1941, Jones 1949; Mackereth 1957
  Perla or *Dinocras cephalotes* Curtis: Hynes 1941; Jones 1949; Carlsson 1962
Perlodidae
  Arcynopteryx picticeps Hanson: Sheldon 1972
  Cultus aestivalis (Needham and Claassen): Fuller

and Stewart 1977
  Dictyopterygella bicaudata: Hynes 1941
  Diura spp.: Carlsson 1962; Sheldon 1972; P. Zwick 1978
  Hydroperla crosbyi (Needham and Claassen): Oberndorfer and Stewart 1977
  Isogenoides zionensis (Hanson): Fuller and Stewart 1977
  Isoperla spp.: Carlsson 1962; P. Zwick 1978
  Isoperla grammatica (Poda): Hynes 1941; Mackereth 1957; Carlsson 1962
  Isoperla obscura Zett.: Kuusela 1978
  Perlodes spp.: Carlsson 1962; P. Zwick 1978
  Perlodes microcephala (Pictet) (as *mortoni* Klapálek): Hynes 1941; Jones 1950; Mackereth 1957
  Skwala parallela (Frison): Fuller and Stewart 1977
Chloroperlidae
  Alloperla sp.: Strnad and Gibbs 1978
  Chloroperla torrentium (Pictet): Hynes 1941
  Suwallia sp.: Fuller and Stewart 1977
  Sweltsa sp.: Fuller and Stewart 1977
  Triznaka sp.: Fuller and Stewart 1977

A caddisfly larva, Hydropsyche *sp., holding a partly-eaten simuliid larva in Ghana. Another simuliid larva is caught in the net-silk (above). (Burton and McRae 1972; courtesy of* J. Med. Entomol.*).*

found that 5% of *Perla carlukiana* and 12% of *Perlodes mortoni* had eaten blackfly larvae. Later studies arrived at figures of 14–18.5% (Jones 1949; Mackereth 1957), and 22–24% (Jones 1950; Mackereth 1957) respectively. Mackereth (1957) considered that these and *Isoperla* nymphs selected simuliid prey above its availability in a stony stream, as larval blackflies never formed more than 3% of the invertebrates sampled. Another perlid nymph, *Acroneuria californica*, is a heavy consumer of simuliid larvae, especially in summer (Speir 1976; Oberndorfer and Stewart 1977) and accounts for 7.5% of all such predatory consumption (Speir 1976).

On the other hand, Zwick (1978), from detailed studies in Germany, considered simuliids to be only fourth or fifth choices in the stonefly diet.

Hemiptera (True Bugs)

Both nymphal and adult hemipterans are predaceous, but most aquatic species prefer still or slowly-moving waters unsuitable as blackfly larval habitats. Only Baranov (1934, 1938) in Yugoslavia and Dalmat (1955) in Guatemala report their attack on simuliids (Table 10). One may wonder, however, whether the extremely active and quite large hemipterans of

the genus *Rhagovelia*, found among rapids in, e.g. New Britain, may prove to be blackfly predators? (Laird, personal communication).

Thysanoptera

Aelothripids have been found in France, sucking the haemolymph of *Simulium ornatum* pupae attached to vegetation floating near the water surface (Grenier 1943).

Megaloptera or Neuroptera
(Dobsonflies, Alderflies) [Table 10]

Corydalid larvae inhabit fast, clean, rocky rapids and are proven predators of simuliid larvae and pupae in Texas, USA (Stewart *et al.* 1973) and Guatemala (Dalmat 1955). In Brazil, 12.5% of 16 larvae examined contained simuliids (Gorayeb and Mok 1978; Gorayeb and Pinger 1978). Speir (1976) attributed 24.3% of all consumption of simuliids in western Oregon streams to larval *Sialis californicus*. Under certain conditions, therefore, megalopteran larvae may exercise considerable control on immature simuliid populations.

Lepidoptera (Moths) [Table 10]

Nymphulinae larvae in Brazil have been reported to crawl on top of *Simulium* pupae, and pupal branchial remains have been found in some of them (Gorayeb and Mok 1978; Gorayeb and Pinger 1978).

Coleoptera (Beetles)

Predaceous aquatic beetles are usually found in slow-moving or stagnant waters where immature Simuliidae seldom occur. In England, 2.0–2.5% of each of three dytiscids (*Oreodytes* spp.) fed on *Simulium* larvae (Jones 1949, 1950). Peterson (1960) observed that in a small Utah stream, two individual larvae of *Agabus* sp. respectively devoured four and eight blackfly larvae in 15 minutes. Members of

Table 8. Reference to Hymenoptera predatory on Simuliidae.

Hymenoptera: Vargas 1945; Dinulescu 1966 (A)	*Crossocerus lentus* (Fox): Miller 1978 (A)
Formicidae: Crisp 1956a (E, L, P)	*Crossocerus megacephalus* (Rossi) as *Crabro (Blephar-*
Formica fusca L.: Peterson 1960 (L)	*ipes) leucostomus* L.: Hamm and Richards 1926
Formica obscuripes Forel: Peterson 1960 (L)	(A)
Lasius neoniger Emery: Peterson and Davies 1960	*Crossocerus pusillus* Lepeletier and Brulle (as *varus*
(A)	L. & B.): Leclercq 1954 (A?)
Myrmica brevinodes Emery: Peterson 1960 (L)	*Crossocerus* (as *Crabro*) *wesmaeli* (Van der Linden):
Myrmica emeryana Forel: Peterson and Davies 1960	Minkiewicz 1932 (A?); Leclercq 1954 (A?)
(A)	*Ectemnius (Hypocrabro) stirpicola* (Packard): Krom-
Vespidae	bein 1960 (A)
Polybia occidentalis (Oliv.): Richards and Richards	*Lindenius* (as *Crabro*) *panzeri* (Vander Linden):
1951 (A)	Sickmann 1893 (A); Miller and Kurczewski 1975
Vespula germanica (Fabr.): Crosby 1978 (A)	(A)
Vespula norvegica var. *albida* (Sladen): Hocking	*Oxybelus* sp.: Dinulescu 1966
1952 (A)	*Oxybelus emarginatus* Say: Snoddy 1968 (A)
Sphecidae: Kolesnikov 1977 (A?)	*Oxybelus pyrurus* Rohwer: Bequaert 1934 (A); Dal-
Crabro sp.: Dinulescu 1966	mat 1955 (A)
Crossocerus n.sp.: Miller 1978 (A)	*Strictia* (as *Monedula*) *signata*: Wise 1911 in
Crossocerus decorus (Fox): Miller 1978 (A)	Pomeroy 1916 (A)

the Dytiscidae and other beetle families take occasional immature simuliids but appear to exert little influence on the population as a whole (Table 10).

Hymenoptera

Most predation by Hymenoptera is on adult blackflies. Several species of crabronine sphecid wasps gather simuliids, along with other Nematocera, to provision their larval cells. In Guatemala, the sand-digger wasp (*Oxybelus pyrurus*) is an efficient enemy of adult simuliids when they are taking blood meals, but this predator is unlikely to reduce the total population (Bequaert 1934; Dalmat 1955).

A Hydropsyche *sp. larva attacking a simuliid larva (Burton and McRae 1972; courtesy of* J. Med. Entomol.).

Snoddy (1968) detailed a similar simuliid-feeding behaviour in *Oxybelus emarginata* in southeastern USA. Peckham (1978), however, said that in hundreds of *Oxybelus* nests which he examined in the USA, no nematocerous flies were seen, except occasional fungus gnats. Krombein (1960) found that two of three nests of the wasp, *Ectemnius stirpicola*, contained five and 17 adults (both sexes) of *Simulium jenningsi* Malloch. Vespid wasps also have been observed preying on adult simuliids (Table 8). Richards and Richards (1951) saw *Polybia occidentalis* collecting simuliids which were biting cows. Several species of ants will attack and carry off larvae or pupae when they are exposed by reduction in water level, and even when still covered by a thin film of water (Crisp 1956; Peterson 1960; Peterson and Davies 1960). None of these Hymenoptera make more than occasional use of simuliids in their diet, and they rarely, if ever, bring about any significant population control of blackflies (and see *Addendum, Hymenoptera,* p. 156).

Diptera (Flies)

The most unusual predators of immature simuliids are three species of African *Drosophila* larvae (Table 9) which prey on

An ant, Formica fusca, *holding its prey, a* Simulium tuberosum *larva in Utah, USA (B. V. Peterson 1960).*

the eggs and newly-hatched larvae (Tsacas and Disney 1974; Disney 1975; Grunewald and Raybould 1978). Tsacas and Disney (1974) view these larvae as having equal or greater potential for biological control of *Simulium damnosum*, than do larvae of the trichopteran *Orthotrichia*. They caution, however, that it might be necessary first to control populations of the parasitoid hymenopteran, *Trichopria*, which may be partially regulating levels of these predatory *Drosophila*.

Larvae of chironomids have been observed feeding on those of simuliids (Wu 1931; Grenier 1945b, 1953; Hocking and Pickering 1950; Dalmat 1955; Peterson and Davies 1960; Dinulescu 1966; Kureck 1978). In 1931–1932, 1200 *Cardiocladius* larvae, a genus with a long-standing reputation as simuliid predators (Zwick 1978), were successfully transported from Australia to New Zealand for simuliid control (Anon. 1932). Larvae of several species of the Pentaneurini feed on those of simuliids (Kuusela 1978), but the latter formed only 0.27% of the diet in Pennsylvania, USA (Coffman 1967). Chironomids, by spinning their silken tubes within and between blackfly pupae, may promote the accumulation of algae around their respiratory filaments, thus suffocating them (Grenier 1943, 1953). Also, by attacking the adhesive on simuliid egg masses, larval Chironomi-

dae cause the eggs to detach and wash away to less suitable habitats (Grenier 1945a, b). This is also true for the adult muscid, *Xenomyia oxycera*, with respect to *Simulium damnosum*, in Ivory Coast (Elsen 1977).

The predatory larvae of *Oreogeton* spp. (Empididae) may have a restraining influence on blackfly larval populations in some Alaskan streams, if they behave in nature as they did in the laboratory, killing many more larvae than they consumed (Sommerman 1962). Larval empidids, and possibly dolichopodids, feed on blackfly larvae and pupae in Algeria (Vaillant 1951, 1953) and in Utah, USA (Peterson 1960). Larval simuliids, especially in crowded conditions, show cannibalism (Table 9). Thus, Burton (1971) observed that, in Ghana, 12 of 100 large *S. damnosum* larvae contained 1st-instar simuliid larvae. Only one of 100 larvae of *S. nigritarse* Coq. in South Africa cannibalized blackfly 1st-instars (Chutter 1972). In Norway, larvae of *Prosimulium ferrugineum* (which are rather large) often consume those of smaller simuliids (Johnson 1969).

Adults of several dipteran families feed on all life-history stages of blackflies. Those of two dolichopodid species seized simuliid larvae almost as soon as they were exposed by a change in water level (Twinn 1939). Adult empidids, *Wiedemannia* spp., dragged blackfly larvae from the edge of a swift stream in California (Wirth and Stone 1956). Adult *Xenomyia oxycera* (Muscidae) sucked juices from simuliid larvae on trailing leaves near the surface (Elsen 1977a). Emerging blackflies may be attacked by adult dolichopodids (Séguy 1925), asilids (Séguy 1925), empidids (Séguy 1925; Peterson 1950; Peterson and Davies 1960), ephydrids (Elsen 1977a), limnophorine muscids (Elsen 1977a) and lispine muscids (Balay and Grenier 1954). While resting near streams before ovipositing or during this act, female blackflies may fall prey to empidids (Peterson and Davies 1960), ephydrids (Balay and Gre-

Table 9. References to true flies (Diptera) predatory on Simuliidae (including cannibalism).

Diptera: Dinulescu 1966
Chaoboridae
Eucorethra underwoodi Underwood: Hocking *et al.*
1950 (A)
Ceratopogonidae: Vargas 1945; Dinulescu 1966
Chironomidae: Wu 1931, Grenier 1943, Peterson
and Davies 1960, Dinulescu 1966, Kureck 1978
Camptocladius sp.: Carlsson 1962
Cardiocladius sp.: Anon. 1932, Zwick 1978 (L)
Chironomus spp.: Hocking *et al.* 1950? (L, P),
Rivosecchi 1978 (P)
Orthocladius sp.: Grenier 1945b
Pentaneurini spp.: Kuusela 1978 (L)
Pentaneura melanops (Meigen) sp. group: Coffman
1967 (L)
Smicridea sp.: Dalmat 1955 (L)
Simuliidae (on small simuliid larvae): Petersen 1924,
Fredeen 1962, Peterson 1962
Prosimulium (Helodon) ferrugineum (Wahlberg):
Johnson 1969
Simulium larvae: Badcock 1949
Simulium damnosum Theobald: Burton 1971
Simulium (Eusimulium) nigritarse Coq.: Chutter
1972
Simulium (Odagmia) ornatum Meigen: Smart 1934
Simulium (Simulium) venustum Say complex:
Peterson and Davies (1960)
Asilidae: Riley 1887, Johannsen 1903, Emery 1913,
Hobby 1931c, Baranov 1938b
Dioctria longicornis Meigen: Baranov 1938b
Leptogaster cylindrica de Geer: Séguy 1927
Empididae: Petersen 1924, Baranov 1938a
Clinocera (Hydrodromia) binotata Loew: Peterson
1960 (L?)
Clinocera near *trunca* Melander: Peterson 1960
Empis snoddyi Steyskal: Snoddy 1978
Hemerodromia melanosoma Melander: Peterson and
Davies 1960
Hemerodromia seguyi Vaillant: Vaillant 1953
Hilaria sp.: Séguy 1925, Peterson and Davies
1960, Davies L. 1978
Hilaria maura F.: Baranov 1938b
Metachela collusor Melander: Peterson 1960
Neoplasta scapularis (Loew): Peterson 1960 (L?)
Oedalia ohioensis Melander: Peterson and Davies
1960
Oreogeton sp. (near *basalis* Loew): Sommerman
1962
Oreogeton cymballista Melander: Sommerman 1962
Platypalpus puerinus Melander: Peterson 1960
Platypalpus xanthopodus Melander: Peterson 1960

Rhamphomyia spp.: Peterson 1960, Peterson and
Davies 1960
Rhamphomyia basalis Loew: Peterson and Davies
1960
Rhamphomyia clauda Coq: Peterson and Davies
1960
Roederioides junctus Coq.: Peterson and Davies
1960
Syndyas dorsalis Loew: Peterson and Davies 1960
Wiedemannia sp.: Wirth and Stone 1956
Wiedemannia (Roederella) ouedorum Vaillant: Vail-
lant 1951; 1953
Dolichopodidae: Petersen 1924, Baranov 1938a
Chrysotus sp.: Twinn 1939, Peterson and Davies
1960
Chrysotus obliquus Loew: Peterson and Davies
1960
Dolichopus affinis Walker (as *splendidulus* Loew):
Twinn 1939
Hydrophorus algens Wheeler: Peterson 1960
Rhamphium effilatum (Wheeler): Peterson and
Davies 1960
Ephydridae
Ochthera sp.: Balay and Grenier 1964
Ochthera insularis Becker: Elsen 1977a
Ochthera mantis (De Geer): Peterson and Davies
1960
Drosophilidae
Drosophila sp.: Grunewald and Raybould 1978 (E,
L)
Drosophila cogani Tsacas and Disney: Tsacas and
Disney 1974, Disney 1975 (E, L)
Drosophila gibbinsi Aubertin: Disney 1975 (E, L)
Gouteux 1976
Drosophila simulivora Tsacas and Disney: Tsacas
and Disney 1974 (E, L)
Anthomyiidae
Scatophaga scybalaria Fab.: Hobby 1931a, b
Scatophaga stercoria L.: Hobby 1931a, b
Muscidae
Coenosiine muscids: Barnley and Prentice 1958
Coenosia sp.: Baranov 1934; 1938a, b
Coenosia sp. (as the anthomyiid *Chirosia crassiseta*
Stein—revision by Crosskey and Davies 1962):
Baranov 1934; 1938a, b
Limnophora (Calliophrys) riparia Fall.: Grenier
1945a, 1945b
Lispe spp.: Balay and Grenier 1964
Lispe nivalis Wiedemann: Balay and Grenier 1964
Xenomyia oxycera Emden: Crosskey and Davies
1962, Elsen 1977a, 1978

nier 1964; Elsen 1977a), limnophorine muscids (Elsen 1977a), and lispine muscids (Crosskey and Davies 1962). Dipteran predators, such as dolichopodids, pursue and capture blackflies in flight (Peterson and Davies 1960), as do empidids (Peterson and Davies 1960; L.

Davies 1978; Snoddy 1978), and asilids (Baranov 1938; Lugger *in* Riley 1887). Considerable biocontrol of simuliids may be exercised by adult lispine muscids in Upper Volta (Balay and Grenier 1964) and by coenosiine muscids in Uganda (Barnley and Prentice 1958).

Table 10. References to other insectan orders predatory on Simuliidae.

Ephemeroptera: (Nymphs): Dinulescu 1966 (L)
 Baetidae
  Baetis spp.: Crisp 1956a (L)
  Centroptiloides   bifasciata   (Esben-Petersen):
   Agnew 1962 (L)
 Heptageniidae: Grenier 1953 (L)
  Heptagenia spp.: Howard in Riley 1887 (L?);
   Cameron 1922 (L)
  Heptagenia dalecarlica Bengtsson: Carlsson 1962
   (L)
  Heptagenia sulphurea Müller: Dinulescu 1966 (L)
 Ecdyonuridae
  Afronurus sp.: Crisp 1956a (L)
 Ephemerellidae
  Ephemerella   mucronata   (Bengtsson):   Kuusela
   1978 (P or E?)
 Potamanthidae
  Potamanthus luteus (L.): Dinulescu 1966 (L)
Hemiptera
 Hydrometridae
  Hydrometra stagnorum L.: Baranov 1938b (A)
 Veliidae
  Velia sp.: Baranov 1934 (A)
  Velia rivulorum Fab.: Baranov 1938b (A)
 Gerridae
  Gerris najas de Geer: Baranov 1938b (A)
 Belostomatidae
  Abedus ovatus St°l: Dalmat 1955 (L, A)
Thysanoptera
 Aelothripidae: Grenier 1943 (P)

Megaloptera or Neuroptera
 Sialidae
  Sialis californica Banks: Speir 1976 (L)
  Sialis fuliginosa Pictet: Hildrew and Townsend
   1976 (L)
 Corydalidae: Gorayeb and Mok 1978 (L), Gorayeb
  and Pinger 1978 (L)
  Corydalus sp.: Dalmat 1955 (L)
  Corydalus cornutus L.: Stewart *et al.* 1973 (L, P)
Coleoptera (aquatic spp.): Riley 1887 (L)
 Dytiscidae: Service and Lyle 1975 (L)
  Agabus sp.: Peterson 1960 (L)
  Oreodytes daviisi (Curtis): Jones 1950 (L)
  Oreodytes rivuli (Gryll.): Jones 1950 (L)
  Oreodytes septentrionalis (Gryll.): Jones 1949 (L)
 Gyrinidae
  Aulonogyrus   striatus   L.:   Vaillant   1951   (L),
   Grenier 1953 (L)
  Gyrinus natator L.: Baranov 1938b (L)
 Hydrophilidae
  Anacaena limbata (Fab.): Peterson 1960 (L)
  Cymbiodyta sp.: Peterson 1960 (L)
 Staphylinidae: Crisp 1956a (L?), Peterson 1960
  (L?), Gorayeb and Mok 1978 (L?), Gorayeb and
  Pinger 1978 (L?)
 Elmidae?: Crisp 1956a (L?), Peterson 1960 (L?),
  Gorayeb and Mok 1978 (L?), Gorayeb and Pin-
  ger 1978 (L?)
Lepidoptera
 Pyralidae
  Nymphulinae: Gorayeb and Mok 1978 (P?),
   Gorayeb and Pinger 1978 (P?)

Addendum, Hymeroptera

Since this chapter was submitted, H. A. Baker (personal communication, 1979) has drawn another two relevant papers to my attention. Lewis (1953) reports a surprisingly high incidence of "accidental" parasitism by chalcid peril-ampid planidia within the heads of *S. damnosum s.l.* adults in Sudan, and Lewis *et al.* (1961) notes a planidium insect larva on the head of an adult *S. damnosum* in Ghana.

Invertebrates other than Insects

Coelenterata

Grunewald (1978) observed *Hydra* sp. feeding on 1st-instar *Simulium* larvae,

with up to 12 head capsules per predator. Such predation, on second and later lar-val instars, was observed by Hocking and Pickering (1954 [Table 11]).

A Hydropsyche sp. larva from a stream in Maine, USA, holding a larval blackfly in its jaws (S. P. Strnad and K. E. Gibbs, personal communication, 1978).

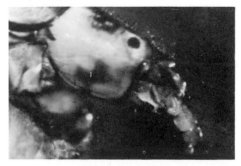

Platyhelminthes

Of several turbellarian species in the rapids below a dam in Maine, USA, only *Phagocata woodworthi* was seen to attack simuliid larvae, suggesting that these planarians were responsible for a marked reduction in the simuliid population (McDaniel 1978).

Annelida

Only isolated instances of leeches preying on simuliid larvae have, so far, been reported (Table 11). Maitland and Penney (1967) found these larvae in the crops of many *Erpobdella octoculata* in Scotland. Chutter (1968) considers this predation to be much less important than that of larval hydropsychids in South Africa.

Arachnida

Spiders of several families spin webs on vegetation, or other supports, e.g. bridges, near running water. Male and female blackflies are commonly caught in these webs (Table 11). In Japan, Saito *et al.* (1978) collected over 400 simuliids of both sexes from 27 spider webs beside rice-paddy irrigation ditches in early summer. However, it is considered that such random predation exerts little influence on blackfly populations.

Crustacea

Cambarus robustus was found to eat larval *Simulium venustum*, but only as 1.06% of its diet (Coffman 1967). Other records indicated that crustacean predation on immature simuliids is low (Table 11). In Palau (Caroline Is), larvae of *Simulium palauense* Stone were observed (G. Bright, personal communication, 1979) to be preyed upon by three species of fish (*Redigobius horiae* [Weber], *Stiphodon elegans* [Steindachner], and *Sicyopus* sp.) and the decapod crustacean, *Atya pilipes* (Newport). This shrimp was observed to eat blackfly larvae in the basaltic waterfall areas of streams.

Table 11. References to animals other than vertebrates and insects predatory on Simuliidae.

Coelenterata, Hydridae	Arachnoidea
Hydra spp.: Hocking and Pickering 1954 (L), Dinulescu 1966 (L), Grunewald 1978 (L)	Spiders: Davies, D. M. 1958 (Unpub. notes ex. Switzerland), Peterson 1960 (A), Peterson and Davies 1960 (A)
Platyhelminthes, Turbellaria, Planariidae planaria: Vargas 1945; Dinulescu 1966 (imm.)	Epeiridae (as Argiopidae): Orii 1975 (A)
Phagocata woodworthi Hyman: McDaniel 1978 (L)	*Meta kompirensis:* Saito *et al.* 1978 (A)
Mollusca, Gastropoda	Tetragnathidae: Davies, D. M. (unpub. notes) (A)
Viviparidae	*Tetragnatha praedonia:* Saito *et al.* 1978 (A)
Viviparus viviparus L.: Pavlichenko 1977a (E)	Dictynidae
Amnicolidae	*Dictyna major:* Longstaff 1932 (A)
Bithynia tentaculata L.: Pavlichenko 1977a (E)	Hydracarina (adults): Hart 1978 (L?, P)
Neritidae	*Typhlodromus (Amblyseius)* sp. (adults) Lewis *et al.* 1961 (A), (larvae) Davies 1959, 1960 (P, A)
Theodoxus fluviatilis L.: Pavlichenko 1977a (E)	Crustacea
Annelida, Hirudinea	Cambaridae
Erpobdellidae	*Cambarus robustus* (Girard): Coffman 1967 (L)
Erpobdella octoculata L.: Kuusela 1978 (L), Maitland and Penney 1967 (L), Pavlichenko 1977a (L, P)	Gammaridae: Howard in Riley 1887 (L), Séguy 1925 (L?), Pacaud 1942 (A), Grenier 1945a,b (L), Pavlichenko 1977a
Hirudidae	*Rivulogammarus* sp.: Dinulescu 1966
Haemopis marmorata (Say): Peterson and Davies 1960 (L)	Potamonidae
Family unknown	*Potamon niloticus:* Hynes and Williams 1962 (L)
?*Salifa perspica* Blanchard: Chutter 1968 (L)	*Potamon berardi:* Hynes and Williams 1962 (L)
?*Rhynchobdella* sp.: Rivosecchi *et al.* 1974 (L)	Asellidae
	Asellus aquaticus (L): Pavlichenko 1977a (E)

Conclusion

Evidence in the literature indicates that predation on simuliids by most vertebrates and invertebrates is generally at a low, or infrequent, level. However, under certain conditions, some birds and fish are able to exercise a considerable influence on blackfly populations. Such larvivorous fish might perhaps merit study as candidate biocontrol agents. This is also true of certain insects, including larval caddisflies and stoneflies, adult odonates, and wasps, as well as such certain dipteran larvae, e.g. those of some African *Drosophila* spp.

Laird (1978) considers that few predators offer the remotest possibility of eventual practical use as biocontrol agents, no matter how avidly famished aquatic omnivores may feed on conveniently available larval blackflies in a laboratory container. Jamnback (1973), while recognizing that predators, including birds, fish and invertebrates, undoubtedly reduce the numbers of blackflies in the field, states that they offer less promise for blackfly control than parasites, which are more easily mass-produced and otherwise manipulated. Nonetheless, the potential of a few predators for simuliid control needs further study.

(References in text dated 1978 and not appearing in References section are "Personal communications".)

Mermithid Nematodes of Blackflies

G. O. Poinar Jr

Of the various groups of nematodes that parasitize insects, none occur in such a wide range of hosts as members of the family Mermithidae. Not only do they attack representatives of several classes of arthropods, but some also develop in members of the phyla Mollusca and Annelida.

The majority of mermithid species parasitize the larval stages of insects and the list of those found in blackflies has been steadily increasing, ever since von Linstow recorded the first case of a mermithid-infected *Simulium* in 1898. These nematodes are also the most common parasites found in blackfly larvae throughout the world. And since they either kill or sterilize their host, they have attracted a considerable amount of attention, especially in regard to their potential as biocontrol agents.

Systematics

Most of the mermithid parasites of blackflies belong to the genera *Isomermis,*

Gastromermis and *Mesomermis (=Neomesomermis)*. Some representatives of the genera *Limnomermis* and *Hydromermis* are also found in blackflies, but only rarely.

There are certainly many undescribed blackfly mermithids throughout the world, and with the renewed interest in this group of parasites, many more species are bound to be described in the next few years. However, when describing a mermithid from any host, care should be taken to include the description of the adults. While, of all the life-history stages, the one which leaves the host (i.e. the postparasitic juvenile) is the simplest to collect, and some taxonomists have erected new species on the basis of this stage alone, such practices should be discouraged. The collective genus *Agamomermis* Stiles was created for juvenile mermithids that could not be identified to genus,[17] and it should continue to be used as such. This problem was recently discussed at the International Parasitology Congress held in Warsaw (August 1978) and similar conclusions were reached (Poinar and Welch 1978). When immature mermithids are submitted for identification, they need not be assigned a specific taxon. They can simply be referred to as "unidentified mermithids" or at most, assigned to the genus *Agamomermis*. When the adults are eventually discovered, then, if new, they can be described and placed in an appropriate genus.

However, if living material is available, the collector can attempt to keep the nematodes in water at natural temperatures until the final moult or moults have occurred and the adult characters are formed. The time for the final moult depends on the temperature and can range from anywhere between three days and two months. Apart from the fact that only the adult nematodes have sex organs, it is often possible to determine whether or not a moult has occur-

[17] See pp. 171–180 for another view . . . Ed.

red by examining the tail; the postparasitic juveniles of most mermithids exhibit a thin cuticular horn or appendage on the tip of the tail which is shed along with the juvenile cuticle.

Specimens can be killed and fixed after the final moult, but if enough material is available, then the nematodes should be given the opportunity to mate and deposit eggs. The pre-parasitic juvenile emerging from the egg may also have diagnostic characteristics that can be included in the specific description.

Before preserving the mermithids in fixative, it is necessary to kill them with heat so they will be relaxed and easy to examine under the microscope. There are several methods of "relaxing" nematodes with heat, but the easiest is to pour hot (80 °C) water over the living specimens. Then the killed worms should immediately be placed in fixative, i.e. 4% formalin, TAF, or 70% alcohol. After several days, they can be processed to glycerin by using the simple evaporation method [see Poinar (1975) for a discussion of this and other methods].

After the nematodes have been processed to pure glycerin, they should be mounted in the same medium on a microscope slide, supports added for smaller species, and the cover slip anchored to the slide by one of the many products on the market. Nail polish will work when nothing else is available.

Then an examination can be made to determine the correct taxonomic position. The following characters should be noted.

The number and arrangement of the head papillae. These should be examined in respect to the mouth opening. All of the mermithid parasites of blackflies have six head papillae, but the mouth opening may vary from completely terminal to almost ventral in position.

The shape and size of the amphids. These chemosense organs are often very large in aquatic mermithids and their size and

shape, as well as their proximity to the head papillae, are important characters.

The shape and size of the vagina. This is one of the basic and perhaps the most stable character in mermithid nematodes. Blackfly mermithids may have an S-shaped, V-shaped, or barrel-shaped vagina.

The shape and size of the spicules. Although very important, the size of the spicules often depends on the size of the nematode, so the ratio, spicule length over body diameter at cloaca is often a much more reliable character.

Cross-fibres in the cuticle. These striae in the cuticle can best be seen in the head or tail region under oil immersion. They are absent in most of the blackfly mermithids.

Number of hypodermal cords. This character, now controversial, was employed by many of the early mermithid systematists and tends to remain constant in most genera. In blackfly mermithids, there are either six or eight hypodermal cords that alternate with the same number, respectively, of muscle fields. To see this character, a cross-section of the body must be made. This section should be made in the mid-body of the nematode.

All genera of blackfly mermithids have either six or eight cords except *Gastromermis*. Most species in this genus have eight cords, but Mondet *et al.* (1977a) described two species from African blackflies which possess six cords. In these latter species, it appeared that the dorso-lateral cords fused with the dorsal cord, a condition indicating the origin of the six-corded forms.

Table 1 lists the diagnostic characters separating the various mermithid genera recovered from Simuliidae.

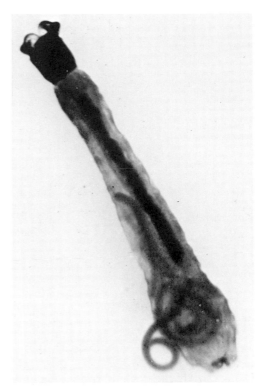

Gastromermis leberri *inside a larva of* Simulium damnosum. *(Collected by J. F. Walsh in Ghana.)*

A larva of Simulium tobetsuensis *parasitized by a mermithid. (Collected by H. Ono near Hokkaido, Japan.)*

A larva of Prosimulium exigens *parasitized by* Mesomermis paradisus. *(Collected by the present author in California.)*

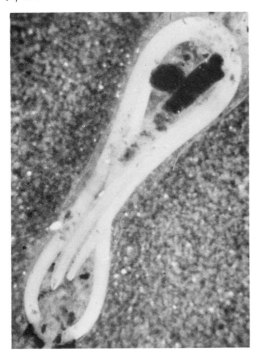

Schematic drawing of the life cycle of a representative blackfly mermithid.

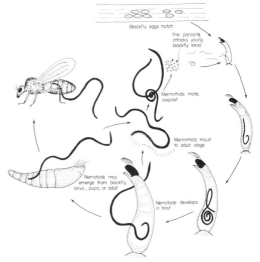

Table 1. Diagnostic characteristics of the genera of mermithids from blackflies.

Genera	Number of Spicules	Size of spicules in relation to cloacal width	Vagina shape	Hypodermal cords
Isomermis	2	1–3 times cloacal width	S	8
Mesomermis (*Neomesomermis*)	2	1–3 times cloacal width	V or barrel	6
Gastromermis	1	2 or more times cloacal width	S	6 or 8
Hydromermis	1	Less than twice cloacal width	S	8
Limnomermis	1	Less than twice cloacal width	S	6

Bionomics

Life-cycle

The complete life-cycle is known for only a few blackfly mermithids. However, the pattern is similar to that of mermithids in general. The pre-parasitic juveniles enter their hosts by direct penetration through the cuticle, then begin to develop in the haemocoel. Depending on the host stage penetrated, and probably other factors such as parasite burden, host nourishment, etc., the nematode may emerge from one of the later larval stages, from a pupa or from an adult host. We know of no blackfly mermithid which emerges from only the larva or only the adult stage of the host.

After emergence, the postparasitic mermithids crawl into the debris in the bottom of the water and there moult, mate and oviposit. The newly-hatched preparasitic juveniles then seek out a new host. It is not known how these minute stages can contact and penetrate young blackfly larvae in fast-flowing streams. It would seem that the current would prohibit adequate contact for penetration, but this obviously is not the case.

Ecology

Free-living Stages

Once the postparasitic juvenile mermithids leave their blackfly host, they begin a free-living existence that may extend up to two years. During this period, no additional nourishment is taken and the worm survives on the reserves built up during its relatively short existence inside the host. These free-living stages (postparasitic juveniles, adults and eggs) remain in the host environment and continue their development, which is influenced by water temperature. In northern latitudes, where the streams flow throughout the year, egg development will continue normally and hatching usually occurs when the preparasitic juvenile is fully formed. However, there is some indication, in populations of blackfly mermithids in California, that some eggs do not hatch "normally" but enter a type of resting period which prolongs the life of the preparasitic juvenile. Since after hatching, the infective juvenile is relatively short-lived (up to one week), it would be to the parasite's advantage if eclosion could be prolonged over an extended period in order to ensure host contact. In such "continuous water" systems, the host and parasite cycles are usually fairly well synchronized.

In tropical and dry-temperate climates where there is a rainy period alternating with a season of drought, the ecology of the parasite is somewhat different. Parasitic development, along with that of the host, is only possible during the rainy season or just afterwards. As the streams dry out, recently emerged, free-living nematodes begin to enter the stream bed and continue their development in the moist soil. Even after the water has completely disappeared, the nematodes can moult, mate and oviposit in the humid stream bed. Some blackfly mermithid eggs can then remain viable (containing the infective stage juveniles) for up to one year. In the tropics, though, the dry season usually does not extend beyond six months.

Soon after the rains begin, the eggs hatch. This occurs at approximately the same time as blackfly oviposition. Thus, the two cycles are regulated by the arrival of flowing water. After the initiation of parasitism, the development of both parasite and host continues as long as flowing water is present. Since, in tropical areas like West Africa, the blackfly–host cycles are continuous and overlapping, further synchronization between host and parasite is probably lacking and contact is made by chance.

There has been much discussion on the "recycling" phenomenon of blackfly parasites. Essentially, in order to explain how the areas of infestation can remain essentially permanent year after year, without the parasites being washed downstream, it was thought that the parasitized adult blackflies returned upstream to release their mermithids, thus replenishing the infestation sites. However, it would appear that a parasitized blackfly might just as well fly downstream with its parasite burden. Also, with our increased knowledge of stream velocity at various distances from the substrate, we learn that it is possible, by

adhering closely to the surface of rocks or debris, for the mermithids to escape the strong forces found even in very rapid torrents. Thus, the recycling effect may be minimal; and, in fact, by moving within protected areas of the micro-environment, mermithid parasites of blackflies may remain for years in their original habitat.

Also unexplained is the spotty distribution of blackfly mermithids. Even in the same stream, parasitized hosts are often found only in certain areas while they are completely absent in adjoining streams. Whether the physical stream bed, aside from chemical factors, plays a role in mermithid distribution is not known.

Parasitic Stages

The actual duration of parasitic development of blackfly mermithids is greatly dependent on temperature. In the case of *Isomermis lairdi* in West African simuliid larvae, it takes only 10–17 days. However, in cold, temperate streams, the mermithids may remain in their larval hosts for months.

Probably, in all cases of blackfly parasitism by mermithid nematodes, it is possible for the latter to be carried into the adult stage of the host. Females of *Simulium damnosum* infected with *I. lairdi* have an urge to oviposit, even though their reproductive system has been atrophied by the parasite. Thus, they can be captured on sticky aluminium plates that are placed near the stream to attract healthy ovipositing females (Mondet *et al.* 1977c).

Pathogenesis

The early investigations of Strickland (1911) revealed that mermithids reduced the fat-body and inhibited histoblast formation of simuliid larvae. However, these effects in the larval blackfly depend on the moment of penetration of the parasite. As Mondet *et al.* (1976) pointed out, seventh stage larvae of *S. damnosum* can harbour one or two young mermithid juveniles and still possess normal fat-body reserves and histoblasts. In the latter case, it is more than likely that the parasites will be carried through the nymphal stage and into the adult host.

Mesomermis flumenalis drastically reduces the host adipose tissue, resulting in a significant reduction of fat-body glycogen reserves. Condon and Gordon (1977) concluded that the depletion of haemolymph and adipose tissue in the host removes precursors required for the synthesis of cuticular proteins during moulting. The same authors also showed that the parasite produces a significant increase in volume of the corpus cardiacum cells, accompanied by an increase in nuclear DNA/RNA activity in the corpus allatum of *S. venustum*. A normal endocrine response of parasitized *P. mixtum/fuscum* indicated that the nematode was not modifying the host's hormonal balance.

In the majority of cases, parasitized female blackflies are sterilized, although they may still take a blood meal and attempt to deposit eggs. In fact, blood may actually be necessary for completion of parasite development.

In a sample of 800 females of *S. damnosum* parasitized by *I. lairdi*, Mondet *et al.* (1976) noted that 98.9% possessed atrophied ovaries, unable to develop even after a blood meal. The remainder (1.1%) had normal ovaries with three to five residual eggs. The above authors also noted that more than 80% of the parasitized females had the fat body greatly reduced or absent. Even when given a blood meal, the parasitized females survived for only six days, whereas non-parasitized females generally remained alive for 12 days.

Geographical Distribution

Found in blackfly populations ranging from the Arctic to the tropics, mermithid nematodes appear to exist wherever their hosts occur. The geographic range has recently been extended to Japan and Central America, although most of the described species of blackfly mermithids occur in Europe.

Representative Species

Included here are brief discussions of four species of blackfly mermithids which show promise as biological control agents. They are the only presently described species that have been carried through at least one complete generation under laboratory conditions. For that reason, these nematodes have potential as biocontrol agents.

Mesomermis flumenalis Welch (1967)

This species is distributed throughout North America and there is one report of its presence in the USSR. In nature, it is considered a univoltine species infecting blackflies usually from February to July. Molloy and Jamnback (1975) confirmed an earlier finding by J. Mokry (Anon. 1973) showing that the infective stages of this mermithid enter their host by direct penetration through the cuticle. At 18 °C, the postparasitic juveniles of *M. flumenalis* moulted in 10–13 days, mated, and completed oviposition 36–75 days later. Eggs held at 12 °C hatched in 35–55 days, but the preparasitic juveniles died two to three days later in the absence of a host (Ebsary and Bennett 1973, 1975).

This species seems particularly adapted to low temperatures and is considered the most important mermithid of Newfoundland blackflies.

Bailey *et al.* (1977) discussed a laboratory method for rearing *M. flumenalis* from postparasitic juveniles through the adult to the preparasitic juvenile stage. A field release of this mermithid was made in New York State by Molloy and Jamnback (1977). After releasing approximately 1.5 million preparasites into a small stream, 71.4% of the early instar *S. venustum* collected immediately below the treatment point were infected. However, the cost of production ($300 per million preparasities) would prohibit the use of *M. flumenalis* as a biocontrol agent at this time.

Gastromermis viridis Welch (1962)

This nematode, easily recognized by its greenish colour in life, has only been recorded from North America. Its biology was studied by Phelps and Defoliart (1964), who showed that parasitic development took only 10–14 days at 23–27 °C, but up to five months in overwintering blackfly larvae. The final moults occurred from five to 18 days, depending on temperature, after emergence from the host and the eggs were deposited within two weeks after moulting. In Wisconsin, an over-all larval blackfly mortality was judged to be about 50% and in one instance, virtual elimination of blackfly populations occurred in one stream, primarily due to *G. viridis* parasitism. Its small host range, however, may restrict its usefulness somewhat.

Isomermis wisconsinensis Welch (1962)

This mermithid has been recorded in Wisconsin and Newfoundland, but is probably found throughout North America. It is commonly found in pupae and adults of *S. vittatum*, infection rates in the latter stage reaching 37–63% (Phelps and DeFoliart 1964). These authors also recorded a development time of 10–14 days at 23–27 °C, although it lasted several months in overwintering blackfly larvae.

This species was also instrumental in eliminating *S. vittatum* larvae from certain streams in Wisconsin, although its limited host range might restrict its usage as a biocontrol agent.

Isomermis lairdi Mondet, Poinar and Bernadou (1977b)

This mermithid is the most common parasite of *S. damnosum s.l.* in the Ivory Coast and probably other parts of West Africa. During the wet season, there are between six to eight cycles of *I. lairdi* each year and the free-living stages can be collected from the moist river beds during the dry season (Mondet *et al.* 1976, 1977c). Rates of infection reached 80% of the nulliparous females and 35% of the entire population later in the season. Of the parasitized female *S. damnosum s.l.*, 98% were sterilized by the mermithids.

The above authors showed that the infective-stage juveniles entered their hosts by direct penetration in the head region of the early stage larvae. At ambient temperatures, the parasitic development lasted from 10–17 days and the entire life cycle was completed between 25 and 39 days. Since the eggs of *I. lairdi*, like the mosquito mermithid, *Romanomermis culicivorax*, can remain in a "resting stage" in the absence of standing water, they could be stored for periods of time which would accommodate their use in mass-breeding programmes or field trials. Of the three mermithids described from *S. damnosum s.l.* at the present time, *I. lairdi* would appear to be the best suited for biocontrol.

Romanomermis culicivorax Ross and Smith

This mermithid can only complete its development in mosquito larvae, but has the interesting ability to penetrate into young blackfly larvae as well (Finney 1975). It was shown in 1976 (Hansen and Hansen 1976; Laird *et al.* 1978) that the infective stages could invade *S. damnosum s.l.* larvae under conditions of water flow at 35 cm/s. Although the nematodes do not develop to maturity in blackflies, they prevent the latter from reaching the

Table 2. Species of blackflies reported to be parasitized by mermithid nematodes.

Host species	Nematode	Reference
Cnephia dacotensis (Dyar and Shannon)	unknown	Webster (1967)
Cnephia emergens Stone	Isomermis sp.	Anderson and DeFoliart (1962)
Cnephia mutata (Malloch)	unknown	Phelps and DeFoliart (1964)
Cnephia pacheco-lunai (De Leon)	Isomermis vulvachila Poinar and Takaoka	Poinar and Takaoka (1981)
Cnephia pecuarum (Riley)	unknown	Webster (1914)
Cnephia sp.	Hydromermis sp.	Anderson and Dicke (1960)
Gigantodax wrighti Vargas	Gastromermis cloacachilus Poinar and Takaoka	Poinar and Takaoka (1981)
Gymnopais dichopticus Stone	Limnomermis sp.	Sommerman et al. (1955)
Gymnopais holopticus Stone	Limnomermis sp.	Sommerman et al. (1955)
Prosimulium alpestre Dorogostajskij, Rubtsov and Vlasenko	Limnomermis sp.	Sommerman et al. (1955)
Prosimulium decemarticulatum Twinn	unknown	Anderson and DeFoliart (1962)
Prosimulium exigens Dyar and Shannon	Mesomermis paradisus Poinar and Hess	Poinar and Hess (1979) (Fig. 3)
Prosimulium fontanum Syme and Davis	unknown	Webster, 1967
Prosimulium fuscum Syme and Davis	Gastromermis sp.	Anderson and DeFoliart (1962)
	Isomermis sp.	Anderson and DeFoliart (1962)
	Mesomermis flumenalis Welch	Phelps and DeFoliart (1964)
Prosimulium gibsoni (Twinn)	unknown	Anderson and DeFoliart (1962)
Prosimulium hirtipes (Fries)	Limnomermis sp.	Sommerman et al. (1955)
	Mesomermis brevis Rubtsov	Rubtsov (1966a)
Prosimulium isos Rubtsov	Mesomermis brevis Rubtsov	Rubtsov (1966b)
Prosimulium magnum Dyar and Shannon	Gastromermis sp.	Anderson and DeFoliart (1962)
	Isomermis sp.	Anderson and DeFoliart (1962)
Prosimulium mixtum/fuscum Syme and Davies	Gastromermis vividis Welch	Ezenwa (1974)
	Isomermis wisconsiensis Welch	Ezenwa (1974)
	Mesomermis flumenalis Welch	Phelps and Defoliart (1964)
Prosimulium multidentatum (Twinn)	unknown	Twinn (1939)
Prosimulium ursinum (Edwards)	Limnomermis sp.	Sommermann et al. (1955)
Simulium adersi Pomeroy	Gastromermis phillipponi Mondet, Poinar and Bernadou	Mondet et al. (1977a)
	Isomermis lairdi Mondet, Poinar and Bernadou	Mondet et al. (1977b)
	Mesomermis sp.	Mondet (personal communication)
Simulium aokii Takahosi	unknown	Poinar and Ono (1979)
Simulium alcocki Pomeroy	Isomermis lairdi Mondet, Poinar and Bernadou	Mondet et al. (1977b)
	Mesomermis sp.	Mondet (personal communication)
Simulium arcticum Malloch	Hydromermis sp.	Peterson (1960)
	Limnomermis sp.	Sommerman et al. (1955)
Simulium argus Williston	Hydromermis sp.	Peterson (1960)
Simulium argyreatum Meigen	Gastromermis boophthora Welch and Rubtsov	Welch and Rubtsov (1965)
Simulium aureum Fries	Gastromermis clinogaster Rubtsov	Rubtsov (1967a)
	Gastromermis gastrostoma (Steiner)	Rubtsov (1963)
	Hydromermis sp.	Peterson (1960)
	Isomermis sp.	Rubtsov (1963)

Table 2 (*contd*)

Host species	Nematode	Reference
Simulium bicorne Dorogostajskij, Rubtsov and Vlasenko	*Limnomermis* sp.	Sommerman *et al.* (1955)
Simulium bidentatum	unknown	Poinar and Takaoka (1979a)
Simulium cervicornutum Pomeroy	*Gastromermis leberrei* Mondet, Poinar and Bernadou	Mondet *et al.* (1977) (Fig. 1)
Simulium cholodkovskii Rubtsov	unknown	Shipitsina (1963)
Simulium corbis Twinn	*Gastromermis virdis* Welch	
	Mesomermis flumenalis Welch	Ezenwa (1974)
		Ezenwa (1974)
Simulium cryophilum Rubtsov	*Isomermis rossica* Rubtsov	Rubtsov (1968a)
	Limnomermis cryophili Rubtsov	Rubtsov (1967b)
	Limnomermis macronuclei Rubtsov	Rubtsov (1967b)
	Limnomermis zschokkei (Schmassmann)	Rubtsov (1963)
	Mesomermis flumenalis Welch	Rubtsov (1963)
	Skrjabinomermis sp.	Rubtsov (1963)
Simulium damnosum Theobald	*Gastromermis leberrei* Mondet, Poinar and Bernadou	Mondet *et al.* (1977) (Fig. 1)
	Gastromermis philipponi Mondet, Poinar and Bernadou	Mondet *et al.* (1977a)
	Isomermis lairdi Mondet, Poinar and Bernadou	Mondet *et al.* (1977b)
	Isomermis tansaniensis Rubtsov	Rubtsov (1972)
	Mesomermis ethiopica Rubtsov	Rubtsov (1972)
Simulium decorum Walker	*Isomermis* sp.	Anderson and DeFoliart (1962)
	Mesomermis flumenalis Welch	Ezenwa (1973)
Simulium defoliarti Stone and Peterson	*Hydromermis* sp.	Peterson (1960)
Simulium emarginatum Davies, Peterson and Wood	unknown	Webster (1967) (personal communication)
Simulium equinum (L.)	*Gastromermis likchovosi* Rubtsov	Lichovoz (1978)
	unknown	Doby and Laurent (1953)
Simulium erythrocephalum DeGeer	*Gastromermis boophthorae* Welch and Rubtsov	Welch and Rubtsov (1965)
	Gastromermis crassifrons Rubtsov	Rubtsov (1967a)
	Gastromermis minuta Rubtsov	Rubtsov (1967a)
	Gastromermis virescens Rubtsov	Rubtsov (1967a)
	Isomermis rossica Rubtsov	Rubtsov (1968a)
Simulium furculatum (Shewell)	unknown	Hocking and Pickering (1954)
Simulium galeratum Edwards	*Hydromermis bostrycodes* Steiner	Rubtsov (1963)

Table 2 (*contd*)

Host species	Nematode	Reference
Simulium griseicolli Becker	unknown	Marr (1971)
Simulium hargreavesi Gibbins	*Gastromermis leberrei* Mondet, Poinar and Bernadou	Mondet *et al.* (1977a) (Fig. 1)
Simulium japonicum Shiraki	*Mesomermis japonicus* Poinar and Saito	Poinar and Saito (1979)
Simulium jenningsi Malloch	*Isomermis* sp.	Anderson and DeFoliart (1962)
	Limnomermis sp.	Wu (1931)
Simulium kerteszi (Enderlein)	*Isomermis rossica* Rubtsov	Rubtsov (1968a)
Simulium latipes (Meigen)	*Isomermis rossica* Rubtsov	Rubtsov (1968a)
	Limnomermis cryophili Rubtsov	Rubtsov (1967b)
	Limnomermis macronuclei Rubtsov	Rubtsov (1967b)
	Mesomermis flumenalis Welch	Levis and Bennett (1975)
Simulium luggeri Nicholson and Mickel	*Isomermis* sp.	Anderson and DeFoliart (1962)
Simulium maculatum Meigen = *Titanopteryx maculata* (Meigen)	unknown	Shipitsina (1963)
Simulium malyschevi Dorogostajskij, Rubtsov and Vlasenko	*Limnomermis* sp.	Sommerman *et al.* (1955)
Simulium metallicum Bellardi	unknown	Dalmat (1955)
Simulium metallicum Bellardi	*Isomermis benevolus* Poinar and Takaoka	Poinar and Takaoka (1979b)
Simulium metallicum Bellardi	*Mesomermis guatemalae* Poinar and Takaoka	Poinar and Takaoka (1981)
Simulium metallicum Bellardi	*Neomesomermis travisi* Vargas, Rubtsov and Fallas	Vargas *et al.* (1980)
Simulium morsitans Edwards	*Gastromermis boophthorae* Welch and Rubtsov	Welch and Rubtsov (1965)
	Gastromermis crassicauda Rubtsov	Rubsov (1967a)
	Gastromermis longispicula Rubtsov	Rubtsov (1967a)
	Isomermis rossica Rubtsov	Rubtsov (1968a)
	Limnomermis tenuicauda Rubtsov	Rubtsov (1967b)
Simulium nishijimai (Ono) = *Odagmia nishijimai* Ono	unknown	Ono (1978)
Simulium nolleri Friedrichs = *S. argyreatum* Meigen	*Gastromermis* sp.	Rubtsov (1963)
	Mesomermis flumenalis Welch	Rubtsov (1963)
Simulium ochraceum Walker	unknown	Dalmat (1955)
Simulium ornatum Meigen	*Gastromermis odagmiae* Rubtsov	Rubtsov (1967a)
	unknown	Bequaert (1934)
Simulium panamense Fairchild	*Neomesomermis travisi* Vargas, Rubtsov and Fallas	Vargas *et al.* (1980)
Simulium paramorsitans Rubtsov	unknown	Rubtsov (1966b)
Simulium piperi Dyar and Shannon	*Hydromermis* sp.	Peterson (1960)
Simulium quadrivittatum Loew	unknown	Garnham and Lewis (1959)
Simulium securiforme (Rubtsov)	*Gastromermis rosalbus* Rubtsov	Rubtsov (1967a)
Simulium subcostatum (Takahasi) = *Eusimulium subcostatum* Takashasi	unknown	Ono (1978)
Simulium tobetsuensis Ono	unknown	Poinar and Ono (1979) (Fig. 2)

Table 2 (*contd*)

Host species	Nematode	Reference
Simulium tuberosum Lundstrom	*Isomermis* sp.	Anderson and DeFoliart (1962)
	Limnomermis sp.	Sommerman *et al.* (1955)
	Mesomermis camdenensis Molloy	Molloy (1979)
	Mesomermis flumenalis Welch	Ebsary and Bennett (1975)
Simulium unicornutum Pomeroy	*Mesomermis* sp.	Mondet (personal communication)
Simulium venustum Say	*Gastromermis viridis* Welch	Ezenwa (1974)
	Isomermis wisconsiensis Welch	Ebsary and Bennett (1975)
	Mesomermis camdenensis Molloy	Molloy (1979)
	Mesomermis flumenalis Welch	Welch (1962)
Simulium verecundum Stone and Jamnback	*Gastromermis rosalbus* Rubtsov	Rubtsov (1967a)
	Isomermis rossica Rubtsov	Rubtsov (1968a)
Simulium virgatum Coquillet	*Hydromermis* sp.	Peterson (1960)
Simulium vittatum Zetterstedt	*Gastromermis viridis* Welch	Welch (1962)
	Hydromermis sp.	Peterson (1960)
	Isomermis wisconsinensis Welch	Welch (1962)
Simulium vorax Pomeroy	unknown	Haeusermann (1971)
Simulium sp. = *Eusimulium* sp.	*Hydromermis* sp.	Anderson and Dicke (1960)

pupal stage. In 1977, about 50 kg of infective stages in damp sand were added to stream water flowing through a metal trough with *S. damnosum s.l.* larvae in the Marahoué River in the Ivory Coast. Some of the blackfly larvae later netted after drifting from the trough were infected.

In 1978, 140 kg of damp sand cultures (produced, like the earlier material, at the USDA's Gulf Coast Mosquito Research Laboratory, Lake Charles, Louisiana) were taken to Bouaké, Ivory Coast, where they yielded some 80 million preparasites which were used in experiments with *S. damnosum s.l.* in a large trough-type recirculating water system in an outdoor laboratory at IRO/OCCGE/ORSTOM. While numerous total penetrations of first-instar larvae were observed, the overall incidence of such invasions was very low (less than 0.5%) and no significant drift of parasitized larvae was observed (Anon. 1978). At a time when it was not yet feasible to mass-produce equivalent numbers of true blackfly mermithids, these experiments at least confirmed the Hansens' (1976) findings with respect to the remarkable ability of a still-water mosquito mermithid to invade a blackfly host in fast-moving water; and provided a model for a future field application methodology based upon siphoning preparasites into streams. Further confirmation was provided by Finney and Mokry (1980) following exposure of Canadian *S. verecundum* and *S. vittatum* to *R. culicivorax* preparasites.

Host Range

Over 60 species of blackflies are known hosts of mermithid nematodes and the actual number is probably several times greater than that. There has never been any case of a blackfly mermithid develop- ing in hosts outside the Simuliidae, however, and most species are restricted to a range of two to 15 blackfly species.

Mermithidae: Taxonomic Criteria for their Juvenile Stages and Blackfly Biocontrol Prospects

I. A. Rubtsov

Nematodes, whether free-living or parasitic in plants and animals, are usually small, transparent organisms with a marked coelomic cavity. For most of them, and for weighty reasons, taxonomic practice in describing and diagnosing the species is based on adult characters. Normally, there is a consistent arrangement of the oesophagus, mid-intestine, rectum and sexual organs, which are all visible in whole mounts. Whenever possible, nematologists examine numerous examples of each species in process of description—this is important for adequate comparison of a significant number of diagnostic criteria for both sexes. These criteria include the measurement of bodily proportions and other structural details. There are, however, additional avenues to the taxonomic description and diagnosis of mermithid species, primarily because these are large nematodes by comparison with the free-living ones and phytoparasites; which

are encountered comparatively rarely, often only as single individuals and in the juvenile stages. Mermithids are often first observed by entomologists rather than by nematologists. Females are dominant in the family, and parthenogenesis is not infrequent. The long, uniform, thread-like shape of the body, the simplified head-capsule, the external isomorphism and overall opacity, create additional problems in defining the species. The trophosome (homologous with the middle intestine) covers the stichosome (homologous with the oesophagus); the latter lies in a groove of the trophosome and has a narrow canal. Stichosome and trophosome lie parallel to one another, although less consistently than in all other nematodes. This enhances opportunities for examining and evaluating the most important taxonomic features—the structure of the stichosome and the sexual organs.

Developmental stages may only be obtained by collecting the hosts themselves and extracting their parasites from them. By this procedure, juvenile individuals at early stages of development are secured. Only in the younger juveniles is the stichosome visible. As does the oesophagus, it conveys much taxonomic information. The structure of the stichosome is unique for each species, important features including: its length (from one-sixteenth to the full length of the body); the number of stichocytes (4–80+); the arrangement of the stichocytes around the oesophageal tube; the correlative arrangement of the epicytes; the presence of absence of gomorocytes; and the isomorphism or polymorphism of the stichocytes. Individual stichosomal variation is not marked in the lowest taxa.

In fully-grown juveniles (postparasites) the stichosome is less easily discerned, so that the structure of the longitudinal cords (especially the lateral ones) acquires paramount taxonomic value. Of importance here are: the number of cords, their width and the number of rows of cells within them; the single-

or double-layered arrangement of the nuclei; the shape and relative measurements of the border and middle cells; and, the nature of variability of all these features along the body from the head to the tail. Not only the various individual species, but also siblings and intraspecific forms, are easily differentiated by the structure of their lateral cords. Individual variability is, of course, a universal phenomenon, but the distinguishing characteristics of the cords of juveniles are, nevertheless, held to be of basic taxonomic importance. However, in mature worms, both the cords and the stichosome are so reduced that, at best, only barely-perceptible traces of them remain.

Also of taxonomic importance, is the structure of the cuticle at the anterior section of the oesophageal tube. This is unique for each genus and varies among the individual genera. Invasive juveniles always have a buccal cavity and stylet, the length of which differs sharply among the species of each genus. There are similar differences in body-shape from species to species. In parasitic and postparasitic juveniles, stichocytes are sometimes observed before the stoma. A quaritile, adjoining the cuticle or penetrating it (e.g. *Mesomermis* spp., *Isomermis* spp.) is embodied in the end of the oesophageal tube. Relevant muscular attachments are evident at the anterior tip of the cuticle, the relative thickness of this zone being characteristic for each species. In most adults, the structure of the anterior end of the oesophageal tube is simpler and more uniform than in juveniles. The cuticle of juveniles performs multiple functions during development in the host. In addition to its supportive role, it is linked with nutrition and growth. During the parasitic phase, its structure is very complex. Thus, up to 12 individually distinct layers can be differentiated within the cuticle of *Gastromermis* spp. In the free-living phases, the cuticle becomes simpler, its supportive and protective functions then predominating.

The structure of the amphids is also complex and variable, the shape and measurements of the amphid pocket differing sharply as between juveniles of *Mesomermis* spp., and *Spiculimermis* spp., moderately in *Isomermis* spp., and only

The parasitic worm, Mesomermis simuliae *Müller, A, schematic representation of the position of the fully-grown larva of the parasitizing worm within the blackfly larva. B, anterior extremity of the fully-grown larva of the parasitic male worm. C, the anterior extremity of the mature male worm. D, the anterior extremity of the mature female worm. E, the posterior extremity of the fully-grown larva of the parasitic male worm. F, the posterior extremity of the mature male worm. G, the middle part of the body of the mature female worm, with the vulva, vagina and uterus. a, amphids; aa, anal aperture; va, vagina; vp, ventral papillae; vlm, ventral longitudinal margins; vu, vulva; dp, dorsal papillae; dlm, dorsal longitudinal margin; pi, posterior intestine; cu, cuticle; lp, lateral papillae; llm, lateral longitudinal margins; mer, mermis; spm, spicule muscle; nr, nerve ring; nch, nerve chain; oc, oscmocytes; pa, papillae; wg, web-line gland of the larva; oe, oesophagous; po, pores within the cuticle; ra, rectal appendages; mi, middle intestine; sp, spicule; tet, tetradonemes; tr, trophosome; t, tail; lta, larval tail appendage.*

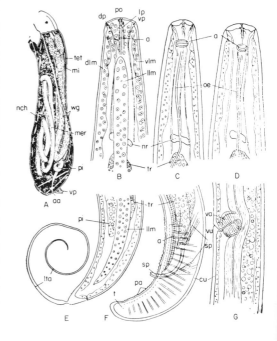

slightly in *Gastromermis* spp. A unique and important feature in the juvenile stage of mermithids parasitizing black-flies, is the presence of the tail appendage, which is, for example, quite distinct in each of the very numerous species of the genus *Mesomermis*. In some cases, it is shaped like a sharp cone, the last trace of which may be discarded towards the close of the parasitic phase of development.

Particularly towards the end of the parasitic phase, and decidedly more so among postparasites, the sexual rudiments (especially the vagina and spicules) are distinctive for all known species of mermithids parasitizing Simuliidae. In microscopic sections of the anterior and posterior thirds of the body, the oligo- and poly-propagation of the ovaries, the measurements of the ovocytes and the number of cords, are all clearly identifiable. The structure of the trophosome, its differentiation and the presence or absence of a canal within it, are held to be best observed from sections and whole mounts of juveniles. The significance of juvenile taxonomic criteria to the identification of adult mermithids is evident from Table 1 (revised, with additional matter, from an earlier version already published by the author).

Table 1 was compiled while investigating the taxonomy of each order of mermithids over a 20-year period, during which about 15 000 individuals, referable to more than 350 species, were studied. It is submitted that this table confirms the taxonomic potentialities of juveniles. Moreover, it emphasizes that practical

Table 1. Taxonomic criteria of juvenile mermithids and their relevance to adults.

Taxonomic criteria	Adults	Juveniles
L, a, b, c, c¹, V	3	3
Greatest diameter of the body, and its diameter at: the head papillae, nerve ring, and anus; and their ratios	3	3
Number and arrangement of head papillae	3	3
Amphids (measurements, shape, position)	3	3
Cuticle (structure and changes of thickness in body)	2	3
Longitudinal cords (their number on sections)	3	3
Longitudinal cords (their cell structure)	1	3
Anterior end of oesophageal tube	2	3
Length of oesophageal tube and structure of stichosome	1–2	3[a]
Structure of trophosome	2	2
Location of vulva	1	1
Shape of vagina or its rudiment	3	2[b]
Number of rows of ovocytes in ovaries (on sections)	3	3
Size of eggs or ovocytes	3	3
Means of egg hatching	2	–
Copulatory apparatus of male or its rudiments	3	1[c]
Sexual papillae	3	–
Tail appendage	–	3
Invading larva (shape, structure)	–	3
Host	–[d]	3
Availability	1	2–3
Total	42–43	50–51

[a] In parasitic larvae, especially in young ones.
[b] In mature parasites and postparasites.
[c] In postparasites.
[d] In the absence of juvenile stages.

differentiation of the juveniles is simpler than that of the adults. At the outset, juveniles clearly exhibit a larger number of taxonomic criteria, both specific and generic, than do adults, and the latter less-clearly exhibit those structures basic to the characterization of species and subspecies, e.g. cuticle, cords, stichosome, trophosome, the reproductive system, etc. Noting that these structures undergo radical change with ontogenesis, such important criteria as the spicule (again, easier to recognize in juveniles) must also be taken into account.

This is not, of course, to say that the characters of adult mermithids should be regarded as of secondary importance, let alone as unnecessary. The species (and intraspecific taxa) must be described in as much detail as possible at every stage of development—i.e. eggs, invasive stages, parasitic and postparasitic juveniles and adults, too. It is important, though, to recognize that taxonomic criteria dominant among juveniles (e.g. the cords, stichosome, the differentiated cuticle, trophosome, etc.) and reduced or entirely cast off in adults. Structures that are completely replaced include the caudal appendage, amphids, cuticle, etc. In the non-feeding, postparasitic worms, the basic function of which is reproduction, all the reserves of the different organs contribute to forming the reproductive products. In contrast, the taxonomic criteria of the juvenile parasitic stages are concerned with the processes of feeding, growth and development. Consideration of both the juvenile and adult characteristics thus makes species-diagnosis easier (e.g. by facilating the designation of taxa under consideration for practical biocontrol use). Proper diagnosis is, of course, important for other practical purposes, too; not only in the case of species that are adequately isolated morphologically, but also in that of their siblings and biological variants—which cannot usually be differentiated on adult characters, despite possible, profound biological differences.

This brings us to the question of the desirability of describing mermithid species from a single, juvenile (usually postparasitic) individual. Such action is obviously neither permissible nor called for in the case of phytoparasitic or free-living nematodes, which can be gathered in large quantities and at different stages of development. Mermithids, however, are rarities by comparison with plant parasities. Also, their juveniles are encountered in the field up to six times as frequently as the adult stages. This simple fact has to be taken into account. The very large mermithids (i.e. *Isomermis rossica* and *Gastromermis boophthorae*) have

Gastromermis boophthorae Welch and Rubtsov. A, the extremity of the fully-grown larva of the parasitic female worm. B, the posterior extremity of the fully-grown larva of the parasitic female worm. C, the middle part of the female body with the following sex organs: vulva, vagina, and part of the uterus. D, the anterior extremity of the mature male worm. E, the male spicule. F, the posterior extremity of the female worm.

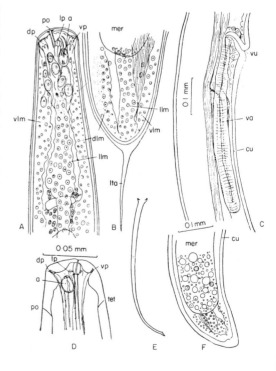

usually been described on the basis of a full series of developmental stages. This is not so with respect to the unique, but undoubtedly atypical, species, *Gastromermis longispicula* and *Mesomermis bistrata*. It must be emphasized too, that several eminent nematologists have already prepared descriptions of mermithids on the basis of juvenile material (e.g. O. Linstow, G. Meissner, A. Hagmeier, G. Steiner, B. Chitwood, J. Christie and I. N. Filipjev). Such action was justified by the need, not only to understand a fauna, but to better comprehend the continuing development of larvae following invasion of the host. The investigations of Welch (1958) and Poinar *et al.* (1976) on *Mermis pachysoma*, a parasite of social wasps, led to a highly pertinent revelation—the elucidation of a distinctive developmental cycle for mermithids having an intermediate or paratenic host.

The best reason that can be advanced for using juvenile taxonomic criteria is that when the necessary studies are carefully conducted, valuable information applicable to the adult stages will also result; as will a variety of valuable data (however cryptic) of a morphological, anatomical and biological nature. If the juvenile stages, especially the parasitic ones, are clearly characterized, recognition of relevant free-living adults is easy. However, the converse does not apply. The reason for this is that the numerous, pronounced criteria of the juveniles are either reduced to traces, or altogether lacking, in adults. The shape of the body and the stylet of the preparasites suggests the manner of their penetrating the host—the long, thin ones enter through the cuticle, while the short, thick ones do so orally.

A paradoxical situation has arisen due to ignorance of the taxonomy and species-identification with respect to juveniles. It has often been overlooked that mermithids are the richest in species of all orders of parasitic nematodes. They have a very wide range of hosts, parasitizing nearly all orders of insects and also such ecologically adjacent groups as spiders, molluscs, leeches and myriapods. Yet, half a century ago, little more than 100 species were known; most of them difficult to diagnose. Rauther (1930) was already aware that many more mermithids (perhaps up to 700 species or so) must exist, apart from the few then described. Filipjev (1934) stated even more categorically: "The number of species, apparently is very high, much higher than could be supposed from our present material".

The author, with the help of limited technical assistance, has, over the past 15 years, described more than 320 new species of mermithids; and, unprocessed material in his hands, will add scores of new ones. Thanks to recent widespread interest in Mermithidae, the number of their species seems likely to far exceed the 700 species postulated by Rauther. A four-volume fundamental work, entitled "Key to Identification of Parasitic Nematodes", illustrates how poorly the mermithids have been studied. In the fourth volume, although the list of parasites and their hosts alone occupies some 400 of the 924 pages, mermithids are not mentioned at all. The reasons for this are two-fold: the scarcity of materials, and the extremely scattered references.

Approximately three-quarters of the 15 000 or so mermithid individuals, processed by the author over the past 15 years, were juveniles (mainly parasitic and postparasitic stages of species affecting blackflies). Others have, of course, recorded many mermithids from these and other hosts, but their investigations have frequently been systematically incomplete because of the general tradition of describing and diagnosing them from adult characteristics only. The above collections of mermithids parasitizing blackflies, were made in various regions of the USSR, from Leningrad to Siberia, and from the northern tundra to Soviet Central Asia. Important material from related hosts was also received from Czechoslovakia, the Federal Republic of Germany,

France, UK and northern regions of Africa and Tanzania. Altogether, about 40 species of Simuliidae have yielded mermithids (some 10 000 in all) in the author's recent studies (he personally collected more than half of this total of specimens). The rather prominent representation of mermithids in blackflies is explained as follows: Simuliidae are particularly subject to attack by mermithids, the massive accumulation of their larvae, in fast currents of streams, being conducive to this; invasive preparasites are stream-borne; and these worms are well-adapted to penetrating the host.

Other contributory factors include susceptibility to mermithid attack on the part of the host, the location of the larval site, and synchronization of the developmental periods of both parasite and host, etc. There is also the obvious fact that the author was, first and foremost, deeply absorbed in questions concerning the taxonomy and biology of blackflies themselves, not to mention their natural enemies, and the tempting prospect of using mermithids for the control of Simuliidae.

Of around 10 000 individual mermithids dissected or extracted from the hosts, more than 90% were juvenile (postparasitic and parasitic). Juveniles were often maintained, alive, in the laboratory so that mature examples could be reared. It should be noted that free-living, mature mermithids were scarcely ever detected by the usual methods of collection. Thus, the searching and analysis of hydrobiological samples of the benthos invariably yielded negative results. Although it is true that such field-searching was limited, it is far simpler to obtain adult mermithids by dissecting out the postparasitic juveniles and by subsequently rearing them for 9–10 days. In the process of rearing these worms to adults, small losses (approximately 20%) usually occur due to fungal disease.[18]

Juveniles (it being useless to fix dead or dried individuals) were relaxed by lightly heating in distilled water or Ringer's solution (around 80 °C for 1 minute), and then fixed in either (1) seven parts 40% formalin, two parts triethanolamine and 91 parts water, or (2) TAF, 20 parts 4% formalin solution to one of acetic acid. Mounting was undertaken from glycerine (after water and other impurities were evaporated, requiring some three days at 30–35 °C) into a glycerine-gelatin slide.

The processing of approximately 10 000 individuals, already referred to, and analysis of their reference data, permitted the conclusion that species belonging to some 10% of the 67 genera of the family Mermithidae, parasitized blackflies. In Europe, Africa, and Asia, representatives of five genera (i.e. 12.5% of the 40 freshwater mermithid genera) have so far been reported from blackflies. Descriptions of no less than 58 species and 21 subspecies have been published. In descending order of incidence, these taxa represent five genera, as follows:

The genus *Mesomermis*: *albicans, arctica, baicalensis, bilateralis, biseriata, bistrata, brevis, canescens, caucasica, comosa, ethiopica, longicorpus, mediterranea, melusinae, minuta, ornata, parallela, patrushevae, polycella, prisjaznoi, sibirica, simuliae, s. acricauda, s. acutangula, s. avrensis, s. brachyamphidis, s. brevipenis, s. caudata, s. latichordata, s. longipes, s. obtusicauda, s. paimponti, s. rotunda, s. simuliae, tumenensis, vashkovii, vernalis.*

The genus *Gastromermis*: *alekseevi, ambianensis, arosea, bobrovae, boophthorae, b. longiscapa, b. distoma, b. coerulescens, b. cinerea, b. glaucescens, b. mutica, clinogaster, crassicauda, crassifrons, kolymensis, latisecta, leberrei, likhovosi, ?longispicula, minuta, obiensis, odagmiae, philipponi, rosalbus, striatella, terminalios, tschubarevae, virescens, v. acutipennis, v. virescens, viridis.*

The genus *Limnomermis*: *crassicauda, cryophili, europea, macronuclei, slovakensis.*

The genus *Isomermis*: *brevis, rossica, r. gallica, lairdi, tansaniensis.*

---

[18] See also p. 329 (Ed.).

The genus *Spiculimermis: fluvialis.*

Many taxa, from the first three genera above, have been considered as subspecies, the validity of which has often been questioned because of both fragmentation of the material and uncertainty as to the reliability of such morphological characters as the length and structure of the stichosome. On the basis of lengthy experience, the author is now inclined to raise subspecies, described by him, to the species-level; and to regard intraspecific taxa as distinctly less common. Taking into account species of which descriptions are currently in press, and also unprocessed material in collections, it is estimated that, in Eurasia alone, at least 80 species of mermithids parasitize some 30 species of blackflies. This estimate is largely based upon juvenile characters. Three worms, of as many genera, are viewed as of major importance as blackfly parasites in North America—*M. flumenalis* Welch; *Gastromermis viridis* Welch; and *Isomermis wisconsinensis* Welch. It is considered that these will probably prove to be species-complexes, with close relatives in the Palaearctic, Nearctic and Ethiopian Regions. Thus, in the USSR, *G. boophthorae* must resemble *G. viridis*, and *I. rossica* seems allied to *I. wisconsinensis*. Central America's first blackfly mermithids have only recently been recorded. Noting, however, that various terrestrial mermithids have already been observed in South America and Australasia (e.g. species of *Amphimermis*, *Ovomermis* and *Hexamermis*), it is held probable that simuliid mermithids will eventually prove to be as widely dispersed globally as these three genera.

While characteristic mermithid breeding sites are restricted for ecological reasons, migration (e.g. of the host) has assured their widespread geographical distribution. However, these worms have not yet been reported from Cuba, despite the examination of some 5000 simuliids from four species; although mermithids were detected on grasses in Jamaica, which is even further from the nearest mainland than is Cuba.

Filipjev (1934) recognized mermithid parasitism of Simuliidae to be common in the USSR, and he recorded the genus *Spiculimermis*—syn. *Amphidomermis* from them. In the author's experience, blackfly parasitism by *Spiculimermis* (numerous species of which develop in chironomids) is only due to *S. fluvialis*. This species has been recorded from many streams, only one of them from the tundra region.

Isomermis sp., parasitizing blackfly larvae. A, a sagittal, optical section of the anterior end of the body of the fully-grown larva of the parasitic female worm. B, the anterior extremity of the fully-grown larva of the parasitic female worm, with the anterior part of the laterla, longitudinal margin. C, the vagina and part of the uterus of the mature female worm, D, a general view of the young parasitic larva. E, the anterior extremity of the mature male worm. F, the posterior extremity of the fully-grown larva of the parasitic female worm. G, the posterior extremity of the mature male worm with copulatory organs.

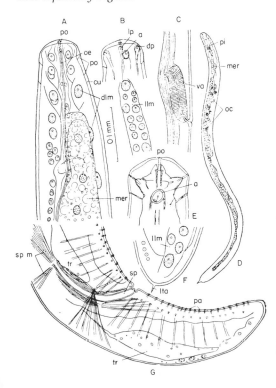

The processing of considerable material of *Mesomermis* spp. from blackflies of France, Czechoslovakia, Leningrad/USSR, Khabarovsk/USSR, Western and Eastern Europe, in areas where blackflies are abundant, has shown that foci of species differentiation are characteristic of this genus. In addition to the species which are clearly distinguishable by taxonomically significant criteria, subspecies (whether siblings or poorly-categorized forms) may be present in similar habitats within a single drainage area.

Little is yet known of mermithid variability, and much of what has been ascertained remains to be published. In this connection, one of the author's articles considered, as manifestations of individual change, the unusually high variability of *Isomermis rossica* with respect to colour, structure of the cords and the stichosome, number of stichocytes, shape and measurements of the amphids, etc. Doubts arose even while the article was in preparation, and now, after critical re-examination of the material and appraisal in the light of growing experience, it is conceded that the author was mistaken in treating variability as an individual matter in his earlier report. Noting that an unusual degree of variability seems associated with certain "mermithid species" in specific aquatic habitats, it is now believed that recognition of the significance of juvenile/adult character combinations, together with the importance of what were previously regarded as intraspecific forms, raises the probability that hybridization and subsequent sharing of some of the characters occurs. Environmental adjustment to particular types of larval blackfly sites, and the occurrence of closely-related taxa and their probable hybridization, suggest future means for applying mermithids to the biocontrol of blackflies. This matter has been touched upon in earlier papers, and is reviewed hereunder with the inclusion of additional suggestions.

First of all, until very recently (see p. 303 [Ed.]) *in vivo* mass production of these mermithids has not been possible to any meaningful extent, because of the lack of laboratory colonies of Simuliidae. *In vitro* culture of blackfly mermithids from the host, or tissue culture, has not yet been developed either. Based on existing knowledge, *Isomermis* spp., followed by *Mesomermis* spp., and then *Gastromermis* spp., would seem to offer the best prospects for candidate biocontrol agents for future field use against simuliids. The reasons why the relatively rarely-encountered *Isomermis* spp., are listed before the relatively common species of *Mesomermis* are: *Isomermis* vigorously penetrates the host, is easily established, and has intraspecific forms which, when collected from different sources, yield fertile hybrids. *Isomermis rossica* is ecologically flexible, and is clearly one of the most promising of the species from a biocontrol standpoint; another is *I. lairdi* (see pp. 165, 327 [Ed.]).

How can the effectiveness of a particular mermithid as a biocontrol agent be increased? The answer is to increase its biotic potential. What is especially important, is to increase its capacity to overcome ontogenetic bottlenecks by, e.g. penetrating the haemocoel of the host without the latter eliciting adverse reaction. In nature, mermithids have proved quite able to overcome such factors. In *I. rossica*, for example, highly variable (and as far as can be judged, hybrid) species have been encountered. This nematode predominated around Leningrad during the 1950s, permitting detailed analyses of the structure of the stichosome, and the colour and structure of the longitudinal cords in 600 individuals, collected from different rivers and streams over an area of 25 km^2.

Initially, it was puzzling to see the extraordinarily high variability of the stichosome, cord structure and colour. However, following lengthy analysis of the material, it was concluded that hybridization and the subsequent splitting off of new taxa from the parent stock, had led to this phenomenon. After all, the universal significance of hybridization (e.g. through heterosis) for plants and animals, is well-known. In agriculture, both increased maize productivity and the improvement of the breeds and the productivity of domestic animals, are due to controlled hybridization. It is submitted that more attention must be paid to hybridization as a means of increasing the effectiveness of biocontrol agents, including mermithids. It is important to realize that, during hybridization, heterosis is clearly manifested by the subject's variability, rate of growth, development, and fertility; and especially, in the strengthening of certain weaknesses affecting survival. Therefore, a practical way to increase mermithid effectiveness may be to hybridize populations of a particular taxon from neighbouring, but ecologically different, sources offering separate sets of circumstances for development. Naturally, what is needed for progress in this area is a means of experimentally monitoring the capability of populations, or ecological forms of the species, for hybridization; also, means of evaluating manifestations of heterosis. Genetically remote forms of the species will naturally yield less viable offspring, if indeed they interbreed at all.

To repeat, future use of deliberately contrived mermithid-hybrids against blackflies is felt to offer better prospects for biocontrol than the mere use of natural species. An impediment, pending the general availability of laboratory colonies of blackflies, is the difficulty in securing enough fertilized mermithid females for the mass production of invasive larvae, as in the case of *Romanomermis culicivorax* which, for years now, has been cultured routinely *in vivo*.

Conclusion

Mermithids comprise the group of nematodes that is richest in species. However, entomologists encounter them in the field more often than do nematologists. This places the mermithid taxonomist in the same position as the palaeontologist—he usually lacks life-history series of individuals exhibiting the various stages of development of each taxon. Moreover, mermithids are most frequently found as juveniles in nature. However, juveniles having more diagnostic criteria than adults, diagnosis of these early stages is held to be more practical (a reversal of the situation existing in most groups of animals).

The most important adult characters evident in the older juveniles are: the shape of the body, the distinctive cuticle and the arrangement of the organs of the head-capsule (the amphids, papillae, oesophagus tube, etc.) and the rudiments of the sexual organs—which are characteristic for each species. In addition, certain anatomical features, which are greatly reduced or lacking in adults, are prominent in juveniles. These are the longitudinal cords, stichosome, trophosome and the tail appendage. Another useful piece of information is that the host is usually known (this not being the case when free-living adults happen to be collected). Invasive preparasites have additional characteristics of taxonomic importance—a head-capsule with a stylet, a body-shape permitting conclusions as to the method of penetration into the host, the distinctive structure of the stichosome, the cellular structure of the trophosome, etc. This additional complex of specifically juvenile criteria (morphological, anatomical and biological) is held peculiar to each species and of great importance for diagnosis and classification. Where living, parasitized blackfly larvae are collected, the adult stages can, of course, be reared too, offering additional species-characters.

Ecologically limited and geographically defined blackfly breeding sites with which various mermithid parasites of these pests and vectors are associated, and the concentration of parasitism, suggests that there is a promising biocontrol

future for controlled selection and hybridization of candidate control agents and
for the practical application of these hybrid agents for control purposes. Therefore, it is urged that wider attention be
given to the desirability of a special taxonomic approach to these mermithids,
based upon the wider use of juvenile
characters, occasionally derived even
from single individuals, while not, in any
way, excluding traditional taxonomic
approaches when feasible, i.e. when
adult worms are available.

fected larvae. Causal agents of blackfly diseases are exposed to constant downstream transportation because of the lotic nature of the habitat. For infections to be transmitted from larva to larva, there must be some means by which pathogens and parasites are returned upstream, thereby countering this downstream movement. Such a reversal may be achieved via diseased adults, when these either deposit infected eggs or simply attempt to oviposit—in the latter instance, the stream being contaminated with infective stages from their decaying bodies.

Diseases of Blackflies

J. Weiser and A. H. Undeen

Diseases are quite common in populations of blackflies, especially in their aquatic stages. They are usually apparent in the larvae because white or coloured cysts of the causal organisms are visible through the host cuticle. As it is easy to select infected individuals from dense populations, a misleading impression can be gained that the percentage of parasitaemia is very high. However, seldom more than *c.* 15% of larvae are infected (sometimes by more than one pathogen). It is exceptional to find one particular entity infecting most of the available susceptible individuals in epidemic proportions.

Disease is most evident in older larvae (4th to 6th instar), mortality thus characterizing the prepupation period. In the case of some diseases, the victims remain in the larval stage for 10–14 days longer than normal; developing through four additional instars, judging by increases in head-capsule dimensions (Maurand 1975). Removal of many uninfected individuals through pupation thus causes a purely illusory increase in the percentage of parasitized larvae in the population. Infections are relatively rare in adults, some of which, however, exhibit pathogens carried over from lightly in-

Virus

The role of viral infections in the natural limitation of blackfly populations is unknown. Indeed, these particular pathogens had not been reported from simuliids at all until little more than a decade ago (Weiser 1968). Since then, other blackfly viruses have been encountered, as detailed below; only, though, in localized, temporary outbreaks, which terminate abruptly, not to recurr for a considerable time. This statement refers only to patent infections—latent ones could obviously persist in the simuliid population (or in reservoir hosts?) for lengthy periods.

Three groups of baculoviruses have now been described from blackflies: iridescent viruses (IVs), cytoplasmic polyhedrosis viruses (CPVs) and densonucleosis viruses (DVs).

Iridescent virus is represented by an entity occurring occasionally in *Simulium ornatum* in Czechoslovakia (Weiser 1968). It has now been reported from other blackflies (*S. variegatum, S. monticola,* or *S. reptans*) in the UK by Batson *et al.* (1976). Also 27% of *S. securiforme* larvae in the Ostravice River, Czechoslovakia, were infected. In all cases, the IV concerned exhibits spherical, non-inclusion bodies, ranging from 120–130 mm in dia-

Viral stromata in the cytoplasm of blackfly fat-body cells infected with IV (×160).

meter. The IV is produced in large quantities in the cytoplasm of cells of the fat-body, muscles and hypodermal tissue. It causes a bluish-green or violet iridescence of infected larvae of the host. Visibly-infected larvae die without pupating. Diagnosis can be confirmed by conventional microscopy, being based on the demonstration of typical viral stromata in the cytoplasm of cells in stained sections.

IVs are known from other Diptera (Culicidae, Tipulidae) and also from Coleoptera, Lepidoptera, etc. Oral infection is difficult to achieve. In the laboratory, IV transmission is usually accomplished by injecting infective particles into a secondary host. *Galleria melonella* (wax moth) larvae are useful in this respect for certain IVs, but it has not proved so for the one from Simuliidae. Experimental transmission of IV to blackflies has not yet been reported.

Densonucleosis virus was recorded recently (Federici 1976; Federici and Lacey 1976) from larval *Simulium vittatum* (Colorado River, California). The infection caused hypertrophy of the midgut epithelial cells and (to a lesser extent) fat-body nuclei. The altered cells are readily seen on dissection—their hypertrophic cells bulging with a mass of spherical virus particles, 180 mm in diameter. Whether or not this virus is pathogenic to blackfly larvae remains to be established—DVs of mosquitoes are only mildly pathogenic, but those of wax moths and crickets show virulence which increases with subsequent passage in susceptible hosts.

CPV was first discovered in Newfoundland blackflies (Bailey *et al.* 1975). Infections developed in *Stegopterna mutata* during the winter; incidence increasing in larvae of this univoltine species, ranging from 10–22%. A few infections were also noted in larval *Prosimulium mixtum/fuscum*, for which Bailey (1977) made similar observations. The latter author also extended the host list to *S. tuberosum*, *S. venustum* and *S. vittatum* (in none of which incidence exceeded 5%), and he indicated that as many as 54% of a larval *St. mutata* population can be infected. Federici and Lacey (1976) reported the same pathogen from *S. vittatum* larvae from the Colorado River. Weiser (1978) noted its presence in Czechoslovakia, where he discovered it in *S. argyreatum* at infection levels of 7–10% (by comparison with 1% in California and up to 54% in Newfoundland).

White spots (in the gut wall, in the region of the cardia and the blind sacs, and at the insertion of the Malpighian tubules) render CPV-infected simuliid larvae easily recognizable to the naked eye. In heavy infections, the midgut is seen through the cuticle as a white band. On removal, the gut has the appearance of white lace, the empty spots indicating where nuclei of infected cells are located. The cytoplasm of hypertrophied cells is completely filled with minute refringent polyhedra, 1.0–2.5 µm in diameter.

Bailey (1977) reported evidence of experimental transmission to 1st-instar larvae. While he was successful in infecting

Cells in the midgut of Simulium argyreatum *infected with CPV (×40).*

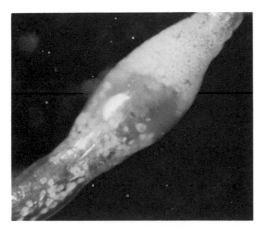

up to 100% of exposed larvae, his experiments were plagued by high control mortality. Weiser (1978) demonstrated that late-instar larvae, harbouring CPV, shed their infected cells during evacuation of the gut prior to pupation, so that patent infections are not evident in either pupae or adult. Nevertheless, the possibility of transmission of CPV to adults and later generations cannot be ruled out, despite the present lack of supporting evidence.

It should be mentioned that the Czechoslovak infections recorded by Weiser (1968) came from the outflow from Darko Lake, a study site where blackfly larval development had been investigated for 30 years prior to this first European appearance of the CPV. Symptoms being clearly apparent, as stated above, it is considered most unlikely that the infection had been present, but overlooked earlier.

Bacteria

Saprophytic

It is known that bacteria occur in blackfly larvae at times when the current is sharply reduced, or ceases altogether, with consequent O_2 depletion (Weiser, unpublished data). Under such circumstances, gut bacteria pass through the intestinal wall, causing septicaemias and larval mortality in 2–3 h. This situation has not been encountered where streams remain dry; larvae then sometimes surviving unharmed (especially on aquatic vegetation) for more than 5 h, regaining activity upon resumption of the flow. Pupae thus exposed to the atmosphere survive under sufficiently moist conditions, eclosion afterwards occurring normally and without bacterial complications. Virulent strains of the heavily-pigmented species, *Serratia marcescens*, are the most common cause of larval septicaemia during periods of fluctuating water velocity.

In the terminal phase of infections due to such pathogens as the fungus, *Coelomycidium simulii*, bacteria may be a secondary cause of death. Saprophytic bacteria also aid the release of the *C. simulii* zoospores by causing rapid decomposition of the body wall of the dead host. Mortality resulting from bacterial infections in natural blackfly populations is difficult to detect, moribund larvae detaching from leaf or stone substrates so readily. To date, none of the bacteria isolated from blackflies has been successfully transmitted back to the host.

In Newfoundland, *Cnephia ornithophilia* adults, emerging in the laboratory, have shown gram negative bacteria in one or both wings (Undeen, unpublished data). A similar bacterium was also found in larval *Simulium venustum*—appearing as a large, white, abdominal cyst—at a Newfoundland pond outlet. Both bacteria were cultured on nutrient agar plates, but attempts to transmit them to blackfly larvae were unsuccessful.

Pathogenic

Bacillus thuringiensis var. *israelensis* de Barjac

Some of the strains of *Bacillus thuringiensis* used with good results in the control of Lepidoptera were claimed (Lacey and Mulla 1977) to be pathogenic to larval *S. vittatum*, although the required concentration of spores in water was too high for practical application.

With the discovery (Goldberg and Margalit 1977) of serotype 14 of *Bacillus thuringiensis* (var. *israelensis*, de Barjac 1978)—highly pathogenic to mosquitoes (the natural hosts)—a new weapon for evaluation against blackfly larvae was at hand. In laboratory tests, using the magnetic stirring system of Colbo and Thompson (1978), it proved toxic to several species of blackflies in Newfoundland (Undeen and Nagel, 1978). LC_{90}s of between 2.5×10^3–5.6×10^3 spores/ml resulted from 30-minute exposures. One-minute dosages of 5×10^4 and 10^5 spores/ml caused 87% and 93% mortality respectively. One small efficacy test, involving a one-minute application of *B.t.i.* (10^5 spores/ml) was conducted in a Newfoundland stream. Two hours after application, larval samples were brought to the laboratory and maintained in circulating water for 48 h along with controls from undosed sites. Corrected mortalities were 100% for localities 5, 10, 15 and 30 m downstream from the dosage point, and 83% at 50 m.

Further tests were conducted during 1979 (Colbo and Undeen 1980). These involved the *S. venustum/S. verecundum* complex, *S. vittatum*, the *S. tuberosum* complex, *Prosimulium mixtum/fuscum*, *Cnephia ornithophilia* and *Stegopterna mutata*. The *B.t.i.* used was cultured on tryptose blood agar base (Difco) for 5–7 days at 30 °C, scraped from the agar surface, suspended in water and stored at 5 °C. Calculated dosages of 10^5 vc/ml for 1 minute of flow were administered from a sprinkling can, calibrated to empty in 1 minute. The treatment was applied at the head of a single channel leading to riffles over the next 100–200 m downstream. During and after dosage, water samples were taken about 10 m below the application point at 15-s intervals. These were brought back to the laboratory for assessment of the "applied dosage" by plate counts. Blackfly larval samples were collected at measured intervals downstream from the dosage point, while control larvae were gathered upstream. These samples were taken back to the laboratory on ice, and sub-samples were placed on a magnetic stirrer rearing system (Colbo and Thompson 1978). Each day, water was changed; 25 mg of Tetra® fish food was then added and, after 48 h, surviving larvae were counted. Percentage mortalities of up to 100% were obtained from dosages of only *c.* 10 ml of "formulation" (10^{10} vc/ml) per m^3 of stream water. This quantity compares favourably with the figures for practical applications of the Abate® formulation used in WHO/OCP (Le Berre, personal communication and see p. 91, and the methoxychlor formulation used in the ARBFRB (Shemanchuk, personal communication and see p. 123). This, combined with its harmlessness to NTOs (Colbo and Undeen 1980) strongly recommends B.t.i. as a possible supplement to, or even replacement for, the larvicidal chemicals on which reliance is presently placed for blackfly control.

Late in 1978, tests at Bouaké, Ivory Coast (Undeen and Berl 1979) indicated that S. damnosum is as susceptible to B.t.i. as S. verecundum in Newfoundland. One-minute exposure tests in concrete troughs, using the same formulation as in the Canadian experiments, produced 95 and 99% mortalities at 5×10^4 and 10^5 spores/ml respectively. The powdered formulation (the international standard provided by H. de Barjac of the Pasteur Institute, Paris) had an LC_{90} of about 3.5 $\times 10^4$ spores/ml in S. damnosum compared with 1.2×10^5 spores/ml in S. verecundum. In trough tests the Pasteur Institute powder required 5×10^6 spores/ml for 84% mortality. Appropriately formulated, though, this material might have potential as a practical blackfly control agent.

Recently, a primary B. t. i. powder produced by Bellon-Biochem (in collaboration with the Pasteur Institute), and some experimental formulations derived from it, were tested against S. damnosum in stream-mounted flumes and in a natural stream. The results from these tests, reported in a series of WHO mimeographed documents (1979–1980) appear very encouraging. Applied dosages of 0.2 mg/l for 10 minutes produced 100% mortality in the flumes. In the stream test an increase in drift of simuliid larvae (but not of the non-target organisms) was observed. Moreover, in the experimental flumes there was no significant kill of non-target organisms.

Laboratory evaluations and field tests were undertaken with B.t.i. in Guatemala (1979) against the onchocerciasis vector Simulium ochraceum (Undeen et al. 1980). The flow volume of the streams in

which tests were conducted ranged from 10–840 l/minute. Larval mortality resulted from the application of between 1×10^5–2 $\times 10^5$ vc/ml at up to 200 m downstream from the dosage site. Preliminary field results indicated that in the relatively small streams utilized by S. ochraceum, higher B.t.i. dosages per unit volume of water might be required than in the case of larger streams. For practical applications, too, dosages (preferably using a formulation needing from 1–10 minutes for dispersal) will have to be applied along the streams at intervals of less than 0.25 km for the larger ones, and as short as 20 m for the smaller ones.

Laboratory experiments upon larval Simulium ornatum and S. latipes, showed that the uptake of B.t.i. and resulting mortality is proportional to the current bringing bacteria to the larval mouthparts (Weiser and Vankova 1978). However, these studies were not undertaken in controlled flowing water circumstances; nor were tests of the response to high dosages of B.t.i. by NTOs captured in blackfly streams. Even under these very artificial circumstances, only low activity of B.t.i. was recorded against larval Chironomidae and Trichoptera (preying upon Simuliidae); and no activity whatsoever was registered against Plecoptera, Ephemeroptera, Coleoptera, Hydracarina. Crustacea, Hirudinea and Mollusca. All these Czechoslovakian tests having been made in still water, and perhaps in conditions inimical to the subjects, further studies are obviously needed in flowing water, along similar lines to those of Colbo and Undeen (1980); who demonstrated, in Newfoundland, in 1979, during the field testing of B.t.i. referred to above, that no ill effects were suffered by Chironomidae or Trichoptera (Hydropsychidae and Philopotumidae), nor by various other aquatic insects of the orders Diptera, Coleoptera, Plecoptera, Ephemeroptera and Odonata.

Finally, over and above assessments of B.t.i. against NTOs, studies are needed of the organism's capacity for surviving

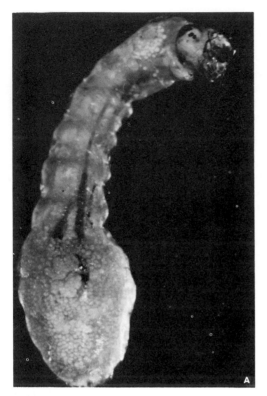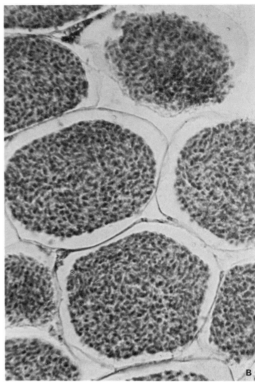

A, Simulium ornatum *larva infected with* Coelomycidium simulii. *B, Thalli of* C. simulii *containing zoospores (×26).*

passage through the gut of filter feeders, on plant and other substrates and in stream sediments.

Fungi

Fungal diseases of blackflies are more commonly encountered than viral or bacterial ones. Invasion takes place during the larval stage and, in some cases, the resultant infections persist into adults. Moreover, the latter exhibit certain fungi which have never been recognized in larvae and may perhaps be specific to the imagines.

Members of the Saprolegniaceae sometimes infect stressed or damaged immature Simuliidae. Thus, Nolan and Lewis (1974) isolated *Pythiopsis cymosa* from a *Prosimulium* pupa, visibly infected by a fungus. Cysts of *P. cymosa* were duly produced on nutrient agar, but attempts to reinfect larvae with them failed. Live blackfly larvae have been taken from Newfoundland streams with fungal hyphae penetrating through the peritrophic membrane and midgut wall. In such larvae the gut does not pull out of the body easily as in uninfected ones. When it does, the fungal mycelia give the midgut a "hairy" appearance (Undeen unpublished). Whether these fungi are only casual invaders, or primary pathogens of blackfly larvae, is not known.

Coelomycidium simulii Debaisieux is the most common, and most widely distributed, fungus of blackflies. It is an aquatic chytrid which invades early larval instars by means of flagellate zoospores (Debaisieux 1919). The developing thalli are incorporated into the fat-body, where they grow, dissolving the tissues with secreted proteolytic and lipolytic enzymes. Finally, all tissues of the host are dissolved except for the gut and cuticle; the space between which is filled with spherical thalli, their multinucleated plasmic content cleaving into hundreds of flagellate zoospores. The sporangial wall is then dissolved, or ruptured, and the zoospores, which have a nuclear cap (appearing as a refringent body apposed to the nucleus), leave the dead host. Sometimes, the sporangia are thick-walled. These "hibernating stages" of Debaisieux (1919) have several release papillae (Weiser and Zizka 1980). They are very seldom encountered and appear to be neither specifically winter forms nor, indeed, to represent an obligatory stage. Details of the transmission of *Coelomycidium* disease are yet to be elucidated. Infections occur throughout the year. Under Central European conditions, there is a prominent peak during the winter and a less marked one 3–4 months afterwards, up to 30–40% of blackfly larvae harbouring the fungus in pockets of infection which characteristically occur in small brooks (Weiser 1966). Many species of Simuliidae (*Simulium*

Sporangia, thought by J.W. to belong to Coelomomyces *and by Laird (1977) to be referable elsewhere, from* Simulium ornatum *(×180).*

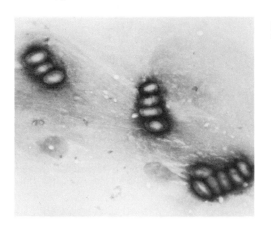

latipes, S. hirtipes, S. pictipes, S. iridescens, S. adersi, S. pugetense, S. reptans, S. neavei, S. ornatum, S. argyreatum and many others) are known hosts for *C. simulii* in various parts of the Northern Hemisphere, South America, and tropical Africa and Asia [see WHO Bibliographies of Jenkins (1964) and Roberts and Strand (1967)]. Morphologically aberrant strains of this, or a closely related fungus, have been reported from the Chingan Mountains of Soviet Central Asia (A. M. Dubitskii, personal communication). Growth of *C. simulii* on artificial media has been reported by Soper and Roberts (1978).

Bodies labelled as "sporangia of *Coelomomyces* sp. in *Simulium ornatum* larvae" were illustrated by Weiser (1977). They had a thick, corrugated wall, measured $60 \times 45\,\mu m$, and were found only once, in a single larva (some term the host *Odagmia ornata*—see p. 17) in a tiny, spring-fed, brook in Czechoslovakia (1943). Laird (1977b) did not agree with the identification of this organism as *Coelomomyces*; the only other suggestion that *Coelomomyces* (a genus of wide occurrence in mosquito hosts and present in some chironomids too) might occur in Simuliidae is that of Garnham and Lewis (1959), who discovered what they supposed to be a *Coelomomyces* in *Simulium metallicum* in Honduras.

Simuliomyces lairdi Dubitskii et al. (1978) is an entomophthoracean fungus described from larval *Odagmia* sp. from mountain streams of the Terskey Alatau, Tien Shan Range, Soviet Central Asia. It is rare to find a member of the Entomophthorales in an aquatic habitat. These fungi are characterized by non-aquatic conidial stages, and their conidia are distributed by active discharge from conidiophores on the surface of the infected insect. In this aquatic example, neither conidiophores nor conidia are present. Instead, large hyphae produce spherical to oval thick-walled zygospores with an exosporium 4–$5\,\mu m$ thick and endosporium 2–$3\,\mu m$ thick. The zygospores themselves are 60–80 µm in diameter. Azygospores are spherical to rhomboidal (17–56 µm × 14–42 µm), with a mucous exosporium swelling in water to 12 µm in thickness at the poles and disappearing when dried (Dubitskii 1978). Attempts to isolate *S. lairdi* were frustrated by the appearance of a contaminating *Sporotrichum* sp. The usual frequency of the infection in the *Odagmia* population was 1–6%, but levels as high as 60% were recorded. The parasitized *Odagmia* larvae were milky white and thus easily recognized in nature. In the same habitat, a larval blepharocerid (*Tianshanella* sp.) harboured *S. lairdi* too, and perhaps transmission of the infection involves both hosts. Dubitskii *et al.* (1978) recovered some resting spores from the submerged surfaces of stream flora.

In addition to the fungi already discussed, several genera and species of Trichomycetes occur in the gut or on the body surface of blackfly larvae. Although up to 100% of the host population may be positive for these organisms, Trichomycetes only appear to harm the host on rare occasions and when other stresses are evident. Examples of these phoretic fungi from larval blackflies are: *Harpella melusinae*, *Penella hovassi*, *Smittium simulii* and *Paramoebidium chattoni* [see WHO Bibliographies of Jenkins (1964) and Roberts and Strand (1967)].

An as-yet undescribed representative of the Phycomycetes has been reported from the ovaries of *Prosimulium mixtum/fuscum* and *Stegopterna mutata*, more than 50% of the females being infected on occasion (Undeen and Nolan 1977). In newly emerged *St. mutata*, the earliest stages of this fungus are elongated hyphae, present in developing ovaries. These hyphae develop into irregular, oval or spherical thalli, measuring 14–17 µm × 17–20 µm. Eventually, the thalli are deposited instead of eggs during attempts, by the host, to oviposit. When free in the egg mass, the thalli protrude in a corona of 2–7 club-shaped sporangia which later becomes navicular. The con-

tents of the thallus flow into these fruiting bodies, which remain dormant, attached to the empty spore case in *St. mutata* (which has a summer-diapausing egg). Evidence of the phycomycete has since been found in *Simulium verecundum* as well. In this host, material was recovered only from egg masses. A long discharge—or germination—tube developed from the sporangia of this species when the temperature was raised to above 10 °C. Transmission has not yet been demonstrated, although it is felt significant that development of the organisms appears synchronous with that of the eggs of the host.

Fungal thalli have frequently been recorded from various adult blackflies (e.g. *S. verecundum*, *S. damnosum*, *S. ochraceum*, *S. metallicum*, *S. callidum* and *S. gonzalesi*)—in some cases from the general body cavity, in others in association with the ovaries. Perhaps some of these thalli may represent stages of a similar phycomycete, extending its range to West Africa and South America (Garms 1975; Collins, personal communication) as well as Newfoundland.

Entomophthora culicis Fresenius is best known from adult mosquitoes, huge numbers of hibernating examples of which become fixed by the hyphae to, e.g. the walls of caves. The genus also parasitizes other Diptera, including blackflies; ovipositing females of which may become attached by the rhizoids to a leaf or stone, adjacent to the deposited egg mass. The adult simuliid is transformed into a 5–8 mm gray mass of hyphae, with conidiophores protruding in all directions. These are single, binucleate, and bear spherical conidia measuring 11–16 µm × 8–12 µm which, when discharged, become fixed to the surface by a halo of cytoplasm. Rhizoids are broad hyphae without any terminal swelling. Resting spores, which are formed inside the insect, are spherical. These are quadrinucleate, averaging 27 µm in diameter. Massive outbreaks of *E. culicis* infection are unknown. Instead, it is usual to find a

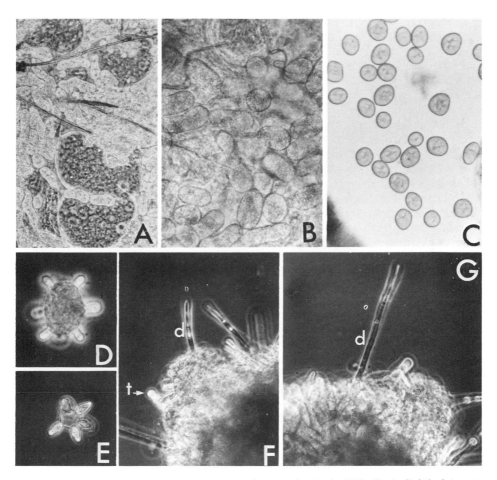

A, Fungal hyphae seen in St. mutata *less than 24 h post-eclosion (×315). B, A slightly later stage in which the hyphal segments are shorter and more rounded (×562). C, Mature "spores" which are deposited by the infected blackfly instead of eggs (×315). D, Development of "secondary sporangia" from the spores (×765). E, The apparent resting stages of the fungus, attached to the now-empty original spore case (×315). F and G, The development of tubes from the resting stage, under the influence of increased temperatures (these last are from spores deposited on* Simulium verecundum *egg masses) (×315).*

few infected adults on, for example, *Phragmites* leaves floating at the surface of small streams. *S. argyreatum* and *S. ornatum* are the most common hosts in Czechoslovakia.

Entomophthora curvispora Nowakowski is the next most common fungus reported from old female blackflies after oviposition. The surface of the host bears many conidiophores with comma-like, long, oval conidia, 25–40 μm × 10–15 μm (Gustafsson 1965). Secondary conidia produced from already-discharged ones, are pyriform and measure approximately 16 μm × 12 μm. Spherical resting spores in the hyphal mass average 37 μm in diameter. When first formed, they have a thin, brown epispore which is later autolysed. The fungus-bearing adult simuliids are found in Central Europe, at the edges of mountain brooks, on stones and vegetation. Lewis (1965) has reported an *Entomophthora*-like fungus from *Simulium damnosum s.l.* in Kumba, Cameroon.

Entomophthora spp. from blackflies can be cultivated on artificial media (Gustafsson 1965); and, as already mentioned, preliminary results concerning the maintenance of vegetative thalli of *Coelomycidium simulii* on nutrient media, have been published (Soper and Roberts 1978). Any practical application of these findings, of course, lies in the future. Dubitskii (1978) reported some possible *C. simulii* infections in larval blackflies reared in the laboratory. However, the variation in numbers of infected larvae in his different samples, suggests the possibility that, rather than being acquired in the laboratory, these cases simply represented the fruition of cryptic infections already present in larvae when brought in from the field. Failure to achieve experimental infection elsewhere, suggests that very specific conditions, as yet not understood, are pre-requisite to successful transmission. It is considered that even once success in culturing *C. simulii*, in the laboratory, has opened the way to mass production, the success of field applications will depend upon careful timing to ensure that the proper environmental circumstances exist.

Protozoa

Tetrahymena-like ciliates have been recorded from adult female *S. damnosum s.l.* in West Africa (Lewis, 1960, 1965; Corliss *et al.* 1979), *S. ochraceum, S. metallicum* and *S. callidium* in Guatemala (Garms 1975) and *S. verecundum* in Newfoundland (J. Mokry, personal communication) Corliss *et al.* (1979) published a preliminary account of such a ciliate from adult males and females of the *S. damnosum* complex from Ivory Coast and were able to assign the organism, with certainty, to the genus *Tetrahymena*. In no case have any external signs of infection been apparent. However, active ciliates are conspicuous in the haemolymph on dissection of the host.

Nothing is known of the mode of transmission or of possible pathogenicity to the Simuliidae concerned.

Lewis (1965) reported parasites resembling Haplosporida from the ovaries of 0.3% of nulliparous *S. damnosum s.l.* females examined in Cameroon. Beaudoin and Wills (1968) described *Haplosporidium simulii* from *S. venustum* in a Pennsylvanian stream. Its spore is uninucleate and has an orifice capped by a tight-fitting lid, very difficult to distinguish when closed. Spores measure 4.8 µm × 3.6 µm, and sporogenous examples contain from one to 64 nuclei. Bodies, larger than the spores, were also observed undergoing division. There is no information on this organism's transmission or its pathogenicity to the host.

Obligate microsporidan protozoan parasites have been discovered in many different species of Simuliidae in many parts of the world (see references in Jenkins 1964, and Roberts and Strand 1977). Some of the 21 species of Microsporida known from blackflies are distributed throughout the major zoogeographical regions. Others have a sharply restricted distribution, comprising foci in suitable microclimatic circumstances and even specific blackfly populations. While certain microsporidan protozoa parasitize a number of different blackfly hosts, others appear confined to a single host-species.

A Simulium ornatum *larva infected with a microsporidan (Tuzetia debaisieuxi) (×6.7).*

Among those Microsporida having worldwide distribution are: *Thelohania bracteata, T. fibrata, T. varians, Pleistophora debaisieuxi* and *P. multispora.* These occur from lowland streams to mountain torrents, and from the tropics to temperate latitudes in both Palaearctic and Nearctic Zones. Species usually found sporadically, but of wide occurrence under both boreal and mountainous conditions, are: *Nosema stricklandi, Stempellia rubtsovi, S. simulii, Caudospora simulii, C. polymorpha, C. pennsylvanica, Pleistophora turgenica, Thelohania djungarica* and *Weiseria laurenti. Pegmatheca simulii* has only been collected from one locality in Florida. Some species (e.g. *Caudospora alaskaensis, C. nasiae* and *Weiseria sommermannae*) appear typical of high northern latitudes. Knowledge of the distribution of blackfly Microsporida is very incomplete. It is, nevertheless, striking that examinations (initiated by WHO/VBC) of preserved Simuliidae in major collections (such as those of the British Museum, Cambridge University's Natural History Museum and the Zoological Institute, USSR Academy of Sciences, Leningrad) have not revealed any additions to the comparatively small number of microsporidan species already known from blackflies. Bearing in mind the large and growing number of these protozoans now known from mosquitoes, it is thought-provoking that the collection of blackfly larvae in nature (for taxonomic purposes), has yielded only further examples of the most characteristic microsporidans of these insects—it is hardly likely that, had massive infections due to new taxa been sampled, they would have been overlooked.

Larvae harbouring microsporidans are characterized by white, irregularly lobate cysts, posteriorly. Their cysts sometimes extend the full length of the body, the fat-body becoming filled with various life-history stages of the microsporidan concerned. Studies of fresh material from living, infected larvae are essential for accurate generic diagnosis. It is often difficult, if not impossible, to determine the identity of microsporida from preserved or dead larvae.

Spores of *Nosema* develop individually. Those of *Thelohania* are produced in groups of eight. Those of *Pleistophora* occur in irregular masses, formed from multinucleate plasmodia. Other microsporidans *Nosema stricklandi* Jírovec (1943), for example, produce single, pyriform spores. In the case of *N. stricklandi,* they exhibit irregular swellings on the surface, are binucleate, and measure 5 μm × 2.5 μm. Occurring in the fat-body of winter-developing larvae of *S. latipes,* they can be found only in early spring. Maurand (1975) identified this European-described species from *Stegopterna emergens* in Canada.

Thelohania bracteata (Strickland 1913) was transferred to the genus, *Amblyospora,* by Hazard and Oldacre (1975). It produces octosporous pansporoblasts, forming almost spherical spores (3–4 μm × 2.5–3 μm), often slightly flattened at the poles. In some hosts, the cysts are brown; in others, white. In the vicinity of Prague, there are two peaks in the incidence of *A. bracteata*; one in January/ February and a second in July/August. Some of the hosts of *T. bracteata* are: *Simulium ornatum, S. pratorum, S. latipes, S. securiforme, S. venustum, S. decorum, S. equinum, S. monticola, S. variegatum, S. pugetense, S. verecundum* and *S. vittatum.* This species has been reported from both North and South America and also in Europe and Asia.

Spores of *Thelohania varians* (Léger 1897) are morphologically very similar to those of *T. bracteata,* except in that the subspherical ones are more elongated (3–5.5 μm × 3.4 μm). Besides parasitizing the fat-body, this pathogen later invades the Malpighian tubules and silk glands too. Infections due to *T. varians* are rarer than by *T. bracteata.* They exhibit minor peaks in April/May and September/ October. The spores sometimes occur as tetraspores, their contents being doubled, so that there are, for example, two nuclei instead of one. Some of its

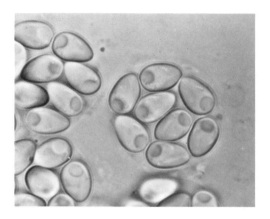

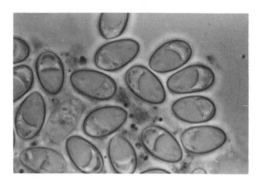

Spores of Tuzetia debaisieuxi from Simulium verecundum larvae (×1620).

Thelohania fibrata spores from Simulium venustum larvae (×1890).

hosts are *S. ornatum, S. latipes, S. securi-forme, S. pugetense, S. tuberosum* and *S. venustum.*

Thelohania fibrata (Strickland 1913) differs from its two congeners in having regularly oval spores, measuring 5–7 μm × 3.4 μm. Before spore formation, granules of lipoprotenic secretions appear in its octosporous pansporoblasts. This microsporidan is known from *S. ornatum, S. caucasica, S. latipes, S. securiforme, S. venustum* and *S. argyreatum* in Europe; and from *S. venustum, S. verecundum, S. latipes, S. tuberosum* and *S. vittatum* in North America. It has proved common wherever blackflies have been thoroughly sampled for Microsporida. Morphological variations may occur locally. Thus, spores of *T. fibrata* in *S. damnosum s.l.* and *S. hargreavesi* in West Africa are larger than normal, measuring 8 μm × 5 μm (Ezenwa *et al.* 1974). Near Prague, *T. fibrata* peaks in July/August, with a second peak in October and (sometimes) a minor one in February.

At sporogony, *Tuzetia debaisieuxi* (Jírovec 1943) produces morula-like, spherical sporoblasts on the surface of the sporogonial plasmodium. In smears, it exhibits rosette-like sporoblasts with the nuclei at the periphery of the finger-like protrusions. Its spores are broadly oval (7–8 μm × 3–4.5 μm). Their clusters separate easily into individual spores. Usually grouped, mature spores are not evident on prepared slides. There are no prominent seasonal peaks for *T. debaisieuxi* populations. In Czechoslovakia, its hosts are: *S. ornatum, S. pugetense* and *S. securiforme*; while, in North America, it parasitizes members of the genera *Stegopterna* and *Simulium*. This widespread organism apparently lacks close host specificity.

Pleistophora simulii (Lutz and Splendore 1908) has the same development sequence as *T. debaisieuxi* and occurs in the same localities. The spores are smaller, ranging from 4.0–5.5 μm × 2.5–3.5 μm (Weiser 1947) to 6.3 μm × 4.0 μm (Vavra and Undeen 1980)—almost the size described for *T. debaisieuxi* by Jírovec (1943). The latter species has about twice the number of polar filament coils, as does *T. simulii* (Vavra and Undeen 1980).

Pleistophora multispora (Strickland 1913) has broadly pyriform spores, measuring 4–5.5 μm × 2.5 μm. Its pansporoblasts are very cohesive and are visible in the cysts under the dissection microscope. The number of spores in these spherical groups varies from 32 to 64 (or perhaps even more). Also characteristic for this species is the presence of a coat of mucus around each spore, detectable with Indian ink. In the Prague region, *P. multispora* produces well-differentiated peaks in May/June, August/September and during

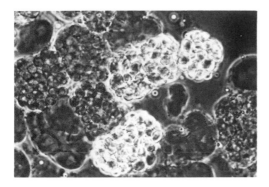

Groups of spores of Pleistophora multispora *from* Simulium venustum *(×600).*

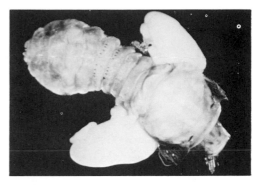

An adult Simulium reptans *packed with spores of* Thelohania columbaczense *(×15).*

the winter months. It is known from *S. ornatum, S. latipes, S. securiforme, S. bezzii, S. corbis, S. decorum, S. morsitans, S. venustum, S. angustitarse, S. pugetense* and *S. argyreatum* in Europe; *S. vittatum, S. tuberosum, S. venustum* and *S. verecundum* in North America; and *S. damnosum s.l.* and *S. hargreavesi* in West Africa.

The undermentioned blackfly microsporidans are much less frequently reported than those discussed so far:

Stempellia rubtsovi Issi, described from the Caucasus (Issi 1968), forms pansporoblasts containing from two to 16 spores, the dimensions of which vary in proportion to the numbers of individuals. Thus, where only two spores are formed, they measure 12–13 μm × 5.4–

Spores of Pleistophora multispora *from* Simulium venustum. *A little India ink in the suspended water reveals its characteristic mucous coat (×720).*

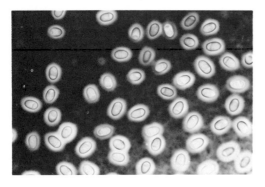

7.3 μm, while individuals of groups of 16 measure 5.4–6 μm × 3.5–4.8 μm.

Stempellia simulii Maurand and Manier (1967) also forms groups of eight to 16 uniform, elongate-pyriform spores (4.5 μm × 1.5 μm) in the fat-body. This species was described from *S. bezzii* in the south of France.

Thelohania columbaczense Weiser (1960) occurs in the fat-body of adult *S. reptans golubaciense* from the Danube. It has pyriform spores, measuring 3–3.5 μm × 1.5–1.8 μm.

Thelohania jungarica Dubitskii (1978) infects *Odagmia* sp. in the Alatau mountains of Soviet Central Asia. Its spores are elongate-pyriform (5–7.5 μm × 2.5–3.8 μm) and occur in groups of eight, in the fat-body.

Pleistophora turgenica Levchenko and Issi (1973) infects *Odagmia* sp. in the mountains of Kazakhstan, Soviet Central Asia. Groups of up to 26 spores are present in the larval fat-body. They are elongate-pyriform, measuring 4.8–6 μm × 1.8–2.4 μm.

Pegmatheca simulii Hazard and Oldacre (1975) from *S. tuberosum* (Florida, USA) is characterized by the formation of groups of satellite pansporoblasts hanging on thin connecting strands from the original sporont. Eight, broadly-oval spores are formed within each satellite, measuring 3–3.7 μm × 1.9–2.5 μm.

Spores of members of the genus

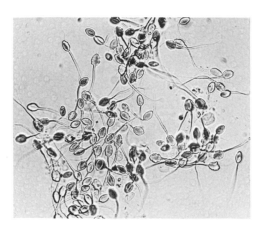

Spores of Caudospora simulii *from* Simulium latipes *(×450).*

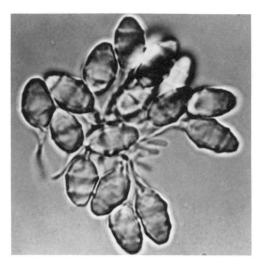

Spores of Caudospora polymorpha *(×2400).*

Caudospora exhibit long appendages from the posterior end. *Caudospora simulii* was described by Weiser (1946) from *S. latipes* in Czechoslovakia. In the type material, the fat-body appeared brick-red due to pigmentation of its outer layer. In Newfoundland, a morphologically indistinguishable species, referred to *C. simulii* (identification confirmed Weiser), was discovered by Frost (1970) and later discussed in detail by Frost and Nolan (1972). This strain, from *P. mixtum/fuscum*, forms white, instead of red cysts. The caudal appendages of *C. simulii* measure 14 μm, or even more, tapering to a terminal strand so fine that it is most difficult to determine its precise endpoint. The spore itself is oval (5.8 μm × 5 μm) and it features two longitudinal, lateral alae. Infections become evident in the spring, in the late instars of winter-developing larvae.

Caudospora pennsylvanica Beaudoin and Wills (1965) has longer spores (5.3 μm × 3.2 μm) and caudal appendages even longer (*c.* 23.5 μm) than those of *C. simulii*. The lateral alae of the latter species are lacking, though. *Caudospora pennsylvanica* infects the fat-body of *P. magnum* in the eastern USA.

Caudospora alascaensis Jamnback (1970) has oval spores (4.6–5.5 μm × 2.9–3.5 μm), with the merest vestige of cauda protruding for less than 1 μm from a spore-body without lateral alae. The only known host is *Prosimulium alpestre* from Alaska.

Caudospora polymorpha (Strickland 1911) forms oval spores (4.8–6 μm × 2.9–3.9 μm), with cauda only 8 μm long. Its spore exhibits two thickenings, girdling the middle and anterior end. The host is *Stegopterna mutata*, and infections have been reported from the eastern USA and Canada (including Newfoundland).

Caudospora nasiae Jamnback (1970) forms oval spores, without alae or thickened rings. They measure 3.7–4.5 μm × 2.4–3.3 μm, the cauda averaging 8.7 μm. The species was described from *S. adersi* in Ghana.

Weiseria laurenti Doby and Saguez (1964) is characterized by a spore having the surface ornamented by sutures, terminating at a collar, protruding laterally. The spores (5 μm × 3 μm) are roughly pyriform. The organism was found in the fat-body of *S. inflatum* larvae in the south of France.

Weiseria sommermanae Jamnback (1970) was described from *Gymnopais* sp. in Alaska. Its oval spore (3.8–6 μm × 3.4–4.8 μm) has a suture at the broadest dimension.

Octosporea simulii Debaisieux (1926) has elongate spores, of tubular appearance, measuring 7.5 μm × 2–3 μm. It was described from Belgium, and parasitizes the gut of the host blackfly larva near the insertion of the Malpighian tubules.

The incidence of Microsporida in blackfly larval populations is generally low, most commonly below 1%, but varying from species to species, and ranging up to 15%. Very little is known about infections in adult blackflies. However, Shipitsina (1963) dissected some of these from the Yenisey River, Siberia. Their microsporidans were not identified, but the rate of infection was highest in *Titanopterix maculata* (22–31%), followed by 2–10% for *S. ornatum*. *Stegopterna reptans galeratum* and *Gnus cholo-*

dkovskii exhibited much lower infections (1–3%). Shipitsina was of the opinion that the infections reduced longevity in females.

When blackfly eggs are brought from the field to the laboratory and used for subsequent rearing purposes, the resultant larvae sometimes exhibit Microsporida. This is presumptive evidence that one mode of transmission among blackflies is the vertical one, via the eggs. The role of the spores in horizontal transmission between simuliids remains unknown, for, despite repeated direct feeding of spores to the various larval instars (in Newfoundland and elsewhere, Undeen), no infections have ever resulted.

Some microsporidan spores have sur-

Table 1. Summary of data on simuliid pathogens.

Pathogen	Stage infected	Pathogenic	Laboratory cultured	Experimentally transmitted	Results
Iridescent virus	larva	+	—	—	
Densonucleosis virus	larva	?	—	—	
Cytoplasmic polyhedrosis virus	larva	+	—	+	
Bacillus thuringiensis (Lepidopteran serotype)	larva	+	+	+	Dosage required impractically high
Bacillus thuringiensis var. *israelensis*	larva	+	+	+	1-minute dosage of 10^5 sp/ml = 100% mortality
Bacteria (unidentified)	larva adult	+	+	—	Found in laboratory-reared adults and field-collected larvae.
Trichomycetes	larva	—	+	+	
Coelomycidium simulii	larva	+	+	+	
Saprolegniaceae	larva	+	+	+	Not known whether these are obligate pathogens of blackfly larvae.
Aspergillus	larva adult	+	—	—	
Entomophthora	adult	+	+	—	
Ovarian phycomycete	adult	+	—	—	Prevents egg production
Microsporida	larva	+	—	—	
Ciliata	adult	?	—	—	
Gregarinida	larva	+	—	—	
Haplosporidium simulii	larva	?	—	—	

face structures which may assist flota-
tion. Such structures may also cause
differential sedimentation or result in
the better filtration of spores from the
stream, by the host. Substantiation of
any of the suggested functions would, of
course, be evidence that horizontal transmission of blackfly microsporidans does,
indeed, take place under appropriate conditions. Until resolution of the enigma
of transmission, the potential of microsporidans as biocontrol agents for use
against Simuliidae cannot be assessed.

Conclusion

Table 1 summarizes data on simuliid
pathogens, including certain parameters
for their future usefulness as biocontrol
agents—pathogenicity, ease of transmission, and potential for mass production.
These three parameters are all favourable
for only two of the pathogens listed,
some *Bacillus thuringiensis* serotypes and
Coelomycidium simulii. The latter, as indicated earlier, has responded well to
preliminary *in vitro* cultivation. However,
this approach has yet to yield infective
stages. Furthermore, experimental transmission data indicate only a low, and
inconstant, percentage of infection.
Although lepidopteran strains of typical
B. thuringiensis kill blackfly larvae, the
dosages required are far too high to offer
any practical prospects. Only in the case
of *B. thuringiensis* var. *israelensis* are all
three parameters fully satisfied. *B.t.i.* is
easily grown on artificial media, is readily
ingested by blackfly larvae, and produces
mortality at adequately low concentrations. Certain other pathogens (e.g. CPV)
are pathogenic and can be transmitted,
but production of enough polyhedra for
testing is not yet feasible. Others (e.g.
Aspergillus spp. which were recorded by
Lewis (1965) from nulliparous *S. damnosum s.l.* in Cameroun) probably lend
themselves to mass production; but,
even where fungi of the genus instanced
to prove highly infective to blackflies in
the field, the risk of infections in vertebrate NTOs would preclude their practical development for obvious safety
reasons.

IV Physiology

Blackfly Physiology

E. W. Cupp

Information concerning blackfly physiology has been restricted because of the notable absence[19] of colonies for continuous, standardized studies. However, where blackflies are easily available from natural populations, several key aspects of morphogenesis and physiology have been investigated. A general framework thus exists for comparison with the physiological–developmental processes of other kinds of biting Diptera. Since large portions of this research have used field-collected material sometimes infected with one or more of a variety of pathogens, the effects of parasitism on biochemical, physiological or behavioural processes have also been noted.

Embryogenesis

Embryogenesis of *Simulium pictipes* is generally similar to that in other Diptera

and does not differ significantly from *Chironomus* sp. (Gambrell 1933). At oviposition, eggs are glued together by a thick adhesive substance. Each egg is covered by a thin, elastic chorion that remains clear and transparent during development of the embryo. A vitelline membrane, which cannot be detected during the initial phases of embryogenesis, surrounds the cytoplasm. Following fusion of the pronuclei, cleavage nuclei appear near the centre of the egg, increase by mitotic division and move to the periphery to form the blastoderm. Pole cells (the primitive germ cells) are formed in the posterior portion of the egg and remain outside the blastoderm until it is fully formed. The germ band is superficial and "extraordinarily long". Soon after it is formed, blastokinesis occurs. Formation of embryonic membranes (amnion, serosa) is more rapid than that reported for *Chironomus*. Gastrulation and organogenesis occur sequentially in a manner typical of nematoceran Diptera. The silk glands, so important later in helping the larva to anchor itself in the lotic environment, are ectodermal derivatives detectable during the second day of embryogenesis.

Duration of embryogeny within the Simuliidae can vary widely, with such environmental factors as temperature, light, and oxygen content interacting to prolong the time required for development to be completed (Ivashchenko 1977) or to prevent eclosion (Kurtak 1974). Also, since the simuliid egg is susceptible to desiccation (Muirhead-Thomson 1957a; Tarshis 1968), survival of the embryo outside the aquatic environment is favoured by substrates having a high relative humidity (Colbo and Moorhouse 1974). The act of hatching is initiated by the swallowing of intrachorionic fluid which causes the embryo to swell (Davis 1971). Expansion of the body results in the penetration of the egg shell by a cephalic hatching spine. Following the rupture of the shell, the larva emerges head-first.

---

[19] Until very recently—see pp. ix, 303 (Ed.).

Larval Nutrition

While little is known about the exact dietary requirements and digestive processes of blackfly larvae, some inferences can be made from general information concerning their method of feeding as well as the type and quantity of materials ingested. Most blackflies are filter feeders, using paired cephalic fans to remove sestonic particles that commonly range in size from 10 to 100 μm in diameter (Chance 1970a). Colloidal particles (Wotton 1976) and bacteria (Fredeen 1964) are also incorporated into the diet, suggesting that oral infection by a variety of pathogens can occur.

The cephalic fans are complex structures, consisting of a variety of sclerites, sensory hairs, and fan subdivisions (Chance 1970b). When completely expanded, the rays of the larger primary fan cover an angle of 200–250°. The number of rays of each fan varies with age and species. The efficiency of feeding is also variable but it has been estimated that blackfly larvae ingest 1–10% (by weight) of the material passing through the cephalic fans (Kurtak 1978). Simuliids that lack these food-gathering organs (e.g. *Twinnia biclavata*) graze the substratum by means of the labrum, maxillae, and mandibles (Chance 1970b). Some filter-feeding species can also browse, and occasionally feed on large algal filaments (Burton 1973). Feeding rates vary with larval age and species, and in relationship to such environmental conditions as water velocity, temperature, and food concentration. Early-instar larvae of *S. vittatum* ingest and retain coarser particles than older instars (Merritt *et al.* 1978). The rate of feeding by this species can also vary, with younger larvae digesting food plugs at a more rapid rate than older ones (Mulla and Lacey 1976). However, older larvae of *S. damnosum* feed slightly faster than younger ones (Elouard and Elsen 1977). The rate of movement of the food plug in this species is dependent upon the absorption rate of new food particles. This, in turn, is a function of current velocity and particle concentration. Generally, replacement of the gut contents of simuliid larvae occurs in 30–60 minutes (Kurtak 1978).

Larval Physiology

The physiological implications of the rapid turnover of food are unknown. These factors may be associated with the types and quantity of enzymes involved in digestion, or the trophic value of ingested material. Nutritional stress due to food deprivation extends development time, disrupts synchrony of development, and reduces the average size of adults; fecundity is also greatly affected (Colbo and Porter 1979; Chutter 1970).

Blackfly larvae occur in freshwater habitats and are probably both ammoniotelic and ureotelic in terms of nitrogen excretion (Gordon and Bailey 1974; 1976). Unfed larvae of *Boophthora* (= *Simulium*) *erythrocephala* and *Wilhelmia lineata* excreted an average total nitrogen output of 0.240 mg N/100 mg wet weight during a 24-h period with ammonia accounting for 76.6–98.6% of the total nitrogen (Grunewald 1978). High levels of ammonia, urea, and arginine, an orthithine cycle intermediate, occur in the haemolymph of *S. venustum*, *S. vittatum* and *Prosimulium mixtum/fuscum* (Gordon and Bailey 1976). Sodium is the principal cation, making up about 60% of the four major cations (Na^+, K^+, Mg^{2+}, and Ca^{2+}) in *S. venustum* and *S. vittatum* and 48% for *P. mixtum/fuscum*. Haemolymph magnesium levels are slightly lower than those in other Diptera, whereas calcium levels are much higher. The osmotic pressure of haemolymph from all three species is slightly lower than those values given for other insects but does resemble the haemolymph osmolarity for larval *Aedes aegypti*.

Larval haemolymph of these three blackfly species also contained high levels of glutamic acid, alanine, proline, glycine, serine, and histidine. Glycerophosphoethanolamine is a major phosphatide. The presence of these important biochemicals in the haemolymph, as well as its general physico-chemical properties, make it a highly suitable environment for parasitic mermithids (Gordon and Bailey 1974).

The larvae of some temperate blackfly species are capable of entering a state of dormancy. This condition, termed oligopause by Mansingh et al. (1972), is characterized by the ability to make rapid biochemical–metabolic adjustments to changing water temperatures (Mansingh and Steele 1973). At temperatures below 4 °C, larvae of Prosimulium mysticum are in a state of hibernation, showing little inclination to ingest food or any evidence of growth. Hibernating larvae also sequester high amounts of polyhydric alcohols (glycerol, sorbitol, mannitol) that contribute to the lowering of supercooling points of insects. Increases in water temperatures significantly elevate O_2 consumption and decrease the levels of polyhydric alcohols, with a concomitant rise in trehalose levels. Sensory adaptations parallel changes in temperatures, with a high degree of locomotion and sensitivity to touch associated with increased temperatures. Since relatively high acetylcholinesterase activity has been detected in other overwintering species of blackflies (Odintsov et al. 1970), the ability to respond to tactile stimulation is not surprising.

Developing larvae can re-enter dormancy when returned to 4 °C, with lowering of respiration to a hibernating level occurring within a few hours. These physiological mechanisms are thus significant in the maintenance of certain parasitoses since overwintering larvae are also able to serve as reservoirs for Microsporida, including Pleistophora simulii, P. debaisieuxi, Thelohania varians, T. fibrata and T. bracteata (Lichovoz 1975), and mermithid nematodes (Ebsary and Bennett 1975).

Larval Physiology as Affected by Parasitism

The alteration of certain features of larval physiology by mermithid parasitism has been noted in Prosimulium mixtum/fuscum and Simulium venustum (Condon and Gordon 1977). When parasitized by Mesomermis flumenalis, S. venustum larvae show unique histopathological changes in the endocrine system, including an apparent enhancement of protein synthesis within the corpus allatum as indicated by an increase in the nuclear DNA/RNA activity of that gland. The corpus cardiacum also increases significantly in volume and retains neurosecretory materials. No effects were seen in the endocrine system of P. mixtum/fuscum parasitized by this mermithid.

In both blackfly species, the neurosecretory system is composed of neurosecretory cells located within the brain (Condon et al. 1976), resembling the brain neurosecretory cells in mosquito larvae. The corpus allatum is a single structure, lacking a dividing membrane. A similar observation was made for the structure of the corpus allatum in Simulium ornatum (Cazal and Maurand 1966), thus suggesting that the simuliid endocrine system is intermediate between the typical nematoceran system and that reported for Brachycera as exemplified by the Tabanidae.

Reduction in fat-body glycogen in parasitized S. venustum larvae is drastic but M. flumenalis does not induce accelerated glycogenolysis. Fat body nutrients are also rapidly depleted in P. mixtum/fuscum. Based on the comparative effects of parasitism on both the endocrine system and the fat body in S. venustum and P. mixtum/fuscum, Condon and Gordon (1977) hypothesized that mermithid parasitism in blackfly larvae prevents further morphogenesis (i.e. pupation, histoblast development) by depletion of metabolites and possible inhibition of fat-body protein synthesis rather than an alteration in basic endocrinology.

Larvae of *S. bezzii* when infected with *Thelohania fibrata* and *Pleistophora debaisieuxi* also show alterations in respiration as evidenced by changes in mean oxygen consumption (Boemare and Maurand 1976). The nuclear envelope of larval fat body cells of *S. vittatum* infected with *T. bracteata* shows changes in both the size and distribution of membrane particles (Liu 1972). The average diameter of the nuclear pore is significantly larger (83.76 nm v. 133.67 nm in infected cells), with a reduction in the number of pores per square micrometre (Liu and Davies 1972a). The biochemical implications of this physical alteration suggest an impairment of RNA synthesis. The cytoplasm is also noticeably affected, with a reduction in the size of mitochondria and a general scarcity of endoplasmic reticulum (Liu and Davies 1972b).

Pupal Nutrition and Cocoon Formation

The pupal stage of the family Simuliidae is quite unique among the Diptera in that it is an active stage capable of feeding for several days before spinning its own cocoon (Hinton 1971). The feeding phase can last up to four or more days after larval–pupal apolysis. The ability of the pharate pupa to feed is due to the fact that when the epidermis retracts from the larval cuticle, many of the muscle connections remain attached to the mouthparts and prolegs (Hinton 1968). After feeding and defaecating, the pharate pupa spins its own cocoon. When spinning is initiated, much of the pupal cuticle is already heavily sclerotized.

Before completion of the cocoon, the plastron of the spiracular gills becomes filled with air (Hinton 1968). A layer of air is held between the vertical struts occur-

ring within the branches of the spiracular gill (Hinton 1976). It has been suggested that variation in resistance to wetting of the gills is, in part, an eco-physiological reason for the distribution of species in such differing habitats as large lowland rivers as opposed to small upland streams.

Adult Emergence

The *Simulium damnosum* complex and several other African species appear to be capable of regulating the time of pupation so that adult emergence can occur during the day (Disney 1968). *Simulium kenyae*, *S. unicornutum*, and *S. damnosum s.l.* pupate mainly in the day, with distinct peaks occurring according to species. As a result of diurnal pupation and a presumed ability to prolong the pharate adult phase of the pupal stage, adult eclosion usually occurs during the day. The timing of adult emergence is influenced by water temperature on the day of eclosion. A similar diurnal pattern has been recorded for *S. ornatipes* (Hunter 1977a). However, the fact that eclosion in the laboratory becomes arhythmic when constant white, purple-pink, and long UV light are used, suggests that adult emergence of this species is strongly influenced by a light-mediated circadian rhythm.

The method of adult emergence in many species of the Simuliidae is noteworthy. Transition from an aquatic habitat to an aerial one by the imago is made by means of an air bubble, which transports the adult to the water's surface. The adult moves immediately to a resting site where cuticular hardening is completed (Peterson 1978).

Mating

Blackflies typically form mating swarms, though for some species, mating occurs when females are encountered while crawling over objects that protrude from the larval site (Downes 1958). Unlike mosquitoes, where mate-detection by males is generally dependent upon reception of auditory stimuli, the antennae of adult blackflies are similar in both sexes (Downes 1969). The holoptic eye of the male is distinctly divided dorsoventrally, with the dorsal aspect specialized for detecting small, rapidly-moving objects in a swarm (Wenk 1965; Kirschfeld and Wenk 1976). The corneal facets in the dorsal eye are considerably larger than in the ventral eye; the screening pigment also differs, being translucent for light of longer wavelengths and light brown in colour as opposed to the dark red-brown pigment in the ventral eye. The retinular cells and their rhabdomeres are extremely elongate (up to 300 μm) in the dorsal eye and penetrate the basement membrane to extend into the ventral region of the head. The retinular cells in the central eye are similar to other Diptera. The entire female eye is also anatomically similar to the ventral eye of the male. In *Cnephia dacotensis*, a species that mates on rocks or plants without forming a swarm, the male eye is reduced so that there are no sharp dorsoventral divisions, suggesting that loss of ocular specialization has occurred as a result of change in the mating location (Downes 1958).

Sperm transfer in many species of Simuliidae takes place by means of a spermatophore (Davies 1965; Wenk 1965). This structure, which can be seen in recently inseminated nulliparous adults, consists initially of a viscous mass which becomes hollowed out to surround the bundle of sperm. The spermatophore wall is subdivided into an inner hyaline layer and a spongy outer layer. While the chemical nature of these laminae is not known, it can be surmised from histological evidence that the male accessory glands produce the granular secretory materials involved in spermatophore formation (Raminai and Cupp 1978).

Adult Nutrition

Nectar feeding by both male and female blackflies appears to be a significant feature in adult biology. The contents of plant nectars and juices apparently satisfy energy requirements for maintenance of mating swarms (Wenk 1965a), and contribute to female longevity (Hunter 1977b)—including that of parous adults (Watanabe 1977). Sugars identified from the diverticulum of blackfly populations having a parity rate of 36% or higher indicated the presence of as many as six kinds (fructose, glucose, sucrose, maltose, melibiose, and raffinose). The frequent use of nectar by both parous and nulliparous *Simulium ochraceum* in Guatemala may be an important factor in host seeking, thus increasing the vector potential of this species for transmission of *Onchocerca volvulus* (Cupp and Collins 1979). Orientation of blackflies to nectar sources appears to rely on olfactory stimulation (Wenk 1965a).

Many species of blackflies are strong fliers and are capable of covering relatively long distances. The occurrence of glycogen particles in the interfibrillar sarcoplasm of the flight muscles (Liu and Davies 1971) as well as glutamate oxaloacetic transaminase in flight-muscle mitochondria suggests that both carbohydrates and amino acids may function as energy reserves for muscle contraction (Liu and Davies 1972c). The presence of this transaminase indicates that the tricarboxylic acid cycle thus occurs in blackflies as in other insects for production of high energy bonds.

The general orientation of females questing for a bloodmeal consists of a series of varied stimuli emanating from the host, including colour and silhouette profile, CO_2, and odour. Bradbury and Bennett (1974a, b) suggested that three hierarchical zones (long, middle and close range) of orientation mechanisms can be described. These zones correspond to various sensory inputs. Long-range orientation is strongly dependent upon host-specific odours; it is probable that these olfactory stimuli initiate host seeking at distances further than that known for CO_2. The most striking example of this is the orientation of *S. euryadminiculum* to the ether extract of the uropygial glands of *Gavia immer* Brünnich, the common loon (Fallis and Smith 1964; Bennett *et al.* 1972). The "forest" form of *Simulium damnosum s.l.* also relies heavily on human sweat as an olfactory stimulus (Thompson 1976c, 1976d). However, the "savanna" form of *S. damnosum s.l.* is not particularly attracted to odour or exhaled breath, but relies primarily on sight for host orientation.

Middle-range orientation is initiated by a CO_2 stimulus that also serves as the apparent principal force in directed orientation. This may also be significant in host finding by nocturnally active blackflies (Fallis and Smith 1964; Raastad and Mehl 1972), though vision can be involved to some degree in middle-range orientation as demonstrated for *S. adersi* and *S. impukane* (Fallis and Raybould 1975). Close-range orientation behaviour is associated with responses to such visual stimuli as shape, colour, and size of the host (Bennett *et al.* 1972). Movement of the host or its parts may also be important (Wenk and Schlörer 1963).

In a comparison of the numbers and types of sensilla associated with the antennae, palps and legs of mammalophilic, ornithophilic and autogenous blackflies, differences between sensilla associated with olfaction, CO_2 detection and contact chemoreception were seen (Mercer and McIver 1973a; Mercer and McIver 1973b; Sutcliffe and McIver 1976). A large number of probable olfactory receptors (sensilla types A1, A2, A3) were found on the antennae of anautogenous species. *Simulium baffinense*, an autogenous form, had significantly fewer olfactory receptors with the exception of A2 sensilla. *S. rugglesi* females had more probable olfactory receptors than did the males, a situation similar to that reported

in anautogenous mosquitoes. Presumed CO_2 and/or olfactory receptors were found in sensory pits of maxillary palpal segment 3 in *S. venustum, S. euryadminiculum, S. rugglesi*, and *S. baffinense* (Mercer and McIver 1973b). These receptors appeared as bulb-shaped, thin-walled sensilla. The ornithophilic species (*s. euryadminiculum, S. rugglesi*) had more bulb organs than *S. venustum*, a mammalophilic blackfly. However, both types of blood-feeders had more of these sensilla than *S. baffinense*, an autogenous species.

Once a host has been selected, blood-feeding by the female may ensue. Peg sensilla located ventrally on the tarsomeres may be involved in contact chemoreception and in initiating probing behaviour (Sutcliffe and McIver 1976). Heat is also an essential factor in stimulating probing behaviour, with both adenosine triphosphate and diphosphate serving as gorging stimulants (Sutcliffe and McIver 1975).

Ingestion of blood is accompanied by the inoculation into the wound of salivary secretions containing several low molecular weight proteins (Poehling *et al.* 1976). One or more of these substances may promote agglutinin activity or serve as an anticoagulant (Yang and Davies 1974). During blood-feeding, 50–70% of the salivary secretions can be injected (Poehling *et al.* 1976; Gosbee *et al.* 1969); the time required for recharging of the glands varies. The structure of the paired salivary glands suggests that the tube-shaped main gland of the female is specialized for blood-feeding (Poehling 1977), with the accessory portion of the gland similar in shape, histology and histochemistry to the male glands (Wachtler *et al.* 1971).

A peritrophic membrane forms in the posterior midgut within 30 minutes of ingestion of avian or mammalian blood (Yang and Davies 1977). The membrane consists of up to seven concentric laminae within 12 hours of formation. Blood digestion begins at the periphery of the blood bolus, particularly at the posterior end. Trypsin, believed to be secreted by the midgut cells, is primarily responsible for blood digestion (Yang and Davies 1968a). However, blood-feeding also stimulates invertase activity, which doubles immediately after a blood meal (Yang and Davies 1968b). Amylase does not appear to be involved (Yang and Davies 1968c). In several species of blackflies parasitized with Microsporida, the digestive process may be impaired so that blood residues remain in the midgut up to 200 h after ingestion (Yang and Davies 1977).

Oogenesis

The physiological patterns of ovarian development in blackflies can vary from strict autogeny to anautogeny (Downes 1971; Davies *et al.* 1977; Davies 1978). In some cases, complete egg development can occur *without* a blood meal, perhaps as an adaptation to severe weather zones (Downes 1965). If an adequate diet is available, these species sequester lipid and protein substrates during the larval stage for oogenesis in the pupal or imaginal stage. Such blackflies usually emerge with the ovaries in an advanced stage of development, mate without swarm formation, and oviposit soon after insemination. The mouthparts are reduced, and each individual completes a single gonotrophic cycle. In other specialized, parthenogenetic species, the pharate adult may remain inside the pupal case; so that eggs are liberated after the dead insect disintegrates (Carlsson 1962). A blood meal may be required after completion of the primary gonotrophic cycle in some blackflies (Davies 1961; Davies 1978), or for the initiation of the first gonotrophic cycle in most temperate and tropical blackflies (Davies *et al.* 1977).

Little is known about the endocrinological–biochemical events associated with oogenesis in anautogenous blackflies. However, several significant qualitative features of oocyte maturation in *S. vittatum*, a common autogenous species, have been described. Fat-body glycogen apparently provides the major energy source for cellular metabolism during key synthetic events associated with oogenesis. As ovarian growth proceeds, the quantity of this stored carbohydrate decreases (Liu and Davies 1972d). Glycoprotein is also presumably synthesized in the fat-body cells as a source of carbohydrate containing yolk. Lipid granules in the oocyte are similar ultrastructurally to phospholipid inclusions observed in yeast cells (Liu and Davies 1972e). Oocyte lipid inclusions differ, however, from those in the fat-body in that the latter are lamellated (Liu and Davies 1972f). Four types of lipid inclusions can be seen in the oocyte by freeze-etching, with a large percentage suggested to be lecithin (Liu 1974). During lipid deposition in the oocyte, mitochondrial divisions occur, perhaps directly contributing to the incorporation of these materials or aiding in energy synthesis for brush border formation (Liu and Davies 1973). Yolk protein granules in the oocyte are crystalline particles bound as long cylindrical filaments possessing a single limiting membrane (Liu 1973).

Differentiation of follicular cells during oocyte maturation in *S. vittatum* involves changes in the dispersal of chromatin material from its central location in the nucleolus to a wider pattern of distribution in the nucleoplasm (Liu *et al.* 1975). Cytoplasmic changes include an increase of rough endoplasmic reticulum as vitellogenesis proceeds, and more frequent occurrence of Golgi vesicles and dense droplets, particularly towards the interfaces of the follicular cell–oocyte surfaces. Following the completion of vitellogenesis, a previtelline membrane substance is deposited between the follicular cells and the oocyte.

Oviposition

Oviposition behaviour appears to be cued by changes in light intensity and ordinarily takes place at sunset for several Nearctic species (Davies 1952). Egg deposition by other species may also occur on cloudy days (Hunter 1977a). Oviposition flight activity for *Boophthora* (= *Simulium*) *erythrocephala* and *S. sublacustre* is closely related to changing light conditions, with a reduction of intensity of at least 1.9%/minute required for the initiation of this behaviour pattern (Reuter and Rühm 1976). Oviposition behaviour may also show a diel periodicity, with some egg deposition occurring at sunrise (Corbett 1967). In blackflies parasitized by both mermithids and fungi, upstream movement by females and subsequent oviposition is not altered so that maintenance of both types of pathogens is insured in a lotic habitat (Wenk 1976; Undeen and Nolan 1977).

The selection of an oviposition site is due largely to colour attractiveness, with many subarctic species preferentially ovipositing on green and yellow objects as opposed to those coloured purple, blue, orange, or red (Peschken and Thorsteinson 1965; Golini and Davies 1975a; Golini and Davies 1975b).

V Ecology

Preimaginal Blackfly Bionomics

M. H. Colbo and R. S. Wotton

The purpose of this chapter is to outline the bionomics of simuliids during their aquatic phase by the synthesis of ideas advanced in, or deduced from, the literature. These ideas include some derived from data concerning other stream insects. It is submitted that the bionomics of preimaginal blackflies cannot be understood in isolation—relevant biotic and abiotic factors must be comprehended for full elucidation of the role simuliids play in stream ecosystems (see pp. 227–235). A broad-spectrum approach is thus prerequisite to development of biological and integrated control programmes along rational pathways.

The Fluvial Ecosystem

Two recent general reviews of the physical nature of fluvial systems (Curry 1972; Beaumont 1975) present the current methods of characterizing their dynamics. Hydrologists view fluvial systems as comprising the whole basin—not just the channel, for much of the silt and chemical loading originate from areas beyond the defined channels. Streams are dynamic systems that are proceeding towards a steady state, balancing input and output all along the channel, although equilibrium is rarely, if ever, achieved. Thus, probabilities and means are dealt with (rather than absolutes), in characterizing the systems. Radical perturbations from the "normal" situations may occur in consequence of climatic, geological and (particularly in recent decades), man-made effects.

The flow along a water course is primarily controlled by two forces: gravity and friction. Equilibrium is thus achieved by the balancing of these forces along the river, which is reflected in the geomorphology of the system. The rate of geomorphic change will depend on the nature of flow, of the substrates, and on the passage of time. From the biological standpoint this means that although statistical estimates will indicate the number of potential larval habitats (i.e. rapids, bends) along a given channel system, stability within the habitat cannot be assumed. Each of these geomorphic features is in a constant state of change; reflecting the discharge pattern of the system and the load of sediment in the stream. Therefore, the location of different habitats will be ephemeral in geological time, even though the fluvial ecosystem may have already existed in a similar form for thousands of years. This is particularly true in zones of erosion where the water velocity is highest, resulting in constant abrasion of exposed substrates. Organisms occupying such zones must thus adapt to a constantly changing environment. Hynes (1970a, b) authoritatively reviewed the ecology of running waters, outlining the various ways in which organisms have adapted to the fluvial ecosystem. The family Simuliidae is almost totally limited to this system and has thus developed many morphological and life-history adaptations, enabling it to compete successfully in specific microhabitats characteristic of most streams.

Eggs, Embryonation, and Hatching

The eggs of simuliids are generally oval to subtriangular in lateral view. They are deposited in masses or chains on substrates (in zones at or just below the water surface) or actually into the water, whereupon, they sink to the bottom (Davies and Peterson 1956; Ussova 1964; Abdelnur 1968; Rühm 1971a, 1975). All eggs are coated with a sticky, gelatinous outer layer readily adhering to any substrate with which it comes in contact (Gambrell 1933), thus resisting the tendency to be swept downstream. Laying of eggs directly onto a substrate has at least two potential advantages. Firstly, eggs on the surface of masses are exposed to well-oxygenated water, facilitating rapid development. Second, eggs can be deposited immediately upstream of favourable sites for development of immature stages. In species known to recolonize temporary streams, such as some members of the S. damnosum complex (Le Berre 1966), S. ornatipes Skuse (Colbo 1974) and S. vittatum Zetterstedt (Fredeen and Shemanchuk 1960), oviposition occurs on the surface of certain substrates.

On present knowledge, eggs laid on substrates near the surface do not undergo a diapause, though hatching may be delayed where batches are several layers thick—this is presumably due to the O_2 deprivation deep within the egg mass (Ivaschenko 1977). Eggs of this type can be stored in the laboratory by lowering the temperature (Fredeen 1959 and routine practice at RUVP, Colbo, unpublished). Some of these species, such as the S. verecundum complex, over-winter in the egg phase. Late in the summer and fall, though, egg masses cannot be found despite the presence of adults, suggesting that there are two types of eggs and oviposition patterns. This is a reasonable assumption, for blackfly eggs have never

been shown to withstand desiccation[20] (Tarshis 1968; Colbo and Moorhouse 1974); although, as water levels frequently fluctuate, the probability of eggs on vegetation being intermittently exposed to drying appears high. It is believed that this explains why eggs, attached to substrates near the surface, characteristically hatch rapidly. There are, of course, several species that deposit eggs on substrates well below the water surface (Ussova 1961) thus avoiding desiccation.

The simuliid characteristic—lack of resistance to egg desiccation—can be interpreted as evidence that the original (and perhaps still the most widespread) mode of oviposition in the family is via the dropping into the water of eggs which then sink to the bottom. In such cases, they will only be exposed if the stream ceases to flow and its bed completely dries out. This is extremely unlikely in most river systems harbouring simuliids. In northeastern Australia, Colbo and Moorhouse (1974) have shown how Austrosimulium pestilens Mackerras and Mackerras has adapted its life-history to avoid desiccation by synchronizing oviposition onto the water surface with the receding phase of a flood in alluvial streams. This results in eggs being buried deep in the bottom deposits. Here they become fully embryonated, remaining quiescent for at least a few years, and only hatching when erosion exposes them in the next flood (Colbo and Moorhouse 1974; Hunter 1978). In this way the species is able to inhabit river systems which may not flow for several years, and in which the surface of the bed becomes completely dry.

On the Canadian prairies, Fredeen et al. (1951) have shown that S. arcticum Malloch oviposits over the water, its eggs also becoming buried in the bottom deposits where they survive at least until

---

[20] Though this might be implied by a statement in Hynes (1970a).

late in the following summer. Recently, Ross and Merritt (1978) have discovered that *Prosimulium mixtum/fuscum* may have a second hatch in mid-winter following disturbance of the stream-bed by flooding. The account of development and hatching of *C. pecuarum* (Riley) by Bradley (1935b) would suggest that this species, too, may exhibit similar behaviour, which may well prove to extend to other blackflies as well. It is thus apparent that the possibility of long-term egg survival must be considered in any control programme.

The ability of the first instar larva within the egg to withstand freezing, has been reported by Ussova (1961). She found certain species capable of surviving temperatures of at least several degrees below 0 °C. Smart (1934) and Kurtak (1974) also reported that eggs of *S. pictipes* (Hagen) hatched after being encased in ice. One of us (MHC) has observed eggs in storage which had quickly frozen by accident when the temperature in an incubator plunged to −15 °C. *Stegopterna mutata* (Malloch) eggs, which were already embryonated, survived and hatched. So did some partially embryonated eggs of the *S. venustum* complex. However, eggs of the *S. verecundum* complex and *S. vittatum*, when deliberately placed in a deep freeze at −15 °C, did not hatch.

The embryology of simuliids has been studied for well over a century (for review see Gambrell 1933; Craig 1969), though there is very little detailed information on factors controlling development. Temperature obviously affects the rate of development. It has also been shown, or inferred, that there are threshold temperatures below, or above, which development does not occur (Fredeen 1959; Davies and Syme 1958; Colbo, unpublished). Ivashchenko (1977) has demonstrated the effect of O_2 and light on the rate of embryonic development in eggs of *S. ornatum* Meigen and *S. latipes* Meigen. However, he was dealing with eggs which immediately began develop-

ment leading to hatching without a quiescent break. Probably, O_2 tension affects the rate of development of most other simuliids as well. Photoperiod may, in some cases at least, trigger the cessation or commencement of development and, perhaps, hatching. However, controlled experimentation is required to establish this with certainty.

Possibly, life-history strategies (as far as the portion spent within the egg shell is concerned) can be postulated from the recorded life-histories of simuliids. The first strategy is where the egg is simply the end product of the reproductive phase, serving no function as a resistant or resting stage. Once laid, development and hatching follow without delay. The rate is of course controlled by environmental factors which affect normal physiological functions. The species concerned all depend on continuous production for survival; living, as they do, in situations always favourable for larval development (with the proviso that development may be prolonged in cold climates during the winter, and that the species may disperse from permanent breeding sites). At least in higher latitudes, numerous species have a prolonged resting phase, under the control of seasonal changes, during the egg stage. Certain "winter" species survive inside the egg from late spring or early summer through to autumn—but this does not imply that the embryo is at the same stage in each species. Other blackflies have only a single spring or early-summer generation, passing the warm summer—and all, or part, of the cold winter-period—within the egg. Yet another variant involves those species which have repeated summer generations and a resting phase within the egg over the winter-period. It is also known that in warm climates some simuliids can survive for long periods within the egg shell, and hatch when exposed during increased flows (Colbo and Moorhouse 1974; Hunter 1978). Therefore it is not always photoperiod and temperature changes that control the duration of the dormant period.

The process of hatching has not been well-studied. The most recent and complete paper is that of Davis (1971), describing the process in a member of the *Simulium venustum* complex in Newfoundland. He refers to previous studies, but overlooks earlier observations on the same blackfly or at least a closely related member of the same complex made by Hocking and Pickering (1954).

Larval Morphology and Instar Number

There is an extensive literature on aspects of larval morphology. Over 135 years ago, Verdat (1822) and Planchon (1844) gave a general description of larvae, pupae and the attachment organs. Although their description of mouthparts was superficial, they recognized their adaptation to filter feeding. Our first detailed description of the complete larval morphology is that of Puri (1925). The next major contribution (Grenier 1949) includes a detailed anatomical account of the preimaginal stages, particularly at the histological and functional levels, with attention to adaptive variations and a review of the relevant literature. More recently, anatomical studies have primarily concerned certain morphological aspects, such as the larval head structures as they relate to feeding (Chance 1970b; Couvert 1970; Craig 1974, 1977b; Davies 1974). The endocrine system has also been explored in studies concerning parasitism (Condon *et al.* 1976), and the nervous system of larvae (with and without cephalic fans) has been studied by Gelbič and Knoz (1972). More detailed investigations of certain aspects of external morphology have been undertaken in relation to taxonomy (e.g. Davies 1974; Gouteux 1977), for instar differentiation (Crosby 1974; Craig 1974; Ruhm and

Sandars 1975), and with respect to pupal respiration (Hinton 1976).

The number of larval instars in the family Simuliidae is not fixed (i.e. in the family Culicidae, where there are always four), varying from six to nine depending on the species (e.g. Pták and Knoz 1971; Crosby 1974; Craig 1975; Rühm and Sandars 1975; Fredeen 1976; Mokry 1976a; Elouard 1978). The problem becomes yet more complex now that Ross and Merritt (1978) have suggested that, for *P. mixtum/fuscum*, the instar number can be either six or seven depending on the duration of larval development. This is, unexpected, for Rühm and Sandars (1975)—who examined S. [*Boophthora*] *erythrocephalum* (de Geer) populations developing at different rates (spring and summer generations)—reported that each had seven instars; the summer generation simply growing faster and being of smaller size than the winter one.

Some species have proved altogether inseparable into instars by the use of morphological criteria. For example, *Cnephia dacotensis* (Dyar and Shannon) exhibits a thoroughly confusing overlap of measurements after the third instar (Ross and Merritt 1978). One of us (MHC) also found that instars of the closely-related *C. ornithophilia* Davies *et al.* could not be recognized from measurements of apotome width. This differentiated only the first four instars in field-collected material. In the laboratory when third and fourth instar larvae, brought back from the field, were individually reared to adults, they passed through five and four additional instars, respectively, making a total of eight for each group. In the laboratory rearings, it was discovered that an individual did not grow evenly between moults (i.e. the increase of apotome varied from 10 to 30% between instars for the same larva). It is obvious that, with this degree of plasticity, instar determination—whether by using Brooks' (Dyar's) Rule or the modification of Crosby (Craig 1975)—will not be possible.

It is thus clear that extrapolation of the number of instars from one blackfly species to another is meaningless. However, for the majority so far examined, it appears possible to separate the instars using methods published by Crosby (1974), Craig (1975), and Rühm and Sandars (1975). However, although the number of instars appears to be constant in most species, the exact parameters to separate them may not be fixed.

It is well-established that simuliid size will vary seasonally (Edwards 1920; Neveu 1973a; Rühm and Sandars 1975), perhaps in response to changes in food supply (Colbo and Porter 1979) and temperature (Colbo, unpublished). This means that each generation, or cohort, will have to be investigated to determine the parameter most suitable for distinguishing its instars. Despite these difficulties, it is necessary to determine the number of instars in order to conduct quantitative biological and chemical studies. Accurate determination is also valuable in testing the efficacy of biological and chemical agents under consideration for potential control use (e.g. Molloy and Jamnback 1975; Bailey and Gordon 1977; Thompson and Adams 1979).

Feeding

Larval feeding is a very important consideration in most blackfly suppression programmes, be they chemical, biological or other potential components of future integrated methodologies—many control agents acting only after ingestion. Two modes of feeding are recognized in blackfly larvae—browsing over substrates, and filter-feeding. Over 135 years ago, Planchon (1844) described the latter as the method of feeding. A few species feed only by grazing well-known examples being in the genera *Twinnia*, *Gymnopais*; and also, the monotypic *Crozetia crozetensis* (Womersley). The lat-

ter species maintains what is assumed to be the more primitive habit; for its home (the Crozet Is, southern Indian Ocean), is believed to have been colonized by long-distance dispersal, the ancestral stock probably finding no filter-feeding competitors upon arrival. Davies (1974) presents an evolutionary sequence for the development of filtering fans from the raking ones used in browsing. The former, being an "improvement" on the latter, must certainly have furthered the predominance of filter-feeding species which, today, indeed, far outnumber all other Simuliidae.

For blackfly larvae to have filtering fans does not of course preclude their grazing at times (e.g. Craig 1977b). Characteristically filter-feeding species may, on occasion browse over the substrate. On these occasions their food then consists of epilithic, and epiphytic organisms which may even include invertebrates such as larval chironomids (Serra-Tosio 1967). In West Africa, simuliid larvae have even been reported actively feeding on algal filaments, *Oedogonium* sp. (Burton 1973). Chance (1970b) suggests that browsing, being mainly confined to the area around the point of attachment, may therefore not be a feeding method *per se* but one which allows larvae to maintain a clear surface on which to attach silk should they require to move. This review will, however, be confined to filter-feeding.

The cephalic fans are complex, sclerotized structures consisting of rays arranged in two main series. The primary rays are very much the largest. Aborally, they have microtrichia, the size and arrangement of which varies from species to species. Within the primary fans (i.e. between the primary rays and the fan base), are small secondary fans having rays with a plumose arrangement of microtrichia ventro-laterally. The structure and functional morphology of the cephalic fans will not be considered further here—those particularly interested in this topic are referred to the meticulous research of Chance (1970b), Craig (1974, 1977b) and Davies (1974).

Craig (1977b) has provided fresh insight into the structure and function of these fans. His observations, and others in the literature, provide an understanding of how they work as food collectors. Larvae being (for the most part) found in fast-flowing water, an adequate current velocity is required to bring food particles to the head fans. It is also evident that the positioning of the body during feeding is important. These two factors are linked, the larval body moving with changes in the direction of the current, as is very noticeable among feeding larvae under observation in nature. Larvae attach to the substrate on "pads" of silk that they secrete, anchorage being maintained by rows of small hooks at the rear of the abdomen. These hooks ensure a firm hold. The larval body, trailing downstream from the attachment site, is turned through 90–180° (usually approaching the latter figure) so that the ventral surface of the head is oriented upwards, and the fans held into the current. In addition to characteristic "flicking" movements (Chance 1970b) the fans are contracted (by folding) from time to time; entrapped particles then being transferred to the cibarium by means of mandibular brushes (Craig 1977b). The fans are folded (flexed) by muscle action, resulting in distortion of the cuticle. It is the elastic action of the distorted cuticle that unfolds the fan (Davies 1974). Feeding shows no diel periodicity (Kureck 1969) although larvae may cease feeding for brief periods.

When the larva is in the feeding position, trailing in the current, its posterior end is, of course, upstream. The fact that the broadest part of the streamlined body is approximately one-third of the overall length from the anterior end, presumably encourages the water to flow more closely along the larval surface and on to the cephalic fans. The proleg, situated ventrally to the head capsule, then serves to "split" the flow of water; which will pass to left and right and sweep over the aboral surface of the fans, as well as through the spaces between the primary fan rays. The microtrichia on the fan rays are directed at an angle into the collecting basket formed by the head fan. However, even the delicacy of these structures cannot account for the retention of the fine particles captured by feeding blackfly larvae. Ross and Craig (1980) have now determined the manner of retention of these fine particles. They have shown that a substance secreted by two cibarial glands is expelled onto the fans in the cibarium, thence being spread over the fans by the mandibular rakes. This sticky secretion enables the simuliids to capture, and retain fine particles.

Several investigators have measured the size of particles ingested by larvae. There is general agreement that the maximum diameter of these particles will be equivalent to the diameter of the cibarium (while recognizing that particles of greater diameter can be ingested if flexible and if they can be "packed" into the gut). Chance (1970b) recorded the size range of such particles as 0.5–$300\,\mu m$ in length and 0.5–$120\,\mu m$ in width, most measuring 20–$100\,\mu m$ in length and 10–60 μm in width. In the gut of larvae from streams in North Wales, Williams et al. (1961) found the majority of particles to be in the range 9.5–$17.0\,\mu m$ dia. Blackfly larvae are capable of ingesting large quantities of small particles, Fredeen (1964) having demonstrated that they can fill their gut with bacteria (c. $1\,\mu m$ dia.). Clearly, therefore, larval filtering capacity is considerable. Indeed, Wotton (1976) has shown that larvae of at least one simuliid species are capable of ingesting particles of colloidal size (0.091 μm). In further experiments (Wotton 1977), larvae were shown to take particles from the water in the proportions in which they were present, the range offered being 0.7–$35.0\,\mu m$. Kurtak (1978) found that only 1–10% of particles present in a suspension were filtered by larvae. It thus appears that larvae are qualitatively efficient at filtering, but not quantitatively so.

As there is an abundance of small particles (<2 μm dia.) in most flowing waters, particles of such size are probably abundant in the gut of blackfly larvae. Examination of the gut contents of larvae from a moorland stream (Wotton 1977) and a lake-outlet (Wotton 1978a) show that at least 70% of all particles are of this diameter. How many of these small particles are ingested as aggregates, or on the surface of other particles remains to be determined.

Feeding-rate measurements, in the field, have shown that blackfly larvae have a gut retention time of 20 minutes to c. 2 h (Ladle et al. 1972; Mulla and Lacey 1976; Wotton 1978a,b). The value varies with flow-rate, numerical density of particles in the water, and probably with temperature, too. Mulla and Lacey (1976) have shown temperature to have a marked effect on feeding rate, a finding that contrasts with those of Ladle et al (1972) who reported little change in feeding-rate at water temperatures between 8–21 °C. Another factor, perhaps of some significance, is the relative abundance of small particles, which may be important in contributing to the retention time of c. 2 h reported by Wotton (1978a) for the dominant species inhabiting a lake outlet in Swedish Lapland. The presence of mermithid nematodes within the body of larvae at this site served to increase the retention time yet further. It should also be noted that differences in rate of passage have even been observed between blackfly species inhabiting the same site (Elsen et al. 1978).

Small larvae have a shorter gut retention time than large ones, yet maintain a similar assimilation efficiency in a lake outlet in Finland (Wotton 1978b). This may well result from the differing sizes of the cylindrical mid-gut, characterizing the life-history stages sampled. The surface area:volume ratio will be higher in small, than in large larvae, and digestive enzymes will have a greater relative sur-

face on which to act should digestion principally occur adjacent to the peritrophic membrane. Large larvae require a greater retention time to achieve the same assimilation efficiency but lower respiratory demands allow more of the assimilated energy to be used in growth (Wotton 1978b). The large particles (of c. 100–150 μm dia.) filtered preferentially by small larvae (Merritt et al. 1978) (and optimally filtered by all larvae, Kurtak 1978) may be acting as "pistons" to push material through the gut. Such a process would seem feasible if very fine particles are the important ones in nutrition. This will result in low assimilation efficiency and short gut retention times. It also serves to explain why small larvae have a faster feeding rate, especially after large quantities of particles (of up to 20 μm dia.) are administered to larvae feeding in the field (Wotton 1978b).

There is still much to be learned about which of the ingested materials are assimilated. This probably varies with both species and habitat. Maciolek and Tunzi (1968) demonstrated that blackfly larvae in a Sierra Nevada stream utilized diatoms. Moore (1977), working on the larvae of streams in Canada, pointed out that most algae are little-affected by passage through the gut, and submitted that they do not contribute significantly to the larval nutrition. Larvae have been grown to pupation in the laboratory on a purely bacterial diet (Fredeen 1960, 1964), and microorganisms may play an essential part in the nutrition of larvae (especially if the bacteria in detrital coatings are considered, as well as free-living forms). While Baker and Bradnam (1976) concluded that bacteria did not play a major role in the nutrition of larvae in a chalk stream, their conclusions were based on calculations where assimilation efficiency had to be estimated. Carlsson et al. (1977) postulated that it was probably detritus particles of less than 2 μm which were important in sustaining growth of blackfly larvae in a lake outlet in Sweden. They were unable to find any large reduction in the population density of algae downstream from the larval assemblages, as had been claimed by Maciolek and Tunzi (1968). Wotton (1979) has subsequently postulated that coprophagy may be an important feature of larval feeding. It is hoped that studies using radioactive tracers will improve our knowledge of the proportions of ingested food actually utilized by larvae. Such studies are, however, extremely hard to conduct in the field, and the labelling of many potential food items will, inevitably, be very difficult.

The assimilation efficiency of simuliid larvae seems low even by comparison with other detritivores (Berrie 1976). Most particulate matter passes through the gut unaffected by the processes of digestion. This has considerable implications when biocontrol methods, based on the ingestion of pathogens, are being considered.

Macrodistribution

The term "macrodistribution" as used herein refers to the distribution of sites where preimaginal populations develop—"breeding sites"—within the overall geographical range. In other words, it covers the type of fluvial system and/or portion thereof where preimaginal population of a given species may be found. In this respect it is evident that the female simuliid, by her oviposition behaviour, selects the sites where populations of preimaginal simuliids will begin to develop. The suitability of a site will determine the survival of the progeny and by this selection pressure, appropriate oviposition sites will have been determined for a given species. In recent times, because of the rapid changes in habitat due to man's activities (Oglesby et al. 1972; Whitton 1975), two new possibilities have arisen. Oviposi-

tion sites appearing suitable to adult blackflies may not be suitable for preimaginal development due to pollution etc., especially if oviposition cues are primarily visual (Golini 1975; Elsen and Hebrard 1977b). Alternatively, where streams are suitable for preimaginal stages, the adult portion of the cycle may have been disrupted due to: failure of mating (e.g. removal of relevant markers, see p. 261; local eradication of the hosts for blood-feeding; destruction of resting sites; and/or removal of the cues for oviposition (i.e. deforesting the stream banks will alter the nature of the stream). Thus preimaginals will be absent from stream habitats perfectly capable of supporting them. Therefore, in examining the present-day distribution of simuliids, it must always be remembered that in most areas the consequences of human activities are likely to result in discrepancies between those areas and others, untouched by man. It is thus increasingly apparent that the total biology of even a single blackfly species cannot be understood without an enormous amount of long-term research.

Simuliid species, like those of other major insect taxa, can be divided into groups which predominate in certain zones along a stream (Grenier 1949; Dalmat 1955; Carlsson 1962; Garms and Post 1967a; Maitland and Penney 1967; Germain et al 1968; Thorup 1974; Williams and Hynes 1971; Konurbayev 1978; Colbo and Moorhouse 1979). The determinants of this zonation are not always clear. An excellent overview of the methods and problems of river zonation is given by Hawkes (1975). Grenier (1949) suggested that current velocity and O_2 tension are important. Davies and Smith (1958) showed that the spring rise in temperature significantly affects the distribution of two Prosimulium species in northern England. More recent studies on this topic (Wotton 1978c) have suggested that temperature differences affect distribution (in this case, discharge was altered too, though).

Williams and Hynes (1971) earlier suggested that temperature is an important factor in determining simuliid distribution on Mt Elgon, Uganda. The fact that certain species are found only at lake outlets, and are capable of filtering fine-particle sizes (Wotton 1978a), may suggest that food supply plays a role in distribution as well. Indeed, all these factors are believed to be of significance in the distribution of Scandinavian simuliids (Carlsson 1962; Carlsson *et al.* 1977). Sheldon and Oswood (1977) published a model to describe the distribution of blackfly populations downstream from a lake outlet. Based in part on the quantity of seston issuing from the lake. Glatthaar (1978) conducted an extensive study, using cluster analysis, of factors affecting simuliid distributions in Switzerland. However, many criteria suggested to explain distribution may in fact have little more than descriptive significance (i.e. areas of steep relief, lake outlets, cool temperatures, etc.), there being little experimental evidence to prove that they themselves are the limiting factors. Exceptions are chemical factors affecting larval development, important laboratory and field experiments concerning which have been undertaken over the past decade (Grunewald 1972, 1973, 1976a; Grunewald and Grunewald 1978). Nevertheless, Glatthaar (1978) has placed greater emphasis on physical factors as major parameters for the distribution of simuliids in Switzerland.

Mountain streams probably exhibit greater diversity of microhabitats (at least with respect to current velocity and substrate type) than lowland ones. Colbo and Moorhouse (1979) suggested this as one possible reason for the diversity of species in the former habitats. However, it has also been noted that shady, heavily-wooded areas have an impoverished blackfly fauna, breaks in the canopy and the edges of the forest showing a marked increase in species (Zahar 1951; Bishop 1973b; Colbo and Moorhouse 1979). Obviously, current velocity differences are not of prime importance in such cases. The diversity referred to being limited to clearings (Bishop 1973b), or zones where streams emerge into the open (Colbo and Moorhouse 1979), making it difficult to see how the nutritional level of the water could be so quickly altered, unless attached algae are, indeed, important as food. Once again, adult oviposition behaviour would seem to be a major determinant for species distribution. The plasticity of the oviposition behavioural patterns of simuliids has yet to be elucidated. Furthermore, many patterns evident today may be the result of selection and consequent trait fixation in ancient populations existing under quite different environmental circumstances. Man's influence has also decidedly altered distribution patterns (Harrison 1958; Wolfe and Peterson 1959; Chutter 1968; Glötzel 1973; Lewis 1973; Rivosecchi *et al.* 1974; Colbo and Moorhouse 1979). These alterations, when compared with unaltered habitats, can provide useful insight into the problem under discussion: e.g. Carlsson *et al.* (1977) explain a massive concentration of blackfly larvae, at a lake outlet, as due to localized food abundance. These authors suggest that females of the populations in question have evolved an oviposition behaviour of flying upstream until reaching a lake, there ovipositing and thus maintaining a large population. This has been confirmed by Wotton *et al.* (1979). However, Glötzel (1973) investigated a river without an outlet from a lake, reporting large larval blackfly populations at certain points along the system where suitable food was also abundant. In this case, the oviposition cues were cryptic, in the absence of so obvious a factor as a lake outflow; chemical and visual components may have been involved. It might be asked whether Simuliidae, developing under certain defined conditions, are perhaps attracted back to oviposit at the original emergence site, paralleling the spawning behaviour of salmon?

Microdistribution

There is much more information available pertaining to the microdistribution of simuliids and movements within the population than is the case for macrodistribution.

One of the earliest factors recognized was that of current velocity. It is readily apparent, by direct observation, that blackfly larval populations are often in, what are readily observed to be, the faster water currents. It has long been known that simuliids are characteristically, filter-feeders (Verdat 1822); and various workers have established that the preimaginal stages are rheophilic organisms (Grenier 1949). Harrod (1965) showed that, for *S. ornatum*, a minimal current of 0.2 m/s is needed to maintain the head fans in extended position for filter-feeding. Furthermore, although a particular species may be found in a wide range of currents, a limited range of velocity is optimal for each species (Phillipson 1956, 1957). Thus, in the field, the recording of the presence of a species in a certain current is less meaningful than coupling the observation with a measurement of the density (i.e plot larval population density against velocity). However, close examination of the preimaginal distribution reveals that it is discontinuous even over limited areas of the same substrate (Grenier 1949; Wolfe and Peterson 1959; Maitland and Penney 1967; Elliott 1971; Colbo 1979a). Thus, clumps of larvae, and areas lacking them altogether, occur adjacent to one another.

This type of distribution is illustrated by Colbo (1979a), in describing the pattern of *P. mixtum/fuscum* larvae on a benth-observatory window, a glass substrate of uniform texture, where microcurrent patterns must control the distribution of larval clumps. In this instance, current pattern was caused by the natural stony bottom of the stream adjacent to the window—the upper level of the water approached a laminar type of flow pattern on the glass, but blackfly larvae were almost exclusively positioned along the lower portions of the window where the current pattern was very complex due to the ruggedness of the stream bed. Larvae were in fact most numerous in turbulent areas, as evident from the vibration of the free parts of their bodies in the current. These observations agree with very detailed laboratory and field studies in France (Decamps *et al.* 1975), indicating that the laminar layer has little influence in nature, but that turbulence plays a very important role. These authors show that while excessive water movement may result in forces strong enough to tear the simuliid larvae from the substrate, optimal turbulence may enhance the availability of their food particles. In support of this, Kurtak (1978) notes that, in one experiment, larvae in turbulent water ingested more of his introduced artificial particles than did those where the flow was smooth. In the laboratory, *S. decorum* Walker prefers somewhat rugose surfaces as was determined by Hudson and Hays (1975), who created the required microhabitat by attaching sand grains to artificial channels (see also Decamps *et al.* 1975).

It is therefore apparent that different types of substrates variously affect the flow patterns over their surfaces. This may account for part, at least, of the preference for a particular substrate exhibited by individual simuliid species. In fact, Edwards' (1920) differentiation of larval habitats in Britain appears to have been based on this idea. Moreover, in sluggish streams, the only zones with an appreciable flow will be near the surface towards the centre of the channel and here emergent vegetation or other protruding objects will be the only suitable substrate. Indeed, trailing leaves (sedges, grasses, etc.) are the preferred substrates of many larval blackflies. In rapid streams having an alluvial bed, only plant or other protruberances will afford attachment sites free from the abrasion of a constantly shifting bed load of mineral particles. It has also been shown that substrates coated with periphyton are not used as attachment sites by larval blackflies, the substrate preferences of which have been considered by many investigations (e.g. Edwards 1920; Zahar 1951; Carlsson 1962; Chutter 1968; Rühm 1970; Colbo and Moorhouse 1979). Also, Smart (1934) and Disney (1972b) noted seasonal changes in such preferences. These were due to the scarcity of vegetation in winter, in the instance reported by Smart (1934).

There are differences, too, in the positioning of larvae in relation to one another. Some simuliid species space themselves over a substrate so that each larva occupies an attachment site and a space surrounding it. Others form clumps with no apparent spacing (Mackerras and Mackerras 1948; Colbo and Moorhouse 1979; Colbo 1979a). The size of the clear area around the larvae varies with size and current velocity (Colbo 1979a).

The depth of the water in relation to colonization was investigated by Lewis and Bennett (1975), who found that most larvae become attached to tiles near the water surface, where the current was much faster than at the bottom. It thus seems that larval distribution is more dependent upon current and suitable substrate than upon depth (c.f. Ulfstand 1967). In very large rivers, with fast-moving water, larval blackflies have been found to depths of several metres (Zivkovic 1955; Elsen 1977b; Depner and Charnetskii 1978). Fredeen and Spurr (1978) confirmed this, reporting that, in the Saskatchewan River, larval *S. arcticum* are plentiful on substrates only 10–36 cm above the stream bed. In Africa, Elsen (1977b) found simuliid larvae and pupae to depths of 3 m, concluding that the current was a major factor in determining distribution. It must, however, be remembered that these larvae are positively phototropic, provided that the currents are equal or the water is temporari-

ly stagnant (Grenier 1949). Positive phototropism would bring larvae towards the surface when deep currents were inadequate, thus exposing them to the highest potential current of the channel. This is probably of most value to newly-hatched larvae emerging from the substrate on the bottom of pools, etc., as noted by Colbo and Moorhouse (1979). Rubtsov (1964) stated that young blackfly larvae can remain suspended in a water column for a considerable period. If young larvae travel up substrates towards the light they will thus approach the water surface, where even weak currents will carry them a considerable distance—thereby optimizing their chances of reaching faster water. The whole matter clearly merits further study.

Larval age influences microdistribution. Initially, early instars are dispersing from the oviposition sites, so that a higher proportion of them will be present at those sites during the actual hatch (Rubtsov 1964; Rühm 1970; Colbo and Moorhouse 1979). In the case of Australian *A. bancrofti* (Taylor), certain sites have only early instar larvae, the older instars always migrating to zones of greater current velocity (Colbo and Moorhouse 1979). However, in the more rapid waters substrates have young larvae too, indicating broader current preferences for these instars than for older ones.

As is well-known, blackfly pupal microdistribution differs from that of the larvae (Maitland and Penney 1967; Colbo and Moorhouse 1979). Changes have also been described in the positioning on the host of simuliid larvae and pupae living in a phoretic relationship with crabs, prawns and mayflies (Disney 1971a, b, c, d).

Interactions with Other Organisms

Interactions between simuliids, as well as between simuliids and other organisms, are important to the understanding of blackfly biology. Perhaps the most fascinating such interactions are the phoretic ones, where larvae and pupae are attached to crabs, prawns and mayflies in parts of Africa and the Himalayan region (Crosskey 1969; Rubtsov 1972b; Lewis 1973). This relationship was probably first recognized by G. E. Hutchinson while collecting in India in 1932. He communicated his findings to Traver who recorded them in a paper on Himalayan mayflies (Traver 1939). Simuliid species having an obligate association with any of the above organisms show certain morphological adaptations to the habitat. These are: increased size of posterior attachment organ; reduced head fans and antennae; and modification of mouth parts and setal structures (Crosskey 1969; Rubtsov 1972b). Corbet (1961b) and Rubtsov (1972b) both suggested that such associations may have arisen as a means of avoiding being crushed in fast mountain streams where riverine substrates are often moved bodily, with consequent destruction of immature Simuliidae attached thereto. Germain and Grenier (1967) describe the relationship with respect to an African mayfly, noting that both parties have similar food habits (they are detrivores). The studies of Raybould (1969), Raybould and Mhiddin (1978) and Raybould et al. (1978) would indicate that the S. neavei/crab association may have arisen as a means of blackfly survival in rivers having periods of little or no flow. Indeed their work suggests that, in this instance, larvae can feed with little or no current as well as in flowing water, although the resultant modification of head structures cannot be definitely related to feeding (Craig, personal communication). Disney (1971a, b, c, d) discusses several relationships with respect to size of host and location of the larvae on them. Among insect substrates for simuliids, are not only Ephemeroptera, but also Odonata (Corbet 1962; Burton and McRae 1972b). However, while the attachments are genuinely phoretic in the case of the former order, they are simply opportunistic in that of the latter. Nevertheless, the occasional records from Odonata suggest the way in which obligatory phoretic relationships were probably evolved.

Besides truly phoretic relationships, there are many more associations between larval blackflies and other organisms which, in one way or another, influence populations of these biting flies. Predators, parasites and pathogens are the most obvious. These were listed by Jenkins (1964) and Strand et al. (in Roberts and Strand 1977), and detailed reviews are presented herein (see pp. 139–158, 159–170, 181–196). Among other interactions influencing blackfly populations are the yet-unelucidated ones between these and literally hundreds of species of plants and animals which may also be present in the "breeding sites".

The reviews of Hynes (1970b, 1975) and Cummins (1975) show that drainage from catchments provides particulate material to streams. The action of microorganisms and other invertebrates upon allochthonous substances also release particulate and dissolved organic matter. Autotrophic production in some streams can also be very high (Minshall 1978); simuliids may utilize some material directly, some as a result of microbiol action and some released by the shredding of other invertebrates (combinations of these routes may occur). Thus, vegetation growth, microorganism activity in the stream or adjacent to it, but also that of other macroinvertebrates, plays an important role in furnishing suitable food for associated invertebrates (Cummins 1975). Determining the individual role of the various stream inhabitants, and their effects upon simuliids, is a matter for further laboratory and field research, in which efforts should be made to break new ground by imaginative, multivariate experimentation.

Competition for space also occurs, although it may be difficult to show conclusive evidence for it. Chutter (1968) showed that larvae of two species of Hydropsychidae were negatively correlated with simuliid distribution, while another was not. He believed that in addition to predation [for hydropsychid larvae are major predators of Simuliidae (see pp. 148–150), although such predation is "passive", in that the blackfly larvae eaten by these caddisflies simply drift into their nets], there is competition for space and food between the two families, because the building of hydropsychid nets and cases can alter the flow pattern over a substrate. Competition for space may also result in blackflies being physically disturbed by a number of other species (particularly certain mayflies, stoneflies and chironomid midges). Among the Simuliidae, at least some species exhibit a marked response to contacting another organism. Often, in such cases, they either loop, or loosen their posterior attachment organ, thereby moving (sometimes even drifting downstream, on a silk thread). Disney (1972b) observed this type of interaction with mayflies in artificial troughs. The drift of stream insects can be (at least in part) related to the results of competition for space (Waters 1972), for interactions for space between simuliids and their associates is inevitable at high population densities. This is true, not only of interactions with animals, but also of those with plants. Algae cover stones in certain seasons in many areas, preventing simuliid attachment altogether (Sommerman et al. 1955; Chutter 1968; Colbo 1974). The growth of emergent vegetation, as already mentioned, will provide substrates for some simuliids and also will alter flow patterns in the stream, which will in turn affect certain simuliid species (Harrod 1964b; Patrusheva 1966; Rühm 1968, 1970).

Dispersal of Preimaginal Simuliids

In this account of simuliid biology, Verdat (1822) described how larvae disperse by looping over the substrate using their proleg and their posterior attachment organ, or descending downstream on a silk thread anchored to a substrate. Verdat viewed the silk as being extruded from the interior proleg. Much later, though, it was recognized that the silk produced by salivary glands, issues from the mouth via a duct opening in the labropharyngeal complex (Puri 1925; Grenier 1949; Craig 1977b). Larvae employ "looping" for short distances to alter their position on an attachment site (Pomeroy 1916; Grenier 1949). Downstream movement from one object to another will often be accomplished by expelling a silk thread and drifting away with the current (Rubtsov 1964; Tarshis and Neil 1970). Large numbers of such threads may become entangled to form webs near batches of hatching eggs from which first instars are actively dispersing (Rühm 1970). Massive displacement of older larvae due to sudden fluctuations in water level may also result in a web of silk on the river bottom (Tarshis and Neil 1970). Larvae can break the silk thread, too, thereafter drifting with the current until catching upon another substrate (Ruhm 1970, Quélennec 1971) perhaps by using the silk as a sticky anchor.

Reviews of drift patterns and the effects of various factors on them in stream invertebrates, including simuliids, can be found in Hynes (1970a, b), Waters (1972), and Müller (1974). Most invertebrates that drift appear to do so at night. Many reach a peak of abundance in the water column just after dark. Some diurnal blackflies are well-represented in daytime drift. The reasons for drifting are equally varied, and the references already cited review these in some detail. Published reports of simuliid drift patterns cover the range of alternatives known from other stream fauna (Reisen 1977; Colbo and Moorhouse 1979). This is partly due to the fact that not all blackfly species or age classes will disperse downstream unless disturbed. For example, those species restricted to lake outlets were reported by Wotton *et al.* (1979) to have a low level of drift by comparison with the overall larval populations at these sites.

Oviposition often occurs in restricted areas (e.g. trailing vegetation, pools, or lake outlets), it is the early-instar larvae—dispersing to habitats more appropriate for later development—which may be expected to drift. However, many of the drift nets used have too large a mesh to retain early instars (Maitland and Penney 1967). Movement from oviposition sites has been recorded by Rubtsov (1964), Rühm (1970) and Colbo and Moorhouse (1979). The latter authors discovered that first-instar larval drift reaches its peak near mid-day, and declines at night; when the older instars predominate in drift samples. They further suggested that early-instar drift constitutes a normal part of the (diurnal) behavioural drift pattern, being the result of immature blackflies shifting from oviposition sites to larval ones. Clifford (1972) had earlier noted the diurnal drifting of simuliids, primarily of early instars.

Older-instar drift is a pattern resulting from disturbances due to the activity of associated organisms. Disney (1972b) showed that a nocturnal pattern of population shift could be induced in experimental troughs, by introducing nocturnally-active mayfly nymphs. There are also simuliid species which continue to disperse downstream throughout their development, an example being *P. mixtum/fuscum*, under winter conditions in Newfoundland (Colbo 1979a). Although the drift pattern of this species has not been studied, it might be expected that all stages would occur in the drift, but that they would be seasonally separated—i.e. first instars predominating in the fall, and late instars in spring. Reisen (1977) considers the possible correlations between drift and various factors. In nature, seasonal drift, with its diel periodicity, can only be properly comprehended against an understanding of the sequ-

ence of movements of early instars to suitable feeding sites, and, afterwards, to sites for pupation. Disturbance by other organisms is also controlled by the light cycle determining the activity of the associated fauna. Large-scale drift due to, for example, floods or chemicals, will occur at the time of that disturbance. Pesticide-induced drift has been discussed by Wallace and Hynes (1975). Full weight must be accorded to the significance of drift in any organisms being considered for use in integrated control of blackflies; for one effect, of whatever control agent is involved, may be to cause larvae to detach from substrates. This, in itself, may be deleterious to the blackfly in question, depending on the specific situation. Thus, if blackflies are consequently swept downstream into slow or stagnant water, they may not be able to reach suitable sites for re-attachment; and are thus controlled as effectively as those killed by the control agent. Where actual establishment of biocontrol agents in the breeding site is the aim, the dispersal of infected larvae downstream will naturally be an important consideration.

Pupal Phase

It is not always realized that, in Simuliidae, the metamorphosis to the pupal stage occurs within the larval body, the cocoon thus being spun by the pharate pupa while still enclosed by the larval cuticle, which is shed after completion of the cocoon (Hinton 1958). The method of spinning the cocoon has been described by various authors (e.g. Burton 1966). Briefly, the pharate pupa attaches to a site around the perimeter of which, and over itself, it spins silk threads. It then coils back under this web and, from within, completes the spinning, then securing itself either to the cocoon or to the substrate. Considerable movement occurs within the larval skin, which soon ruptures dorsally. The thorax of the pupa emerges, freeing the plastron covered respiratory filaments (Hinton 1964), which are promptly extended. The next step is the freeing of the head by rupture of the epicranial sutures, and the withdrawal of the abdomen from the old larval skin. The pupa moves its abdomen well down into the cocoon, among the threads in which it engages the abdominal hooks. The cocoon is so oriented that its opening is at the downstream end. However, not all simuliids form well-defined cocoons. This is particularly true of the generally more primitive Prosimuliinii and (to a lesser extent) Cnephii. Also, pupae may occur either singly or clumped together.

Smart (1934) indicated that pharate pupae of *S. ornatum* spin cocoons wherever they happen to be at the time. Maitland and Penney (1967), however, stated that larvae migrate to the downstream side of stones before pupation takes place. Colbo and Moorhouse (1979) found that, in Australia, *A. bancrofti* behaves in the manner described by the two previous authors, its cocoons occurring on the downstream side of stones; while, in aquaria, *Simulium ornatipes* migrates from a stream of air bubbles to spin cocoons. They also discovered that a definite diel cycle can be induced by light, although the cycle is disrupted if the larvae are disturbed. Again, they found, under natural conditions and at times of peak population density, that *A. bancrofti* occur as conspicuous patches (all members of which are of the same age) on the downstream side of stones, suggesting cyclic pupation as well. As is well known, the eventual emergence of adults from the pupae is cyclic, and predominately diurnal (Davies 1950; Colbo 1977), although the pattern may be influenced by temperature (Disney 1969; Kureck 1969).

Preimaginal Growth and Population Dynamics

Numerous authors have dealt only with the temporal occurrence of populations

(Edwards 1920; Puri 1925; Zahar 1951; Dalmat 1955; Rubtsov 1956; Orii *et al.* 1964; Lewis and Bennett 1974b). In the past two decades or so, various investigators have begun to study larval growth patterns (Davies and Syme 1958; Davies and Smith 1958; Obeng 1967; Maitland and Penney 1967; Thorup 1974; Zwick 1974; Reisen 1975; Carlsson *et al.* 1977; Ross and Merritt 1978; Wotton 1978c).

In temperate regions, temporal distributions and growth patterns vary. For example, numerous species have only one generation annually; but the larval phase may occur in spring, summer or winter, and the length of the development may vary with temperature. Winter conditions have been shown to severely retard growth and to delay pupation (Thorup 1974; Ross and Merritt 1978), this state being referred to by Mansingh (1971) as oligopause; in which there is retardation of growth and build-up of certain polyhydric alcohols, presumably to protect cells against super-cooling (Mansingh *et al.* 1972; Mansingh and Steele 1973). Some univoltine species with only a winter larval population have been shown not to tolerate high temperatures (Rubtsov 1940; Davies and Smith 1958); although one such species, *Cnephia ornithophilia*, has been found to develop at temperatures above 20 °C in the laboratory, progressing from the first instar to the adult at the same rate as summer generations of the *Simulium verecundum* complex and *S. vittatum* (Colbo, unpublished). The reverse is also true, for Mokry (1976a) found that a member of the *S. verecundum* complex[21] would not tolerate low temperatures in at least the older instars. This species is a summer-developing one with overwintering eggs. On the other hand, the larvae of certain simuliid species (e.g. *S. vittatum*)

---

[21] Mokry recorded it as *S. venustum* but Colbo and Porter (1979) state that it was probably one of the *S. verecundum* complex.

are able to develop at all "normal" temperatures throughout their areas of distribution. Therefore, larval tolerance to a given temperature cannot be assumed. In tropical areas, the effect of temperature on blackfly larvae has not been studied for any species, although Hynes and Williams (1962) suggested that temperature may be a factor in the zonation of species in rivers draining from Mt Elgon, Uganda; as did Germain et al. (1968) with respect to a river in West Cameroon.

Neveu (1973a) has shown how adult size varies seasonally, indicating that temperature is probably the chief determining factor. The same conclusion was reached by Ladle et al. (1977). Decrease in size often being accompanied by a reduction in fecundity (Rühm and Hechler 1974; Hechler and Rühm 1976), it might be expected that temperature will affect the population size of succeeding larval generations.

The autogenous egg production in S. vittatum is approximately halved by increasing rearing temperature from about 19° to 20 °C (Colbo and Porter, unpublished). Similar relationships have previously been demonstrated in experiments with mosquitoes (Van Den Heuvel 1963) and ceratopogonid midges (Akey et al. 1978). Therefore, it follows that, in nature, those species produced under varying temperature conditions should exhibit natality changes.

Food supply also plays a role in survival and natality (Colbo and Porter 1979). This could have been a factor in the seasonal changes noted by Neveu (1973a) and Hechler and Rühm (1976). Colbo and Porter (1979) also found that low food availability retards the time, and decreases the synchrony, of development. Thus, larvae would spend longer in the stream, increasing probabilities of predator attack and parasitic infection. The same argument could also apply to retardation of development caused by reduced temperature, though compensation through increased natality will occur.

Food and temperature are important factors influencing the population dynamics of preimaginal simuliids, although their level of influence, and precisely how they interact with stream morphology and velocity, competition, predation, pathogens, etc., are not clear. It must be remembered, too, that in the tropics and warm-temperate zones, all simuliid life-history stages are found throughout the entire year. This, inevitably emphasizes interspecific and intraspecific competition, other stream organisms being subject to the same stimuli. Wotton (1978c) suggested that intermittent hatching of eggs may reduce interspecific competition, while Chutter (1972b) felt that cannibalism by older instars may limit the populations of newly-hatched larvae, thus reducing the membership of the next cohort.

Ladle et al. (1972), Neveu (1973b) and Reisen (1975) all investigated simuliid production in stream systems. The first-named authors demonstrated variation in production between generations of individual species, also noting that maximum production occurs during periods when suspended solids are at a minimum. Reisen (1975) attempted to correlate production for each instar and to derive a life table thereby.

It is essential to have proper appreciation of competition between species and within individual species. As discussed under "Microdistribution", interaction between blackfly larvae limits the population of a species on a substrate and, consequently, density. It is equally true that species interact. The latter needs more attention, for control measures aimed at reducing one species may affect the abundance of another, which may or may not be desirable.

Greater understanding is needed of how various potential food materials and temperature affect simuliid populations, through studies of variation within given species, linked with the relevant seston and temperature profiles. Many species exhibit distribution changes between generations (e.g. species dispersing from permanent streams to temporary ones). Comparisons of population growth patterns and available food sources might clarify this complex problem. Laboratory studies, using various individual food organisms (including cultured bacteria and algae, leaves of specific plants, etc.) would obviously contribute to this understanding, especially if coupled with field studies. In the warmer regions, where simuliids are multivoltine with short generation times and a complex of potential foods, such studies will be particularly difficult. Nevertheless, rapid development permitting several replicates, during a relatively brief time-span, represents a bonus for such research, which also suggests a multi-variate experimental approach—as yet, a scarcely touched field.

Key Factors in Planning Control Programmes

The various means by which blackfly eggs develop, so as to ensure that the preimaginal phase occurrs at the appropriate time for a given species, have already been outlined. While blackflies vary with respect to the number of generations annually, survival from one breeding season to another is only *via* the egg or larval stage; pupae and adults are not known to survive between breeding seasons. The pupal stage lasts from two days to 2–3 weeks, the actual duration being closely related to temperature (e.g. 2–3 days at 30 °C and about three weeks near 0 °C). Adult females of certain species have been shown to survive for several weeks. In Guatemala, Dalmat (1955) used mark-recapture methods, and found that over 50% of recaptured *S. metallicum* Bellardi survived for longer than three weeks, 25% for over a month and one example for 85 days. On the other hand, most of two other species that he marked, *S. ochraceum* Walker and

S. callidum Dyar and Shannon, did not live beyond one week, although 25% of the former survived for at least three weeks. Davies (1953) kept two North American *Simulium* sp. in the laboratory for up to about three weeks, but the majority were dead within 7–10 days. It seems, therefore, that adult blackflies do not live long enough to survive between larval development phases, in areas with discontinuous cycles.

However, adults are highly important for dispersal and recolonization, especially in species depending on larval survival in refugia. Fredeen and Shemanchuk (1960) indicated that *S. vittatum* recolonized irrigation canals from permanently flowing main channels. In Australia, *Simulium ornatipes* is distributed from the coasts to the dry interior. It survives in small, permanent streams, dispersing out from these during spells of heavy rain. In fact, at some time in the past, *S. ornatipes* colonized two islands, some 1400/1500 km east of the Queensland Coast, respectively becoming ancestral to a new subspecies and a new species, taxonomically close to the parent stock (Colbo 1974; Bedo 1977). This demonstrates a wide dispersal capability. In Africa, members of the *S. damnosum* complex also recolonize streams from permanent breeding sites, the adults dispersing up to several hundred kilometers[22] (Le Berre 1966; Noamesi 1971; Thompson 1976b). In an exhaustive study in Canada, Baldwin *et al.* (1975) demonstrated a mean dispersal from 9.3–13.1 km (max. 35 km) for a member of the *S. venustum* complex, the distribution of which is known to be patchy. Thompson (1976b) reported greater dispersal of *S. damnosum* s.l. along large rivers in Cameroon and Bennett and Fallis (1971) captured most of their blackflies near lakes and ponds

---

[22] This phenomenon led to serious problems in the early years of WHO/OCP (see pp. 96–100 . . . Ed.).

(i.e., near the habitat of the waterfowl that this species feeds upon).

Those species depending on larval populations in isolated refugia for survival appear to have a tenacious larval phase, able to endure poor conditions. An example is *S. vittatum*, which survives in the larval state throughout the year. In the laboratory, it was found (Colbo and Porter 1979) to survive almost twice as long, when deprived of food, as a member of the *S. verecundum* complex, which survives in the egg. However, *S. vittatum* was found to develop as fast as the members of the *S. verecundum* complex, when adequate food was provided.

It is also of interest that some phoretic simuliids can apparently survive for very long periods, even in still water, while browsing for food (Raybould and Mhiddin 1978; Raybould *et al.* 1978). They may thus tide over periods when the stream is dry and the host consequently carries them into subterranean crab burrows which thereby provide additional refugia under adverse conditions. Furthermore, although simuliids which disperse extensively, broaden their access to a wide range of sites, offering suitable cues for oviposition, not all of these will be equally suitable for larval development. For example, food supply may be low with decreasing size, fecundity, and, indeed,

survival prospects; while larval development will be prolonged and asynchronous (Colbo and Porter 1979). Davies (1963) describes such variations in duration of larval size, emergence time and adult size, for the *S. venustum* complex in Canada. The questions must therefore be asked: are all these populations viable; and (in the case of vectors or pests) do they contribute significantly to the adult biting population (i.e. can they disperse as far as others, are they more voracious biters, etc.)? Colbo and Porter (1979) also questioned their ability to sustain parasitic infections when applied as biocontrol agents.

As at least some simuliids are rapid colonizers of new habitats (e.g. *S. damnosum s.l.* in the African savanna and *S. vittatum* in North America) their exposure to biocontrol agents should be implemented during the least favourable period for the target blackfly (i.e. when it is confined to a limited habitat). This may be particularly important under temperate conditions, where the building up of strong populations of such species may depend upon only two or three generations while conditions are optimal. Nevertheless, it would seem reasonable to believe that, even in the tropics, this strategy may lead to, at least temporary, amelioration of blackfly attack.

Hydro-chemical and Physical Characteristics of the Larval Sites of Species of the *Simulium damnosum* Complex

J. Grunewald

All natural fresh waters contain dissolved and particulate inorganic and organic compounds, and gases which vary greatly depending on place (geology, topography, climate), season, time of day, depth, and biota. These compounds are vitally important to the existence of aquatic organisms and there is an intensive and continuous exchange of ions and gases between the latter and the surrounding water. It is therefore likely that the immature stages of Simuliidae, which colonize various kinds of waters all over the world, are influenced by their physico-chemical environment. For the Palaearctic, Nearctic and Ethiopian Regions, the literature covers a number of factors affecting the larval habitats of various blackfly species (Table 1). However, correlation between physico-chemical parameters of the aquatic environment and its blackfly fauna is rarely fully documented; and virtually no laboratory experiments have been conducted, except as regards temperature (Mokry 1976a; Ross and Merritt 1978), pH, conductivity and nitrogen compounds (Grunewald 1978; Grunewald and Grunewald 1978).

For *Simulium damnosum s.l.*, some physico-chemical data, from stream sites in west and east Africa, are presented by Carlsson (1968, 1970). Disney (1975b) gives some information on the water chemistry of rivers harbouring *S. damnosum s.l.* populations in the Kumba area of West Cameroon. Neither author, though, differentiates between breeding sites of the various *S. damnosum*-complex species. As members of this complex differ among themselves in their distribution, biology, ecology and behaviour, it would not be surprising if the larval stages differed in their habitat requirements with respect to hydro-chemical and physical conditions. Dunbar and Grunewald (1974) and Grunewald (1974) conducted investigations concerning the hydrogen-ion content, the conductivity and the ionic composition of the water at larval sites of some *S. damnosum*-complex species. More detailed records are presented by Grunewald (1976a, b) for 13 species of East and West Africa and by Quillévéré *et al.* (1976c, 1977a) for six species of this complex in Ivory Coast. Grunewald *et al.* (1979) provide detailed information on the physico-chemical parameters of breeding sites of the *S. neavei* group, which includes the most important vectors of human onchocerciasis in East Africa. Little information is yet available for central South American blackfly species, especially for vectors of onchocerciasis.

Physico-chemical Parameters of the Larval Sites

Water Temperature

Water temperature is an important ecological factor in lotic habitats and its effect

Table 1. A summary of references concerning hydro-chemical and physical parameters of breeding sites of Simuliidae.

Author	Date	Altitude	Water current	Water temperature	pH	Conductivity	Salinity	Water hardness	Alkalinity	Chloride	Chlorine	Sulphate	Sulphide	Calcium	Magnesium	Potassium	Sodium	Iron	Chromium	Lead	Aluminium	Zinc	Copper	Ammonia	Nitrite	Nitrate	Nitrogen total	Nitrogen Kjeldahl	Phosphate	Silicate	Consumption of KMnO₄	Oxygen, BOD	Free carbon dioxide	Water colour	Turbidity
Carlsson, G.	1962	+	+	+	+	+		+		+				+				+						+							+	+	+	+	
Carlsson, G.	1967		+	+	+					+				+				+						+							+	+	+		
Carlsson, G.	1968	+	+	+	+					+				+																		+			
Carlsson, G.	1970	+	+	+	+	+				+				+																					
Carlsson, M. et al.	1977		+	+	+									+																		+			
Chutter	1968	+																														+			+
Disney	1975b						+	+															+					+	+							
Glatthaar	1978	+	+	+	+	+			+			+		+	+		+	+						+	+	+			+	+		+			
Grenier	1949	+	+	+	+	+	+					+		+	+		+							+	+	+						+			
Grunewald	1972			+	+	+	+	+	+				+		+		+	+	+						+	+	+			+	+	+	+	+		
Grunewald	1973			+	+	+		+	+	+		+		+	+	+		+						+	+	+			+	+	+	+	+		
Grunewald	1976a		+	+	+	+		+	+	+		+		+	+	+		+						+	+	+			+	+	+	+	+		
Grunewald	1976b		+	+	+	+		+	+	+		+		+	+		+							+	+	+	+					+			
Grunewald	1978			+	+	+		+	+	+		+		+	+		+							+	+										
Grunewald & Grunewald	1978	+	+	+	+	+		+	+	+		+		+	+	+	+	+						+	+				+	+	+	+	+		
Grunewald et al.	1979	+	+	+	+	+		+	+	+		+		+	+	+	+							+	+	+			+	+	+	+	+		
Lewis & Bennett	1975		+	+	+								+																						
Maitland & Penny	1967	+	+	+	+		+				+																								
Muirhead-Thomson	1970				+			+																											
Quillévéré et al.	1976c		+	+	+	+	+					+		+	+	+	+	+			+			+	+		+		+	+	+	+	+		
Quillévéré et al.	1977a		+	+	+	+	+					+		+	+	+	+	+											+	+	+	+	+		
Reisen	1974	various parameters		+	+		+																	+											
Rivosecchi et al.	1974		+											+										+								+			
Syed Hyder Ali et al.	1974	+		+														+	+	+		+				+			+			+			
Williams & Hynes	1971	+	+	+	+	+																													
Wotton	1978a	+	+	+				+																										+	
Zahar	1951	+	+	+																				+	+	+						+			

on benthic organisms is very complex. For the various *S. damnosum s.l.* species, different ranges of temperature have been recorded.

In the rivers and streams of the savanna zone of West Africa the immature stages of *Simulium damnosum* and *Simulium sirbanum* develop at water temperatures ranging from 24°C to 33°C, with mean values of 27°C. Similar values have been obtained from large rivers of the tropical rain forest zone where *Simulium sanctipauli* breeds and from the larval habitats of *Simulium soubrense*. In the rain-forest zone, the water temperature of the small, heavily-shaded streams harbouring larvae and pupae of *Simulium yahense* have been found to be lower with a mean value of 25°C (23–30°C). Similar values (mean, 25°C; range, 22–29°C) were recorded for *Simulium squamosum* (Grunewald 1976a, b; Quillévéré *et al.* 1976c, 1977a).

Quillévéré *et al.* (1976c, 1977a) showed that at the larval habitats of *S. damnosum*, *S. sirbanum*, *S. sanctipauli* and *S. soubrense* in Ivory Coast, water temperature was much lower during the rainy season than in the dry one. For the larval sites of *S. yahense* and *S. squamosum* the difference in temperature between the dry and rainy season was less pronounced.

Grunewald (1976a) recorded equivalent temperatures from *Simulium damnosum* complex larval sites of the East African taxa, Kiburenzi and Kisiwani, and West African ones, *S. damnosum s.s.* and *S. sirbanum*. He found that larval habitats of the three East African mountain forms, Nyamagasani, Sanje, and Nkusi, had much lower water temperature (means, 18.9°C, 19.8°C and 21.1°C, respectively), reflecting the different altitudes concerned. To date, the lowest temperature recorded for any member of the *S. damnosum* complex is 16.8°C (Sanje). Very little is known about daily fluctuation of the stream temperature at larval sites of the various members of the *S. damnosum* complex. Sanje offers an exception, Grunewald having found that the daily fluctuation at a habitat in the Eastern Usambara Mountains, Tanzania, was 5°C in the cold season (June, July) and 1–2°C in the warm season.

Water Current

Another major parameter affecting organisms in lotic habitats is water current itself. Blackfly larvae depend upon a particular current velocity not only for their feeding activities but also for preventing the accumulation of a region of depleted water around them in consequence of their metabolism (Ruttner 1926). Optimum velocities differ among larval blackflies, including those of the *S. damnosum* complex. Taxa of this complex have been recorded from larval sites where the current velocity ranged from 0.4 m/s (Carlsson 1968) to 2.40 m/s (Elsen and Hébrard 1977b). With respect to Kibwezi, Grunewald (1976a) reported his highest velocity as 2.18 m/s. He recorded similarly high current rates for Kisiwani; which was, however, also found to develop in currents as slow as 0.68 m/s. Under laboratory conditions, Grunewald reared Kisiwani from egg to adult at only 0.43 m/s. At the larval sites of Sanje, though, the range in water velocity was somewhat less (0.82–2.15 m/s). Nevertheless, Grunewald succeeded in rearing Sanje larvae in the laboratory in slower currents. In contrast to the two taxa just mentioned, Kibwezi, were never found in the field where the current was below 1.72 m/s. They clearly required higher velocities, for it was not possible to rear them where the flow rate fell below 1.40 m/s.

Since the current decreases towards the substrate to which the simuliids are attached, and since there is considerable variation in velocity and turbulence over very short distances (Ambühl 1962), most of the relevant measurements made from rivers and streams are of little significance for any particular individual of a blackfly species. These data define the

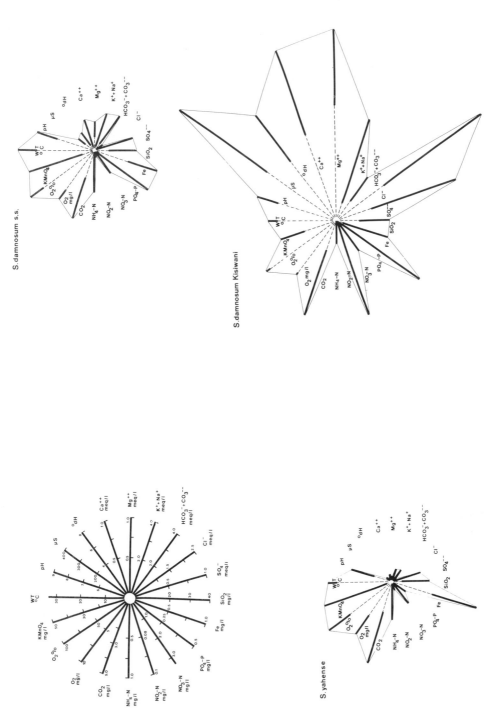

Physico-chemical diagram of the breeding sites of the aquatic stages of S. yahense (rain forest zone of West Africa), S. damnosum s.s. (savanna zone of West Africa), and Kibwezi (East Africa). The various parameters radiate from a common point in all directions. The range of a parameter within which the aquatic stages of a species were encountered is indicated by a bar. Connections of the maximum values of all parameters yield a polygon, the shape of which is characteristic for each species (°dH = degrees German hardness) (after Grunewald 1976a, b).

overall situation as regards stream velocity, but not the precise flow rate at the spot where the larvae are attached. Laboratory experiments defining the water velocity requirements of larvae of the *S. damnosum* complex have yet to be extrapolated to a range of actual field conditions.

Hydrogen Ion Concentration and Conductivity

As background to the distribution of the immature stages of this complex, the hydrogen-ion concentration and conductivity are as important as water temperature and water velocity. On the basis of differences in pH and conductivity, Grunewald (1976a, b) divided the species of the *S. damnosum* complex into three main groups.

GROUP I (*S. yahense, S. sanctipauli* and menge): larval habitats characterized by acidity, hydrogen ion content ranging from pH 5.7–6.2, and conductivity always below 50 μS.

GROUP II (*S. damnosum s.s., S. sirbanum, S. squamosum, S. soubrense*, Sanje, Nkusi, Nyamagasani, Jovi): larval habitats characterized by weak acidity to weak alkalinity, conductivity attaining values of 150 μS.

GROUP III (Kibwezi and Kisiwani): larval habitats characterized by high alkalinity (pH 7.7–10.0), with conductivity values from 400–950 μS, because of a very high content of dissolved solids.

In laboratory experiments by Grunewald and Grunewald (1978) both Kibwezi and Kisiwani were reared from egg to adult in water with pH and conductivity values similar to those found in the field; the rate of survival was 20% and 53%, respectively. They were unable to rear the larvae of either form in water near neutrality, and with conductivity values near 70 μS (as measured in the field for Sanje, Group II). It seems likely that neither Kibwezi nor Kisiwani have osmotic or ionic regulatory abilities

enabling their survival in water with low salt concentrations and of (approximate) pH 7.0. In this respect studies of larvae of other taxa of the *S. damnosum* complex are much to be desired.

The findings of Quillévéré *et al.* (1976c, 1977a) fit well into the group system of Grunewald except for *S. sanctipauli*. Studying 10 larval sites in Ivory Coast they found *S. sanctipauli* larvae developing in alkaline water with conductivity values higher than 100 μS. *S. sanctipauli* larvae seem to tolerate a wider range of pH and conductivity than Grunewald found at three larval sites in Liberia, where the water always showed an acid reaction[23] and low conductivity values. Therefore, *S. sanctipauli* should be included in Group II.

In Ivory Coast breeding sites of species of the *S. damnosum* complex, Quillévéré *et al.* (1976c, 1977a) found lower conductivity values during the rainy season than during the dry season. There were distinct differences as between the rivers and streams of the savanna and the larger rivers of the rain forest; but the smaller streams of the latter zone compared more closely with Savanna larval habitats. In the Ivory Coast studies, rainy-season conductivity values were very similar in the larval sites of all members of the *S. damnosum* complex investigated. For this reason Quillévéré *et al.* did not attach special importance to the electrolyte concentration with respect to the distribution of these taxa.

---

[23] Quillévéré *et al.* (1977a) point out that, because of the low concentration of free carbon dioxide, the water of these rivers would be expected to be more alkaline than measured by Grunewald. The acid reaction of rivers and streams of the tropical rain forest zone, however, is not due to the concentration of free CO_2 but to humic and fulvic acids as shown by Shapiro (1957) and Golterman (1975), and to intensive ion exchange processes which take place with organic matter (Gessner 1959).

In the Usambara Mountains in Tanzania, Grunewald *et al.* (1979) working with two members of the *S. neavei* group, showed that the conductivity values were lower in rivers and large streams harbouring immature stages of *S. nyasalandicum*, than in small streams yielding *S. woodi* (the local vector of onchocerciasis). During the rainy season, when the conductivity values were lowest, *S. nyasalandicum* invaded the small streams which then had electrolyte concentration levels similar to the river production sites during the dry season. When the conductivity values of the small streams increased during the dry season this species withdrew to the larger streams and rivers. *S. woodi*, on the other hand, remained in the small streams only during the dry season when the conductivity values were highest.

Major Ions

While the total concentration of the dissolved solids is important for the larvae of at least some members of the *S. damnosum*-complex species the composition of the major ions in the water seems to have no effect upon them. Disney (1975b) found that calcium and carbonate were the dominant cations and anions in the larval sites of *S. damnosum s.l.* in the Kumba area of West Cameroon. Quillévéré *et al.* (1977a) found that in Ivory Coast calcium and traces of sodium, and carbonate were the dominant cations and anions in the relevant larval sites (this was subject to some seasonal variation). Grunewald (1976a) found sodium (+potassium) and sulphate as the main ions in the breeding sites of *S. yahense* and *S. sanctipauli* in Liberia. For the other species in West and East Africa, sodium (+potassium), calcium, magnesium and carbonate were the characteristic cations and anions. Although Kisiwani was always found in sodium-(+potassium) and chloride-containing water in the field, it was reared in the laboratory with good success in water containing sodium (+potassium) and carbonate (Grunewald and Grunewald 1978). The latter authors assumed some slight effect of the ionic composition in the case of Kibwezi larvae. However, these findings remain to be verified by detailed studies.

The oxygen requirements of the immature stages of Simuliidae are not fully known. The general opinion that blackfly larvae are always limited to water courses with a high concentration of O_2 has not been verified by field investigations. Carlsson (1967) suggested that many authors overestimate the importance of the oxygen content of the water to blackfly larvae. He (1968) found larvae of *S. alcocki*, *S. alcocki occidentale*, and *S. unicornutum* in Nigerian rivers with an O_2 saturation as low as 21.5%.

Although the larval sites of members of the *S. damnosum* complex are usually well supplied with dissolved O_2 (the mean saturation values ranging from 85 to 105%—Grunewald 1976a), some of the species are occasionally found at much lower O_2 concentrations. Grunewald (1976b) collected larvae of *S. yahense* at an O_2 concentration of 66% (5.06 mg O_2/l) and Quillévéré *et al.* (1976c) recorded remarkably low O_2 levels (about 2.5 and 4.7 mg/O_2/l) from larval sites of the same species (unfortunately they do not quote the O_2 saturation figures). The lowest reported saturation value of O_2 in larval habitats of other members of the *S. damnosum* complex is 75% (5.49 mg O_2/l) for the East African form Sanje and 78% (6.00 mg O_2/l) for the West African savanna species *S. damnosum s.l.* and *S. sirbanum*. Comparing O_2 contents for study sites during the dry and rainy season, Quillévéré *et al.* (1977a) found no significant differences. Kisiwani is the only form of the complex repeatedly (seven times) found in water fully or oversaturated with O_2 (Grunewald, 1976a).

Carbon Dioxide

The content of free carbon dioxide does not directly affect the larvae of the *S. damnosum* complex. It may affect them indirectly by its influence on the pH value. In West Africa the lowest concentration of free CO_2 has been reported by Grunewald (1976b) and Quillévéré et al. (1977a) in rivers harvouring larvae of *S. sanctipauli*. The highest values have been determined at larval sites of *S. yahense*. In East African larval habitats of Kibwezi where 34 analyses were undertaken, 33 altogether lacked free CO_2. In one analysis, though, a rather high concentration (5.94 mg CO_2/l) was found.

Ammonium, Nitrite, Nitrate

The three compounds of nitrogen analysed (ammonium, nitrite, nitrate), were always in low concentration at larval sites of the *S. damnosum* complex. The highest concentration of ammonium was 1.158 mg NH_4-N/l at a Sanje site in East Africa. The mean value of ammonium for this form, however, was 0.230 mg NH_4-N/l (Grunewald, 1976a). Quillévéré et al. also found ammonium concentrations exceeding 1 mg/l at a larval habitat of *S. soubrense* in Ivory Coast. However, the mean values of the relevant ammonium content of the various species concerned, did not exceed 0.5 mg NH_4-N/l.

The influence of ammonium on the distribution of various blackfly species was reported by Grunewald (1976a). In the Kibwezi River, Kenya, he found larvae of *S. adersi* and *S. hargreavesi* at a pH value of 7.5 and an ammonium content of 2.296 mg NH_4-N/l caused by the inflow of polluted water. Five kilometers downstream in a similar habitat he found larvae of *S. mcmahoni* and Kibwezi at pH 8.9 and 0.154 mg NH_4-N/l. Most of the ammonium had already been oxidized to nitrate and thereby rendered harmless to the larvae of Kibwezi and *S. mcmahoni*. The larvae of *S. adersi* and *S. hargreavesi*

tolerate higher concentrations of ammonium than those of the various species of the *S. damnosum* complex. Since *S. mcmahoni* larvae also utilize larval sites with water at pH 7 the variation in hydrogen-ion content may not be the only reason for the different distribution in the Kibwezi River.

With respect to the larval sites of two species of the *S. neavei* group in East Africa, Grunewald et al. (1979) determined significantly higher ammonium concentrations in the rivers and larger streams inhabited by *S. nyasalandicum* than in the smaller ones harbouring immature stages of *S. woodi*. This suggests that the larvae of *S. nyasalandicum* may tolerate higher ammonium concentrations than those of *S. woodi*. In rivers providing otherwise suitable conditions for blackfly larvae, but with more than 2.3 mg NH_4-N/l, neither larvae nor pupae were found by Grunewald (1976a). Ammonium as ionized NH_4+ is not acutely toxic to aquatic animals due to its apparent inability to enter cells. However, un-ionized ammonia (NH_3), being able to enter cells, appears to be very poisonous (Milne et al. 1958). The ratio NH_3:NH_4+ in water depends mainly on water temperature and pH value—the higher the temperature and the pH value the higher the proportioon of toxic ammonia in the water. In laboratory experiments with larvae of the Palaearctic species, *Boophthora erythrocephala*, which were reared in water containing 0.6 mg NH_3/l, Grunewald (1978) found mortality rates of 48.3% after two hours and 100% after four hours.

In unpolluted, oxygenated rivers only trace quantities of nitrite are found. In water harbouring immature stages of the *S. damnosum* complex, the concentration of nitrite seldom exceeds 0.009 mg NO_2-N/l; although the two East African forms, Kibwezi and Kisiwani, have been found in water with a maximum concentration of up to 0.039 and 0.125 mg NO_2-N/l, respectively (Grunewald, 1976a, b).

As with nitrite, Grunewald (1976a, b)

found the content of nitrate, to be very low in the water of all preimaginal sites investigated. In West Africa, values around 1 mg NO_3-N/l have been determined. The highest concentration 3.164 mg NO_3-N/l was found for *S. squamosum* in the rain forest zone of Cameroon. In East Africa the quantity of nitrate was even lower than in West Africa; the highest nitrate concentration being recorded in a larval habitat of Kisiwani containing 1.578 mg NO_3-N/l. Neither nitrite nor nitrate seem to affect the immature stages of the *S. damnosum* complex in concentrations analysed in natural larval sites. In "closed" rearing systems in the laboratory, however, an increase of mortality of Palaearctic blackfly larvae was found when the concentration of nitrogen accumulated to 10 mg/l due to decaying food particles and larval metabolic by-products (Grunewald 1973).

Phosphate

In West Africa the water of rivers and streams investigated by Grunewald (1976a, b) and Quillévéré *et al.* (1976c 1977a) was very poor in phosphate. Phosphate as ionic orthophosphate, did not exceed 0.5 mg PO_4-P/l. The highest values have been determined in waters producing the savanna species *S. damnosum* and *S. sirbanum* and in the rivers of the rain-forest zone harbouring *S. sanctipauli* and *S. soubrense*. In the small streams of the forest zone where immature stages of *S. yahense* and *S. squamosum* occur, the phosphate content hardly reached 0.2 mg PO_4-P/l. Quillévéré *et al.* found the phosphate concentration in the rain-forest zone to be lower in the rainy season than in the dry one, while in the rivers and streams of the savanna zone, the phosphate concentration did not vary with the seasons. Although in East African rivers harbouring Nyamagasani, Nkusi, Jovi and Sanje the content of phosphate seldom exceeded 0.2 mg PO_4-P/l, Kibwezi and Kisiwani larvae were

found at phosphate concentrations up to 0.63 mg PO_4-P/l and mean values over 0.2 mg/l. In natural, unpolluted watercourses phosphate and the nitrogen compounds did not directly affect the immature stages of the *S. damnosum* complex. Being essential nutrient salts for growth of plants including algae (which form part of the larval nutrient), phosphate, ammonium and nitrate may affect larval populations indirectly.

Organic Matter

The organic matter in rivers and streams, determined by the oxidation of potassium permanganate, show variation in West Africa (Grunewald 1976a, b). During the dry season, Quillévéré *et al.* (1976c, 1977a) found higher values of organic matter in larval sites of *S. yahense* and *S. squamosum* than in those of *S. sanctipauli* and *S. soubrense*. In the rainy season, however, there was an increase of organic matter in the sites of the two last-mentioned species, the concentration reaching a similar level to that recorded for *S. yahense* and *S. squamosum*. At larval sites of the two species of the savanna zone (*S. damnosum* and *S. sirbanum*), Quillévéré *et al.* found higher levels of organic substances in the dry season than in the rainy one. The organic matter in the blackfly larval habitats, as determined by the consumption of potassium permanganate, gives no real indication of the amount of nutrients available to the larvae. This is because dissolved organic matter, like humic acids, is included, and not all forms of organic matter are oxidized by permanganate. Organic matter may affect *S. damnosum*-complex larvae by reducing the hydrogen ion concentration by organic acids (Shapiro 1957) and by intensive ion exchange processes (Gessner 1959).

Analyses of iron and silicate in blackfly larval habitats (Grunewald 1976a, b; Quillévéré *et al.* 1976c, 1977a) and of aluminium (Quillévéré *et al.* 1976c, 1977a) did not indicate any relation between these parameters and the distribution of the immature stages of the *S. damnosum* complex. In rearing experiments iron and silicate had no influence on the survival rate of the larvae of *B. erythrocephala* (Grunewald 1973).

Conclusion

Our knowledge of ecological parameters influencing the immature stages of the *S. damnosum* complex and of other blackfly species is based mainly on field observations. Relatively few laboratory experiments have been conducted. Nevertheless, it is possible to make a general statement concerning the hydro-chemical and physical conditions of the environment and their importance for the aquatic stages of the *S. damnosum* complex. The four physico-chemical parameters most affecting the distribution of the immature stages of this complex in natural streams and rivers are water temperature, water velocity, hydrogen ion content and the concentration of dissolved salts (conductivity). The larvae of the *S. damnosum* complex are unable to develop in polluted stretches of rivers and streams because these exhibit high concentrations of ammonium and have O_2 saturation rates below 60 to 50% (occasionally as little as 30%). However, the limits of tolerance for the various parameters, differ for the various *S. damnosum*-complex species.

Investigations of larval sites in other geographical regions are needed to show the tolerance limits of parameters within which the various blackfly species are found. Nevertheless, such field data necessarily characterize a momentary situation only. Long-term studies are needed to determine diurnal or other fluctuations and the influence of the duration of particular parameters at given levels, especially when conditions are extreme.

Furthermore, the effect of a given parameter on the tolerance of the immature stages for extreme levels of a second factor or combination of parameters, cannot be determined by field evaluation. A correlation between temperature, water velocity and O_2 consumption of invertebrates was reported by Ruttner (1926), who found that when water temperature increases, faster currents are required to supply the organisms with a sufficient amount of O_2. Ambühl, too (1959), found that an increasing uptake of O_2 correlated with an increasing velocity of water current in many aquatic insects. Investigating three *Corixa* species (Hemiptera/Heteroptera) Claus (1937) found a positive correlation between O_2 uptake and salinity. More laboratory experiments are therefore required to resolve the importance of various physico-chemical parameters for the aquatic stages of the *S. damnosum* complex. Nevertheless, these field data give an indication of the conditions that are likely to be required by the immature stages, for the establishment of laboratory colonies of species of the *S. damnosum* complex (pp. ix, 307–315).

The Effect of Chemical Treatments Against Blackfly Larvae on the Fauna of Running Waters

R. R. Wallace and B. N. Hynes

Efforts to control populations of Simuliidae have been made principally through the use of insecticidal chemicals on the larval and adult stages. Larviciding has been an attractive approach to control as larvae are sensitive to low concentrations of many poisons and control exerted at breeding sites in rivers is an economical, efficient procedure.

Although various chemicals were used in early attempts at larvicidal control (Weed 1904), effective chemical control was not demonstrated until 1945[24] when Fairchild and Barreda successfully used DDT in Guatemala. Their work was soon confirmed in Canada (Arnason et al. 1949; Davies 1950; Fredeen et al. 1953a, b) and Alaska (Gjullin et al. 1950). For the next two decades, research centred on evalua-

---

[24] But see p. 71, also note, p. 81, re 1944 work of R. R. Langford in Ontario, Canada. (Ed.)

tions of various formulations of a growing list of chemical larvicides (Chance 1970a; Muirhead-Thomson 1971; Jamnback 1973). There was, at the same time, a growing realization of the consequences of such pesticide use, e.g. the accumulation of DDT and its isomers, impact on so-called "non-target" insect communities, and the development of resistant strains of simuliids (Suzuki et al. 1963; Jamnback and West 1970; Walsh 1970a). There is an important distinction between the aims of temperate (generally nuisance-abatement) and tropical (vector) control programmes. In temperate areas, only limited injury to NTOs is generally acceptable, but in the tropics, where severe medical consequences may result from outbreaks, a strong case can be made for intensive larviciding (Jamnback 1976) although questions have been raised about the rationale of widespread African insecticidal operations (Asibey 1975). However, Mellanby (1976) succinctly notes that:

> if millions of citizens in a developing country would die from malaria unless there was a fairly generous use of DDT, it would probably be right to risk some temporary effects on wildlife. In a wealthy western country, it may be sensible to use a more expensive and less efficient pesticide which is less likely to have side-effects. The danger is that if the western world bans persistent insecticides before adequate substitutes are available, less developed countries may follow their examples with disastrous results.

In 1970 the virtual ban on DDT in Canada accelerated the search for alternative larvicides (Wallace et al. 1973) and the examination of their adverse environmental effects (Gardner and Bailey 1975). It is well known that chemical control of insect populations can occasion far-reaching responses by species and, indeed, communities (Ripper 1956; Newson 1967; Ferguson 1969; Metcalf and McKelvey 1976).

From the 1950s to the late 1960s there was a gradual recognition that eradication of *Simulium* was unlikely to succeed and control of populations became the accepted objective, the important benthic interactions between predator and prey species gaining recognition as natural regulators of *Simulium* populations. This, coupled with the valid concern for monitoring the consequent side-effects of pesticidal treatments, has led to a wide acceptance of holistic assessments of benthic communities.

It is misleading to state that any insect community necessarily has a "non-target" segment (Ferguson 1969). The term is not based on profound biological differences between insects that comprise "targets" and those that do not but it reflects human attitudes toward control operations and ignores non-selective effects or the longer-term consequences. However, the term "NTO" does allow for a more precise definition of the ideals of pesticidal applications.

It is not our intention, here, to present an exhaustive review of the growing literature on the various consequences of blackfly larviciding, but rather to submit an overview, highlighting key areas of past experience and future promise.

Methods for Assessing Insecticidal Impacts in Running Waters

Field and laboratory techniques have been used in the assessment of the impact of pesticides on stream fauna. Laboratory studies are valuable in that detailed observations of the individuals of one species can be made under rigidly controlled conditions. However, such studies are complicated in that the results are of limited applicability, especially in formulating predictions of adverse consequences to benthic populations. Also, such factors as water quality, may alter the impact of pesticides (Muirhead-Thomson 1971). This complicates the interpretation of laboratory tests which often use standardized waters. Nevertheless, laboratory toxicological tests do determine specific toxicant effects and accurately compare precise responses to specific chemicals and their various formulations (Muirhead-Thomson 1973).

On the other hand, field studies allow for a rapid estimation of the overall consequences of pesticidal applications and, whether observationally or experimentally oriented, furnish the only "real-life" reflection of those consequences. Such studies, however, are limited as to comparability by temporal (seasonal) and spatial (location and stream type) variabilities and by our common inability to sample quantitatively. As a result of these problems, we still have nothing more than gross qualitative tools with which to predict the effects of stress on lotic benthic communities.

The key is to bring these two widely disparate approaches to toxicology (laboratory *v.* field studies) closer together, allowing for quantifiable, experimental approaches to situations which reflect the actual conditions of stream communities. Although some experimental channels have been constructed for use in the laboratory (Wallace *et al.* 1975), this approach suffers from many drawbacks, not the least of which is the expense and effort needed to maintain such systems. There has been considerable success in establishing channels in the field which, while reducing the formidable problems of quantitative sampling and natural variability of stream habitat types, allow for experimental testing. Such an approach was pioneered by Wood and Davies (1964), Wilton and Travis (1965), Travis and Guttman (1966), and Travis and Schuchman (1968), who constructed stream troughs for rearing or testing various pesticides and formulations against *Simulium*. However, these methods are not suitable for the many other lotic species which frequent a wide range of aquatic habitat types. Fredeen (1969) was among the first to attempt replicable, comparable toxicological testing in a large river, by treating small plots with DDT and DDD. This technique, of using smaller artificial channels within rivers, has been improved upon by Dejoux (1975) and others (Wallace 1973). The technique of using portable prefabricated channels, which may be set up quickly, has been employed with considerable success (Dejoux 1975) and is probably one of the most widely applicable means yet devised for toxicological testing in rivers.

Field Methods

The methods for determining the effects of pesticides on benthic fauna in lotic ecosystems, include the use of Surber, drift, emergence, kick, and artificial substrate samples (Gorham 1961; Hitchcock 1965; Ide 1967; Hatfield 1969; Muirhead-Thomson 1971; Fredeen 1974; Flannagan *et al.* 1980).

However, these methods, when used in estimating biomass or the standing crop of benthos, are often crude and subject to large errors (Gaufin *et al.* 1956; Needham and Usinger 1956; Allen 1959; Macan 1962; Chutter and Noble 1966; Eriksen 1968; Hynes and Coleman 1968; Hamilton 1969; Chutter 1972a; Wallace *et al.* 1973). Indications of the extent and distribution of benthic fauna in the intergravular (hyporheic) areas of streams, makes accurate estimations of biomass even more difficult (Coleman and Hynes 1970; Bishop 1973a; Hynes 1974; Hynes *et al.* 1976).

Samples of drift taken before, during and after treatments is a valuable, but still unquantified, measure of larvicidal impact. Studies of drift give us a quick, general indication of the relative impact of pesticides on the species present. The increase in numbers of drifting invertebrates after treatments is clearly associated with subsequent reductions in standing crop and is consequently a useful barometer of at least the relative impact. The exact relationship between standing crop and the amount of drift remains unsubstantiated, limiting the quantitative usefulness of such techniques.

Reviews on the effects of pesticides on NTOs in running waters (Chance 1970a; Muirhead-Thomson 1971) note that, in addition to the problems associated with estimating biomass in rivers, comparisons are further complicated by geographical differences and the annual variability of the types of streams or rivers in which Simuliidae occur. In addition, the interrelationships between benthic species, as exemplified by studies on aquatic trophic levels (Cummins 1973; Kaushik and Hynes 1971), appear very complex and are poorly understood.

The fundamental problem of estimating the biomass or the numbers of a species in small rivers continues to be a major impediment to the quantification of pesticidal, or any other, impact in running waters. We know much less about how to sample in larger, silty rivers such as the MacKenzie, Athabasca, Saskatchewan, or Volta.

The methods for assessment of blackfly larval populations include Surber samples (Wallace 1973), cones (Wolfe and Peterson 1959), polyethylene tapes (Williams and Obeng 1962), glass tubing samplers (Noamesi 1964; Burton 1964), floats (Fredeen 1975) and tiles (Lewis and Bennett 1974a). Although research on the natural factors affecting the dynamics of simuliid populations has been done (Carlsson 1967; Maitland and Penney 1967; Chutter 1968; Ladle *et al.* 1972), there is comparatively less information published relating blackfly larviciding operations to the dynamics and interrelationships of aquatic invertebrate populations. One of the first, and unfortunately last, long-term studies of simuliid population dynamics was that of Davies (1950).

Until standard, quantifiable methods are devised for estimating simuliid benthic populations, aquatic biologists will be at a considerable disadvantage in reaching definitive conclusions about "target" or "non-target" effects of control operations. These problems are further aggravated by the lack of definitive keys for the identification of the Simuliidae and other benthic insects.

Laboratory Methods

There is an enormous literature on the techniques, both static and flow-through, used for testing toxicants. Many of these techniques have been, or could be, modified so as to allow for testing with aquatic invertebrates from running waters (Anon. 1971; Muirhead-Thomson 1971). An example is the magnetic-stirrer, larval blackfly rearing system (Colbo and Thompson 1978).

The approaches used in laboratory testing make predictions of community dynamics of benthos in running waters difficult, if not impossible. For example, we suffer from a lack of laboratory cultures of running-water organisms, including blackflies (but cf. Collins *et al.* 1976).[25] This lack of availability of many species for laboratory testing, ties such programmes near to breeding places and to the natural cycles of benthic populations. The strict use of standardized waters (for comparative evaluations of formulations and dosages) and the variable susceptibilities between species limit applicability to field situations where great differences may occur between rivers (or within a river from one season to another) as regards hydrochemistry.

This is not to deny the utility of such testing programmes. Many types of pesticides may be rapidly "screened" through such a testing system, which greatly facilitates the ability of field researchers to reduce the numbers and

---

[25] Since the draft of this chapter was prepared *Simulium decorum* has been established in self-maintaining laboratory cultures (Simmons and Edman 1978).

types (including combinations) of pesticides for critical field evaluations. Such programmes, when combined with field experiments, are useful, especially when evaluating "target" responses (Muirhead-Thomson 1973).

The Impact of Pesticides in Running Water

Overview

Fortunately, many of the principles of the impact of pesticides in running waters were elucidated by early research on spraying programmes, using DDT and other pesticides, to control forest insects. The use of aircraft made it inevitable that some chemicals would not only reach target watersheds directly, but would also drift or flow out of them.

Linduska and Surber (1947), Hoffman and Surber (1948, 1949), and Savage (1949) made some of the first surveys concerned with stream NTOs. They reported heavy losses and drift of bottom fauna after spraying with DDT. These early studies first raised the questions of the rates and types of re-colonization of simuliid and other invertebrate (and indeed vertebrate) populations—matters still poorly understood, because of the lack of quantifiable data.

Although DDT is no longer widely used, examples of its environmental impact are instructive. We know that pesticidal operations may cause substantial declines in the standing crop of stream invertebrates (Hoffman and Drooz 1953; Ide 1956; Webb and MacDonald 1958; Gorham 1961; Hastings *et al.* 1961; Coutant 1964; Hitchcock 1965; Dimond 1967; Ide 1967; Wallace and Hynes 1975). There is often an initial heavy loss of aquatic insects in streams inadvertently treated in forest pest control, but the genera lost and the dynamics of recovery vary among streams. Hitchcock (1965) demonstrated considerable specific dif-

ferences in the responses of Ephemeroptera, Plecoptera, and Trichoptera to
various dosages of DDT. Such losses of
insects may extend to other stream
populations. Filteau (1959) reported large
reductions in benthos in Quebec streams
after applications of DDT, and presumed
that a subsequent increase in the standing crop of algae was a result of the loss
of many aquatic herbivores.

Gorham (1961), Hitchcock (1965), and
Dimond (1967) dealt with the recovery of
benthos in streams treated with aerial
DDT sprays. Ide (1967) studied changes
in populations of aquatic insects in
streams unintentionally sprayed with
DDT. After spraying, there was an interval of several weeks when no emergence
of aquatic insects occurred. The fauna
recovered qualitatively in two or three
years, except for the Trichoptera, which
required four or more years. A second
spray, applied within three years, produced a reduction in numbers equal in
severity to that due to original spray.

Such early sprayings with DDT probably involved extremely heavy[26] pesticidal dosing of rivers. However, the literature is confusing and easily misleading
unless it is remembered that the pesticide
types and dosages often varied considerably. Only recently has it been possible to
make comparative studies of adverse
effects of larviciding on NTOs, recognizing the inherent differences between
streams in the various seasons and latitudes where control is carried out. This is
thanks to the accumulation of results of
studies where dosages and formulations
were carefully monitored, applied and
recorded.

Without clear differentiation between
the major types of rivers in which blackflies breed, the environmental consequences of such treatments cannot be compared with one another. All too often in
the past, different types of rivers and the

waters that flow in them have been considered as equivalent in discussions of environmental impact. To ignore the
tremendous faunistic or ecological differences which may exist between rivers, is
to impair our ability to improve the
accuracy (in terms of timing and selectivity) of such treatments. Not only may the
quality of waters effect pesticidal efficacy
(Cabejszek *et al.* 1966; Muirhead-
Thomson 1971), but the standing crops
and productivity of rivers may vary widely. It is important to consider such basic
hydrological factors in control programmes, but even more so, in assessing harm
to NTOs, particularly if such studies
range over different seasons. It is unfortunate that environmental impact studies
rarely begin more than a few months
before the treatment(s), and rarely extend much beyond the period of control.

Of the many factors which may affect
the impact of the pesticidal formulations;
rate of flow, turbidity (particularly the
presence of fine silts), and temperature
are probably of paramount importance.
Apart from these physico-chemical factors, it must be realized that such water
quality parameters do not merely influence the chemicals used, and the ways in
which they are dispersed; they also fun-
damentally influence the composition,
distribution, and standing crop of the
fauna and flora. A bewildering series of
potential variables must therefore be considered in formulating a balanced view of
resultant environmental impacts.

As noted elsewhere (Jamnback 1976)
there have been numerous large-scale
projects to control blackflies from Africa
(McMahon 1967; and see pp. 85–103) to
Central America (Lea and Dalmat
1955a, b; Mallén 1974; and see pp. 105–
111) to North America (West 1971; Fredeen 1975; and see pp. 117–132) and
Japan (Uemoto 1971). Undoubtedly, as
people move into new areas, whether in
the tropics or in northern Canada, for
instance, more and better control programmes will be mounted. The WHO/
OCP in the Volta River Basin of West

---

[26] See footnote [10], p. 81 (Ed.).

Africa, is particularly interesting, in this respect, as it is the first massive programme (*c.* 654 000 km²) which will be continued for up to 20 years.

WHO has also expended considerable effort in ensuring that adverse effects of Abate® larviciding (see pp. 101–102) are closely monitored. This provides a unique opportunity to study many fundamental, poorly-understood problems of the dynamics of populations of at least simuliids.

Phases of Environmental Impacts of Pesticides in Rivers

There are three distinct responses of benthic communities to acute exposures to poisons—catastrophic, reversion and recovery phases. The amplitude of the responses is proportional to the type of pesticide (formulation, amount, and exposure period), the type of river and its faunistic composition. Each phase is usually discernable by sampling drift and bottom-dwellers.

The types of drift of benthos have been classified as "behavioural, constant, and catastrophic" (Waters 1965, 1969). The causes and consequences of each type may be quite different. The term "catastrophic" was originally applied to drift arising from some physical disturbance to the stream, such as floodwaters (Waters 1969). The term has also been used to describe that resulting from a chemical disturbance, as in the control of blackflies (Wallace and Hynes 1975).

Catastrophic Phase

When a poison, such as a pesticide, enters a watercourse, there is often sudden, acute mortality among benthic communities. The poisoned insects cause enormous increases in the numbers and types of animals being swept downstream. As rivers are commonly treated during the day, the pre-treatment drift found is usually low as diurnal increases resulting from behavioural characteristics of benthos occur during the night (Waters 1969).

This catastrophic drift may begin at, or shortly after, the time of arrival of the pesticide and may last for up to five hours (Hatfield 1969; Wallace and Hynes 1975; Wallace *et al.* 1976; Flannagan *et al.* 1980b; Dejoux 1977a, b). Form and magnitude of the response are governed by the amount, type, and formulation of the pesticide and the mixing processes in the river. The latter determines the dispersion and time of passage of the treated column of water. The degree of toxicity will also be influenced by many other independent chemical, biotic, and physical (hydrologic) factors, all of which may have tremendous seasonal or latitudinal variability. Other factors, such as the size and type of the river and standing crop (abundance and species composition), will determine the degree of contact between the poison and the benthos. The amplitude of the resulting catastrophic drift may vary from low total numbers caught (Wallace 1973) to literally hundreds of thousands of specimens (Flannagan *et al.* 1980).

The study of drift of stream invertebrates is fascinating. It has been suggested (Waters 1969) that behavioural drift is a function of the rate of production with excess numbers being removed and distributed downstream. Waters (1969) extended this hypothesis to the work of Dimond (1967), who found that high rates of drift recurred only after streams, sprayed in forest insect control, returned to previous stable levels.

The measure of catastrophic drift therefore constitutes a potentially valuable barometer of the degree of stress in treated benthic communities. However, until such time as we are able to quantify the relationship between benthic standing crops and the amount of drift, such measures will constitute qualitative indicators only.

The literature is unclear as to the fate of insects which drift during or after a treatment. Individual species may exhibit widely differing responses to pesticides, and, drift measured carefully (with nets emptied and replaced regularly at short intervals) may reflect susceptibilities. Species that are physiologically susceptible or that inhabit exposed river substrates will be more likely to be contaminated, killed and swept away into drift. Drift sampling after treatments gives us a rapid view of the more susceptible animals, whereas post-treatment bottom samples indicate survivors or recolonizers.

It is often assumed that animals which drift away after pesticide treatment are badly injured or moribund; or at least destined to be eaten downstream (Muirhead-Thomson 1971). In a clear-water stream in Quebec, treated with methoxychlor (0.075 mg l⁻¹ for 15 minutes), most of the drifters, captured and retained in clean water, died. *Hydropsyche* spp. (Trichoptera) were particularly susceptible. Many drifting Plecoptera (genera *Paragnetina* and *Acroneuria*) eventually died; however, some recovered and became hyperactive, exhibiting symptoms similar to those reported by Sanders and Cope (1968) and Jensen and Gaufin (1964a) for *P. californica* and *A. pacifica*. Nevertheless, within 76 hours, most of the Plecoptera that "recovered" had died (Wallace 1971).

Survival among drifters treated either with 0.1 mg l⁻¹ of Dursban® or Abate® for 15 minutes was very low (Wallace *et al.* 1973), although the total mortality was probably increased by packing of the animals in the nets for 24 hours.

Fredeen (1975) hypothesized that some drifters which were "stunned" by sublethal exposure to methoxychlor in the Saskatchewan River, could have recolonized downstream. However, there was no direct evidence for this. Haufe *et al.* (1980a) list relevant genera as biological indicators for the monitoring of environmental impact of blackfly larviciding with methoxychlor.

Differences in the times of appearance of various constituents of drift (Simu-

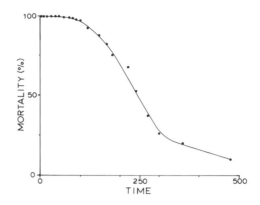

Cumulative mortality of Ephemeroptera drifting after a treatment with Abate®. The curve indicates the % mortality of insects which were collected from the drift at the points shown, and held in uncontaminated tanks for up to 180 minutes (after Dejoux 1977b).

liidae, Trichoptera, Ephemeroptera, Plecoptera), indicating differing sensitivities, were noted by Wallace and Hynes (1975), in small, clear streams in Quebec, and by Flannagan *et al.* (1980b) in Alberta.

Dejoux (1977b) elegantly demonstrated that some Ephemeroptera and chironomids drifting soon after a treatment (Abate® at 0.1 mg l⁻¹ for 10 minutes) of a tropical African river, eventually died. By the use of small, uncontaminated holding tanks set in the river, he found that the insects which drifted up to 90 minutes after the treatments all died, but thereafter, the rate of mortality steadily declined. In this case, the larger part of those drifting in the river would have died since most of the drift occurred soon after the treatment.

The picture is complicated, however, by specific responses of benthos to different chemicals. Fredeen (1969) found that cased *Branchycentrus* and *Hydroptila* (Trichoptera) in the St Lawrence River were more tolerant to DDD than netspinning *Hydropsyche* and *Cheumatopsyche*. Disabled individuals of the former, when captured in drift nets after treatment and transferred to clean water, had a higher rate of recovery than the

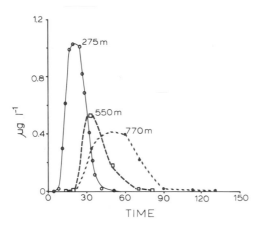

Concentrations of Rhodamine B dye in the Chalk River as monitored at 275 m, 550 m, and 770 m downstream from the point of treatment (15-minute addition). Time shown in minutes (from Wallace et al. 1976).

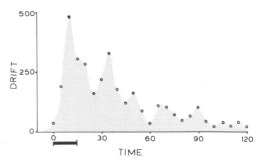

The catastrophic phase of pesticidal impact. Total numbers of benthic animals caught in drift nets after a treatment with methoxychlor (0.075 mg l^{-1} for 15 minutes). Time in minutes, dark bar indicates the time of treatment. Sampling interval 5 minutes in a modified channel constructed in the Speed River, Ontario (from Wallace 1973).

net-spinners. Fredeen speculated that the cases may have been protecting the larvae from handling or the effects of the pesticide. Additionally, the net-spinners may have captured the pesticide in their nets and been contaminated with it, possibly by ingestion. Over a period of 48 hours, recoveries for those caught drifting ranged from 2–19% for species of *Hydropsyche* (fewer with higher doses) and from 6–46% and 34–74% for *Hydroptilidae* and *Branchycentrus* spp. respectively (Fredeen 1969).

There is a substantial dilution of chemicals in rivers as they move downstream, initially caused by the physical processes of mixing and dilution. With distance travelled downstream, concentrations decrease, and the time of passage of the pollutant increases (Dale *et al.* 1975; Wallace *et al.* 1976). This process of dilution makes it inevitable that at some point, in a river which is long enough, a sublethal concentration will be reached for at least some exposed benthic species. However, we presently have few good indications of the consequences of such sublethal exposures to the many species in benthic communities, including Simuliidae.

In two studies involving techniques initially pioneered by Fredeen (1969) using small sections of a river for experimental applications, Wallace and Dejoux separately studied the catastrophic drift shortly after treatments. Sampling intervals as short as five minutes revealed that the form of the post-treatment drift with a Canadian application of methoxychlor (0.075 mg l^{-1} for 15 minutes) was very similar to that shown for a tropical stream treated with Abate® (0.1 mg l^{-1} for 10 minutes) (Dejoux and Elouard 1977; Wallace 1973). The duration of the response by benthic insects was shown to extend far beyond the time of passage of the chemical.

Concentrations of methoxychlor (bars), Rhodamine B dye (dark circles), and numbers of drifting blackfly larvae (open squares) in the Chalk River after a 15-minute treatment at 0.79 µg l^{-1}. Time in minutes. Station located 275 m downstream from the point of treatment (from Wallace et al. 1976).

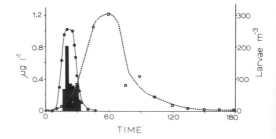

During the catastrophic phase there is a maximum probability of injury to lotic animals, as the highest concentration of the pesticide in the stream is followed by the drifting of large numbers of contaminated organisms. It has been shown that many insects so exposed and drifting may very quickly concentrate the pesticide (Hatfield 1969; Fredeen et al. 1975; Wallace et al. 1976; Flannagan et al. 1980b). Predation on the pesticide-bearing drifters, by invertebrates or fishes, may rapidly spread the contamination and related toxic effects.

Reversion Phase

The second major phase in the response of lotic benthos to pesticides commences from c. 90–450 minutes after the passage downstream of the toxicant and may persist for days or weeks. The onset of this phase is marked by a pronounced decline in the numbers of animals drifting downstream, an increasing proportion of which are alive, while the cumulative rate of mortality progressively falls. The chaotic time during the catastrophic phase is eventually succeeded by the period of reversion when, as sick or dying animals are crowded out or consumed by aggressive survivors, a relative calm returns to the benthic community.

The duration of the reversion phase depends on the impact of the toxicant and the persistence of its action. Certainly, the degree of the initial disruption of the benthic community will influence predator/prey relationships as will the effects of persistent chemicals. Within several days after treatment, the survivors may retain their positions or move into less populated areas.

Animals which have survived the pesticide accumulations will be left to compete with unexposed ones, including those that move in from upstream areas. Here we enter the poorly-understood area of sublethality and chronic, toxic effects. Community responses to such effects are so little understood that

reasonable discussion is difficult. We have many interesting clues as to gross community responses during this post-treatment time. Elouard et al. (1974) and Dejoux and Elouard (1977) found that the normal, nocturnal rate of drift in treated, tropical streams may be greater the evening after the application of Abate® even though the catastrophic phase of drift had ended hours before; contrary to results of Flannagan et al. (1980b) who showed considerable reductions in normal nocturnal drift after methoxychlor treatments of the Athabasca River, Alberta.

Recovery Phase

The phase of recovery in rivers begins when massive recolonization (through hatching or drifting) commences in treated areas. The onset and duration of this phase is a function of the degree of the impact, the persistence and toxicity of residual amounts of the poison, and the degree to which hatching occurs and/or drifters from upstream move into the area. The possible routes for recolonizers are discussed below. However, the rates of such movements are poorly defined and probably highly variable, certainly from the viewpoint of population or community interactions.

Benthic communities may exhibit a remarkable degree of resilience to stress, and the repopulation of denuded or severely stressed environments is based on complex mechanisms. The drift of animals from untreated areas (Fredeen 1977b) and the movement of animals living within the hyporheic, or deep substrata, of a river bottom are thought to constitute the major routes of recolonization. In the case of pesticides passing quickly downstream, animals living in this habitat area would not be exposed unless residues were retained on the upper layers of substrate.

Recolonization by drifters from upstream is undoubtedly an important mechanism and has been studied extensively. Waters (1964) found that drift could restore benthic populations to normal levels in a short time. Rivers with hard or solid substrates will obviously not harbour many organisms at the hyporheic level and, as such, may not recover quite as rapidly as those which have both intragravular and upstream untreated areas. We know that the deeper strata of certain types of stream gravel may harbour large numbers of benthic invertebrates (Coleman and Hynes 1970; Bishop 1973a; Radford and Hartland-Rowe 1971; Williams and Hynes 1974). The potential importance of water-borne toxicants which percolate down (also carrying particulates and O_2) (Vaux 1962) to the hyporheic level, remains unquantified. However, these factors probably strongly influence the vertical, benthic distributions (Bishop 1973a). By using small channels constructed in a stream, Wallace (1973) found that although drift constituted the major route of recolonization after a treatment (methoxychlor $0.075\,\mathrm{mg\,l^{-1}}$ for 15 minutes), there was some evidence to indicate that either hatching or diapausing eggs and/or movement of insects up from deeper intragravular spaces may contribute significantly to repopulation. If drift may be correlated with rates of benthic production (Waters 1969), density-dependent competition would eventually force some individuals to drift or to move into deeper strata of gravel. However, such a mechanism would be operative only if the substrate were fully occupied. This could account, in part, for the failure to correlate drift with standing crop (Chaston 1972).

When very long stretches of a river are treated, diapausing or unhatched eggs may provide for the eventual repopulation. Although little is known about ovicidal toxicity, there is some evidence of adverse insecticidal effects to aquatic insect eggs. Ide (1967) found that no emergence occurred from a river treated with DDT for several weeks. This indicated that either eggs were affected

adversely or any young, newly-hatched animals came into contact with sufficient residual amounts of the pesticide to kill them. From this point of view, chemicals used against blackfly larvae should certainly have a short residual life—any poisoning of competitors or predators is counter-productive.

Uncontaminated upstream areas are vital to recolonization and, therefore, should not be treated. However, such considerations present a dilemma in control operations, since fast headstream areas, particularly those just below lakes, often provide an excellent habitat for blackfly larvae.

To summarize, little is understood about the sequence of events in benthic communities in the interval between the evacuation of a treated area and its repopulation. As a generalization, rivers may quickly assimilate toxicants by dispersion or dilution, and thereafter recover. The mechanisms of such a recovery are traceable to the nature of the evolution of benthic communities. Stresses (often extreme) from flooding, ice scour, and shifting substrates associated with changeable hydrologic regimes, make repopulation an essential facet of benthic survival. Indeed, the basic cause of natural nocturnal drift may be related to a requirement for mobility among benthos.

Recovery in Large, Temperate, Silty Rivers

Fredeen (1975) used artificial substrates (anchored fishnet floats) to assess the impact of a methoxychlor larviciding in the Saskatchewan River. There was an enormous reduction in populations of Simuliidae and Plecoptera (*Isoperla*) up to 161 km downstream from the treatment (Trichoptera, Ephemeroptera, and Chironomidae were not as severely affected). It remains to be established that population estimations, based upon data from net floats, accurately reflect benthic population trends. Also, the comparisons of recovery at treated sites with levels found at the post-treatment control sites might be more appropriate than consecutive measurements at a single treated site.

Increased populations of Ephemeroptera, found 121 km downstream after the treatment, suggested to Fredeen (1975) that the nymphs may have been able to re-establish themselves after drifting in from treated areas upstream. Recolonization, from untreated areas, and egg hatching, probably repopulated the treated areas, too.

Fredeen (1975) noted that 10 weeks after pesticiding, re-established populations on artificial substrates equalled or surpassed the pre-treatment densities. Blackflies, however, were among the most aggressive recolonizers, reaching pre-treatment densities, at some locations, within one week. Also, it must be recalled that although populations of several orders may be replaced by temporal succession, aberrations may persist for several years in severely affected populations of particular species.

The data of Fredeen (1974, 1975) indicate that treatments for blackflies with methoxychlor in large, silty rivers, has an immediate, severe impact on NTOs; the Plecoptera and Ephemeroptera being the most severely affected. Burrowing species, which may inhabit deeper, less exposed substrata (such as the Chironomidae) may escape immediate deleterious effects. In many cases, the impact is just as, if not more, severe on the Plecoptera than on simuliid species. The action of the pesticide in such waters indicates a mode of contamination by contact and ingestion. The Plecoptera would receive fatal dosages from contact-type pesticides, whereas filter-feeders such as Simuliidae, *Hydropsyche* and *Cheumatopsyche*, which suffer enormous losses (Fredeen 1974) would, presumably, consume contaminated particles as well as experiencing direct cuticular contact.

This indicates that emulsifiable concentrates in silty rivers are not as "selective" in their impact as previously thought, despite the acknowledged affinities of such pesticides for suspended solids. Further evidence for this non-selectivity is provided by the work of Flannagan (1976) and Flannagan *et al.* (1980b) (see also pp. 128–130).

Recovery in Small, Temperate, Clear-water Streams

In clear-water streams, pesticidal formulations must be adjusted so as to maximize dispersion. Wilson and Snow (1972) found that wettable powder formulations (0.058–0.091 mg l^{-1} for 18–23 minutes) of Abate® severely affected Baetidae and Chironomidae. Interestingly, at the lower concentration the blackfly larvae were not killed, although they detached, while the mayflies and chironomids were harmed. At higher concentrations, a wider range of NTOs were affected. Repopulation (apparently by hatching of eggs) took more than three weeks to occur.

After two ground treatments with DDT in Labrador (0.1 mg l^{-1} for 15 minutes) Hatfield (1969) found that benthic populations were sharply reduced and that a catastrophic drift occurred for up to 90 minutes after treatment.

Wallace *et al.* (1973), indicated that 15 minute treatments with 0.1 mg l^{-1} Abate®, 0.075 mg l^{-1} methoxychlor, and 0.1 mg l^{-1} Dursban® caused considerable increases in drift. No recovery data were provided beyond 24 hours. However, the stream fauna was not eradicated by the treatments. Wallace and Hynes (1975) indicated similar, heavy, drift after ground-level and air spray treatments with methoxychlor in Quebec. Burdick *et al.* (1974) found that air sprayings with methoxychlor reduced the standing crop by about 20%, with recovery occurring within one season.

It is interesting that while ground-level applications inject rather more pesticide into a river at the point of application than does aerial spraying, the latter usually is equally effective in killing larvae, suggesting that processes (e.g. aerial spraying) which disperse the pesticide into small droplets, enhance contact with the larvae. It is unfortunate that this phenomenon has not been more thoroughly investigated as it could result in more efficient, less damaging ground treatments. An interesting innovation by Burdick et al. (1968) was the treatment of streams with several formulations of methoxychlor administered at ground level with a handsprayer. However, a tremendous variation in environmental impact was demonstrated in that simuliid larvae comprised from 8.8–88.9% of the drift of invertebrates in the various streams so treated. In drift samples taken after experimental aerial sprayings, simuliid larvae comprised from 40.1% to less than 10% of the drift in the streams monitored. Wallace et al. (1976) discuss the possible use of dyes such as Rhodamine B to enhance the effectiveness of ground treatments. They demonstrated a complete mortality of larvae from the Chalk River for up to 770 m after a treatment with methoxychlor (0.79 µg l^{-1} for 15 minutes), the lowest effective concentration reported to date for ground treatments.

In an intensive study of short-term effects of treatments with methoxychlor (0.075 mg l^{-1} for 15 minutes), Wallace (1973) found that insect biomass was reduced from 20% to 50% within 3.25 h after treatment in an open channel. Drift from upstream, untreated areas, constituted the major route of recolonization by the insects. Generally, samples indicated that the maximum environmental impact occurred within three days of the treatment, but that recovery was well under way within seven days.

Recovery in Tropical Rivers

Tropical rivers, particularly those of West Africa, are marked by year-round warm temperatures and the enormous changes in flow which occur between the wet and dry seasons. Lack of rain causes these streams to become small, clear-running brooks in the dry seasons and in the wet season, they are transformed into swollen, silty, torrential rivers. Such extremes of flow cause considerable difficulties in choosing formulations and modes of treatment (Lewis 1974; Quillévéré et al. 1976a,b).

From the point of view of pesticidal impact, the rate of production of benthos (affecting replacement) is probably a crucial consideration (Hynes and Williams 1962). By contrast with Canada, where control of blackflies is usually achieved with two aerial applications annually (West 1961), in Africa, during prime seasons, weekly treatments may be required. This is an indication of the potential rapidity with which the tropical rivers may recover, although cases of severe adverse environmental affects of pesticides have been reported (Corbet 1958a, b) for larger rivers. Factors such as the rate of reproduction and the relatively high primary productivity often permit rapid recovery of the benthos from periods of stress.

As a result of the WHO/OCP now under way in West Africa, there have been intensive studies on the effects of larvicidal control. Ironically, tropical Africa was the scene of work that drew early attention to adverse effects of pesticides upon NTOs (Hynes 1960). Presently, the literature on tropical river ecology related to such treatments, is steadily and rapidly increasing. Jamnback (1976) and Lewis (1974) have reviewed the many different types of pesticides evaluated in West Africa to date. Relevant papers generally concern the rationale which led to the selection of Abate® as the pesticide of choice for West African blackfly control operations. The success of WHO/OCP rests on interruption of onchocerciasis transmission for up to 20 years. As such, it is vital that strenuous efforts be made to ensure that the rivers subjected to that level of treatment are carefully monitored.

Lauzanne and Dejoux (1973) and Dejoux and Troubat (1976) have reported the effects of intensive treatments of West African streams with Abate® (0.5 mg l⁻¹ for 15 minutes, and 0.1 mg l⁻¹ for 10 minutes, respectively). Although there was a decline in benthic populations of treated streams, the effect was less severe than would be expected for temperate streams so treated (17 times over four months).

No evidence has been forthcoming of serious, catastrophic fish kills as a result of the WHO/OCP treatments in Africa, although injury has been reported (Philippon *et al.* 1973; Elouard *et al.* 1974). Corbet (1958a) found that following blackfly larviciding with DDT in the Victoria Nile, although some groups of insects seemed unaffected, almost all species of Ephemeroptera and Trichoptera were eliminated. As a result, a small fish (*Mastacembelus* sp.) which was abundant prior to treatment, became very scarce for up to five weeks following treatment, possibly because of starvation.

Elouard and Forge (1977) studied a tropical stream which, after having received intensive treatments with Abate®, was left untreated for two months. They found that the fauna declined in density and diversity in relation to the progressive drop in flow of the river in the dry season, and that the time for which treatments were suspended was too short for a full return to densities found in nearby, untreated streams.

Dejoux (1977b) found that after treatments with Abate®, reductions of up to 75% of the NTOs had occurred after a modest overdosing (1.0 mg l⁻¹). He recommended that, for the initial treatment in a long series, the dosage be reduced to half that usually employed, so as to reduce the overall adverse effect on NTOs.

Vidy (1976) analysed the stomach contents of fish from treated and untreated rivers in West Africa and could not detect profound changes in the benthic insect dietary composition of the fish.

Dejoux and Troubat (1976) noted that, in studies of rivers in Ivory Coast, treatments with Abate® for 18 months did not appreciably alter the faunal composition in treated rivers, although they estimated that the latter then had a faunistic density of 25–30% below that in untreated streams.

The astonishing resilience of tropical rivers to stress was demonstrated in 1975 (Elouard) when an aircraft crashed into the Black Volta and released 170 litres of Abate® (resulting in a river dosage estimated at 55 times the normal, accepted dosage). Shortly after the accident, the benthic fauna was considered to have been very seriously affected. However, samples taken two days after the crash indicated that no insect group had been eliminated. Repopulation by drift and hatching were considered to be the principle routes for the rapid recolonization.

Susceptibilities of Aquatic Invertebrates

Muirhead-Thomson (1971) gives an excellent summation of the literature on the susceptibility of so called "non-target" species to larvicides used against blackflies.

Species of Crustacea (including the microscopic forms) were rapidly adopted as useful indicators of stress from freshwater pollution due. The ease with which they are maintained in the laboratory has made them desirable test organisms. However, as Muirhead-Thomson (1971) points out, there are wide differences in susceptibilities among many aquatic species: *Daphnia* spp. are, for instance, much less susceptible to Endrin® than is *A. pacifica* or *P. californica* (Plecoptera). Muirhead-Thomson (1971) cited numerous other instances where variable susceptibility has been demonstrated. In short, the search for an ideal "indicator species" has been largely frustrated. Indeed, the concept is probably an illusory one, unless a highly susceptible species is used as the basis on which to establish standards of safety. The problem with this is that many stream insect communities are far more susceptible than are blackflies to many insecticides.

As noted earlier, interspecific differences in susceptibilities are known. Intraspecific factors may also be important; as a general rule younger or smaller aquatic insects are generally more susceptible than are more mature ones (Gjullin *et al.* 1950; Jensen and Gaufin 1964a, b; Hitchcock 1965; Sanders and Cope 1968; Wallace *et al.* 1976). Gjullin *et al.* (1950) suggested that older simuliid larvae are less susceptible to DDT than younger ones. Sanders and Cope (1968) found that small individuals of *Pteronarcys californica* were consistently more susceptible to some insecticides than were larger examples of that species.

Jamnback and Frempong-Boadu (1966) found that after a 5-minute exposure to $0.4\,mg\,l^{-1}$ methoxychlor, slightly fewer mature blackfly larvae had detached than had smaller ones. Fredeen (1974) noted that treatments of methoxychlor in the Athabasca River affected more smaller larvae than larger ones. Jensen and Gaufin (1964a), working with two species of Plecoptera and several insecticides, found that the resistance of one to DDT was greater than that of the other, and that larger specimens were more resistant to several compounds. Wallace *et al.* (1976) discussed aspects of intra- and interspecific susceptibilities to methoxychlor by *S. venustum* and *S. vittatum*, where smaller individuals of both species appeared more susceptible than larger ones.

It is known that accumulation of pesticides is often related to a high ratio of surface area: mass of absorbent (Kenaga 1974). As such, the apparent enhanced toxicity of pesticides to smaller aquatic insects may be due, not so much to physiological susceptibility, as to a higher potential rate of absorption per unit volume than for larger specimens.

Muirhead-Thomson (1970) cautions against the direct application of laboratory toxicity tests to the field, by noting the differences in susceptibility that may result from identical testing with static and flow-through apparatus. In order to generate realistic answers to questions of toxicity, the testing apparatus should resemble, as closely as possible, the natural conditions of habitat preferred by the animal. In addition to reducing stress on the subject, and thereby not giving rise to accentuated mortalities and unrealistic analysis, conditions which are comfortable to the organism will allow one to maintain healthy test specimens longer and permit longer-term studies of recovery or sublethal effects.

Another important aspect of the culture of laboratory species is that by having benthic organisms on hand in the laboratory, it is possible to make detailed observations on their behaviour and habitat preferences. Such data could be invaluable in exploiting behavioural or physiological characteristics in enhancing the kill of Simuliidae or minimizing harm to NTOs.

It is unfortunate that so little effort has been, and continues to be, spent on such basic culture work. Certainly it is difficult to convince funding administrators as to the utility of such expensive (in terms of dollars and time) basic research, especially in programmes directed towards control operations. However, its importance in advancing the state of the science cannot be stressed enough. It is to be hoped that more texts of the stature of Hart and Fuller (1974), further extending our understanding of limits of tolerance, food habits, and chemical parameters for aquatic species, will be forthcoming.

In discussions of toxicology, susceptibility and NTO-safety, field researchers often overlook the true, physiological basis for these effects. Field ecologists tend to observe insect exposure to toxicants and then quantify the results as mortalities within the benthic community. Brooks (1976) succinctly points out that selective toxicity can be broadly subdivided ecologically and physiologically. The former is due to behavioural characteristics which bring the organism into contact with the pesticide, whereas the latter occurs when "marked differences in acute toxicity follow an equivalent contact between different organisms and the same toxicant" (Brooks 1976). As such, morphology, biochemistry and mode of action (of the pesticide) are the determinants of physiological differences.

Our designation of a "target" has a misleading connotation and often overlooks the true physiological site of action (i.e. the *true* "target") among the Insecta. Ecologists and vector control workers may be able to accentuate or tailor pesticidal formulations so as to better exploit habitat or feeding preferences of aquatic target species. However, we must ultimately return to physiological considerations of the chemical mode of action. The availability and accumulation of the toxicant at the site of action and the physiological susceptibility of that site to it, will determine its toxicity.

These complex interactions are affected by innumerable factors (O'Brien and Yamamoto 1970; Sun 1970; Welling 1977) including excretion, detoxification, and the kinetics of penetration and activation. This is a highly involved subject and has an immense literature associated with it (Brooks 1976; Welling 1977).

Welling (1977) gives a comprehensive review of the approaches to describing insect–insecticide interactions, including quantitative modelling which can greatly assist clarification of the complicated processes and factors contributing to final lethal action.

Although subject to the qualifications noted by Welling (1977) (the distinction of rate and amount is unclear), the relatively simple mathematical/graphic model of Sun (1968) does serve to illustrate the complexity of the subject. The principles of such interactions may be broadly applicable to the fauna of running waters. An insecticide, to be effective, must reach its biochemical target in sufficient, potent quantities to produce a lethal effect. The figure illustrates the

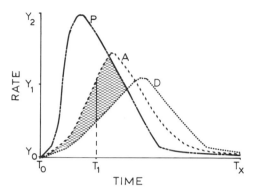

An hypothetical curve for a phosphorothioate insecticide showing that the rate of cuticular penetration (P), activation (A), and detoxification (D) increase rapidly and then decline. The area shaded represents the total effective, absorbed dosage (the S-analog). The amount at a particular time (T_1) during the exposure is shown (adapted from Sun 1968).

relationship between the rates of penetration, detoxification, and activation of the toxicant. As an aside, the use of aquatic species in such physiological studies is exceedingly rare, doubtless because of the difficulties of maintaining such animals in the laboratory.

Other, better, multicomponent models have been developed (Hollingworth 1971; Welling 1977), allowing graphical illustration of the three-dimensional relationships: time-, dosage- and target-inactivation. The application of such findings to actual field circumstances is dauntingly difficult. Species (susceptibility) and habitat (exposure) are two major factors basic to NTO impact. Insecticidal chemicals may act on aquatic species through ingestion and/or contact mechanisms. Ingested pesticides may enter the insect body by absorption from the gut. As for contact poisons, the details of the mechanisms of entry, although a matter of intensive research, remain controversial (Welling 1977).

Whether for "target" or "non-target" species, the accumulated dosage at the biochemical target may reach the threshold of lethality. If it does not, we must face the inadequately-understood area of sublethal toxicology. Consideration of pertinent factors, such as mechanisms of feeding, habitat (exposure) preference, and species-susceptibilities, emphasizes our lack of knowledge of aquatic pesticidal effects, and the imprecision of many, if not most, designations of "targets" and "non-targets". That these considerations are not merely academic is illustrated by the work of Travis and Wilton (1965), who found that many simuliid larvae eventually recover from exposures to Lindane and Sevin. These pesticides apparently induce a (temporary) paralysis of the mouth brushes, the authors hypothesizing that a rapid paralysis would interrupt feeding during exposure, hence preventing the consumption of lethal doses of toxicant. Obviously, under such circumstances, detoxification and/or excretion mechanisms must also be operative.

Wallace *et al.* (1976) discuss possible mechanisms for the different rates of uptake of various formulations of methoxychlor in "target" *Simulium* and "non-target" *Hydropsyche*. These mechanisms lead to discussions of resistance or residue accumulations. Some cases are known for Ephemeroptera (to DDT— Grant and Brown 1967), *Hexagenia* (Newsom 1967) and possibly others (Ferguson 1969).

Oppenoorth (1976) makes the interesting point that toxicants which act on a wide range of species are less likely to induce resistance than those which are more selective, stating:

This is a very unpleasant situation in view of the need for selective insecticides that interfere as little as possible with natural enemies and other useful species of insects. It should be emphasized that selective insecticides . . . have thus two contradictory influences on the development of resistance: a favourable one, since undue selection of species to which the insecticide is not directed is avoided, and an unfavourable one, since the susceptible species can develop resistance by mimicking the properties of naturally tolerant species.

Croft and Brown (1975) have extensively reviewed the literature on the responses, to insecticides, of terrestrial arthropods which are natural enemies of pests. Their data certainly show that such interactions are much better understood with respect to terrestrial than aquatic arthropods. However, many of the principles so derived may be applicable to aquatic fauna. Similarly, sublethal levels of pesticide residues have been found to stimulate some populations of terrestrial arthropods. Luckey (1968) termed the process whereby chemicals at sublethal concentrations exerted a stimulatory effect as "insect hormoligosis". Croft and Brown (1975) discuss examples of terrestrial insects exhibiting this physiological response.

The latter authors deal exhaustively with the factors which govern development of pesticide resistance among natural enemies of pests. Clearly, much less is known about this topic with respect to such resistance on the part of pests themselves. This is true even for terrestrial arthropods, resistance problems concerning which are generally better understood than those of aquatic ones. If a differential frequency in resistance exists between pests and their enemies, it is most probably due to other factors such as exposure or population dynamics; for, usually, physiological mechanisms affecting susceptibility are similar for both.

Croft and Brown (1975) conclude that a wise pesticidal control strategy should include the monitoring of susceptibilities of not only pests but also their natural predators, declaring that:

The greater overall understanding of selection in both groups thus obtained could result in an optimization of spray practices leading to a greater balance in the adaptation of both groups to insecticides and ultimately to a greater balance in the ecosystem; thus avoiding our past error of destroying natural enemies and inducing resistance in the pest.

They note that selection pressures

from surviving natural predators delay resistance in pests. At the same time, the survival by sufficient prey (target) individuals allowed by chemical control programmes aimed at less than complete pest suppression, may allow natural enemies of the "target" to maintain viable populations, perhaps even to develop pesticide resistance.

Certainly, the better practices of terrestrial pest control are far advanced over comparable programmes against aquatic species, particularly those in running waters. There is no reason not to expect that the latter cannot eventually reach the level of sophistication of the former.

Pesticidal Accumulations in Lotic Animals

Intricate mechanisms for pesticidal accumulations in aquatic insects may vary with the species and the habitats. Much of the early experimental work was done with DDT or its isomers (Wilkes and Weiss 1971; Derr and Zabik 1972).

The widely-held view of food-chain concentration (bio-magnification) of chemicals has been questioned (Södergren et al. 1971; Rosenberg 1975). Södergren et al. cautioned that:

> only diffuse relationships between residue levels and the age and trophic levels of organisms in a stream ecosystem were found. This suggests that transport routes and accumulation processes are much more complicated and inter-specifically variable phenomena than often assumed.

Hamelink and Waybrant (1971) hypothesized that biological accumulation (of chlorinated hydrocarbons) "depends on adsorption and solubility differences" and they rejected the idea that biomagnification of pesticides depends on passage of the residues through a food chain. Johnson et al. (1971) also found rapid magnification of DDT and Aldrin® in invertebrates and noted that "aquatic invertebrates influence both the

quantity and quality of insecticide residue passed via the fish food chain." The accumulation of methoxychlor in invertebrates is equally complex. Metcalf et al. (1971) found that in a model ecosystem it was readily metabolized to mono- and di-OH derivatives. However, contrary to the findings of Burdick et al. (1968), Metcalf et al. (1971) claimed that the snail Physa sp. stores substantial levels of methoxychlor.

Pesticides may enter lotic food webs either during a blackfly larvicidal treatment or during the later phases of lower-level residual exposure. Organisms so contaminated may be eaten by predators which, in turn, may become poisoned. There is growing evidence that the principle route of pesticidal uptake for Simuliidae during treatments is by ingestion (Fredeen et al. 1953a, b; Noel-Buxton 1956; Fredeen 1962b; Travis and Wilton 1965; Kershaw et al. 1965, 1968; Helson 1972; Ladle et al. 1972). The rate of uptake of particulates, such as pesticides, by simuliids may be very rapid. Wallace et al. (1976) found, in laboratory tests, that blackfly larvae concentrate particulate $(0.075 \text{ mg l}^{-1})$ and ethanol (0.1 mg l^{-1}) formulations of methoxychlor to 2.31 and 0.68 mg kg^{-1} respectively, within 30 minutes; and that larvae collected after field exposures to as little as 0.79 μg l^{-1} for 15 minutes had accumulated methoxychlor residues from 0.24 mg kg^{-1} to 2.57 mg kg^{-1}. In showing that benthic insects which drifted during a methoxychlor treatment $(0.309 \text{ mg l}^{-1}$ for 15 minutes) contained, on average, 17.5 mg kg^{-1}, Fredeen et al. (1975) introduce the possibility that such fauna may accumulate appreciably higher levels of pesticidal residues in the field than in the laboratory.

Among the fauna collected by Fredeen et al. (1975), 8–10 days after their treatment, the important food fish, Hiodon alosoides (Goldeye), contained 0–1.5 mg kg^{-1} methoxychlor. Such residues obviously constitute a hazard to predaceous fish feeding on the drifters, either directly or via bio-magnification. The fact that no methoxychlor was detected in fish collected 17 weeks later is understandable in view of the known elimination of this compound from fish and the wide-ranging habitats of the latter. Recent work on Athabasca River fishes (Bond and Berry 1980) indicates just how such pesticidal accumulations take place. Goldeye, for example, were found, by these investigators, to have 95% and 75% of the food volume composed of insects (Plecoptera comprised 46% and 34% of the respective totals). On 20 May 1976, 96% of the food volume of Walleye (*Stizostedion vitreum*) was composed of insects (63% Plecoptera, 33% Odonata).

It is clear, therefore, that at the time of the year that treatments for blackflies occur in western Canada, a preferred fish food is an insect (Plecopteran) most severely affected by such treatments. Ingestion, as well as adsorption (Fredeen *et al.* 1975; Lockhart *et al.* 1977) appears as a probable mechanism of contamination of the fish in such rivers.

Major kills of fish have not been reported from the Saskatchewan River; however, little research has been done to quantify contamination due to such treatments, whether directly (via ingesting contaminated drifters) or indirectly (via reduced food). As Lockhart *et al.* (1977) note, the risk of fish poisoning was systematically underestimated in their experiments with caged fish. Again, capturing of free-living fish selects for survivors and fish dispersal makes conclusions difficult too. Except for casual observations, there are no field data on the long-term effects of blackfly larviciding upon natural populations of fish. Flannagan *et al.* (1980b) found that all the invertebrates sampled within seven days of the methoxychlor treatment ($0.3 \, \text{mg} \, \text{l}^{-1}$) contained residues, levels being highest at the stations nearest the point of treatment. Maximum residues for the Ephemeroptera, Plecoptera, and Trichoptera were all near $11 \, \text{mg} \, \text{kg}^{-1}$.

Linduska and Surber (1947) warned of the potential danger to fish through feeding on pesticide-bearing insects. One of the earliest actual observations of such behaviour was made by Hoffman and Surber (1948). Poisoning in this fashion certainly contributed to the fish kill reported by Hatfield (1969) after a ground treatment with DDT ($0.1 \, \text{mg} \, \text{l}^{-1}$ for 15 minutes) in Labrador Following treatment, many dead fish [*Salvelinus fontinalis* (brook trout)] were collected. These fish contained residue levels of DDT tenfold higher than pre-treatment ones. Hatfield (1969) concluded that fish mortalities were caused by contamination of fish food by DDT, above maximum tolerance levels.

Burdick *et al.* (1968) indicated that low-level exposures of brook trout to methoxychlor in hatchery ponds resulted in less accumulation of the insecticide than that of DDT. Methoxychlor was not found in the food chain in as short a period as 36 days, whereas DDT residues were present in samples of fish, and dragonfly nymphs, as well as pond water. Kapoor *et al.* (1970) noted that in a model ecosystem, methoxychlor was found in Mosquitofish (*Gambusia affinis*) at a level 1500 times that of the water. Nevertheless, this was a much lower rate than that found in comparative studies using DDT. The authors noted that there was considerable evidence that "methoxychlor in fish is in a dynamic equilibrium rather than in a storage state as with DDT". However, Merna *et al.* (1972) noted that the 96 h TL_{50} for perch is about $20 \, \text{mg} \, \text{l}^{-1}$, and indicated that there is a very low tolerance level below which perch are able to metabolize methoxychlor with no mortality. Waiwood and Johansen (1974) found that $0.1 \, \text{mg} \, \text{l}^{-1}$ methoxychlor increased O_2 consumption and activity, but all the test fish [*Catostomus commersoni* (White Sucker)] died within 85 h. It is interesting to note that this is the same fish species that Flannagan *et al.* (1980b) found drifting in large numbers shortly after the treatment with methoxychlor in the Athabasca River. Oladimeji and Leduc (1974) found significant reductions in the growth of *Salvelinus fontinalis* and markedly increased maintenance requirements when fed low levels of methoxychlor for 30 days.

Mount (1967) concluded that:

> Concentrations of pesticides in the water that appear to be acceptable under continuous exposure may not be adequate to protect the fish when pesticides are present in their food. Also, unacceptable residues in edible portions may result from exposure of fish to the water concentrations that are not directly harmful to the fish themselves.

Any larvicidal operation for blackflies must take these considerations into account, especially in cases where new formulations or insecticides are being evaluated.

Techniques which have been used to "screen" chemicals through the use of model ecosystems have been reviewed by Metcalf (1977). These systems are useful in that they provide relatively rapid evaluations of pesticidal degradation of many chemicals in many different organisms. Such studies outlined the degradative and bioconcentration pathways for methoxychlor (Metcalf *et al.* 1971).

Studies using models of running waters have been undertaken (Warren and Davies 1971; Metcalf 1977). While having only limited applicability to the field situation, they do offer potential for devising means of rapid detection of larvicides before widespread use, and for possible amelioration of undesirable side effects.

It is known that fish food habits alter with changes in the availability of benthic insects caused by pesticidal applications (Elson 1967; Ide 1967; Keenleyside 1967; Kerswill 1967). Also, Simuliidae may comprise an important segment of the available food of brook trout (Power 1966) besides, perhaps, that of some invertebrates (Peterson and Davies 1960).

Pesticidal Formulations as Affecting Selectivity

One of the vital keys to improving the efficacy of larvicides against Simuliidae (and presumably therefore minimizing harm to NTOs), is formulation. As pesticidal active ingredients (technical material) are usually far too potent to be applied without dilution, they are often combined with other materials into a more dilute "formulation". The two most common types of formulation are emulsifiable concentrates and wettable powders. The former are dissolved in solvents which produce an emulsion in water. In the latter, a finely-divided, inert carrier, together with wetting agents, promotes aqueous dispersion. Oils, and/ or synergists, feature in some formulations and, of course, the manner of application is highly important too (Farquharson 1976).

Early research centred on finding ever more effective blackfly larvicides. At a later stage, attention was focused upon tailoring individual pesticides so as to enhance adsorption or ingestion. One approach was to use chemicals that would be adsorbed by particulates, which were either present in the rivers (Fredeen *et al.* 1953a, b; Travis and Wilton 1965) or mixed in with the chemical (Noel-Buxton 1956; Fredeen 1962b). This ultimately led to detailed considerations of larval feeding preferences (Williams *et al.* 1961) and mechanisms (Harrod 1965; Ladle *et al.* 1972; Chance 1970b). It was hoped that larvicides would eventually be specifically tailored to size-ranges optimum for ingestion by the target blackflies. Such formulations have been prepared and tested in the field (Kershaw *et al.* 1965, 1968; Helson 1972).

Emulsifiable concentrates and oil solutions are, by far, the most widely used formulations. Any formulation enhancing the dispersion of the pesticide into actual blackfly-breeding zones will both increase the effectiveness of the compound and reduce the total amount required for control. However, such successes must be balanced by a consideration of concomitant "non-target" effects. Obviously, techniques which reduce the total amount of larvicide used will reduce the probability of undesired effects.

It is interesting that aerial spraying of oil formulations is so effective, estimated at as low as $8 \mu g l^{-1}$ for 5 minutes (Jamnback 1976). This is probably due to the dispersion of the oily particles in the water. It is known, too, that oil solutions and emulsions are more effective in more turbid waters (Fredeen 1962b). Unfortunately, though, research is needed on the hydrodynamics of oil emulsions, and their concomitant availability to blackfly larvae. Relevant investigations would shed light on contamination problems in other habitats.

Turning to particulates, it is true that such formulations are available to many filter-feeding species (Helson 1972). Further research, in this connection, should permit enhancement of selectivity by capitalizing on behavioural or feeding characteristics. Such techniques are, however, not without problems. Helson (1972) found that particulate methoxychlor used in experimental larviciding of streams severely harmed larvae of Philopotamidae and some types of Chironomidae, as well as Baetidae and Heptageniidae [Ephemeroptera]. Particulate formulations, while representing an advance in control practise, cannot thus be viewed as a chemical panacea for the selective elimination of Simuliidae from running waters.

Other, more subtle, physio-chemical interactions between solvents, particulates in the water, and other chemicals, such as dyes, may be usefully employed to enhance kills and increase selectivity. Wang et al. (1972) found that Rhodamine B, for instance, greatly enhanced the adsorption of parathion onto clay particles in water, and Wallace et al. (1973) discussed the possible significance of such a mechanism in their field work

with methoxychlor. Lacking full understanding of the consequences of such subtle interactions, we thus cannot maximize the advantages of selectivity. It is curious that in West Africa, methoxychlor solutions ($0.1 \text{ mg} l^{-1}$ for 30 minutes) have proved insufficiently effective against blackfly larvae (Le Berre et al. 1976; Jamnback et al. 1970; Philippon et al. 1976; Jamnback 1976). This highlights the exacting nature, and extreme importance, of formulation; which is further influenced by such factors as seasonal change (especially in tropical areas) (Quillévéré et al. 1976a) and application methods (Quillévéré et al. (1976b).

The physico-chemical relationship between methoxychlor and particles is of importance (Fredeen et al. 1975; Merna et al. 1972). It is possible that, in silt-laden rivers, methoxychlor forms a dynamic equilibrium between the suspended silt and the water. Aquatic insect larvae may thus be poisoned by contact with the contaminated water or silt. Such a mechanism would explain the high mortality of Plecoptera (contact) and Simuliidae (ingestion) in the Saskatchewan (Fredeen 1974, 1975) and Athabasca Rivers (Flannagan et al. 1980b). However, as Fredeen (1975) noted, "the exact relationship between turbidity and the larvicidal effects of methoxychlor have not been determined". Beltaos (1977) proposed that the rate of adsorption of methoxychlor on the particles (hence loss from the "water") could be described by a linear sorption-desorption equation:

$\Delta m_s / \Delta V \Delta t = K (C_\omega - \alpha C_s)$ where Δm_s is the net amount sorbed from a water volume ΔV over time Δt. C_ω and C_s are pesticide concentrations in the water and adsorbed, respectively, in water volume ΔV. K is a coefficient with dimensions of time^{-1}, and α is a dimensionless coefficient.

Despite the shortcomings of this equation (it assumes the rates of sorption and desorption are proportional to the pesticide's concentrations in water and on particles), it is reasonable to expect the rate of adsorption to be proportional to the contact between the particles and the water. It is possible to begin to apply such calculations predicting the eventual dosages received by target species and NTOs in such waters. This, in turn, may permit predictions of the relative efficiencies of various formulations and concentrations. Such an approach, when combined with laboratory and field evaluations of the biological effects, could lead to the eventual development of a more quantitative approach to monitoring the environmental impact of pesticides in running waters.

Conclusion

Several large-scale projects for larval blackfly suppression have been conducted in different parts of the world (see pp. 75–132). Since these first began, the concept of pest "eradication" has been modified to one of "control". More and more, it was recognized that consideration of the effectiveness of such treatments must be weighed against any undesirable consequences. A major problem, as noted by Jamnback (1976), is that once demonstration control projects are terminated, the associated research efforts (if indeed they ever existed) cease or are reduced to routine monitoring programmes. Invaluable information on the long-term aspects of insecticidal treatments on streams has thus been lost.

Also, of course, the urgency of achieving adequate blackfly control (whether to interrupt onchocerciasis transmission, to protect a labour force, or to save livestock from major harrassment by blackflies) may, in itself, preclude the orderly gathering of data on the wider implications of the chosen control methods. Even worse, relevant research which has been undertaken has sometimes failed to reach scientific journals, thereafter remaining generally inaccessible in internal documents (whether of national or international agencies).

Kershaw et al. (1968) aptly noted that "The use of insecticides on land is now accepted as a much more discriminatory affair than before, and the time has come for a similar approach to the use of insecticides in water." Others, before and since (Hynes 1960; Chance 1970a) have called for more attention to the basic ecology of benthic animals and to the development of increasingly selective pesticide formulations. It is most unfortunate that the very promising work of Kershaw et al. (1968) and Helson (1972) on particulate formulations has not been followed up. Recent developments in the techniques for microencapsulation of pesticides present the theoretical possibility of specifically tailoring both capsule size range, and the chemical contents with a view to achieving optimum ingestion by target blackfly larvae and maximum lethality. Synthetic juvenile hormone mimics developed, in the first instance, against mosquitoes, offer promise against blackflies too (Thompson and Adams 1979). However, the potential both for this approach and for the practical use of ovicidal chemicals remains to be explored. Moreover, next to nothing is known of the effects of such chemicals on benthic communities.

So far, biological control has not registered conspicuous successes in lotic situations (Van Den Bosch 1971). Nevertheless, over the past few years the urgent need for supplementary and alternative measures to control Simuliidae has directed increasingly widespread attention to the prospects for biocontrol on which this book is focused. Enthusiasm for such prospects must, of course, be tempered by the recognition that all control agents—whether chemical, biological, or other—must be assessed against the broader implications for total stream communities. It is in this area where so much more needs to be done. We shall be greatly limited in our ability to assess control operations until we know more about measuring benthic populations, and the cycles of community responses to natural and man-made stress. Studies on the interactions of benthic species with populations of blackfly larvae have been few, and our inability to understand such predator/prey relationships is a severe hinderance to meaningful experimental modelling. Carlsson (1967) recognized this and stated that valuable results in the biocontrol of blackflies could be anticipated from the use of several carefully selected predators. Clearly, our concept of "integrated control" methodologies must take cognizance of biological "allies". It is regrettable that the work of Davies (1950) on several annual cycles of blackfly larval populations before and after a chemical (DDT) treatment was not followed up, despite his exciting observation of considerable increases of blackflies (apparently through destruction of their longer-lived larval predators) in years following the treatment. Our knowledge of the details of the routes and rates of recolonization after treatments remains inadequate, too. While it may sometimes be true that rivers can recover from larvicidal treatments within 2–8 weeks, far too frequently such recovery is due more to the ability of streams to assimilate larvicides and the fauna to recover from their effects, than to our having employed selective, "safe" control measures.

Eventually, it may be possible to achieve formulations which are, indeed, "selective" for blackflies by exploiting habitat- or feeding preferences. It is unlikely that this will be realized, though, without deeper understanding of the interactions among benthic communities and the natural mechanisms of governing their population limitation. One reason for the lack of relevant research is the impossibility of conducting it without assured long-term funding, allowing for continued sampling over a number of years; preceded and followed by experimental manipulations. Such studies are rarely supported by government or university granting agencies—the former being constrained by political priorities and the latter by inadequate research resources. Until we have the results of such studies, we must continue to suffer the consequences of ignorance. If we continue to accept the argument that pesticidal larviciding as now practiced, is the only viable means of achieving blackfly control, we shall be unable to develop better techniques for such control with minimum adverse environmental impact.

Finally, it must be realized that various benthic populations of different rivers may each respond in quite distinct ways to the same control methodology. Even "enlightened" methods of control may produce inimical effects. Balanced views of control operations, paying due regard to river types and seasonal variabilities, may not necessarily assure the safety of benthos; but they will at least provide a sensitive basis of operation which is open to improvement.

Bionomics of Adult Blackflies[27]

P. Wenk

After leaving the submerged, living pupae, the newly-emerged adults must mate and feed. Development of eggs and oviposition now necessitate distinct patterns of orientation behaviour. Only the female feeds on blood, and in locating a suitable source, and relocating the preimaginal site for oviposition, it may have to change its main flight direction several times; initially, away from the emergence place to find its vertebrate host, and after blood-feeding, in the contrary direction. The process has to be repeated for each ovarial cycle, up to six times in all. Since mating takes place only once, all subsequent individual feeding and oviposition cycles are correspondingly simplified. The possibility of transmission of parasites among the vertebrates in question, exists in this blood-feeding/oviposition cycle. Other parasites (specific to blackflies) mainly occur in the preimaginal stages. If they pass through the imaginal stage, the behaviour of the latter may be changed

accordingly (as happens with mermithids—see p. 163).

Emergence

Submerged pupae, from which adults are ready to emerge, are darker than younger ones. Also, the red pigment of the eyes of the imago is visible through the cocoon anteriorly (i.e. the end directed downstream to the opening of the cocoon). The act of emergence takes about one minute. The abdominal segments show slight peristaltic movement, and appear silvery due to the gas released between the imaginal and pupal cuticles. Soon afterwards, the thorax bursts, leaving a T-shaped slit through which the thorax of the adult emerges, followed by the head and abdomen. As soon as the forelegs are free, they are stretched aside and the water, streaming past, tears the blackfly out of the exuvia. The insect does not become wet, but reaches the water surface enclosed in a gas bubble, and ready for flight. Some blisters of haemolymph may be evident in the wings of adults that have emerged in a cage. Nevertheless, such simuliids are able to fly just as well as the others. In the event of desiccation of the pupa within its cocoon, the adult may still be able to emerge (although sometimes more than a

Biological cycle of Simulium *sp. transmitting* Onchocerca volvulus *to man.*

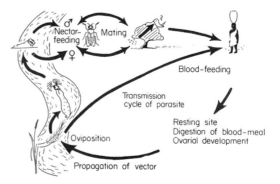

---

[27] See Fallis (1964) for a comprehensive literature review, mainly concerning Palaearctic Simuliidae.

week later) and to complete its life-cycle.

Large numbers of newly-emerged blackflies may be obtained from field-collected aquatic vegetation, placed in a cage and kept slightly moist. This may be used as a test to demonstrate the control of a larval population by an insecticide application, as well as the subsequent restoration of the larval population. Emergence rates are measured by simple counts of the numbers of adults emerging from standard portions of such vegetation. Proportional differences between rates before and after applying insecticides will allow quantitative comparison of the effects of different compounds (Wenk and Schulz-Key 1974).

The diurnal emergence pattern of blackflies seems to be initiated mainly by light. It is also influenced by water temperature. Scarcely any simuliids hatch during the night. In the case of *S. damnosum s.l.*, 60–90% appear by midday. In Cameroon, during especially warm weather (midday temperature of the river, 24.1–28 °C), forest zone *S. damnosum s.l.* males and females (also some associated species), show a peak of emergence between 06.00 h and 09.00 h. On relatively cool days (midday temperature of the river, 20.1–24 °C), the peak is delayed until 09.00 to 12.00 h. It is interesting to note, that under artifically cold conditions (refrigerator temperature, 16–20 °C), the peak was delayed until late afternoon. Under laboratory conditions, savanna-zone blackflies exhibited a time-lag in the peaks of emergence relative to forest-zone ones. The latter, when subjected to the same pattern of cold nights and warm days, as is experienced in the Guinea Savanna Zone, showed a time-lag in emergence greater than that exhibited by blackflies characteristic of this zone (Disney 1969).

Similar results were obtained in Ivory Coast with the Bandama cytotype of *S. damnosum (S. sanctipauli)*, by hourly field collections (Edwards *et al.* 1976). However, this may not hold true for all geographical regions and for different seasons. Thus, the Australian *Austrosimulium bancrofti* shows three consecutive peaks, each higher than the last, in the emergence of males and females; 09.30 h, 13.30 h and (sunset) 17.30 h, when water temperatures of 16–25 °C were measured. When temperatures increased to 23–30 °C, the main peak was around midday at 15–20 °C, 62% of the males and 81% of the females emerged at sunset. In the same region, *S. ornatipes* showed a broad morning peak between 06.00–12.00 h, with a maximum at 09.30 h (Colbo 1977). Lapland *Simulium* species also emerge only during daylight. However, the rhythm persists throughout the continuous light of the subarctic summer, also under permanent artificial light and during permanent darkness (Kureck 1969).

Mating

Swarming

Mating usually takes place quite close to the emergence site. However, the behavioural pattern needed to bring the two sexes together in nature may operate effectively even at some distance from this site. Swarming of males takes place in *S. damnosum s.l.* as well as in many other Simuliidae. The blackflies orient to visual markers standing out against the open sky, e.g. the extended branches of prominently-situated trees (Wenk 1965b). In the UK, even female *S. austeni* have been observed to accumulate near a visual marker (automobile parked in a field of short grass) just prior to mating. Males then arrive and soon comprise an increasingly high proportion of the swarm. No in-flight mating was observed and 50 females, caught by net, proved to be unfertilized and nulliparous (Service 1972).

Sometimes, male swarms have been observed orienting to the vertebrate hosts of the females of their species. The females of *Boophthora erythrocephala* bite the ears of cows, while those of *Odagmia ornata* land on the belly around the naval skin-fold. The males of *B. erythrocephala* may form small swarms (5–10), orienting to the ears immediately after these have been flapped in the cow's attempts to avoid the troublesome females. The latter are then seized in flight by the males, and the pairs fall to the ground, insemination taking place within a few seconds. Sometimes the male pursues a female even into the ear of a cow, accounting for the occasional (*c.* 1%) collection of males (which do not, of course, bite) from this site. The males of *O. ornata* swarm some distance from the cow, catching females flying to the belly of their host (Wenk and Schlörer 1963).

When white blankets are spread out beside the cow, the copulating blackflies may be detected as soon as they fall. Sometimes, the paired simuliids belong to different genera, a fact which strongly supports the theory that, in mating swarms, the males rely on visual stimulii in contacting females. This has been proved by experiments with artificial, visual markers (female *B. erythrocephala*) being replaced by black shot (Kirschfeld and Wenk, in preparation). The highly-adapted dorsal sector of the compound eye of the male is specialized to detect blackfly females in flight, against the blue sky, at distances of up to 50 cm (see p. 277). The duration of copulation differs from species to species, ranging from up to 40 minutes (Friederichs 1922) to seven minutes (Moorhouse and Colbo 1973), 2–3 minutes (Davies 1965), or as little as 1–5 s (Wenk 1965b). Paired *S. salopiense* were observed briefly (1–2 s) drifting on the surface of a stream (Rühm 1971b).

In *S. damnosum s.l.* copulation has never been observed in nature; and it has been held doubtful that mating takes place in flight, whether by swarming with the males oriented to a visual marker (e.g. tree branches), or at some distance from the blood source (e.g. humans). However, there are many indirect

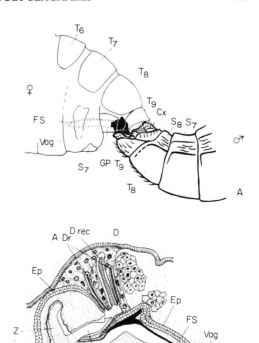

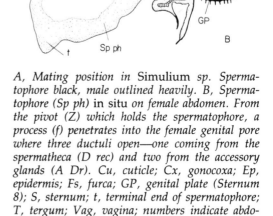

A, Mating position in Simulium *sp. Spermatophore black, male outlined heavily. B, Spermatophore (Sp ph) in situ on female abdomen. From the pivot (Z) which holds the spermatophore, a process (f) penetrates into the female genital pore where three ductuli open—one coming from the spermatheca (D rec) and two from the accessory glands (A Dr). Cu, cuticle; Cx, gonocoxa; Ep, epidermis; Fs, furca; GP, genital plate (Sternum 8); S, sternum; t, terminal end of spermatophore; T, tergum; Vag, vagina; numbers indicate abdominal segments (after Wenk 1965b).*

observations in support of the above hypothesis. For example, swarms of blackfly males oriented to visual markers (specifically trees) have been observed in the West African savanna (Le Berre and Wenk 1966). Again, in forested regions, males are sometimes captured together with females actually attacking humans, indicating that the former are active close to the blood source.

During copulation, the male attaches a spermatophore (a 2-chambered sperm-containing structure secreted by the male—Davies 1965) to the genital plate of the female. The latter opens the spermatophore (probably by enzymatic secretions of the additional genital glands), squeezing the sperm into the spermatheca via the corresponding duct (Fig. 2, Wenk 1965b). The spermatophore is easily seen, under the dissecting microscope, in many nulliparous females. The inexperienced observer will locate spermatophores more readily when about 50–100 females are bulk-stained with acid fuchsin, aniline blue, etc., in 40% alcohol, and washed in 95% alcohol (Wenk and Raybould 1972). The structures will be stained selectively, since the chitinous cuticle of blackflies protects their internal organs from absorbing the stain.

At certain hours of the day, a high percentage of S. damnosum s.l. females, attacking man, have been found to carry a spermatophore. In the West African savanna, during the rainy season, the author has found many females with a spermatophore during the afternoon (31% at 17.00 h, increasing to 45.5% at 18.00 h, when the blackflies ceased to bite). In the same region, during the dry season, the rate increased to 80% (Renz, personal communication). In earlier investigations, all Simuliidae (including S. damnosum s.l.) with a spermatophore, have proved nulliparous. It thus seems reasonable to assume that copulation takes place near the blood source, or perhaps at a short distance from it—after all, swarming of males seems rather unlikely in the savanna, characterized by so many widely-scattered bushes and trees of various sizes.

Even when a blackfly collector is seated on a white blanket (c. 3 m × 3 m) he may fail to observe any paired simuliids. By comparison, about 5–10 copulating blackflies may be caught, in the hour before dawn, in the immediate vicinity of a cow with about 100 females in each of its ears and another 100–200 biting its belly and udder. However, these represent only a few of the many pairs present—probably most of them will not fall directly onto a blanket only 9 m^2, situated 1–2 m distant from the cow. The copulation of S. vittatum has frequently been observed by this technique (Mokry, personal communication). When the person attacked by S. damnosum s.l. was positioned between a net and a ventilator blowing into it, small numbers of males and many females were caught. While participating in this experiment, in the West African savanna, the author noticed that males tended to escape from the net (which had quite a wide mesh) quite early; however, the bulkier abdomen of engorged females facilitated their retention. In this investigation, the percentage of nulliparous females with a spermatophore attached increased during the day; from 18.3% (09.00–12.00 h) to 25.0% (12.00–15.00 h) up to 48.5% (15.00–18.00 h). In similar experiments, in the forest region of Cameroon, the percentage of females with spermatophore remained approximately the same throughout the day (Le Berre and Wenk 1966).

Mating in Captivity

Mating of S. (Wilhelmia) salopiense in glass tubes, was first observed by R. W. Dunbar (Davies 1965). Another Palaearctic species, B. erythrocephala, and afterwards, S. damnosum s.l., proved to mate in cages (c. 40 cm × 40 cm × 60 cm), covered with black cloth on all sides but one—this featured a triangular window of gauze, the blackfly within being attracted towards it phototropically, and then scrambling upwards towards the peak of the triangle. The adults used in these experiments emerged from pupae on plant material earlier placed in the cages. Regularly, more than 80% of the females carried a spermatophore 48 hours after emergence. Mating took place while the blackflies were concentrated on the gauze triangle, through which it was easy to observe the process. The adults remaining numerous for about 10 days, blood-feeding and oviposition of viable eggs duly took place (Wenk 1965c). The experimental cages also facilitated detection of separate taxa of the *S. damnosum* complex, as well as of other species amenable to mating in cages.

Even when a cage contains a mixture of several species of Simuliidae, only a small percentage of which is made up of the taxon actually mating, the relatively small numbers of inseminated females are readily detectable through staining the spermatophore (see p. 262). By this technique, more than 80% of adults of the Kibwezi form proved to mate in the cages, and blood-feeding and oviposition of viable eggs were achieved (Wenk and Raybould 1972). The author obtained mating of *S. damnosum s.l.* in such cages, in Ivory Coast (Leraba River) and northern Cameroon (Boki River, near Garoua). In both instances, some 10% of the females carried a spermatophore despite there being a mixture of *Simulium* spp. present. Near Lomé, Togo, a cage-copulating strain (a probable onchocerciasis vector of the *S. damnosum* complex) has been detected by Grunewald (personal communication). The "black cage and spermatophore-staining technique" thus lends itself very well to field use in discovering readily-mating strains of blackflies suited to further evaluation for laboratory colonization.

Feeding

Patterns of Feeding Activity

Diurnal

Mammalophilic blackflies probably all bite in daylight, but at least some ornithophilic ones (e.g. *Eusimulium latipes*, *S. tuberosum* and *S. reptans*) have been caught throughout the night, by means of light traps (Williams 1962; Raastad and Mehl 1972). While mammalophilic strains of *S. damnosum s.l.*, *S. griseicolle* and *S. unicornutum* have also been taken in light traps (Lamontellerie 1963, 1967; Walsh 1978), the relevant collections were mostly made within the two hours following sunset; and as all three are characteristically evening biters, it seems likely that they had simply been reactivated by the artificial light of the traps.

Simuliidae, like other biting flies, exhibit an endogenous rhythm of blood-seeking activity (appetitive behaviour). The chief stimulus of this among black-flies is sunlight (Wolfe and Peterson 1960). Thus, both African species (during the dry season) and Palaearctic ones (during the summer), attack their hosts soon after sunrise and again, somewhat less severely, in the late afternoon. However, in African forests, biting activity remains relatively constant throughout the day (Giudicelli 1966). In the savanna zone, nulliparous females are more numerous in the evening, especially in the wet- or intermediate season (Bellec 1974). *S. woodi*, a phoretic species and vector of onchocerciasis, in the Usambara Mountains, Tanzania, exhibits a bimodal biting cycle, especially in the warmer months, but its nulliparous females are more numerous in the morning (Raybould 1967b). However, Crosskey (1958b) reported, from Nigeria (on the basis of dissections demonstrating a small fat-body and the presence of residual eggs and cleared Malpighian tubules) that older *S. damnosum s.l.* bite most actively at the height of the day (10.00–14.00 h). Most probably, the rapid change of light intensity—whether increasing in the morning, or decreasing in the afternoon—stimulates the adults. Supportive evidence for this is that no obvious peaks of activity are evident in shaded areas with little light fluctuation while, nevertheless, peaks occur in nearby sunny places (Kaneko *et al.* 1973). Also, at the approach of a thunderstorm, or when sudden clouding occurs for other reasons, blackfly biting intensifies whatever the time of day. Accompanying increase or decrease in relative humidity may also influence blackfly feeding behaviour (Häusermann 1969), although this is of lower significance in rain forest. Emergence rhythm, too, may influence the afternoon peak of nulliparous blackflies (Marr 1962).

Seasonal

Year-long observations disclose major differences in blackfly population density as between the dry and wet seasons, as Häusermann (1969) noted in Tanzania. Correlations between (1) larval density and the incidence of preimaginal sites, and (2) the number of females attacking man, were made (in relation to rainfall) in the savanna and forest zones of Upper Volta and Ivory Coast (Le Berre *et al.* 1964, Le Berre 1966).

Synchronous variation of larval and adult simuliid population density has been observed in West Africa, in forested regions where (during the rainy season) the level of large rivers increases steadily. The number of emergence sites increasing correspondingly, biting density also peaks. From November to February, river levels are low there, the preimaginal sites then consisting of rocky shelves in the low-water channel. Levels then rise gradually, until the September–October peak, when countless emergent trees and bushes then enormously increase potential preimaginal sites. Variation in density of *S. damnosum s.l.* females follows the same unimodal curve: less than 100 bites/man/day at low river levels, several thousand at high ones. During the dry season, when water-courses are non-existent in the savanna zone, breeding sites are, of course, absent. Female blackflies become evident again at the beginning of the rainy season and their population thereafter increases steadily, as already mentioned, until river levels peak.

Inverse variation occurs in low-lying regions. During the dry season, intermittent rapids provide numerous, relatively small, preimaginal sites, which can be correlated with high biting density at the end of the dry season or at the start of the rains. From then onwards, the rivers overflow their flat banks, resulting in such loss of water velocity that few larval sites remain available. However, blackfly breeding in such regions is never completely halted. The annual curve for daily biting density shows a maximum from February to July, at low water levels (up to 100 bites/man/day) and a minimum from August to January, until complete disappearance of the adult blackflies.

Bimodal variation of larval and adult blackfly density is characteristic of savanna zones, where the vegetation is denser than usual, and the rivers have deep channels. High biting density occurs during both dry and wet seasons. However, rapid changes in water levels, whether at the start of the rains or the beginning of the dry season, results in decrease of preimaginal sites and consequent low biting densities. The annual curve for daily biting density shows two maxima, one at low water (up to 300 bites/man/day), and the other at high water (up to 500 bites/man/day). Between these peaks, the biting density falls sharply (Hughes 1952).

Dispersal, Survival, Flight Range, Longevity

Blackfly "dispersal" may be understood as a displacement dominated by the flight capability of the insect; whereas "migration" may be considered as mainly windborne event, following a non-appetitive initiation. Migration is active only in so far as the adults keep themselves airborne by persistent wing beating (Magor *et al.* 1975). It may be difficult, in the field, to distinguish "dispersal" from "migration". Actual flight distances may vary seasonally, too, and radial dispersal may be separated on the basis of changes in plant cover and relative humidity from the dry to the wet season and *vice versa* (Le Berre 1966).

Linear dispersal. In savanna grasslands, the permanent breeding of blackflies is restricted to a few continuously flowing rivers margined by gallery forests. Such lines of scattered bushes and small trees may extend along the banks of rivers flowing only seasonally, too. During the dry season, *S. damnosum s.l.* may be active within these gallery forests, but with the first rains and the recommencement of flow in seasonal rivers, they disperse along the water-surface under protection of the overhanging vegetation, thereby extending the breeding zone. Riffles associated with fords provide suitable larval sites. Accordingly, the biting rate may be high at fords and bridges where travellers necessarily concentrate. The same type of linear dispersal may take place in savanna with scattered bushes, becoming evident in the dry season when blackflies disperse along the rivers.

Radial dispersal is observed in bush savanna during the wet season and in forests throughout the year. Early in the day, when humidity is high, blackflies traverse the grasslands by flying from one clump of bushes to another. They may then bite man in open fields or even in villages, many kilometres distant from their emergence sites. In the forest zone, blackfly dispersal will not be limited by dryness due to solar radiation but rather by the availability of sufficient light for flight orientation. Simuliids may not only disperse along the edges of broad rivers, but also travel rather long distances through deep forest, following small rivers and attacking man in natural clearings and plantations. A quantitative comparison of biting density in different bioclimatic zones (e.g. savanna and forest zones of West Africa) reveals a multiple of the number of attacking (biting) flies/

man/year in the latter zone by comparison with the former one. Twelve-hourly catches have been undertaken weekly at selected stations for a full year or even for several years. The estimation of nulliparous and parous females (parous rate) and of microfilaria of *Onchocerca volvulus* in the first, second or third (infective) stage, permits calculation of the annual transmission potential (ATP) of onchocerciasis (Duke 1968c). It has been estimated that a man exposed to infection over all the hours of daylight for a year will receive 14 000 infective bites potentially transmitting 92 000 infective larvae (Duke 1968a). In reality, of course, the figure is subject to great variation, since not all third-stage larvae are truly infective, while some do not leave the vector during its blood meal, and the hypothesized full year's total exposure to bites is hardly likely to occur for any one individual (Garms 1973). Moreover, much confusion may be caused by the presence, in *S. damnosum s.l.*, of infective larvae of worms other than *O. volvulus*, but hard to distinguish from this species as third-stage larvae (Duke 1967; Garms and Voelker 1969).

The *differential dispersal* of nulliparous and parous blackflies has much significance for the epidemiology of onchocerciasis. In the Sudan- and Guinea Savanna Zones, nulliparous female *S. damnosum s.l.* usually disperse inland, away from their riverine emergence sites. Parous blackflies (especially older ones) characteristically remain close to the banks of the rivers where they developed. However, they may still fly far from the emergence sites along the course of the river. Consequently, the risk of infection is highest close to the breeding rivers. In the forest zone, the overall biting density is much lower inland than along source rivers. Duke (1975) found proportions of parous and infective blackflies higher inland (and onchoceriasis more widespread there) than along the river banks.

The *survival* of *S. damnosum s.l.* populations during the dry season may depend

upon the availability of permanent breeding sites and long-distance migration of
the adults (Lamontellerie 1964; Ovazza *et
al.* 1965, 1967). Such migration may be
facilitated by the harmatan winds, as
proved to be the case in reinvasions of
the Volta River Basin after suppression of
the local blackfly populations by larviciding. At the end of the dry season, blackflies sometimes appear quite suddenly
and in remarkably high numbers, shortly
before the rivers begin to flow or at that
time. Such simuliids are parous and may
even carry infective larvae. It was presumed earlier that they survive as adults
for at least part of the dry season (Marr
and Lewis 1964). However, during a receding flood, there may be survival of
eggs incorporated in the bottom sediments, which remained moist during the
dry season—this was established for *Austrosimulium pestilens* (Colbo and Moorhouse 1974). It could partly hold for *S.
damnosum s.l.*, too (e.g. in seasonal rivers
of the Guinea Savanna, whence migration into the Sudan Savanna would in
volve shorter flights than from permanent breeding sites).

Flight-range during migration has been
estimated by marking, release and recapture experiments in the Forest Zone of
Cameroon. One day later, *S. damnosum
s.l.* females were recaptured, 56 km and
79 km, along the river in question from
the release site. In these instances, the
air-speed probably exceeded the flight
capability of the blackflies themselves,
the displacement being due to an active
take-off, followed by a period during
which the insects simply had to remain
airborne until an involuntary landing.
Thus, the altitude of such flights may
decide the distance of migration. Along
small rivers, far less dispersal takes place,
and Thompson (1976b) reported that biting density declined even more rapidly
along a road leading away from a large
river. In the Mehenge Mountains, Tanzania, about 0.5–1.3% of marked females
were recaptured within 0.5 km of the
release point. The greatest distance re-

corded there was 5 km for one female, six
hours after release at 08.00 h (Häusermann 1969). In Guatemala, females were
recaptured up to 62 and 85 days after
release, having covered distances of
10–15 km (Dalmat 1950b, 1952, 1954).
Techniques for mass-tagging by radioiso-
topes (e.g. P32 or S35) were reported by
Fredeen (1953c) and Gillies (1958). The
intrinsic physiological flight range for unfed simuliids in still air can be estimated
by assessing the energy available to the
insect and the power required for its
flight and metabolic efficiency. The maximum range for *S. venustum* has been
calculated as 112 miles and the flight-
endurance of various biting flies has been
measured from 7–35 h (Hocking 1953).

The *longevity* of blackflies in the field is
difficult to estimate directly. However, in
captivity, survival times have been reported equivalent to those obtained in
the recapture experiments of Dalmat,
already mentioned. Davies (1953) found
that blackflies, especially when fed on
sugar alone, could be maintained for up
to 63 days. The age of wild *S. damnosum
s.l.*, caught coming to feed on man, can
be estimated from the stage of development (based on a six-day scale) of *O.
volvulus* parasitizing them. From such
data, models have been constructed to
illustrate the pattern of biting activity of
forest *S. damnosum s.l.* populations in
relation to transmission of *O. volvulus*. If
nulliparous blackflies are assumed to
attack on day 0, most parous ones appear
on day 3 for their second or subsequent
blood meal, and to their third or subsequent meal on day 6 (Duke 1968b). These
calculations suggested that the last reappearance of simuliids attempting to bite
is on day 13–15.

Orientation to the Host

Distant and Near Orientation

Blackflies often attack their source of
blood (man or other vertebrates) at a dis-

tance from the emergence site. It is hardly likely that only a single sensorial stimulus leads them so far. It has to be assumed that endogenous diurnal rhythms of appetitive behaviour, and exogenous stimuli (light, temperature, odour, humidity) activate the blood-seeking flight. Its direction may be influenced, over long distances, by air movement (anemotaxis). Flight orientation to a host at some distance is determined by qualities of the blood host itself (e.g. odour, movement, contrast), which bring the blackfly into close proximity with it. Final location of the host may be determined by its CO_2 output or colour; and sometimes the shape of the host's body attracts the females to certain regions of its body (Wenk and Schlörer 1963; Peschken and Thorsteinson 1965; Fallis *et al.* 1967; Bellec 1974). Bradbury and Bennett (1974b) proposed a corresponding model for long-range, middle-range and close-range orientation to which blackflies respond during orientation to attractive targets. 100–200 ml CO_2/minute, released by a fixed tube on adhesive cardboard surfaces, at ground level, attracted numerous chicken-biting *S. adersi* (Fallis and Raybould 1975). After a blackfly has settled, further sensorial stimuli are necessary to provoke biting or at least probing. These include temperature (convection and radiant heat), odour or sweat, and similar chemical stimuli of the skin surface. In each blackfly species, this chain of events will be linked in a particular way, for host preferences often differ markedly even between subspecies (they may do so between cytotypes, too). In species where mating is observed, or presumed to happen, in more or less close proximity to the blood source, it must be concluded that the first links (i.e. those relating to far and medium distances, and even subsequent ones at close distances), are the same for males as for females.

Until now, the kind of laboratory experiments that have been conducted with mosquitoes, by means of olfactometers, have failed with blackflies. In captivity, even slight differences in the direction of intensity of light, together with negative geotropism, outweigh any other sense stimuli in attracting blackflies. Accordingly, our knowledge of the orientation behaviour of Simuliidae is based solely on field trials carried out with many species, mostly in the Palaearctic or Nearctic Regions. Field experiments have the advantage of being undertaken in conditions natural to the particular blackfly species under investigation. However, in such experiments, essential and accidental events cannot be separated. In the laboratory, the investigator has the advantage of being able to choose the conditions governing his experiments. However, it is not always clear to what extent these conditions are natural.

In field experiments with caged birds and small mammals, females of different Palaearctic blackflies were captured at wind velocities of 2–5 m/s, their preferences for mammals or birds generally relating more closely to host size than to any other apparent factor (Anderson and Defoliart 1961). Bennett *et al.* (1972) showed that when ornithophilic blackflies were attracted to experimental site by an ether extract of the uropygial gland of the common loon (*Gavia immer*),[28] they could be caught by sticky silhouettes of different shapes and colours; *S. euryadminiculum* landing primarily on the head-neck portion of duck decoys. Black cylinders proved distinctly more attractive than others of various colours. In similar experiments (Bradbury and Bennett 1974a), two-dimensional geometric forms, reflecting the least amount of incident light (black, blue and red) proved most attractive and a matte surface was found superior to a glossy one. Females of *S. euryadminiculum* are able to discriminate between nearby targets on the basis of colour, independent of the amount of CO_2, up to 180 cm downwind

---

[28] Also see p. 289. Ed.

from the source. In medium-range orientation, CO_2 is the main attractant; and in long-range orientation, host-odour predominates (Bradbury and Bennett 1974b).

When 15 cm squares of coloured material were placed on human bait, in pairs, blood-seeking female *S. venustum* also selected certain colours more often than others (Davies 1972). The landing frequency varied inversely with the intensity of reflected light, whether with neutral (white, greys and black) or with colours. With respect to UV, white paper of higher UV-reflectivity proved less attractive. If intensity was constant, green, yellow and orange were less attractive than white and much less so than maroon, purple or grey of the same intensity. The purer the colour, the fewer blackflies landed on it. Attractiveness of the coloured materials varied directly with the ratio of the reflectance from 450–$500 \mu m$ over that from 500–$550 \mu m$ (Davies 1972).

In forest areas, *S. damnosum s.l.* females rely heavily on odour as an attractant and, to a lesser extent, on sight and the exhaled breath of a moving person. Odour appears to be the only obligatory attractant. For "savanna" *S. damnosum s.l.*, neither odour nor exhaled breath appear to be important attractants; some other factors (perhaps sight?) are more significant in this zone (Thompson 1976c). This was confirmed by experiments with animal-baited traps containing a chicken or a sheep (Thompson 1977). Sweat-soaked clothing, worn for several days (Thompson 1976d) was tested for attractiveness to *S. damnosum s.l.* The positive response of which, to trousers as compared with shirts, seemed to agree with the known preference of these blackflies for biting the lower parts of the human body.

Attacking Behaviour

As indicated, at the end of the previous section, most *S. damnosum s.l.* bites are on parts of the human body closest to the substrate. A relevant factor here is the matter of vegetation cover. In the savanna zone, the height from the ground at which biting takes place is influenced by the presence or absence of high, dense grass (Duke and Beesley 1958).

In experiments with lightly-clad Africans in the savanna zone, standing or sitting near an emergence site, 93% of *S. damnosum s.l.* first landed on the ankles, but did not begin to feed until reaching the lower parts of the calves. In the case of people sitting on the ground, some blackflies bit the underside of the thighs. In people harbouring *O. volvulus*, microfilariae have their highest concentration around the hips—which *S. damnosum s.l.* hardly ever bites, even when the person concerned is sitting on the ground. It thus seems that attacking behaviour and the bodily distribution of microfilariae do not fully correspond; a matter of decided epidemiological importance which requires further study (Renz, in preparation).

The pattern of attacking behaviour, described above, appears to be optically oriented. However, this remains to be confirmed by experiments with dummies. It should be remembered that female Simuliidae have been shown capable of orienting to specific parts of a wooden dummy (the silhouette of a horse) by sight alone; *B. erythrocephala* attacking prominently protruding parts, and *O. ornata*, the flat underparts (Wenk and Schlörer 1963).

The timing of the first blood-meal is difficult to determine. In forest areas of Cameroon, parous *S. damnosum s.l.* bite earlier in the day than nulliparous ones. Only a small proportion of adults emerging on a particular morning, bite that afternoon; most of them do so on the following afternoon. On the other hand, the parous rate of man-biting *S. damnosum s.l.* is higher in the West African savanna than in the Forest Zone. Le Berre (1966) postulated that this was due to differences in the average longevity of the blackflies of various zones. However, there are regional differences in the degree of zoophily, especially ornithophily, in the Sudan Savanna of Cameroon (Duke 1967; Disney and Boreham 1969) as well as in the Forest Zone of Liberia (Garms and Voelker 1969). Most probably, the longer a female seeks a blood-meal without success, the more readily will it accept alternative hosts. This has been demonstrated in field experiments where blood was equally available from a chicken and a man. From 10.00–14.00 h, a statistically significant lower parous rate was observed attacking the chicken (Disney 1972a). It may be that, in nature, the first blood-meal of *S. damnosum s.l.* is taken from a bird if a man is not available.

As soon as a female blackfly has alighted on a blood-source, it begins to probe if temperatures are suitable (32–38 °C, optimum 37 °C). Engorgement is stimulated in laboratory experiments, with membrane feeding by 10^{-4} ATP resp. 10^{-4} ADP (Sutcliffe and McIver 1975). This finding was found to hold good for a variety of blood-feeding arthropods other than blackflies (Galun and Margalit 1969).

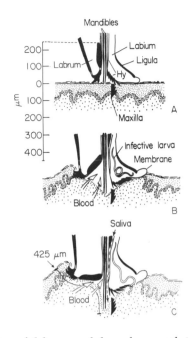

Action of labrum and hypopharyngeal teeth in Simulium *sp. penetrating human skin. Lateral view of distal third of proboscis. A, position before feeding (Hy, hypopharyngeal teeth moving repeatedly forth and back). B, intermediate position during penetration. C, final position after maximal penetration achieved.*

Feeding Behaviour

Function of Mouthparts

In *Simulium damnosum s.l.*, the duration of feeding from humans averages 4–5 minutes (Crosskey 1962). The body of the insect lies close to the surface of the skin, above which the fore-tarsi are kept slightly raised. The manner of action of the mouthparts may be inferred from the anatomical structures, supported by direct observation of living blackflies, mounted ventral side uppermost, under a dissecting microscope (Wenk 1962).

As soon as the proboscis is applied to the skin, the latter will be tautened by the combined action of the labrum and hypopharynx. The labrum anchors to the skin by a pair of small hooks, curving anteriorly when the surface of the hypopharynx (which is furnished with many small hooks) glides backwards. The latter may be observed directly from life (see previous paragraph) when a female, under microscopic examination, is touched with a dissecting needle.

The two maxillae are protruded alternately, due to the flexing of their basal (occipital) sections. The mandibles exhibit rapid scissor-like activity. Actual skin penetration is achieved by flexion of the chitinous components of the mouthparts as is evident from the following anatomical features:

The penetrating components (mandibles, maxillae) are compressed between the forward-curved labrum and the activated hypopharynx with its tip directed backwards. The maxilla penetrates downwards, as a result of flexion of its base, which is linked to the hypopharynx by a small rod. Both rows of hooks on each maxillar stylet, being directed backwards, now anchor the proboscis in place. Next, the mandibles (each of them articulating separately with the labrum and the hypopharynx) cut into the skin. This may be repeated several times until a capillary is reached and blood can be taken from a pool in the wound. Strong muscles act as feeding pumps in the cibarium and pharynx (behind and in front of the brain). Ingested blood passes immediately to the midgut, only sugar solutions being taken into the crop. Saliva may be directed into the wound along the hypopharynx *via* the salivary duct. Since the two labella are spread sideways, the very thin membrane between them (ligula) is stretched and lies close to the tissue exposed by the wound. It is highly probable that the infective larvae of *Onchocerca* spp. penetrate this membrane and move immediately into the open wound.

How far the mouthparts penetrate into the skin of the host may be estimated by measuring the relevant parts of the labrum and the mandibles. In *S. damnosum* s.l., the author measured the length of the bending part of the labrum as 237 μm and the protruding parts of the mandible as another 237 μm. However, the penetration depth may be somewhat less than the sum of these figures (c. 425 μm). Microfilariae are taken up with the blood. However, in several species there are hooks inside the pharynx which may damage the worms when the cibarial pumps are in action and, alternately, constricting the sclerotized paryngeal parts. Damage to microfilariae was observed in *O. ornata, S. metallicum, S. ochraceum, S. veracruzanum, S. vorax* and *S. damnosum* s.l., feeding on human volunteers, carrying onchocerciasis (Omar 1976; Omar and Garms 1975, 1977; Reid 1978).

Digestion of the Blood-meal

As soon as blood has entered the posterior part of the midgut (which is located in the abdomen), the gut cells begin to secrete a peritrophic membrane, which is completed in 30 minutes. Six to 12 hours later, the membrane displays up to seven concentric laminae. Within the first 48 hours of blood digestion, the posterior part of the peritrophic membrane (where digestion is most advanced) begins to decompose. This membrane disappears completely by the end of blood-digestion. If a blood–sucrose mixture goes directly to the midgut, a thin membrane forms around it; but this does not happen if it goes by way of the crop. A "prefeeding section" in the anterior midgut (surrounding a small, but variable, amount of material, namely the meconium) is considered to be the peritrophic membrane of the pharate adult (Yang and Davies 1977). Enzyme activity of invertase increased in the midgut of *Prosimulium fuscum* and *S. venustum* after a blood-meal (Yang and Davies 1968b). *In-vitro* trypsin activity was mainly confined to the midgut, and did not occur in the salivary glands when blood–sucrose mixtures were fed (Yang and Davies 1968a).

Identification of Blood-meal by Precipitin Tests

The blood-meals of haematophagous arthropods may be identified by serological techniques elaborated by Eligh (1952) and Weitz (1952, 1956). However, the digestion rate must be taken into consideration (Weitz and Burton 1953). Some Palaearctic species, e.g., *S. tuberosum*, contained blood from man, cow, horse, dog and bird. On the other hand, *E. latipes* had fed mainly on birds, although it sometimes harboured composite blood-meals of mammalian and avian blood (Davies *et al.* 1962). Some females of *S. damnosum* s.l., caught on vegetation near the emergence sites, proved to contain a blood-meal of primate or avian origin. *S. schoutedeni* and *S. griseicolle*, which were caught in larger numbers, mostly contained avian blood, although some exhibited blood of primate origin (Disney and Boreham 1969).

Salivary Gland

The salivary gland of the adult female consists of a proximal gland formed by large cells with nuclei containing giant chromosomes and an accessory gland with normal cells, leading to the salivary duct. The basal part of the main gland close to the accessory one is mostly pigmented, and of variable size in different species and genera (Bennett 1963). Obviously, the main and accessory glands produce different proteinaceous components of the saliva. Twenty minutes after a blood-meal, cisternae are observable in the cytoplasm of the accessory gland cells, indicating that the structure is more than just a reservoir (Welsch *et al.* 1968). In males, only the accessory gland is present. The main gland in females seems to be a specialization concerned with blood-feeding. In addition to the main- and accessory gland cells, a small variant of main gland cells has been found by histological and histochemical techniques (Wächtler *et al.* 1971). Both glands are fully differentiated at the time of pupa/adult ecdysis. Soluble proteins from the one-day-old pharate female show a very similar pattern to those of the pupal and male glands. With growth of the female main gland and secretion of saliva into its lumen, increasing amounts of proteins—which have been characterized by Poehling *et al.* (1976)—become evident (Poehling 1977). In tests with mammalian and avian blood, agglutinin activity was found in saliva from females of *S. damnosum*, *S. venustum* and *S. vittatum*, which were at least 12 hours old, but not in saliva from 2-hour-old females of *S. decorum*. The anticoagulant activity of the saliva varied among individuals. Heat destroyed this activity (Yang and Davies 1974). After blood-feeding, a PAS-positive substance is lost and synthesized again within the next 24 hours (Gosbee *et al.* 1969).

Uptake of Sugar and Juices

Blood ingested from a host immediately passes to the midgut, while plant juices and sugar solutions are first taken into the crop. Artificial solutions of blood and sugar are partially absorbed in both locations (Wenk 1965c). However, studies of wild-caught female *S. damnosum s.l.*, as well as simuliids reared from pupae, proved that sugar may be dispatched directly to the stomach in a high proportion of adults, and that, in a significant number, none enters the crop at all. Thus, sugar in the stomach of wild-caught blackflies does not indicate autogeny (Disney 1970a).

Little information exists on the *flowers* from which juices are taken by most African Simuliidae. Some Palaearctic species feeding on man, cows, horses and ducks, occur regularly on specific flourishing plants; namely, the White Willow (*Salix alba*) in spring, the yellow blossoms of Parsnip (*Pastinaca sativa*) in summer, and the green blossoms of Ivy (*Hedera helix*) in autumn. Many other Umbelliferae are disregarded, even when abundant nearby. The blossoms on which feeding takes place, are generally inconspicuous, of yellowish to green colour, and with open nectaries. They are most probably located by olfaction senses (Wenk 1965a). In one instance, 30 female *O. ornata* were found in the Cuckoo-pint [*Arum conophalloides*] (Knoll 1926).

Resting Sites

Since blackflies present on flowers are mostly inseminated, and nulliparous as well as parous (some of them exhibiting a blood-meal or fully-developed eggs), a sugar meal may be taken up at any time—but mostly during the nocturnal resting period. Considerable numbers of females and males of *S. schoutedeni* and *S.*

griseicolle, and some of *S. damnosum s.l.* have been collected from the underside of leaves in vegetation bordering rivers of the savanna zone, Cameroon; some of the females were gravid or engorged with blood (Disney and Boreham 1969). However, compared with the numbers of blackflies attacking man, sometimes at a considerable distance from breeding sites, the recorded collections from plants are small. The actual origin of ingested juices remains virtually unknown for *S. damnosum s.l.*, although large numbers of males, and nulliparous and gravid females of this complex, have been collected from vegetation (of unidentified plants) in the vicinity of an emergence site (a dam), following sunset. Near the river, they were found resting low on clumps of grass. The resting zone followed a definite upward gradient away from the river, 37 m from which the blackflies hovered above trees before alighting at the side of leaves up to 5–6 m from the ground. They then crawled to the underside and settled at the tip. No specific plant preferences were noticed, other than a general preference for vegetation with drooping, rather than erect, leaves; the adults were seen on clumps of a variety of grasses, bushes, trees, guinea corn and millet (Marr 1971).

Extrinsic Cycle of *Onchocerca volvulus*[29]

O. volvulus microfilariae reach the midgut of feeding *S. damnosum s.l.* together with the ingested blood. However, few of them actually penetrate the wall of the midgut, whether anteriorly (where no peritrophic membrane is formed) or posteriorly (before formation or hardening of this membrane). Entering the haemocoel, the microfilariae move to the large flight-

muscle fibre, the syncytium, for further development.

The long and slender microfilaria, a first-stage larva, shortens and thickens to the "sausage stage". Then it moults, to become a second-stage larva or "intermediate stage" in which the cells of the hypodermis and gut are differentiated (Duke 1968b). The shed skin of the microfilaria remains visible for some time, especially the caudal details, which are taxonomically significant. In so far as is known, all filariae develop as intracellular parasites in their arthropod intermediate hosts. The cells that they parasitize are polyploid, such as those of muscle fibres and Malpighian tubules, the fat-body, and the salivary glands.

The fully-developed third-stage larva is infective to the vertebrate host. It leaves the flight-muscle of the blackfly and enters the haemocoel. Wriggling through narrow apertures, these larvae may be recognized in the abdomen of the host and, more seldom, in its legs, but most of them reach its head, where they move into the labium. When the blackfly next takes a blood-meal, third-stage larva penetrate its body-wall.

Infective larvae can probably survive in the vector for a considerable time and not all of them, present in a blackfly, will leave it during any one blood-meal—some of them will still be evident in a newly-engorged female. On the other hand, some larvae may leave the simuliid when it feeds on sugar solutions. Once their development has commenced in *S. damnosum s.l.*, the mortality of *O. volvulus* larvae is relatively low (Duke 1962b, Philippon 1976). The survival rate of the female is not adversely affected by the parasite except where more than 20 microfilariae are ingested (Duke 1962b).

The parasitic cycle of *O. volvulus* in female *S. damnosum* averages seven days. The second larval stage is reached on the fifth day and the third, on the sixth (Blacklock 1926, Muirhead-Thomson 1957b; Laurence 1966; Duke 1968b). The "epidemiologically dangerous age" be-

[29] With acknowledgements to WHO (1973, 1976).

gins when the vector is able to transmit onchocerciasis for the first time. Where the vector became infected at its first blood-meal, the second meal will take place 3–4 days later—when the first gonotrophic cycle has been completed, and oviposition has taken place for the first time (then, in such a case, *O. volvulus* larvae are at the "sausage" or "intermediate" stage, and incapable of being transmitted). Thus, a female infected during the first blood-meal can only retransmit larvae at the third meal, i.e when it is 8–10 days old. Therefore, the parasitic cycle of *O. volvulus* cannot be completed between two successive blood-meals of the vector (Duke 1962b; Thompson 1976a).

Certain *parasite-vector complexes* are specific for different biogeographical zones (Duke 1968a, c; Philippon 1976). Intake and development of microfilariae of a local strain of *O. volvulus* is more successful in *S. damnosum s.l.* from the same area (e.g. forest zones) than from elsewhere (e.g. savanna zones). Experiments with microfilariae of a Guatemalan strain of *O. volvulus* have shown similar host-related acceptability as regards forest and Sudan-savanna forms in West African *S. damnosum s.l.* (de León and Duke 1966; Duke *et al.* 1967a, b).

As in the case of most filarial transmissions by arthropod vectors, the passage of *O. volvulus* through an *S. damnosum s.l.* female has the effect of reducing the number of parasites (Duke 1962b; Bain 1971; Philippon and Bain 1972). Thus, while one of these females can ingest several hundred microfilariae (maximum 1000), during a single blood-meal, the average number of infective *O. volvulus* larvae per infected blackfly host is only two in the savanna and five in the forest zones (the maxima are rarely more than 10 and 20, respectively). Many microfilariae are presumably damaged by cibarial and pharyngeal armatures (Omar and Garms 1975, 1977; Reid 1978; and see p. 270). There is no direct relationship between the number of microfilariae in-gested and the number of infective larvae completing the cycle. In the case of simultaneous meals taken on the same onchocerciasis victim, the number of microfilariae ingested is extremely variable from one female *S. damnosum s.l.* to another; although it is not possible to single out any factor upon which these variations might depend (age of females, place of bite, length of blood-meal, time of day). In a group of females fed on a person with onchocerciasis, there will always be a small proportion (i.e. of *S. damnosum s.l.*) which has not ingested any microfilariae. Finally, only 40–50% of a whole population of females which fed on an onchocerciasis victim proved infective on the seventh day of survival, at the end of the parasitic cycle (Duke 1962a, 1973; Laurence 1966; Garms and Weyer 1968; Lamontellerie 1965; Philippon 1968; Philippon and Bain 1972).

Apart from the limiting factors already mentioned, the existence of several other phenomena, reducing the intensity of transmission of onchocerciasis by the vector, should be noted:

Since the *S. damnosum s.l.* female takes meals of *sweet juices*, loss of infective larvae may occur at this time, particularly when there is a heavy parasite load, as in certain forest foci (Wanson and Henrard 1945; Duke 1962b; Le Berre 1966).

It has been shown, moreover, that only 80% of the infective larvae harboured by the vector are transmitted during the blood-meal (Duke 1973; Philippon 1976).

Females parasitized by *O. volvulus* have, at least in certain regions, a *mortality rate* greater than that of uninfected females. This is evident from comparing the survival curves for females from the same batch fed on an uninfected person with those fed on an onchocerciasis victim (Wanson *et al.* 1945; Duke 1966; Omar 1976).

The infective stages of *O. volvulus* transmitted to man will give rise to either male or female filariae. The chances of encountering one another in the human host are certainly remote, in view of the small number of larvae inoculated at each bite.

These chances are further reduced, the lower the number of infective bites. The *length of exposure* is, doubtless, important, too-short such exposures favouring encounters between male/female microfilariae.

Finally, blackfly/human host contact, depending on the biology of the former and the behaviour of the latter (e.g. way of life, housing, clothing habits, daily and seasonal activities of both sexes and of different age-groups, etc.) is of the key epidemiological significance.

Under field conditions, the percentage of female *S. damnosum s.l.* infected by, and infective for, *O. volvulus* is directly related to their physiological age—clearly, nulliparous females cannot harbour the worm. *Infected females* (i.e. those containing immature and therefore non-transmissible *O. volvulus* larvae) must be distinguished from *infective females* (i.e., those containing fully-developed third-stage *O. volvulus*). Theoretically, all females can be infected, although only parous ones (at least 8–10 days old, i.e., about to take a third or subsequent blood-meal) can be infective. Adult populations, dispersing from emergence sites, consist essentially of nulliparous females, so that the chance of human infection diminishes with increasing distance from blackfly streams, where the risk is highest in early morning and at midday (parous females biting in larger numbers then); the degree of infection of associated human populations also, of course, being of major epidemiological importance (Wanson *et al.* 1945; Crosskey 1954, 1957; Lamontellerie 1965; Garms and Weyer 1968; Philippon 1968; Duke 1968c; Duke and Moore 1968). Because of the limiting factors listed above, the natural infectiveness of female populations is generally low. It is highest in the savannas, near emergence sites where it reaches an average of 3–5% of the total population, and 6–10% of the parous population. Exceptionally, the theoretical maximum of 25% infective females among the parous ones may be reached when all have fed on onchocerciasis victims (Philippon 1976).

Ovarial Development

After eclosion from the pupa, until just after the first blood-meal, all the ovarioles remain at the same stage of development. Three sections are visible: the germarium at the proximal end, followed by the partially developed second oocyte and the larger, further-developed, first one. Immediately after the commencement of oviposition, a small follicular relict or "rest body" becomes evident below the largest oocyte.

S. damnosum s.l. females show complete gonotrophic concordance, i.e. a single blood-meal (possibly preceded by a plant-juice meal) is necessary and sufficient to ensure maturation of an egg-batch. Autonogenesis—maturation of eggs without a blood-meal—is unknown; while interrupted or double blood-meals are exceptional.

The gonotrophic cycle (i.e. the interval between two successive ovipositions) comprises three main periods:

(1) that between an oviposition and the following blood-meal, during which the meal of sweet juices is taken and a blood-source sought. In nulliparous females this period (av. 24 h) corresponds to the interval between eclosion and the first blood-meal, during which dispersal may occur;

(2) that between digestion of the blood-meal and maturation of the ovaries, when *c.* 3–4 days are passed at resting places (see p. 271) about which little is known; and

(3) that (up to 24 h) occupied by the search for an oviposition site.

The total duration of a gonotrophic cycle is thus 3–4 days for the first cycle (nulliparous females) and 4–5 days for the following ones (parous females) [Wanson

Table 1. Characteristics used for age-grading *S. damnosum s.l.* females.

	Nulliparous	Parous
Follicular relicts	Never present	Present
Ovaries	Tight, disrupt as soon as stretched	Loose, and easily stretched
appearance	Translucent	Granular
retained eggs	Never present	Present
Spermatophore	Sometimes present in "young nulliparous"	Never present
Fat-body	Large and numerous cells present	Reduced, or completely absent
Malpighian tubules	Rather dark	Progressively clearing with age
Meconium	Sometimes present in "young nulliparous"	Never present

and Lebied 1948; Edwards 1956; Duke 1968b; Le Berre 1966].

The ovarial development stages are the main characteristics used for *age-grading* the females (Table 1).

The follicular relict or "rest body", below the terminal, larger oocyte is present in parous blackflies only. It may appear as a more-or-less distended and empty part of the membrane of the ovariole, when the oviposition has taken place within the previous two hours or so. Later on, it shrinks and becomes difficult to detect. Other secondary characteristics are then useful, especially when large numbers of simuliids have to be dissected.

The ovaries have considerable elasticity in parous blackflies; those of nulliparous ones being fragile. Furthermore, ovaries of parous blackflies have a granular appearance and are less translucent than those of nulliparous ones. Sometimes retained eggs may be present in parous blackflies.

A spermatophore (a small, two-chambered transparent ball with two short, lateral projections)[30] may fall from the genital opening of a freshly-mounted specimen for dissection. This is indicative of a young, nulliparous blackfly that had

mated 12 hours, or less, previously. Found in 40–80% of females, it indicates that mating takes place close to the capture point.

The fat-body, stored during pupal development, is rather large and voluminous in nulliparous blackflies. In parous ones, only a few small cells are evident, if, indeed, the fat-body has not completely disappeared.

In young adults, the Malpighian tubules, being full of excretions from the pupal stage, appear rather dark. In parous examples, the proximal sections are empty, older such blackflies exhibiting tubules empty through their whole length.

A meconium—residual gut-contents of the pupal stage—may be present in freshly-hatched nulliparous flies.

However, the presence or absence of follicular relicts is the only completely reliable criterion, the fat-body being a useful supplement in routine dissections (e.g. in *S. ochraceum*). It is not yet possible to distinguish uniparous and multiparous females.

The first oviposition comprises 400–600 eggs, the number depending on the size of the female. The average size of eggs is larger in forest than in savanna-taxa of *S. damnosum s.l.* The number of ovarioles decreases at each subsequent oviposition (Wanson and Henrard 1945; Lewis 1958;

[30] See pp. 261–263 (Ed.).

Le Berre 1966). Statistical analysis of the number of ovarioles in parous *S. venustum* has revealed that there are up to 5–6 ovarian cycles (Mokry personal communication). This corresponds with the calculations of longevity for *S. damnosum s.l.* and may be true for this complex, too.

The *longevity* of the females can be estimated by calculation, or by direct study, of the regression and aging of the population after the application of larvicidal measures. The survival probability is calculated from the age-composition of the population. *Simulium damnosum s.l.* females have a maximum longevity of one month (the greatest actually reported being 23 days). Consequently, a female cannot complete more than six gonotrophic cycles during its life-span. Variations in longevity are revealed by the average age of the populations. *Simulium damnosum s.l.* longevity is much greater in the savanna than in the forest; and in the savanna in the dry season as compared with the rainy one. That is why, out of 100 eclosions on a given day, in different bioclimatic regions, the number of females surviving on the ninth day has been recorded as three in rain forest, 33 in the Guinea Savanna, and 47 in the Sudan Savanna (Le Berre *et al.* 1964; Le Berre 1966).

Oviposition

Orientation to oviposition sites is highly complex. The female simuliid must first locate running water suitable for larval development, somewhat upstream of her own emergence site, thereby compensating for larval drift (Davies 1962). Positive anemotaxis leads the female upstream, flying immediately above the water (Balay 1964). Perhaps the glittering river-surface is itself a marker for running water, for aluminium panels, 1 m², attract gravid females (and also newly-emerged ones, and males, too) in great numbers (Bellec 1976). Transparent interception

traps, placed above the surface, across the entire stream, have captured large numbers of *S. damnosum s.l.* (of which 94% were migrating females, and 72% gravid) just before sunset, in the section upstream of the main emergence site (Bellec and Hebrard 1977). In Guatemalan rivers, *S. ochraceum* were found to oviposit mainly at midday or in the early afternoon (Garms 1975).

S. damnosum s.l. females oviposit, at the water surface, on vegetation or a rock, so that the mucus secreted at that time is immediately wetted. Each egg-batch comprises 100–600 eggs, deposited together. The colour of the substrate has been shown to influence selection of the site by *S. verecundum* and *S. vittatum* females, which prefer yellow above all other colours, and white over black (Golini 1974, 1975). In one particular river area, various species of Simuliidae may be separated through their individual behavioural patterns with respect to substrate, water turbulence, etc. Some species oviposit directly onto the water surface, their eggs sinking to the bottom (Rühm 1971a; Moorhouse and Colbo 1973). *S. argyreatum* even selects the artificial outlets of fish ponds and roofed, wooden troughs, from the exterior of which the actual water surface cannot be seen (Rühm 1975).

Artificial oviposition may be induced in gravid females, in the laboratory, by exposure to CO_2 (Dalmat, 1950a). However, decapitation may be useful (Lewis *et al.* 1961) or simply a light "tap" on the head. When a blackfly, ovipositing on wet filter paper, is gently but continuously moved, its eggs may be aligned in a row. This facilitates counting the number of fertile and infertile eggs a few days later (Wenk and Raybould 1972). When single females are kept in small vials containing a strip of filter paper, it often suffices to half-fill the vial with water and shake vigorously. In the laboratory, *S. columbaczense* oviposits readily even on dry surfaces (Zivkovic 1958).

Long-term laboratory storage of eggs

has been demonstrated, up to two and one-half years, for *Austrosimulium pestilens* (Colbo and Moorhouse 1974). However, this may be exceptional. Eggs of *Boophthora erythrocephala*, separated from substrate material and submerged in water containing the bactericide Desogen Geigy® 0.01%, were kept in petri dishes for six months with close to full survival (Wenk 1965c). About 90% of *S. damnosum s.l.* eggs hatch after 16 days. In one test (Raybould *et al.* 1973) only 14% of the batch survived. *S. damnosum s.l.* may be airmailed on moist filter paper or cottonwool, covered with plastic foil, with low, but adequate, survival for up to five days (Grunewald, personal communication).

Sense Organs

The Complex Eyes of the Male

Simuliid males are easily distinguished from superficially similar Diptera by their subdivided complex eyes. These consist of two kinds of ommatidia and cover almost the entire surface of the head, except the "face" of the blackfly—namely, the proboscis, the clypeus and a small frons where the antennae are inserted. Whereas the ventral parts of the complex eyes show cornea-lenses of normal size (10–15 μm dia.) and characteristic ommatidia, there are cornea-lenses of a remarkably greater diameter (25–40 μm) above the "equatorial line" of the head-capsule. The zones covered by these large cornea-lenses meet at the median dorsal line in a slightly-lowered suture. The retinular cells and their rhabdomeres are very long (approximately 300 μm), penetrating the basement membrane and continuing for about two-thirds of their own length (sometimes on a slightly curving path) to the optical lobes. The retinular cells and rhabdomeres of the ventral eye resemble those of other Diptera (length, 50 μm). The light-brown screening pigment in

the dorsal eye is translucent for light of longer wavelengths. The ventral eye has dark, reddish-brown screening pigment. The construction of the male dorsal eye is unique among insects. A detailed histological description was published by Dietrich (1909). The eye of simuliid females resembles the ventral one of males.

The function of the male dorsal eye is to detect the small (3 mm × 1 mm) females against the blue sky as they fly above the swarm of males. On detection, the females are then pursued by the males and seized for mating (Wenk 1965b). At dawn, the eye functions close to the physical limit imposed by the quantum fluctuations of light. Under bright illumination, light quantum variation is not a limiting factor; as can be shown by behavioural experiments (Kirschfeld and Wenk 1976).

The Antennae

In simuliids, each antenna is compact in both sexes. It consists of nine flagellar segments, an urn-shaped pedicel containing a simple Johnston-organ and a small basal scape which has a distal, slightly projecting, edge on the outer surface. The flagellar segments are cylindrical, the distal one (the ninth) conical. In *S. damnosum s.l.*, the length of the entire female antenna may vary considerably in relation to body-size (about 400–750 μm). This was one of the criteria used by Quillévéré *et al.* (1977b) and Quillévéré and Sechan (1978) to distinguish their cytotypes.

The types, numbers and distribution of the antennal sensilla of male and female *S. rugglesi*, *S. baffinense* and female *S. venustum* and *S. euryadminiculum* have been determined by Mercer and McIver (1973a). Seven sensillar types, including olfactory receptors, contact chemoreceptors, and mechanoreceptors, were distinguished in all cases. Although significant differences exist with respect to type, number and density of sensilla, as between mammalophilic and ornithophilic blackfly males and females, and autogenous or anautogenous species, it is not understood which stimuli pertain to the various simuliid behavioural patterns.

The Maxillary Palp

The shape of the maxillary palps is similar in both sexes. The palp is very small and is scarcely visible before dissection. The first two segments are tiny. The basal one is provided with muscles, and the second is cylindrical. The third, the broadest, exhibits a sensory peg, the so-called Lutz' organ [this name having been proposed by Hechler (1978), as the organ was first described by Lutz (1910)]. A somewhat smaller segment than the third, resembles the latter. From it, originates the long, slender, terminal segment, which may be longer than all the others combined. In the resting position, the terminal segment of both palps is held backwards, the fourth and third segments forming upward curves, so that the openings of the Lutz' organ are directed forward. When the blackfly is activated (e.g. by bringing an open hand to the netting of the cage), the terminal flagellar segments are turned anteriorly as the insect clambers nearer.

The types and numbers of palpal sensilla have been determined for the females of *S. venustum* and *S. euryadminiculum*, and for both sexes of *S. rugglesi* and *S. baffinense* (Mercer and McIver 1973b). Four types of sensilla occur on the palps: (1) sensilla chaetica (probably mechanoreceptors); (2) the thin-walled, capitate pegs, or bulb-shaped bodies, clustered on the floor of the Lutz' organ; (3) thick-walled pegs (probably contact chemoreceptors); and (4) campaniform organs (probably mechanoreceptors).

Most of the sensilla chaetica and all the thick-walled pegs are on the three distal segments. Circular cuticular depressions surrounded by a ring of raised cuticle, including a convex area, are located distally on the ventral surface of segment two and the outer ventro-lateral surface of segment three. The external appearance and location of these structures suggest their being campaniform organs.

The size and number of sensilla in Lutz' organ show considerable variation during the period of plant growth in the Palaearctic; however, the length of the sensilla shortens continuously (Hechler 1978). Between March and October, the number of sensilla per Lutz' organ was counted monthly in *O. ornata* (95-102-x-53-60-70-78-93). During the same period, the length of the sensilla decreased from 1.5 to 1.2 scale-units. When five different species were compared with one another, at the beginning of plant growth, the females of the ornithophilic *E. latipes* exhibited twice as many sensilla (369.5) as the mammalophilic *W. lineata* (166.5), which is of comparable body-size. *S. argyreatum*, which is autogenous in the first cycle, has the lowest number (54.5) at that time; the mammalophilic *B. erythrocephala* (82.0) and *O. ornata* (101.0) fit neatly into the hypothesis that ornithophilic blackflies have the greatest number of sensilla. The males of all species tested have a low mean number of sensilla (22.0–38.0), without significant differences among them. The size of the Lutz' organ in relation to body size, gives almost equal ratios in the three mammalophilic species. The relative figure is significantly higher in the ornithophilic *E. latipes*, and lower in the autogenous *S. argyreatum* females. Males have lower numbers of sensilla than females. Finally, the length of the sensilla increases with body size up to a certain limit. This must be taken into consideration, too, when measurements of a complex sense organism are used for speculations concerning its function on a comparative basis (Hechler 1978). Syme and Davies (1958) suggested differentiating some Canadian simuliid species on the basis of the structure of the Lutz' organ.

The ultrastructure of the bulb-shaped sensilla shows that the head and the stalk surface is covered with about 3000 pores, the density of which is highest in the head (50 pores/um^2). The construction of the pores resembles a sensillum basiconicum. A tunnel leads into a cavity from which 7–11 tubuli have contact with the dendrite. The latter is evident in the stalk as an undivided line provided with many lengthwise tubuli. Entering the head, the dendrite divides into several layers. In sections, it appears as spirals surrounding a central fluid-filled space. Many neurotubuli are found between the layers, always in single rows.

Normally, each sensillum has only one bipolar sensory cell. Below the basal part of the stalk, the proximal segment of the dendrite is separated from the distal one by a constriction with two ring-like basal bodies. Distally, there is a zone of nine double tubuli. These extend into the outer region of the dendrite, where further (triple) tubuli are seen. There are two tormogenous cells which send microvilli to the base of the dendrite, which is covered by extensions of the trichogenous cell. The latter surrounds the sensory cell, the neurites of which form the nervus palpus maxillaris (Hechler 1978).

Finally, Elizarov and Chaika (1975) have shown by EM that in female (?) *B. erythrocephala*, a labellar sensillum has 4–6 neurones of which one is probably a mechanoreceptor. There are, moreover, three types of hair-like olfactory antennal sensilla: one has thick walls and two neurones; the second has 3–4 neurones; and the third, which has thick, perforated walls and microtubules associated with pores, has 5–6 neurones.

The Sensilla of the Legs

Eleven different sensillar types are present on the legs of (Canadian) *S. rugglesi*, *S. arcticum*, *S. baffinense* and *S. euryadminiculum*. They include seven probable mechanoreceptors, two probable contact chemoreceptors and two possible olfactory receptors. For the most part, males have the same types of sensilla. In the case of the bifurcate type, however, males consistently have fewer sensilla on the legs than do females. This is so for both blood feeding and autogenous species. However, the peg sensilla of autogenous blackflies differ morphologically from those of anautogenous ones (Sutcliffe and McIver 1976).

VI Trapping Technology

Trapping Technology—Larval Blackflies

G. Carlsson, P. Elsen and Karl Müller

Investigations on preimaginal blackfly bionomics face many difficulties. Where, when, and how do the different species of larvae and pupae live, develop and emerge—for example, in large rivers where attachment sites are more or less inaccessible for purposes of observation? To counter such difficulties, and to obtain detailed information on preimaginal simuliids and their ecology, methods have been devised and, indeed, used for many years.

When constructing traps for use in investigations of larval and pupal blackflies, every effort must be made to ensure that the devices should: cover or compensate natural larval and pupal habitats, ecological niches, etc.; when activated, permit and preimaginal stages to continue their development, unimpeded; give qualitative and quantitative results yielding valid information on the natural distribution of the different species at the investigation sites; allow direct comparison of results among different biotopes; provide for both active and passive displacement of the blackfly larvae; furnish data on interrelationships between the

preimaginal stages and key environmental factors, i.e. current velocity, depth, light, nature of substrate, etc.; reveal patterns caused by intrinsic factors, such as biological clocks; be sturdy enough to withstand abrupt changes in water velocity; and not be vulnerable to algal growth, etc., when in use.

Unfortunately, a trap fulfilling all the above conditions has yet to be designed. However, many presently in use, give results satisfactorily reflecting the abundance of preimaginal stages of various species of blackflies in running waters.

Natural Substrate Traps

Over the years, a variety of "natural" substrates have been employed to catch blackfly larvae. Leaves, stones, and branches of trees have been removed from the study site, cleaned, and replaced in the stream in different ways. This has rendered it possible to determine the rapidity and rate of blackfly colonization, the prevalence of larvae, their preferences for particular substrates, and their relationship to environmental factors, such as current velocity, food supply, depth, light, surface of the substrate, physicochemical conditions, etc (Müller 1953; Carlsson 1962). Although such traps give valuable information on larval distribution and the dependence of the preimaginal stages on the factors referred to, natural substrate traps are seldom used today; for it is difficult to standardize them and comparisons between results from different localities may be misleading.

Artificial Traps

Boards

Boards of varying sizes, secured in the stream in different ways, were ex-

Plastic strip with larvae and pupae. Ivory Coast. Photo by P. Elsen.

perimented with over a decade. Large ones, covering many square metres, were used to obtain large numbers of blackfly larvae immediately downstream from lakes, to provide food for salmon hatcheries. This method proved a useful means of sampling over relatively shorter periods. After several weeks, though, they often became coated with algae and fungi, which precluded satisfactory results.

Plastic Strips

Different kinds of plastic strips have been in use for at least 20 years. Their size, colour, and manner of placement in the stream vary greatly (Marr 1962; Rühm 1975, 1976; Elsen and Hébrard 1977b; Carlsson and Müller 1978).

Experimentation has revealed how different species of Simuliidae react with respect to colour, size, shape and surface conditions of the substrate, current velocity, depth, light, food supply, and physico-chemical conditions. It has also been possible to determine where, when and how various blackflies oviposit, also the duration of embryogenesis, dispersal of different stages of larvae, pupation, etc. (see Rühm 1975, for exact methodology). It is preferable to use a methodology lending itself to direct comparisons when studying blackfly larvae in widely-separated areas and in different micro-habitats. Excellent results have followed the use of small plastic strips, 9 cm × 2 cm × 0.2 cm (total area 40 cm²), securing in running waters in the Palaearctic regions. However, in Africa (and probably elsewhere in the tropics), it seems that the radically different simuliid fauna poses a need for longer plastic strips (*c*. 1 metre). The latter methodology gives adequate information on the status of preimaginal stages of blackflies at a particular time; and the results obtained are directly comparable among different localities. Moreover, the trap is economical and easy to construct and handle.

For longer-term investigations, many alternatives to the plastic strips have been devised; particularly good ones being the "piquet gradué" and "pan-

The "panneau-drapeau" device, dismantled and mounted in situ. *Ivory Coast. Photo by P. Elsen.*

neau-drapeau" devices of Elsen (1977b) and Kureck's (1969) "Schlüpfrinne".

(1965), Dejoux and Elouard (1975), and Elouard and Leveque (1977).

Nets

Drift-nets of various kinds have been used to investigate the importance of blackfly larvae to aquatic ecosystems. These are stretched across the stream in such a way that they neither extend above the surface not touch the bottom. The nets can be fixed in the water with the aid of wires and/or stones, in such a manner that they remain in position despite high current velocity. Their mesh ranges from the usual plankton diameters (Müllergauze 25 and Griesgauze 20) to 1 mm. Plankton nets cannot be left in most lotic situations for longer than 10–30 minutes, but 1 mm mesh nets sometimes function for at least 24 hours before silting up. A special "paddle-wheel apparatus" which measures the actual flow-volume, gives more precise information on the density of larval drift (Carlsson 1962). The "net-methodology" has been further developed by Müller

Direct Observations *In Situ*

The above methods are those most commonly used. However, other techniques have been employed to obtain information on preimaginal Simuliidae. It is highly desirable that our understanding of blackfly larvae and pupae should, wherever possible, be secured under natural conditions. For this reason, "stream-aquaria" (Carlsson 1962) and benthobservatories (Laird *et al.* 1974; Mokry 1975; Biswas and Mokry 1978) have been constructed, to enable direct observations of actual stream sites of pre-imaginal blackflies to be undertaken at very close range. While such an approach provides a visual record of benthonic life in running waters, the facility is both stationary and costly.

Another useful complement to trapping, is the water-telescope, which allows direct observation of the installed traps themselves and of surrounding parts of the stream.

Conclusion

Many trapping methods are thus available. However, the choice of which to use must be made with due consideration of the particular biotope under study, and the species represented. It is thus submitted that no single, standardized method for trapping larval blackflies will prove universally applicable. Rather, various of those briefly discussed herein, should be tested in each study area in order to determine which of them are most appropriate to local conditions and the particular simuliids present.

Sampling Methods for Adults

M. W. Service

Different types of traps have been used to catch blackflies, but trapping technology has not achieved the same sophistication as it has for mosquitoes (see Service 1976, for a review) or tsetse flies (Hargrove 1977; and see Jordan 1974 and Challier 1977 for reviews). Moreover, few traps have been designed specifically for catching blackflies.

The selection of any sampling method depends on the type of information required. Most trapping methods give biased samples, but this may not always be a disadvantage if large numbers of the desired species are obtained. On the other hand it may be important to collect representative samples of all simuliid species in an area, or to obtain meaningful estimates of contacts between adult blackflies and humans. This information is more difficult to obtain. The limitations of any sampling technique must be appreciated. Frequently, recourse must be had to more than one such technique in order to obtain reliable results.

The different means described here for sampling blackfly populations are divided into two categories: (1) attractant procedures employing human or animal bait, light and other visual stimuli, CO_2, etc. and (2) non-attractant methods such as catches from suction traps and vehicle-mounted ones. A few traps have been designed more to study blackfly host orientation and behaviour (Thompson 1976c) than to sample field populations. These are not discussed herein.

Attractant Methods

Human Bait Collections

Direct human bait collections undertaken during daylight hours are likely to result in more representative samples of anthropophilic blackflies than other methods. Frequently, catches last only 1–3 h, but because biting times vary among different species it is important to ensure that these short-duration collections are made when the insects are actively biting. Occasionally, although catching extends over long periods adults are caught for only a short time (e.g. 15–30 minutes) every hour; alternatively, continuous catches are performed from about 06.00–18.00 h. In stationary bait catches all blackflies attracted to one or more baits are collected at a single station, but collections can also be made at several sites along a transect, reminiscent of fly rounds catches for tsetse flies.

Certain species, such as *S. callidum* and *S. metallicum* are termed nervous or shy feeders because they tend to hover around their hosts before settling; even after they have landed, they are easily disturbed by slight movements and fly off again. These shy feeders are best collected by descending nets (Berzina 1953; Minář 1962; Monchadskii and Radzivilovskaya 1948).

Several variables arise in using human bait catches to sample anthropophilic blackflies. For example, in common with human bait catches for other haematophagous vectors, both aptitude and attractiveness of the collectors may vary considerably. Dark-skinned people are generally more attractive to blackflies than light-skinned individuals (Crosskey 1955; Hocking and Hocking 1962; Hughes and Daly 1951), and the colour of clothing may also influence the numbers caught (Bellec 1974; Hocking and Richards 1952). When comparing biting rates, it is usually important to sample adults at the same place and time, because most species exhibit periodicity of biting which may alter according to vegetation zone (Duke *et al.* 1967), types of habitat (Giudicelli 1966; Wolfe and Peterson 1960) or season (Crosskey 1955). The degree of anthropophily of a species may also change with altitude (Peterson 1959; DeFoliart and Rao 1965) and time of day (El Bashir *et al.* 1976).

When the number of blackflies biting exceeds about 1000 per hour, it becomes impossible to catch them all. Only samples of those arriving at bait can be caught in such circumstances and, unless care is exercised, this may lead to sampling bias. Nevertheless, despite these and other limitations, human bait collections, when conscientiously performed, can give useful information on temporal and spatial changes in relative biting rates, and help evaluate the effectiveness of control operations. It is not believed, however, that results can be extrapolated to provide realistic estimates of the numbers biting man daily (Duke 1970; Service 1976, 1977), although this has been attempted (Mills 1969).

Animal Bait Collections

Several blackfly species considered highly anthropophilic, including important vectors of human onchocerciasis, also bite other mammals and birds. Because of this and because several species are of veterinary importance, various animals have been used as bait. One procedure consists of a collector either remaining with the bait animals, or periodically inspecting them at about 15-minute inter-

vals, to collect blackflies attracted to them (Abdelnur 1968; Breyev 1950; Davies 1952, 1957; Crisp 1956a; Giaquinto 1937; Golini *et al.* 1976; Guttman 1972; Lewis and Ibañez de Aldecoa 1962; Shewell 1955). A disadvantage of this method is that it may be difficult to differentiate between blackflies attracted to the bait and those attracted by the collector. Moreover, man's presence may either increase or decrease the numbers and species of blackflies attracted.

Bed-nets have been employed in the USA to enclose a variety of bait animals (Wright and DeFoliart 1970) and large descending nets (drop-nets) have been used, particularly in the USSR (Breyev 1950; Ussova 1961), to catch simuliids attracted to large mammals such as horses and reindeer. However, the use of animal-baited traps to collect blackflies has been rather uncommon. In North America, however, because of the veterinary importance of certain blackflies as

Bird-baited trap recovered from tree canopy site and with box-like cover removed. (Photo supplied by G. F. Bennett.)

vectors of avian haematozoa (e.g. *Leucocytozoon* spp.), various domestic and wild birds have been placed as bait in small traps.

Commonly the bird is enclosed in a mesh cage which is placed on a baseboard, either at ground level or raised to various heights. At intervals, a cage is lowered over the bird so as to enclose it and all blackflies attracted to it (Anderson and DeFoliart 1961; Anderson *et al.* 1962; Bennett 1960; Disney 1972a; El Bashir *et al.* 1976; Khan and Fallis 1970). Another arrangement involves periodically enclosing the mesh-bait cage with a box having a hole cut in the top near a corner. Blackflies attracted to the bait pass through this opening and are trapped in a small wire cage covered with cheesecloth. A sliding panel at the base is closed prior to collection and the top cage removed and replaced by another.

In Africa, bird-baited traps have been used to attract man-biting species. In Ghana, for example, Odetoyinbo (1969) removed the baffles and light tube from a Monks Wood battery-operated light trap, and placed a chicken wrapped in wide mesh wire over the cylinder housing the fan. Over 11 trap-days, 931 blackflies were caught, of which 88.9% were of the *S. damnosum* complex. Using a CDC light trap with the bulb removed 50.1% of the 24 088 blackflies caught during 68 trap-days were of the *S. damnosum* complex. Also in Ghana, Crisp (1956a) placed a chicken in a cage, the mesh of which enabled small species such as *S. griseicolle* and *S. adersi* to crawl through to feed on the chicken, but prevented their escape after engorging. In East Africa, Raybould (1965) placed an intermittently-activated fan over a chicken and caught several hundred *S. adersi*; while, in Cameroon, Thompson (1977) used fan traps to catch the *S. damnosum* complex attracted to chicken and sheep.

Diethyl ether extracts of the uropygial glands of the Common Loon (*Gavia immer*) attract *S. euryadminiculum*, but unfortunately other blackflies are rarely caught.

Similar extracts from other birds do not generally attract blackflies. However, the addition of urophygial gland extracts from ducks and loons to traps employing CO_2 can enhance the catch of *S. euryadminiculum* (Bennett *et al.* 1972) and other species of blackflies (Fallis and Smith 1964; Golini 1975). Uropygial gland extracts have a very limited application for catching blackflies, and there has been little success in obtaining other attractants from hosts, although A. W. A. Brown (cited by Davies 1978) found that lysine and alanine, warmed to 37 °C and combined with CO_2, increased the attraction of blackflies more than the gas alone.

All animal-baited traps have inherent difficulties. For example, it is likely that some blackflies will escape capture by descending nets or boxes, while traps incorporating restrictive one-way openings will deter some adults from entering and at the same time will fail to prevent all of those that have entered from escaping. Ability to escape capture will vary according to both trap design and the species of blackfly. Visual stimuli are important in host orientation in some species (Monchadskii 1956; Thompson 1976c; Smith 1966), whereas odours may be more important in others (Hocking and Hocking 1962; Thompson 1976c). Consequently, the efficiency of baited traps will vary according as to whether the baits are easily seen and whether odour dissipates freely from them.

Anderson and DeFoliart (1961) and Bennett (1960) concluded that in some situations trap location was more critical in trapping ornithophilic simuliids than the type of bait used. Golini (1975), working in Norway, also found that trap location was critical (see below).

CO₂ Traps

CO_2 emitted from cylinders or by sublimation of dry ice has been successfully used (especially in North America) in a number of traps to catch blackflies. In the

USA, a New Jersey light trap with the bulb removed and with gas emitted under the roof at a rate of 0.45 kg/h caught 11 species of blackflies, including *Cnephia pecuarum, Prosimulium magnum* and *S. vittatum*. The largest catch was 669 simuliids in 1 h (Snoddy and Hays 1966). Also in the USA, 82 042 blackflies were caught over 32 days in two CDC miniature light traps with the bulbs removed and gas emitted at the rate of 2 l/min (Frommer *et al.* 1974). In later trials with discharge rates varying from 50–1900 ml/min seven CDC traps operated for eight days caught 4808 blackflies. The optimum discharge rate seemed to be about 500 ml/min (Frommer *et al.* 1976). In Lapland, blackflies have also been caught in light traps with the bulb removed and baited with CO_2 (Kureck 1969).

In Italy, Rivosecchí (1972) used small cylindrical traps into which 100 or 260 ml/min of CO_2 was released, and blackflies which were attracted were sucked down by a fan into a collecting bag. He found that the height (0.5–6 m) above the ground, and the type of vegetation into which the traps were placed, greatly influenced the species caught. Similarly, in Norway, Golini (1975), found that raising the height of CO_2 traps from just above the water surface to 1–3 m, or alternatively placing them at ground level but some 10 m away from the water, greatly diminished the catch of *S. rendalense*. This emphasizes the great importance of both trap location and design in catching blackflies.

Using four tent-type traps baited with CO_2 and having an internal vertical cloth baffle, more than 30 000 females of *Austrosimulium bancrofti* were caught over 117 days in Australia (Hunter and Moorhouse 1976). Trap location again proved critical, but the numbers caught were also influenced by temperature, wind, light intensity and precipitation. Curiously, the presence of humans inside the traps drastically reduced the catch of *A. bancrofti*. Similar traps were used by Hunter (1978) to study the daily

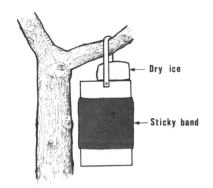

A lump of dry ice placed on top of a cylindrical sticky trap. (M. A. Johnson, del.)

fluctuations in population size of *A. pestilens* in Australia.

In Canada, blackflies have frequently been caught in very simple traps, comprising a dark-painted (usually blue) metal cylinder, coated with adhesive, and having about 1 kg of dry ice wrapped in paper, placed either in, on top of, or at the base of the trap (Baldwin *et al.* 1966; 1975; Moore and Noblet 1974; West *et al.* 1971). As many as 55 560 blackflies have been caught over a three-month period from 23 sticky cylindrical traps baited with a 22.6 kg block of dry ice placed in a styrofoam box at the base of each trap (Moore and Noblet 1974).

Besides being used alone, CO_2 has been combined with other attractants. In Norway, the catch of *S. rendalense* was increased fourfold when extracts from the uropygial glands of domestic ducks were added to CO_2 traps (Golini 1975). In Canada (Fallis *et al.* 1967) and Upper Volta (Bellec 1974) CO_2 emitted from cylinders has been combined with visual stimuli. For example, the catches of blackflies in Canada (both cylinders and silhouettes, resembling humans, coated with "Tanglefoot"), were increased 2.3–33.3 times by the addition to the traps of 400 ml/min of CO_2. In Upper Volta, Bellec (1974) concluded that a suitable silhouette augmented with CO_2 was likely to catch about half the numbers of *S. damnosum* complex attracted to human bait. In

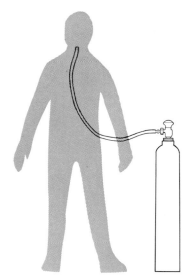

Various sticky two-dimensional silhouettes. (Photo supplied by G. F. Bennett.)

A life-size wooden silhouette of a man supplied with carbon dioxide gas and coated with adhesive. (M. A. Johnson, del.)

Tanzania, few simuliids were caught on round, rectangular or bird-like silhouettes, coated with "Tanglefoot". However, the addition of CO_2 (100–200 ml/min) resulted in attracting large numbers of blackflies, including *S. adersi* and *S. impulsane* or a closely related species (Fallis and Raybould 1975).

Visual Attractant Traps

Both two- and three-dimensional silhouette-type traps have been used to collect blackflies. Some silhouettes are made to resemble natural hosts, others make no pretence of host resemblance. For example, in Canada, Bradbury and Bennett (1974a) used two-dimensional silhouettes such as squares, rectangles, triangles, stars and circles, coated with adhesives, and of different colours, to attract simuliids. The principal species caught were *P. mixtum/fuscum* complex, *S. venustum* (possibly mixed with *S. tuberosum* and *S. verecundum*) and *S. mutata*. Most were caught on surfaces reflecting the least incident light. Also in Canada, Peschken and Thorsteinson (1965) caught

blackflies on cardboard figures, some rectangular, and others cut out to resemble the letters X and Y.

In Upper Volta, Bellec (1974) displayed silhouettes resembling a man, a cow and a heron. He also used various-sized square panels painted different colours. Few simuliids were caught on any of these silhouettes, and it was concluded that by themselves they were of little value for trapping adults of the *S. damnosum* complex. However, when CO_2 was released alongside them, they proved the most efficient of all the trapping devices that he tested. In trials in Tanzania, Fallis and Raybould (1975) used silhouettes, shaped as circles, rectangles and birds, but concluded they were of little value in catching ornithophilic blackflies. With the addition of CO_2 (100–200 ml/min), though, they caught substantial numbers. In Germany, silhouettes constructed to resemble horses and birds (with moveable ears, head and wings) attracted considerable numbers of blackflies. Moreover, the catch of ornithophilic species, mainly *S. latipes* auct., to the bird-like silhouettes, increased when these were slowly pulled along the ground (Wenk and Schlörer 1963). In the USSR, blackflies (including *S. tuberosum* and *S. pusillum*) have been caught on circular, rectangular, pyramidal and con-

A three-dimensional silhouette trap about the size of a sheep. (M. A. Johnson, del.)

ical silhouettes (Breyev 1950; Popapov and Bogdanova 1973; Vladimirova and Popapov 1963).

Manitoba fly traps, designed principally for catching Tabanidae (Thorsteinson *et al.* 1965), containing either cylinders or spheres, have caught small numbers of blackflies including *S. venustum, S. vittatum* and *S. decorum*, when tested in Canada (Peschken and Thorsteinson 1965), and *S. latipes* auct. in England (Service 1977). Three-dimensional traps, resembling either a cow, a sheep or a pyramid raised on legs, caught *S. arcticum* in Canada in the ratio of 30:1:10 respectively (Fredeen 1961). In Upper Volta (Bellec 1974), Ghana (Walsh and McCrae *in* Bellec 1974) and Uganda (McCrae and Manuma 1967), silhouette traps, similar to those of Fredeen (1961), were ineffective in catching blackflies. In Tanzania, however, Haüsermann (1969) used rectangular traps covered with black cloth to catch as many as 100 *S. damnosum s.l.* daily during seasons of high population densities; but when densities were low, few blackflies were caught.

Light Traps

Because most activities such as emergence, blood-feeding, sugar-feeding, dispersal and oviposition are diurnal in most species of blackflies, light traps have seemed inappropriate for sampling adult populations. However, the discovery, in Scotland, that large numbers of several species of blackflies can be caught in Rothamsted light traps (Davies and Williams 1962; Williams and Davies 1957), stimulated interest in the possibility of their use for sampling blackflies.

Since these highly successful results in Scotland, light traps have proved of varying efficiency elsewhere. Sometimes they have been of little or no value, but, on other occasions, they have caught considerable numbers of blackflies. A difficulty in comparing light-trap collections is the multiplicity of variables such as degree of night-time illumination, temperature and other environmental conditions, differences in light intensity and spectral emission of the light source, differences in trap design, and trap location, etc. Unsatisfactory trials with light traps have been reported by Lamontellerie (1963) and Bellec (1974) in Upper Volta, Crisp (1956a) and Marr and Lewis (1964) in Ghana, Walsh (1970a) in Nigeria, and Abdelnur (1968) in Canada. In Colombia, Monks Wood light traps with either daylight or UV tubes that were operating continuously, or repeatedly flashing on and off, were placed adjacent to larval habitats or where adults were biting. However, they failed to catch any blackflies, despite the presence of large numbers of both anthropophilic and nonanthropophilic species in the immediate area (Service, unpublished). More promising results were obtained by Lewis and Stam (1968) in Zaire, Lamontellerie (1967) in Upper Volta, Raybould (1966) in Tanzania, Peterson (1959) in Canada, and Marr (1971), Odetoyinbo (1969), Walsh (1978) and Service (1979) in Ghana. Odetoyinbo (1969), using a Monks Wood trap with a daylight fluorescent tube operated for 27 nights on the banks of the Red Volta River, caught 6325 blackflies; of which 95% belonged to the *S. damnosum* complex. During other trials, he demonstrated that *S. damnosum s.l.* were caught throughout the night and not just after sunset. Similarly, in Ghana, Walsh (1978) operating a Monks Wood type trap, showed that members of the *S. damnosum* complex (almost certainly *S. sirbanum*),

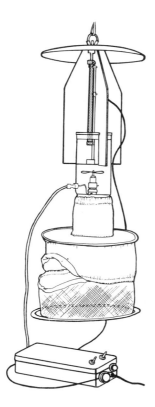

A Monks Wood battery-operated light trap.

A Monks Wood light trap with flashing light situated by a breeding site of S. squamosum *in Ghana.*

were active throughout the night. There was evidence to suggest too that adults of *S. griseicolle* and *S. adersi* were also active at night. In one series of trials, 5401 females, most probably *S. sirbanum*, were caught from 18.30–23.00 h in seven trap-nights. The presence of many gravid females collected during the first two hours after sunset indicated that egg-laying females may be strongly photo-positive. The suggestion (Service 1977), that it might be worthwhile to evaluate light traps near larval habitats to catch newly-emerged blackflies, is supported by the capture of many newly-emerged individuals in other trials.

Recently, as many as 6520 female *S. squamosum* were caught in a single night in a Monks Wood trap with a flashing UV light (Ross and Service 1979) placed adjacent to larval habitats in Ghana (Service 1979). From seven trap-nights

at this site, with daylight and UV fluorescent lights, 14 644 female and two male *S. squamosum* were caught. The catches consisted of gravid and parous females, which had oviposited very recently. In contrast, during 16 trap-nights near other sites known to be producing *S. damnosum s.l.* adults, the traps caught only four females of this complex and one female *S. adersi*.

Further trials with light traps are merited, and it is possible that other types of light sources, such as chemical lights (Service and Highton 1980) may prove useful. In addition to catching newly-emerged adults and those orientated to oviposition, light traps placed near hosts might well catch host-orientated females. This procedure has already proved useful in Africa for catching mosquito species not normally attracted to light traps (see Service 1977 for review).

Attractive Sticky Traps

Commercial tree-banding resins, bird repellents and concoctions of greases and
oils have been used to catch host-seeking
blackflies attracted to odours, CO_2, visual
stimuli and, occasionally, bait animals. In
Canada, for example, nylon gauze coated
with castor oil or "Tanglefoot" and
placed near birds' nests succeeded in
catching a few blackflies including
S. "aureum"[31] (Abdelnur 1968); and dark-
blue cylinders coated with adhesives
have trapped blackflies, mainly S. venustum, especially when the traps were
baited with 1 kg blocks of dry ice (Baldwin and Gross 1972; West et al. 1971). In
Ghana, plywood squares painted with
different colours and smeared with an
adhesive have indicated that S. damnosum
s.l. may be resting high up in trees during the day (Walsh 1972).

In Ivory Coast, Bellec (1976) caught
large numbers of simuliids on 1 m² aluminium panels coated with equal parts of
95% alcohol and "Tween-20" which were
placed on rocks alongside streams. In
one trial 2430 gravid and 19 007 non-
gravid females and 211 males, of S. damnosum s.l. were caught from 57 trap-days.
It seems that the shiny reflecting surfaces
of the panels catch females just prior to or
after oviposition. Similar sticky panels,
placed away from streams, caught few or
no blackflies.

Non-attractant Methods

No single type of attractant trap is likely
to catch representative samples of all
blackfly species in any area. Moreover,
they usually catch predominantly host-
seeking females. Because more representative faunal samples are sometimes

needed, and because information may be
required on activities such as dispersal,
oviposition, swarming, nectar-feeding,
etc. non-attractant traps have occasionally been used to sample blackflies. These
traps sample the aerial population. Consequently, the numbers caught will depend both on their population densities,
and flight activities.

Different non-attractant traps, such as
Malaise traps, rotary traps, suction traps,
vehicle-mounted traps and sticky traps,
have been devised to catch insects. The
first three types have not generally
proved useful in catching blackflies,
although a few have been caught in
Malaise-type traps in Uganda (McCrae
1966) and in Sweden (when sited over
streams—Roos 1957).

Non-attractant Sticky Traps

These traps should neither attract nor
repel insects, but catch them by wind
impaction. Cylindrical traps, coated with
adhesive, are generally more efficient for
sampling windborne insects than flat surfaces because air eddies are produced
around the latter. If, however, flat surfaces are used, the best procedure is to
employ fine mesh netting, lightly-coated
with adhesive, so that air blows through
but the insects are retained on the netting.

In Scandinavia, transparent sticky
cloths (0.5 m²) placed across streams,
caught 426 blackflies dispersing upstream and 231 dispersing downstream
(Carlsson 1962). In Ivory Coast, glass
panels (100 cm × 50 cm) and transparent
plastic sheets (200 cm × 30 cm and 200 cm
× 40 cm) coated with equal parts 95%
alcohol and "Tween-20", placed across
rivers, have proved very successful in
catching S. damnosum s.l. dispersing over
the water (Bellec and Hébrard 1977). A
total of 8296 S. damnosum s.l., 3790
S. adersi, 425 S. unicornutum and 167
S. griseicolle were caught from 32 trap-
days with the plastic sheets, and 59 trap-

---

[31] This specific name is given in inverted
commas because it cannot be equated with the
true S. aureum species found in Europe.

days with the glass panels. Female black-flies comprised 94–98% of the catch, most of which (71–91%) were gravid individuals. Substantially more blackflies were caught from the glass traps than from the plastic ones. Most were caught when the traps were placed just above (1–30 cm) the water surface.

In England, when 30 cm long metal cylinders covered with plastic covers coated with adhesive, were suspended horizontally about 5 cm above the water of a stream, they caught ovipositing blackflies. A maximum of 368 *S. ornatum* were caught on three traps within a week. Gravid females were usually caught around sunset (Davies 1957).

Suction Traps

Suction traps, particularly those of the Johnson-Taylor design (Service 1976),

A "Johnson–Taylor" suction trap. (M. A. Johnson, del.)

probably give better estimates of relative and absolute aerial population densities of insects than do other sampling methods. However, because they are non-attractant they usually have to be operated for long periods (especially when population densities are small) to get meaningful samples. For example, in Scotland, only 693 blackflies, mostly *S. latipes* auct. were caught in a trap operated from 1956–1959; whereas a nearby Rothamsted light trap operated from 1955–1959 caught 17 500 blackflies (William and Davies 1957). In contrast, these suction traps in northern Sudan caught an average of more than 150 *S. griseicolle* from 17.00–18.00 h (El Bashir *et al.* 1976; see Service 1977, for discrepancies in this paper).

Recently, suction traps 30- and 46-cm in diameter have been used in West Africa in endeavours to study the dispersal of *S. damnosum s.l.* into the sprayed area of the WHO/OCP. However, only a single adult of this complex was caught, although over 9000 adults of about eight or nine other species were trapped (R. W. Crosskey, personal communication, 1981).

A disadvantage of these suction traps is that they require a mains-supply of electricity or a portable generator capable of operating non-stop for 12 or 24 hours before refuelling.

Vehicle-mounted Traps

Vehicle-mounted traps (truck traps) are usually conical or pyramidal in shape and are fixed to the roof of a vehicle to catch insects swept into them while travelling. They thus differ from other non-attractant traps in that they sample insect populations along a transect—the route of the vehicle—not at one fixed point. Their usefulness is limited to terrain over which the vehicle can be driven.

Although these traps have rarely been used to sample blackfly populations, the best design of a vehicle-mounted trap was, in fact, made specifically for sam-

A vehicle-mounted trap with anemometer in opening. (Photo supplied by L. Davies and D. M. Roberts.)

pling British simuliids (Davies and Roberts 1973). The trap consists of a pyramidal structure with an opening, 91.5 cm wide and 61 cm high, tapering to 10.2 cm diameter at the rear end. It is covered with polyester netting (13.3 mesh/cm) giving about a 50% open area and ensuring a smooth flow of air through the trap. The vehicle is driven at 48 km/h. The volume of air sampled depends not only on vehicle speed and cross-sectional area of the trap opening, but also on wind speed and direction. Because of the variability of the latter, an anemometer is placed in the entrance of the trap to record the volume of air actually sampled. Inside the back of the van, insects trapped are delivered into small collecting tubes mounted on a turntable which can be advanced electrically so that the catch is segregated for specific lengths of the run. Five circuits, over a distance of 220 km, caught 4681 blackflies, most being *S. reptans*, *S. equinum* and *S. ornatum*.

In northern Sudan, a sweep net was held vertically about 150 cm from the ground through an opening in the roof of a car driven at 20 km/h along a 750 m length of road about 250 m from a river. Collections from 17.00–18.00 h resulted in an average of more than 150 *S. griseicolle*

(El Bashir *et al*. 1976; see Service 1977 for discrepancies in this paper).

Conclusions

Whereas there have recently been interesting developments in trapping technology for sampling populations of tsetse flies (Challier 1977; Jordan 1974; Rogers and Randolph 1978; Rogers and Smith 1977; Vale 1974a, b), and mosquitoes (Service 1976), leading to a proliferation of traps, little progress has been made in sampling adult blackflies. There is clearly a need to focus more attention on the development of better trapping techniques for these insects, and to devise reliable alternative methods to human bait collections. Moreover, there is a need to develop techniques that will allow the capture and study of the behaviour of non-blood-feeding stages of adult simuliids.

A serious difficulty (not unique to blackflies), is that if a trap proves unsuccessful in one situation, it might still be useful in a different area or habitat, under different environmental conditions, for catching the same or different species of blackflies.

It is essential to recognize the limitations of any chosen trapping methods and to appreciate the bias they may have in sampling populations. The procedures selected will largely be dictated by the information required. In control operations, there is need of a cheap and simple standard surveillance method, giving reliable pre- and post-control population indices. This type of sampling, however, may be too imprecise in epidemiological studies of onchocerciasis transmission. It is important to understand what the catches from various sampling techniques represent, and how the data can be correctly interpreted.

VII Colonization

Laboratory Colonization of Blackflies

J. E. Mokry, M. H. Colbo and B. H. Thompson

The laboratory colonization of blackflies has long been a subject of interest to medical entomologists and has received renewed impetus due to WHO/OCP. The availability of any such colony, even of a non-vector species, would greatly enhance our knowledge of blackfly biology in the matters of genetic inheritance and variability as applied to cytotaxonomic studies, the rate of evolution of resistance to insecticides, and the testing of new control agents (both chemical and biological). As well, blood-feeding experiments, which can elucidate host-preferences, biting rhythms and vectorial capacities of many simuliid species, would be much better served by the establishment of laboratory colonies of simuliids.

In order to achieve such colonization, it is necessary to be able to hatch the eggs, rear the larvae and pupae, and then induce the subsequent adults to mate (unless the species is parthenogenetic), blood-feed (where necessary), and oviposit. Each stage of the cycle is critical and has presented great difficulties in the

past. Since the last major review (Muirhead-Thomson 1966), however, many of these problems have been overcome. Most species can now be made to complete a large part of their life-cycles in the laboratory, with three species passing through their entire life-cycle there. The purpose of this chapter is two-fold: firstly, to report the significant advances of recent years, and secondly, to outline areas of research that it is felt might prove productive in greatly enhancing the prospects of colonizing additional simuliid species. In this regard, it should be noted that the life-histories of probably less than 10% of the described species has been elucidated even on a limited basis, while only a handful have been studied in detail. Therefore, there is a vast array of species which may be suitable for laboratory colonization but which have simply never been tried. In addition, the probability of colonization of any species will be improved by a better understanding of the several facets of simuliid biology, outlined hereunder.

Egg Collection and Storage

Fertile eggs laid by gravid females are a potential source of material for a laboratory colony. Eggs of some species of *Simulium* such as *S. verecundum s.l.*, *S. decorum* and *S. damnosum s.l.*, are often laid on trailing stream vegetation by many females simultaneously. These egg masses are readily observable in the field, simplifying collection. Field experiments such as those by Golini and Davies (1975a) in Canada, and Elsen and Hébrard (1977b) in West Africa, have shown that blackflies have colour preferences for oviposition substrates (pp. 206, 276). The use of certain coloured plastic strips or other suitable substrates (pp. 284–285) set in the rivers at natural oviposition sites has provided much information about blackfly biology and, in the present context, introduces an improved technique for egg collection in colonization efforts.

There remains the problem of egg collection from species ovipositing directly over still or running water, a mode typical of a large proportion of simuliids. Extraction of eggs from the stream bottom is possible and has been achieved (Fredeen 1959), but it is tedious and uncertain. Alternatively, certain substrates, if searched at appropriate localities, will yield a high density of eggs (Colbo and Moorhouse 1974). Also, larvae have been hatched out in the laboratory by aeration of aliquots from such substrates (Tarshis 1968). Another approach is to obtain eggs through providing field-collected females with blood-meals. Even then, however, many adults prove reluctant to oviposit. Two established laboratory techniques for inducing oviposition in blackflies are: to expose gravid females to CO_2 (Dalmat 1955); or, to decapitate them (Wenk and Raybould 1972). Again, however, there is significant risk of failure in applying these techniques, at least to North American species. Oviposition in two blackflies (*S. damnosum s.l.* and *S. vittatum*) has been successfully induced by putting the gravid females into vials half-filled with water, which are then vigorously shaken—eggs then being laid almost immediately. Oviposition in several (though by no means all) species, is known to occur during low light intensities (Davies and Peterson 1956; Hunter 1977a), which should be kept in mind when attempting laboratory oviposition, as was recently shown for *S. decorum* (Simmons and Edman, personal communication).

In some cases, eggs can be stored for future use, at least for limited periods, by cold storage (Fredeen 1959; Tarshis 1968; Mokry 1976a), thus providing a source of material for laboratory rearing experiments. However, prolonged cold storage of eggs is, so far, limited to what are referred to as "summer species", in the Northern Hemisphere, i.e. those species laying eggs which overwinter and then hatch in the spring. These belong to the subfamily Simuliinae. A number of them are multivoltine, completing 2–4 generations per summer. Many of these species overwinter in the egg stage, and, as laboratory experience has shown that low temperature is the critical factor in preventing hatching, it is evident that no obligatory cessation of embryonic development, or diapause, need take place. However, a blackfly such as *S. vittatum* overwinters as larvae, and the eggs of this species will only store for 2–3 months at temperatures very close to 0 °C (Fredeen 1959; Colbo unpublished). It has also been found, by Colbo (1974) and Ivashchenko (1977) that a reduction in O_2 will delay development. This can also be inferred from field observation of masses of eggs, where those deep in the batch do not develop and hatch until exposed to the air.

Univoltine species have an obligatory diapause in the egg stage. The Nearctic spring-breeding species, e.g. *S. venustum s.l.*, must pass through a warm period, followed by temperatures near freezing. Some of them hatch relatively early in the year, when water temperature is near 0 °C; others, after water temperatures have risen in the spring. One of us (MHC) has observed that eggs of a univoltine member of the *S. venustum* complex, which oviposits over water, commenced hatching in the spring before water temperatures had risen above 1 °C. In the laboratory, when eggs were stored for many months at room temperature, the embryos developed to the "eye-spot" stage only. Hatching of some occurred when the temperature was lowered to 5 °C. Additional ones hatched a year after oviposition, following an accidental freezing of the egg batch. Winter-developing univoltine species must pass only the warm summer in the egg, hatching when the temperature drops in the fall. Bradley (1935b) showed that *Cnephia pecuarum* eggs hatched after being left in either still or aerated water, in the laboratory, from April until December. However, as his conditions were not specified, it cannot be determined what the cue for development was. *Stegopterna mutata*, a winter species, produces eggs autogenously, requiring six weeks to two months conditioning at temperatures above 15 °C, before being transferred to cool temperatures where embryonic development and hatching will take place (Mokry 1978). These eggs can be stored for at least one year at temperatures near 20 °C, though they also remained viable for a few months when held at temperatures below 5 °C (Colbo, unpublished).

Clearly, the factors controlling diapause, as well as egg development and hatching, need to be defined, so that egg-storage of more species can be realized. Such research could greatly expand our knowledge of simuliid biology and add to the number of species available for colonization experiments.

Larval Rearing

A great variety of techniques has been employed for the successful rearing of simuliid larvae. As these have been reviewed in detail by Muirhead-Thomson (1966) and, more recently, by Raybould and Grunewald (1975), only the salient information is included herein. Before reviewing the various types of rearing systems in current use, however, it would be well to consider what is actually required from such a system with respect to colonization. The two most important considerations are: first, and foremost, a consistently high rate (at least 50%, preferably 70–80%) of survival from 1st-instars to the adult stage; secondly, it is essential that adult emergence takes place over as short a period as possible (ideally, 90% of the adults should emerge no more than 4–5 days apart, allowing maximum opportunities for mating and for efficient management of a colony).

Rearing systems are usefully divided into two basic types: those using water running along a trough, and those in which larvae are reared in some variation

of an aquarium with a system of agitation to create a current. Water for the system may be from either a natural source, or the public supply. In most cases, trough systems have been closed (i.e. water is recirculated through the trough by means of a pump). Such closed systems, depending upon an electric pump, face the difficulties of power interruptions, pump failure, and the frequent need to change or filter the water. Nevertheless, systems of this type have been successfully used to rear the Kibwezi form of *S. damnosum* *s.l.* (Raybould and Grunewald 1975), and *S. decorum* (Taylor, personal communication), as well as several other species of North American simuliids (Tarshis 1973; Wood and Davies 1966). A good example of an "open" trough system is that developed by Jamnback and Frempong-Boadu (1966). Water is gravity-fed from a lake and run through a series of troughs containing larvae, thus avoiding cessation of flow due to mechanical breakdowns. Such a system has the advantage of providing a natural food supply, and continuous renewal of fresh water, preventing build-up of waste products. Grunewald (1978) has stressed the deleterious effect of a build-up of excretory products on preimaginal development (and see p. 312). In fact, for successful rearing of at least certain species, it is necessary to control the concentration of a number of chemicals in the water besides those released by excretion (Grunewald 1972, 1976a, b; Grunewald and Grunewald 1978).

Rearing devices which utilize an aquarium, or modification thereof, providing water current by means of compressed air-jets or mechanical means, have by and large been the mainstay of larval rearing efforts. Among the many references to larval rearing systems using compressed air-jets is that of Simmons and Edman (1978), in raising larval *S. decorum* for their closed laboratory colony. Stir-bar systems have not yet had wide use, though they undoubtedly represent the highest level of proficiency in larval rearing to date. Grunewald (1973), using a stir-bar apparatus in combination with gravel and charcoal filters as well as the roots of a *Monstera* sp. plant to de-ionize the water, regularly reared up to 50% of 1st-instars of *S. erythrocephalum* to the adult stage. Later refinements apparently led to survival rates of up to 80% (Wenk, personal communication). A simplified and modified version of the stir-bar apparatus was developed by Colbo and Thompson (1978). This was designed primarily for laboratory investigations of preimaginal biology and as a bioassay tool. The new design has successfully reared several Newfoundland species of simuliids at survival rates of 65–80%, egg to adult, with a low degree of variance among replicates.

The temperature of the rearing medium strongly affects the rates of growth and survival, at least in the species so far examined. Mansingh and Steele (1973) found that *Prosimulium mysticum* larvae were unable to tolerate temperatures of 20 °C and quickly succumbed to the heat. *Stegopterna mutata*, also a "winter species", showed similar sensitivity to overheating and thus had to be reared at relatively low temperatures [9 °C] (Mokry 1978). Larvae of *S. venustum/verecundum* have been shown to have a tolerance range between about 15 °C and 24 °C, with the best survival above 18 °C (Mokry 1976a). The rate of growth also improves with temperature. Clearly, then, larval rearing temperatures must be considered when choosing a species for colonization studies, not only for the rates of survival of 1st-instar larvae to adult, but also in relation to the larval growth rate which will determine the "turn-over rate" of the colony. However, experimentation is necessary to determine the optimal temperature for the colony; for, a shortening of development time with increasing temperature has been found to markedly reduce potential natality in *S. vittatum* (Colbo and Porter in preparation) and is known to have

similar effects in *Aedes* (Van den Heuvel 1963) and *Culicoides* (Akey *et al.* 1978).

Food supply is another important factor influencing survival, rate of development, synchrony of adult emergence, and female size and fecundity (Colbo and Porter 1979). At present, high-plant-content fish foods are the most successful (Grunewald 1973; Colbo and Thompson 1978; Colbo and Porter 1979). However, the build-up of other microorganisms and fungi in this nutrient-rich diet can create problems, especially if development is prolonged, such as is the case in low temperatures (Colbo *et al.* unpublished). The use of residual antibiotic solutions added to the rearing media may help overcome this problem as was found for *Anopheles stephensi* Liston (Reisen and Emory 1977). Therefore, more experimentation should be undertaken to find additional readily-produced diets. The present possibilities include certain bacteria and algae, besides defined artificial and microencapsulated diets, development of which might well reduce unwanted microorganism production in the water.

Mating

In common with many other Nematocera, most blackfly species form mating swarms in nature (Downes 1958, 1969). The failure to induce swarming behaviour, and hence mating, under laboratory conditions has been the greatest single obstacle to the colonization of blackflies. Laboratory matings have been observed in a few species of blackflies, however, and are worthy of note.

Wenk (1965a) observed that laboratory-emerged *S. erythrocephalum* mated readily, either in aspirators or on windows where they gathered after emergence. Through the use of black cloth cages, with a triangular opening for light,

Wenk was able to concentrate adult blackflies, the increased contact between which led to an increase in mating. Remarkably, *S. erythrocephalum* is known to mate in swarms under natural conditions (Wenk 1965b). Wenk and Raybould (1972) applied the same technique to East African simuliids and found two species which mated in the cloth cages (see pp. 308, 262–263 for comprehensive coverage of this technique).

In North America, there are at least two species which have been observed to mate on the ground without the prerequisite of mating swarms, *viz. S. decorum* and *Cnephia dacotensis* (Davies and Peterson 1956). Laboratory-matings in aspirators have now been reported for *S. decorum* (Simmons and Edman 1978), the behavioural display sequence that both males and females exhibited before mating being suggestive of possible pheromone-like attractants. Perhaps the changes in air pressure, associated with aspirating blackflies from one container to another, serve as an olfactory releaser. On the other hand, Taylor (personal communication) apparently observed mating of the same species in cages, without preliminary displays. In any case, this discovery of a laboratory-mating strain of a North American simuliid was a major breakthrough. It quickly led to the establishment of a colony which was duly taken through 16 generations before it failed (May 1980) due to unexplained high larval mortality in the late instars and very poor egg hatches (Edman, personal communication). A sub-culture brought to RUVP at the 10th generation has now, June 1980, reached the 16th generation.

Another approach to the laboratory-mating problem has been that of the hand-mating technique used so successfully on mosquitoes. Field *et al.* (1967) were able to mate laboratory-reared *S. vittatum* in this fashion, but their work does not seem to have been followed up. Attempts at hand-mating Newfoundland simuliid species, including *S. vittatum*,

did not meet with the same success McDaniel and Mokry, unpublished).

On the other hand, *Stegopterna mutata* exists in both diploid and triploid populations (Basrur and Rothfels 1959). The triploid females do not require mating to lay fertile eggs and are, therefore, tempting subjects for experimental studies. In fact, *S. mutata* has already been taken through complete generation rearing, several times, at RUVP (Mokry 1978).

In general, there has been a distinct failure to appreciate and duplicate the environmental and/or physiological factors stimulating swarming and mating among blackflies. Although there are many reports of males orienting to optical markers in the field (Davies and Peterson 1956; Wenk 1965b; Le Berre and Wenk 1966), no one has yet been able to induce such behaviour in the laboratory. One factor, likely to prove of major importance, is light, or rather change in light intensity; for several species have been shown to swarm only at dawn and dusk (Davies and Peterson 1956; Wenk and Schlörer 1963; Wenk 1965b). Indeed, a non-biting midge, *Chironomus riparius* Meigen, can be stimulated to swarm and mate in cages by changes in light intensity (Downe and Caspary 1973). An often neglected "meeting place" for mating activity is in the proximity of the host. Wenk and Schlörer (1963) and Wenk (1965a) showed that males of *S. salopiense*, *S. equinum* and *S. erythrocephalum* orient to the host and intercept females *en route* to blood-feeding.[32] Observations on Newfoundland blackflies suggest that males orient to the host to a far greater degree than previously expected. Field trials have shown that male *S. verecundum/venustum* are attracted to, and swarm downwind of, a CO_2 source (Mokry, unpublished). Females of these species, attracted to the same CO_2 stimulus, are intercepted in flight and the pair fall to the ground *in copula*. In addition,

---

[32] And see pp. 260–261 (Ed.).

male swarms of the bird-feeding *Cnephia ornithophilia* have been found only at tree-top level at dusk (Mokry and Colbo, unpublished), the time and place of blood-feeding by the females (Bennett 1960).

If general stimulants to appetitive flight and feeding in females are indeed also stimulants to swarming and mating in males of some species, then these factors may also be crucial in inducing laboratory mating. This is particularly pertinent to the case of *S. euryadminiculum*, which is attracted to, and feeds on, the common loon (*Gavia immer*) as a result of specific host odours (Lowther and Wood 1964). In this context, it is suggested that males will also prove to be present nearby and downwind to intercept attacking females. There is, then, the possibility that the highly specific reaction of this species to its avian host is as potent a stimulant to elicit mating activity as it apparently is to feeding activity.

Pheromones or other artificial chemical stimulants also remain an unexplored possibility for inducing mating activity in simuliids. Indeed, the recent extraction, from female *Culicoides melleus*, of a hexane-soluble substance which causes sexual aggressiveness on the part of males, is very encouraging in this regard (Linley and Carlson 1978). As this is the first confirmed report of a contact-mating pheromone in the Nematocera, there is clearly room for more such experimentation upon blackflies.

Blood-feeding and Egg Production

Most simuliids require at least one blood-meal before oogenesis can occur. Even in those species which are autogenous for the first gonotrophic cycle, blood is needed for the development of subsequent egg batches. Blackflies have been provided with blood by soaking pre-heated (38–40 °C) heparinized or citrated blood into cotton-wool pads and making these accessible to them (Wenk 1965a; Yang and Davies 1968a). However, the rate of engorgement is low unless sucrose is added to the blood—a procedure which very greatly reduces the egg-yield.

Recent advances in knowledge concerning many aspects of the physiology and behaviour of blood-sucking insects generally have been applied to good effect in blackfly research. The development and application of artificial membrane systems has been especially encouraging. McMahon (1968) was the first to use a membrane device to provide blood for simuliids. The skin of a two-day-old chick was found to be the most satisfactory membrane when used with either ox- or human blood. Feeding rates of up to 55% were recorded with laboratory-emerged *S. ornatum*, an *Onchocerca gutturosa* vector in England.

There are difficulties, however, in the preparation and handling of animal skins for use as membranes, so more recent systems have employed latex rubber membranes. Sutcliffe and McIver (1975) investigated some factors associated with probing and gorging in *S. venustum s.l.* Temperature was shown to have a marked effect on the number of adults probing the membrane, with a maximum of such activity at 37 °C. Phagostimulants such as ATP and ADP were required to stimulate gorging, confirming the results obtained with other blood-sucking insects (Hosoi 1958; Friend 1965; Galun and Margalit 1969). Among other factors which influenced the rates of feeding was the age of adult blackflies used in the trials (McMahon 1968; Mokry 1976b, 1980a). McMahon showed that laboratory-emerged *S. ornatum* fed at much higher rates (55%) when five days old, as compared with one-day-old blackflies (4.5%). Mokry found similar results with *S. verecundum*, fed on a rabbit, although *S. vittatum*, an autogenous species, fed better when one day old (82%) than when five days old. In Europe,

Grunewald and Wirtz (1978), using a membrane system incorporating an electrical blood-stirring device, very successfully fed both *S. erythrocephalum* and *S. lineatum*. *S. erythrocephalum* has also been fed through a membrane at the London School of Hygiene and Tropical Medicine (Reid, personal communication).

As a research tool, artificial feeding devices provide an avenue for much useful and interesting experimentation such as effects of different host bloods on feeding rates and fecundity, the behaviour and physiology associated with feeding, and experimental parasite transmission studies. However, in the context of blood-feeding for colony reproduction, live hosts have also been used to provide a blood meal. Wenk (1965a) first developed the technique of feeding blackflies (*S. erythrocephalum*) on the ear of a rabbit. Up to 80% of laboratory-emerged females of this species would take blood, about two-thirds of them eventually developing and laying eggs. Wenk and Mokry (unpublished) also used a rabbit to feed laboratory-emerged *S. venustum*, *S. verecundum*, *S. vittatum*, *S. decorum* and *S. tuberosum*, in Newfoundland. Tarshis (1973) developed a successful technique for the bloodfeeding of several ornithophilic species *viz. Cnephia invenusta* (56%), *S. rugglesi* (38%) and *C. ornithophilia* (55%). Blackflies were aspirated into a fine-mesh sleeve covering a shaved portion of the leg of the avian host, or were placed in a gauze-covered cup directly on its breast. In both cases, relatively high rates of feeding were recorded, even for laboratory-reared females. The above author's immediate interest was the study of potential vectors of avian malaria, but his technique has obvious application to colonization studies of ornithophilic biting flies.

The need for blood-feeding, however, is obviated by the selection of an autogenous species for laboratory colonization. The list of such candidates is extensive. Most of the "winter species" in these colder climates are autogenous. Among the "summer species", at least *S. vittatum* and *S. decorum* are autogenous in many localities. Autogenous blackflies produce only their first egg-batch without the benefit of a blood meal, therefore requiring blood if a second oviposition is to be expected.

Two final notes on blood-feeding, autogeny and egg production are of some interest to colonization efforts. Firstly, laboratory studies on *S. vittatum*, at RUVP, have clearly proved that fecundity of the autogenous adults is related to larval nutrition. Very simply, nutritionally-deprived larvae give rise to smaller females, which in turn produce fewer eggs than better-fed ones. Therefore, close consideration must be given to the ability of the rearing system to produce large, healthy females of the species in question, as well as to its efficiency. Secondly, blood-feeding of autogenous females within three days of emergence augments the number of eggs ultimately produced, at least in the case of *S. vittatum* (Mokry 1980b). Such an increase in fecundity, due to blood-feeding, is of obvious significance for colonization studies. Similar autogenous, but haematophagous, mosquitoes are well-known (Trpis 1977, 1978). Like *S. vittatum*, these mosquitoes (*Aedes scutellaris* and local populations of *A. aegypti*) are autogenous, but take blood during the first cycle, if given the opportunity.

Conclusion

The laboratory colonization of any insect species obviously depends upon the successful completion of all stages of its life-cycle. It is equally clear, however, that in order for the colony to be self-propagating, the number of individuals resulting from each cycle must be at least equal to the number with which the cycle began. For example, starting with 50 females and 50 males, should the consequences of mating, feeding, oviposition and larval rearing only produce 10 females and 10 males in the next generation, the colony will soon die out. In addition to the "normal", or expected, mortalities evident at each stage in the cycle, the colony should produce a sufficient excess of individuals to: (1) account for unforeseen mortalities due to accidents or equipment failure, etc., and (2) allow some latitude for errors while perfecting handling techniques, as well as providing material for experimental purposes. Wenk (1965a, and personal communication) suggested an equation for determining the "propagation factor" of a closed colony. In the example below, hypothetical figures have been inserted at each stage of the life-cycle, showing the percentage of the total population in a colony needed to complete that stage. In addition, the number of larvae subsequently surviving to the adult stage is given, and then halved, to calculate the number of females expected.

blackflies in the colony than eggs laid. Therefore, it is essential to maximize the oviposition potential of the females in the colony, whether by larval diet or other means. Again, the selection of an autogenous species for colonization means an equivalent blood-feeding rate of 100%. In the hypothetical example, this would give a propagation factor of over 16. It was noted previously that making blood available to *S. viitatum* significantly increased the number of eggs laid—this would give an even greater margin of insurance. *Stegopterna mutata*, being both triploid and autogenous (equivalent to 100% mating and blood-feeding), seems an excellent prospect for colonization (Mokry 1978). However, high losses associated with storage of diapausing eggs and slow larval development reduce the overall propagation factor to a relatively small figure.

S. decorum, then, remains the only simuliid species to have become well-established in the laboratory, i.e. truly "colonized", without any need of rein-

$$\begin{array}{ccccccc} & \text{blood} & & & \text{eggs to} & \\ \text{mate} & \text{feed} & \text{oviposit} & \text{eggs/}\female & \text{adult} & \text{females} \\ 40\% & \times\ 50\% & \times\ \ 50\% & \times\ 250 & \times\ 65\% & \times\ 50\% & = 8.13 \end{array}$$

Theoretically at least, an eight-fold increase in the number of females, resulting from each cycle, is to be expected. Even with a mating rate of only 10%, the number of females would double after each generation.

Further examination of the equation shows several other interesting features. All factors taken into consideration here, with the exception of the "number of eggs/females", are numerically less than unity. It is important to realize that, at every step of the cycle, there is a degree of mortality—the sum total of which must be compensated for from the total number of eggs laid. Of course, it is obvious that there will never be more

forcement from field-collected material (Simmons and Edman 1978). Although precise figures are yet uncompiled, it is evident that the propagation factor is more than adequate to permit the continuous culture of this species. Two of the most likely European candidates are *S. erythrocephalum* and *S. lineatum*, both of which will mate and blood-feed readily in the laboratory (Wenk 1965a, Wirtz 1976). According to the data given by Wenk, *S. erythrocephalum* could have a propagation factor of as high as 11, more than sufficient to establish a colony. Hence, there presently appear to be no major obstacles to the successful colonization of these species.

Present Progress Towards the Laboratory Colonization of Members of the *Simulium damnosum* Complex

J. N. Raybould

Some of the advantages likely to accrue from the laboratory colonization of Simuliidae, have been pointed out (pp. 299–306). If self-perpetuating colonies of onchocerciasis vectors could be established, they could then be investigated away from natural breeding sites and at times of year when little or no material is naturally available. Complete colonization has yet to be achieved for any vector of *Onchocerca volvulus*, but even single-generation rearing has proved to be very useful.

As *Simulium damnosum s.l.* comprises a sibling species-complex (pp. 45–56), new biological information concerning this complex is of little value unless it is known to which taxon it refers. Clearly, the identification of the adult female blackfly attacking man is of fundamental importance to the medical entomologist. Until now (pp. 45–56), despite advances in morphological taxonomy, chromosomal differences in the larvae have remained the most reliable criteria for species-determination within the complex (WHO 1977). For this reason, the laboratory maintenance and rearing of species of the *S. damnosum* complex (starting with wild-caught adult females and embodying the cytotaxonomic determination of reared late-instar larvae) has proved to be an important research tool, especially in relation to the WHO/OCP (pp. 85–103).

If larvae, pupae and adults of *S. damnosum s.l.* are reared from a single egg batch, and one of the reared larvae is identified cytotaxonomically, the female parent and her progeny can be used in the search for morphological differences between the species. Considerable progress in morphological taxonomy has already resulted from studies of material provided in this way[33] (Garms 1978). In addition, reared larvae can be used for testing insecticides, and the adults, initially free from infection, can be used in transmission experiments with *Onchocerca volvulus* and other parasites.

Progress towards the laboratory colonization of African Simuliidae has been reviewed by Raybould and Grunewald (1975). Prior to 1976, most investigations in this field were carried out in East Africa, especially northeastern Tanzania. Since then, investigations have been centred in the Volta River Basin area of West Africa in connection with the WHO/OCP. At Tübingen, Federal Republic of Germany, considerable progress has also been made in rearing *S. damnosum s.l.* from eggs sent from West Africa (Grunewald, personal communication).

Some important onchocerciasis vectors in Africa belong to the *Simulium neavei* Roubaud group. However, as no further progress towards their laboratory culture

---

[33] See pp. 45–56.

has been made recently, members of this group have been excluded from the present account, which updates the review of Raybould and Grunewald (1975) with respect to the *S. damnosum* complex only. This account should therefore be read in conjunction with the earlier review. Suitable modifications of some of the techniques recently developed for *S. damnosum s.l.*, urgently require testing on species of the *S. neavei* group.

Pages 283–296 should be consulted for methods of collecting material for use in rearing and colonization studies.

Laboratory colonization requires the completion of all stages in the life-cycle in captivity. Each stage is considered in turn below, commencing with the adult.

Completion of the Life-cycle Under Artificial Conditions

Maintaining Adults

Adults of *S. damnosum s.l.* appear to survive better when kept singly in small tubes (Lewis 1960; Duke 1962a; Raybould 1979), or holes bored in a block of expanded polystyrene (Raybould and Mhiddin 1974), than when grouped together in larger containers. Glass tubes are unsatisfactory because water condenses on the glass, trapping the blackflies and causing high mortalities. The tubes are usually kept in the dark. To stabilize conditions, they are sometimes placed in trays floating on water in a drum (Duke 1962a). In any case, a water barrier is required to prevent attack by ants.

Survival rates improve at reduced temperatures, but oogenesis is delayed. Schulz-Key (personal communication) has recently obtained excellent results by confining about 20 *S. damnosum s.l.* females at the bottom of a small tube containing a plug of tissue paper and then storing them on ice in a refrigerator. It is important, however, to provide the correct humidity by adding two drops of water to the tissue.

Adult simuliids require sugar, which is usually provided in solution absorbed in cotton-wool or foam rubber. Alternatively, separate sucrose crystals and water can be supplied. The blackflies may also be fed on fruits, such as sultanas and sliced apple, or flowers on which simuliids naturally feed (Raybould and Grunewald 1975).

Techniques for maintaining adult blackflies require further investigation and improvement. Although a few may survive for a number of weeks, the majority succumb much earlier. Survival rates vary considerably for no obvious reason.

Mating

Failure to induce onchocerciasis vectors to mate in captivity remains the biggest stumbling block to their laboratory colonization. Nevertheless, one East African member of the *S. damnosum* complex, the non-anthropophilic Kibwezi form, will mate readily in cages (Wenk and Raybould 1972). West African *S. damnosum s.l.* from a locality in Upper Volta, have also been recorded mating in captivity. However, the species concerned (although almost certainly anthropophilic) has not been identified (Wenk and

Anterior and posterior views of emergence cage used for mating experiments (J. Grunewald).

Schulz-Key 1974). In addition, apparent mating behaviour by *S. damnosum s.l.* adults emerging in cages has occasionally been observed elsewhere in West Africa (Grunewald, personal communication). Actual spermatophore transfer under these conditions has recently been demonstrated for *S. soubrense* from the River Volta near Akosombo in Ghana (Raybould, unpublished). Certain members of the complex (such as the East African Sanje and Kisiwani forms) do not display mating behaviour in captivity.

Specially designed emergence cages have been used for mating experiments (Wenk and Raybould 1972; Raybould and Grunewald 1975). The cages (40 cm × 40 cm × 60 cm) are made of black cloth with a triangular opening about 40 cm high, on one side.[34] This opening is covered with netting through which light enters and on the surface of which the blackflies move and copulate. The majority of Kibwezi females emerging from pupae placed in such cages are subsequently inseminated by the males.

Blood Feeding

A major obstacle to colonization has been that of inducing adequate numbers of laboratory-reared female simuliids to take a blood-meal. There are, however, some early records of *S. damnosum s.l.* either biting in an outdoor cage (Crisp 1956a; Marr 1962) or being induced with difficulty to take blood in the laboratory (Wanson *et al.* 1945; Muirhead-Thomson 1957a).

McMahon (1968) successfully fed several species of African simuliids (including anthropophilic *S. damnosum s.l.* from East Africa and Cameroon) using an apparatus incorporating a feeding tube containing warmed human- or ox-blood covered by a membrane. Both wild-caught and laboratory-reared adults en-

gorged through the membrane. The best results were achieved with the skin of two-day-old chicks. The same author also induced *S. damnosum s.l.* from Cameroon to feed on anaesthetized chimpanzees and obtained good results with laboratory-reared blackflies.

Raybould and Yagunga (1969) fed wild-caught females of an anthropophilic form of *S. damnosum s.l.* (and other actual or potential onchocerciasis vectors) on rabbits and man by releasing them into sleeves of fine netting placed around the ears and arms respectively. The proportion of blackflies feeding was frequently above 40%. Laboratory-reared adults may also be fed in this way but (unlike McMahon's results using chimpanzees) usually fewer individuals become engorged (Wenk and Raybould 1972). Similar results may be obtained by releasing the simuliids into a sleeve of netting fitted around a leg of a chicken, guinea-fowl (*Numidia* spp.) or other gallinaceous bird (Wenk and Raybould 1972; Raybould and Grunewald 1975). This simple technique was used for studies on the transmission of avian haematozoa (Fallis *et al.* 1973a,b).

An important advance was recently made when Grunewald and Wirtz (1978) showed that results using a rabbit vary greatly with the individual animal utilized: feeding rates obtained with Kibwezi and Kisiwani females were respectively 41 and 51% on a brown rabbit but only 14 and 12% on a white one. Different skin temperatures in the ears may be responsible for this phenomenon. When the same authors offered a mixture of heated blood and saturated sucrose solution (absorbed on filter paper) to four African *Simulium* species (including two in the *S. damnosum* complex), nearly all individuals took blood but subsequently produced very few eggs.

Most of the above-mentioned investigations were carried out in Tanzania or Cameroon. Results have been generally similar for the various *S. damnosum* complex species investigated, with one

---

[34] And see pp. 262–263 (Ed.).

Left, terminal part of rearing apparatus and adult collecting device (Raybould 1967). Right, outlet pipe from rearing tank showing larvae attached to a plastic-strip (J. N. Raybould).

exception—the East African, non-anthropophilic Sanje form rarely took blood irrespective of the feeding method employed. Although most techniques remain to be thoroughly tested in countries west of Cameroon, preliminary results suggest that West African onchocerciasis vectors are unlikely to prove especially difficult to feed.

Refinements of technique lately developed outside Africa with Palaearctic blackflies need testing with the *S. damnosum* complex. Recently, a few *S. damnosum s.l.* females (carried from Ghana to Germany) were successfully fed using a modification of Wade's (1976) membrane feeder (Wirtz and Grunewald, personal communication). Further investigations along these lines are required.

Oviposition and Egg Development

S. damnosum s.l. females only rarely lay eggs in captivity unless induced to do so by immersion in water or decapitation (Lewis *et al.* 1961). The great majority of gravid females will oviposit when thus stressed. A convenient technique is to immerse them in water in a tube lined with polythene sheeting. Eggs laid on the polythene can then be cut out and suspended in a rearing trough (Raybould 1979). Oviposition can be brought about two or more days after blood-feeding according to temperature. A single member of the complex, the non-anthropophilic Kibwezi form, has been successfully induced to oviposit after mating and taking blood in captivity (Wenk and Raybould 1972).

Frequently only a minority of laboratory-laid eggs give rise to larvae. Some complete egg batches fail to develop even when laid by an inseminated female. Nevertheless, on some occasions most of the eggs develop, especially in the case of West African *S. damnosum s.l.* Reasons for this variation require further study. Eggs that develop do so rapidly. Those of West African members of the complex hatch after two days at temperatures of about 30 °C, while those of Kibwezi take four days at 20 °C. The eggs may be stored for limited periods at low temperatures. Experiments with Kibwezi eggs showed that at 1–4 °C survival is about 80% after a week, falling to 20% after a month (Raybould and Grunewald 1975). Hansen and Hansen (personal communication) found 17°C to be optimal for storing the eggs of West African *S. damnosum s.l.* for up to a week—longer periods were not tested. At 20°C hatching occurred during storage, and at 15°C and below hatching was later delayed, even at ambient temperatures.

Rearing Techniques

Much ingenuity by investigators in various countries has gone into developing techniques for rearing larval simuliids (see previous section). Only methods already tried in Africa will be considered here, although techniques developed elsewhere may yet be found to work well with African species.

A method of rearing *S. damnosum s.l.* by modifying a stream habitat and using an emergence cage, was successfully developed by Marr (1962). Laboratory rearing, however, has proved to be more difficult. Larval development requires suitable water-flow and turbulence, correct chemical conditions, and the right kind of food. Some African blackflies, such as *S. adersi*, are tolerant of a wide range of conditions, but those in the *S. damnosum* complex have much more particular requirements. West African members of the complex, however, can all be reared in the same culture water obtained from an aquatic site of any one of them. This is not the case in East Africa where at least the non-anthropophilic Kibwezi and Kisiwani forms need excep-

tional conditions (Grunewald and Grunewald 1978).

The type of rearing system required depends on the circumstances. The mass production of material for investigations connected with biocontrol and other purposes, sometimes necessitates the development of a very large-scale rearing apparatus. A system on such a scale, using large concrete troughs and giving rapid larval development, has been constructed at IRO, Bouaké, Ivory Coast (Berl and Prud'hom 1978, 1979). Research has been generally directed towards the development of compact, small-scale rearing systems, convenient to an ordinary laboratory, and in which particular populations can be reared separately.

As was pointed out by Muirhead-Thomson (1966), laboratory techniques for rearing blackflies may be divided into (a) methods based on agitation and oxygenation of water in a closed system and (b) methods depending on a continuous flow of pumped water.

The former category includes methods in which water in a simple container, such as a glass cylinder, is circulated and aerated by a compressed-air jet (Muirhead-Thomson 1957b; Doby *et al.* 1959) or an electro-magnetic stirrer. These techniques work well with some African simuliids, such as *S. adersi* and *S. unicornutum*, but have so far proved to be unsuitable for certain East African members of the *S. damnosum* complex. Somewhat better results have been obtained with West African *S. damnosum s.l.* Nevertheless, development is slow, mortalities are high, especially in the late larval instars, and only very few under-sized adults are obtained. This is so even when the water is frequently changed and the food supply carefully maintained.

The most successful rearing techniques for the *S. damnosum* complex have been based on the continuous circulation of pumped water over an inclined trough through pipes or tubes. The first record of *S. damnosum s.l.* being reared from egg to adult in the laboratory is apparently that of Wright (1957), who obtained seven adults using an apparatus which circulated water in which the plankton had been concentrated. About 10 years later, the non-anthropophilic Sanje form was reared in large numbers in an apparatus comprising two large tanks between which water was circulated by a centrifugal pump (Raybould 1967a).

Water from the upper tank flowed through four narrow outlet pipes through which strings or plastic strips were suspended. Larvae congregated inside the pipes and on the strips where the water flow was maximal. The pipes fitted into a plexiglass chamber surmounted by vertical collecting tubes into which the emerging adults were attracted when the rest of the chamber was covered with shading cloth. On one occasion when quantitative records were kept, 5023 adults of Sanje were obtained from about 28 640 eggs representing about 17.9% survival to the adult stage. As these results indicate, the technique works well and has yielded valuable information on eclosion and other aspects of development. Nevertheless, the apparatus requires a large volume of water from a natural breeding site and is too large and cumbersome for normal laboratory use.

Most blackfly larvae are very sensitive to water pollution, especially to a high ammonia content (Grunewald 1978). Therefore, one of the problems encountered in the development of suitable rearing systems is that of pollution by metabolic by-products of the larvae and decaying food particles. There are two main approaches to overcoming this problem. One is to either use a large volume of water (as was done with the technique described above) or to regularly change the water. The other is to introduce a self-purification system.

A technique based on the latter approach was successfully developed for rearing B. erythrocephala in Europe (Grunewald 1973), and later modified for use with S. damnosum s.l. (Raybould and Grunewald 1975; Grunewald and Grunewald 1978). The main feature of the method is that the ionic balance and pH of the culture-water (obtained from a blackfly stream) are adjusted and maintained at similar levels to those in the natural habitat (Grunewald 1976a). The apparatus comprises two circulatory systems which operate independently. A centrifugal pump rapidly circulates water between a tank and an inclined rearing tube or trough (or oval channel—see below) where the larvae develop. The water is purified by its slow circulation through a gravel filter, an activated charcoal filter and a container with the submerged roots of Monstera sp. (Araceae) which remove excess phosphates and nitrogen compounds. This technique has been successfully used for rearing both East (Grunewald and Grunewald 1978) and West African members of the S. damnosum complex. Development is rapid if the larvae are adequately fed. Suitable foods includes Tetra® Conditioning Food (this is a widely-used vegetable diet for tropical fish—Ed.) and cultures of algae. The technique is highly satisfactory for use at a permanent laboratory base, but the presence of the purification system renders the apparatus rather difficult to transport. It also takes time to assemble and the Monstera plant requires a period of stabilization.

A readily-transportable rearing apparatus that could be carried great distances across Africa and used immediately on arrival, was recently required for inves-

Rearing apparatus incorporating a self-purification system used for rearing S. damnosum *complex species (Raybould and Grunewald 1975).*

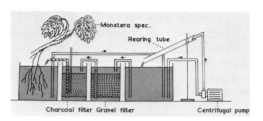

tigations connected with the WHO/OCP. These requirements have been met by using a considerable volume of water (from a blackfly stream) and feeding the larvae with living algae which obviates the need for a separate water-purification system (Raybould 1979). The apparatus used is, as far as possible, constructed from locally-available materials that are easy to pack, transport and assemble. It comprises a simple closed system in which water is circulated by a centrifugal pump between a container and an inclined trough in which most of the larvae develop (essentially the same simple circuit as is shown above). The larvae are fed on microscopic green algae (Chlorophyceae) which are cultured in Nakamura's medium (Nakamura 1949) and removed by filtration. Polythene strips hung into the rearing trough enable attached pupae to be cut out with scissors. An adult collecting device can be fitted to the trough. It excludes light apart from that entering a collecting tube into which the flies are attracted.

The technique has given better results than any obtained hitherto with members of the *S. damnosum* complex using a relatively small-scale apparatus. Recent investigations have shown, however, that Grunewald's technique (as modified for use with *S. damnosum s.l.*) is equally successful if a similar diet of algae is supplied. At a temperature of about 25–30 °C larval development is very rapid, and there is less variation in growth rates than occurs in water circulated by compressed air. Old algal cultures contaminated with bacteria and other organisms appear to accelerate growth. Pupation normally commences between about 10 and 14 days after hatching and full-size adults sometimes develop in under two weeks. Survival rates have not been precisely determined, but several hundred late-instar larvae may be obtained from a few egg batches in one rearing trough. Most of those not removed for identification survive to the pupal and adult stages, although high pupal mortalities

do occasionally occur. Because of their large size and the condition of their chromosomes, the reared larvae are particularly suitable for cytotaxonomic work (Vajime, personal communication).

Some refinements in rearing-trough design have recently been introduced either as part of a new apparatus or as a modification to a pre-existing one. Hansen and Hansen (personal communication) obtained promising results (in Ivory Coast in 1976) using two interconnecting channels sloping in opposite directions in which water turbulence was increased with carefully-designed baffles and recirculated from a balanced aquarium. A compact oval rearing channel, incorporated into Grunewald's (1973) apparatus, is now in use for rearing *S. damnosum s.l.* at Tübingen (Grunewald, personal communication).

All the above methods were designed for rearing numerous larvae together. The only technique enabling individual *S. damnosum*-complex larvae to be observed throughout their development, is that of Elsen and Hébrard (1977a). Their method is based on the controlled flow of water through a series of bifurcating tubes, each larva being reared in a separate tube.

Completion of all Stages in the Life-cycle

Only one member of the *S. damnosum* complex, the East African Kibwezi form, has so far been successfully induced to complete every stage of its life-cycle in the laboratory. In mating trials with this species, Wenk and Raybould (1972) obtained insemination rates ranging from 50–92%. Kibwezi larvae, although particular in their requirements, survive well if suitable conditions are provided (Grunewald and Grunewald 1978). Feeding trials with this species have also been satisfactory—Grunewald and Wirtz (1978) induced 41% of females tested to take blood from the ears of a brown rabbit. The chief problem remaining with the rearing of Kibwezi is that most eggs laid in captivity fail to develop (Wenk and Raybould 1972). If this obstacle could be overcome, then colonization would be feasible.

The Transportation of Living
Simulium damnosum s.l.

Until self-perpetuating colonies of *S. damnosum*-complex species are established, the long-distance transportation of living material for research will continue to be of vital importance.

Eggs may be transported at low temperatures (usually in a thermos flask) or simply kept moist in sealed plastic bags and mailed in thick, light-excluding envelopes. Consignments dispatched from Africa to Europe by the latter method contained viable eggs after three days and sometimes after five or six days. There is evidence that embryonation and hatching are delayed in deoxygenated water in the dark, and the possibility of transporting eggs under these conditions requires investigation (E. Hansen, personal communication).

S. damnosum-complex larvae may be transported in closed plastic bags or in water in small containers with a battery-run oxygenation system. They must be kept cool *en route*. If required for cytotaxonomic determinations, they are best carried on ice (Dunbar 1972). Pupae are normally transported in closed plastic bags at reduced temperatures and in the dark. More information is needed on the effect of temperature and light on pupal survival and development and on adult emergence. It is possible that pupae may be more amenable to long-distance transportation than is at present realized.

S. damnosum s.l. adults are normally transported separately in small plastic tubes (glass ones being unsuitable) wrapped in damp cotton-wool and placed in a cool-box or thermos flask. Schulz-Key's technique of storing the adults (see p. 308) could possibly form the basis of a new method of maintaining them during transportation, provided that they can be kept on ice throughout the journey.

Discussion

The fact that *S. damnosum*-complex species may be taken through various stages in their life-cycle in the laboratory, has made possible many investigations that could not otherwise have been carried out. Nevertheless, colonization has not yet been achieved and progress in this direction has been uneven. Most recent advances have been in the improvement of rearing and blood-feeding techniques (Grunewald and Wirtz 1978; Raybould 1979).

The development of a rearing technique using an apparatus that is easily transported and usable under makeshift laboratory conditions, has been of particular advantage to WHO/OCP (Raybould *et al.* 1979). The high rate of larval development achieved by this method, which approaches that in nature at similar temperatures, is a considerable advance. The success of the method in the apparent absence of a self-purification system, may have been at least partly due to the relatively large volume (60 litres) of water used (Grunewald 1978). It is also possible that the algae themselves provide an alternative self-purification system by removing nitrogen compounds. Preliminary analyses support this hypothesis, although further investigations are required.

In addition to the algae supplied as food, bacteria and other microorganisms contaminating the algal cultures and developing in the rearing system, might also have made a major contribution to the diet of the developing larvae. The importance of bacteria as food for blackfly larvae has been well demonstrated by

Fredeen (1964). Further studies are needed in which not only algae, but also cultures of particular species of bacteria, are offered as food to larvae of *S. damnosum s.l.*

The non-anthropophilic Kibwezi form and *S. soubrense* remain the only laboratory-mating species so far identified in the *S. damnosum* complex, and are therefore the only likely candidates for colonization at present. As was pointed out by Raybould and Grunewald (1975), it is imperative either to develop new techniques for inducing laboratory mating or to discover species, or intraspecific populations, which will mate without special inducement. While the former approach is by no means certain of ultimate success, the latter is much more hopeful. Investigations are urgently required, therefore, to determine exactly which members of the complex will mate in the laboratory and under what conditions.

Acceptable rates of laboratory mating should remove the remaining obstacle to colonizing onchocerciasis vectors in the *S. damnosum* complex. However, it seems probable (Raybould and Grunewald 1975) that even if some members of the complex are successfully colonized, the process may prove too difficult and time-consuming for all but a few specialized research establishments. If this proves to be the case, certain non-vector species may well prove to be more suited to colonization, as has been suggested by Muirhead-Thomson (1966). Despite the marked differences between species, experience with colonies of non-vectors might nevertheless provide useful guidance as to the most fruitful methods of investigating the vectors. Species such as *S. ruficorne*, which breeds in slow-flowing or even still water, and *S. adersi*, which tolerates an unusually wide range of environmental conditions, might prove particularly amenable to colonization if laboratory-mating forms could be found. The fact that *S. adersi* is a good laboratory host for *Onchocerca volvulus* (Wegesa 1970b), is of further interest in this context.

VIII Mass Production of Pathogens and Parasites

Mass Production of Pathogens

R. A. Nolan

Because of recently intensified interest in the testing of pathogens of diversified origin against blackflies and in identifying new pathogens from them, several microorganisms have attained positions of possible future potential for biocontrol (see pp. 181–196). The following discussion will, however, be limited to those organisms having a more immediate potential as blackfly biocontrol agents and offering good prospects for large-scale production by mass-fermentation technology. The organisms of this category are: *Bacillus thuringiensis* var. *israelensis*, *Entomophthora culicis*, *Entomophthora curvispora*, *Coelomycidium simulii* and *Metarhizium anisopliae*. Those candidates which at present require *in vivo* technology are not considered practical either on economic grounds or for reasons of the comparative lack of control during most phases of production. For a discussion of these alternate methods of pathogen production, see Simmonds *et al.* (1977).

Bacillus thuringiensis var. *israelensis*, serotype 14 (ONR-60A/WHO 1897), is pathogenic to larvae of the mosquitoes *Anopheles sergentii, Uranotaenia unguicula-*

ta, *Culex univitattus* and *Culex pipiens* (Goldberg and Margalit 1977) and *Aedes aegypti* (Goldberg and Margalit 1977; de Barjac 1978) in laboratory studies, and also pathogenic to blackfly larvae in both laboratory and field (stream) studies (Colbo and Undeen 1980; Undeen *et al.* 1980). This isolate is pathogenic to larva of *Simulium damnosum s.l.* (Undeen and Berl 1979) and *Simulium verecundum, Simulium vittatum, Cnephia mutata, Cnephia ornithophilia* and *Prosimulium mixtum* (Undeen and Nagel 1978). The ease of cultivation of this particular isolate (ONR-60A/WHO 1897) is greatly enhancing rapid transition to large-scale field studies.

Goldberg and Margalit (1977) used "N2X" [a proprietary medium obtained from Nutrilite Products Inc. and used commercially for large-scale fermentation production of *Bacillus thuringiensis* (HD-1)] and nutrient agar for propagation of isolate ONR-60A without loss of larvicidal activity against mosquitoes. Undeen and Nagel (1978) produced the bacteria, used in their efficacy tests against blackflies, on plates of Tryptose blood-agar base. Two media available for use in mass-fermentation work are the B-4 and B-8 media developed by Dulmage (1970, 1971) and Dulmage and de Barjac (1973) for good δ-endotoxin yields with several *B. thuringiensis* isolates. The B-4 medium contains 1% "Proflo" (a partially de-fatted, cooked cotton-seed flour obtained from Traders Protein Division of Traders Oil Mill Company, Fort Worth, Texas), 1.5% glucose, 0.2% yeast extract, 0.2% Bacto-peptone, 0.1% $CaCO_3$ and low levels of $MgSO_4$, $FeSO_4$ and $ZnSO_4$. Lacey and Mulla (1977) used thirteen strains of *B. thuringiensis* grown on the B-4 medium in bioassays against field-collected *Simulium vittatum* larvae. The B-8 medium contains 2% Proflo®, 1.5% glucose, 0.2% yeast extract, 0.2% peptone, 1% (w/v) corn steep liquor, 0.1% $CaCO_3$ and low levels of $MgSO_4$, $FeSO_4$ and $ZnSO_4$.

Because of previous experience with

Production-scale seed tank operations. Courtesy Dr T. L. Couch, Abbott Laboratories, Chemical and Agricultural Products Division, North Chicago, Illinois.

other varieties of *B. thuringiensis* at the mass fermentation level, the general technical procedures used to achieve operational status, in which the seed tanks and the final production fermentation vessel and medium components are prepared prior to inoculation, have been well defined. Dulmage and Rhodes (1971) and Simmonds *et al.* (1977) discuss some factors which should be considered in selecting fermentation media components. In harvesting the production vessel and recovering the spore–crystal complex, or a mixture of the two as separate entities, the technique(s) used must take into account the fact that the volume of the desired product(s) is relatively small in comparison with the volume of the components of a large-scale production vessel. The initial step in recovery generally utilizes a large capacity, high-speed, continuous-flow centrifuge to concentrate the spore-crystal complex into a

"paste-like" mass while eliminating most of the liquid. The product is then further processed in a temperature-controlled crystallizer before reaching the stage at which the desired formulation is achieved by the use of a combination blender-drier. *Bacillus thuringiensis* formulations have long been available as dusts, wettable powders and granular preparations (Briggs 1963). Dulmage and Rhodes (1971) should be consulted for a discussion of the historical development of fermentation technology in relation to *B. thuringiensis*.

The formulations which have been used until now for testing *B.t.i.* against blackflies have been simple air-dried or acetone-dried spore-crystal mixtures (Undeen, personal communication). Goldberg and Margalit (1977) reported the development of a microencapsulated formulation which maintained the level of toxicity of isolate ONR-60A against

High-speed centrifugal recovery equipment. Courtesy Abbott Laboratories.

mosquito larvae and concentrated the spore–crystal complex just below the air–water interface where larval feeding predominated. The use of microencapsulation (Anon. 1973a) in microbial formulations for use against larval blackflies would have definite advantages in giving greater control over the rate of availability of the spore–crystal complex to the filter-feeding larvae at different depths and distances from the area of application and allow for compensation for water flow velocities by adjusting capsule density (i.e. specific gravity). Also, by controlling the capsule chemical composition, one could enhance the level of bacterial spore–crystal complex release in the specialized environment provided by the appropriate blackfly larval gut. Unfortunately, very little information is available on the bacterial flora of the blackfly larval gut (Malone and Nolan 1978) and, consequently, on the metabolites (including enzymes) released by the

resident or transient gut flora (Ryan and Nolan, unpublished data). Neither do we have information on the chemicals released by the larvae into the gut lumen.

In assaying and standardizing the product prior to final formulation, there is an obvious need for the establishment of colonies of blackfly "tester strains" of known genetic background against which all isolates/mutants can be tested under standard conditions. It is only in this way that the virulence of different strains can be validly compared. It is also necessary to have access to similar "tester" colonies of NTOs (aquatic and non-aquatic) of known genetic background for safety-testing. The problems and techniques involved in *B. thuringiensis* bioassays and standardization have been previously discussed (Burges 1967; Burges and Thomson 1971; Dulmage 1973; Dulmage and de Barjac 1973). In working with new varieties and isolates of *B. thuringiensis*, one area which should be kept in mind in quality control is the possible presence of bacteriophages and their effect(s) on productivity (cell, spore and parasporal crystal yield) and on intra- and extra-cellular metabolite production. Ackermann and Smirnoff (1978) found that, for 48 strains of *B. thuringiensis* tested, at least seven different phage particles with long, non-contractile tails were produced. The frequencies of lysogeny and polylysogeny were 83% and 25%, respectively.

Entomophthora culicis has been found parasitizing *Simulium venustum* (Thaxter 1888; Brumpt 1936; Shemanchuk and Humber 1978) and *Simulium vittatum* (Shemanchuk and Humber 1978). Growth and sporulation studies have been carried out with *E. culicis* by Gustafsson (1965b, 1969). He found that, for the media tested, conidia production was best on media containing coconut milk, while resting spores were sometimes produced on Sabouraud agar (Gustafsson 1965b). The optimum temperature for growth in a liquid medium (as based upon dry weight yield) was approximately 20 °C with a good yield in the 18–24 °C range. The optimum pH for growth was not determined. However, based upon his results with representatives of other species, pH 7.0, after sterilization, was adopted. For the seven nitrogen sources tested (final concentration of 0.013% N), *E. culicis* grew best on neopeptone with successively less growth occurring on casamino acids, L-asparagine and glycine when glucose was the added carbon source. When starch was substituted for glucose, neopeptone still gave the highest yield; but, the relative yields for casamino acids, L-asparagine, glycine and the inorganic nitrogen sources containing ammonia were increased. Thirteen sole carbon sources were tested at a level corresponding to 2% in the final medium. The greatest amount of growth (as based upon dry weight yield) occurred in the presence of glycogen and starch; whereas, fructose, maltose, galactose, sucrose, glucose, casamino acids,

Tween 80, L-asparagine, Tween 60, Tween 40 and glycerol produced successively-reduced yields. A study of the effects of the addition of various substances on the enhancement of yield of an asparagine-starch medium, indicated that yeast extract and neopeptone were the most stimulatory.

Additional studies revealed that, after various combinations of growth factors were considered, either starch or glycogen was a good carbon source (concentration range 0.5–1.5%) and milk powder supplemented with yeast extract was an efficient nitrogen source (concentration range 0.5–1.0%). According to Gustafsson (1965b) media for the growth of *E. culicis* should be neutral or slightly acidic, and the trace elements Mg, Fe and Zn should be added. In the earlier study, Gustafsson (1965b) concentrated on the general conditions required for mycelial growth and conidia production, whereas he later (1969) examined the possible application of the earlier results to biocontrol programmes. Of the four solid media tested, growth and sporulation (conidia production) were best on medium M8 consisting of 2.6% milk powder, 0.1% yeast extract, 2% maltose and 1.5% agar. On the other hand, sporulation was slightly reduced and growth greatly so on medium M9 consisting of 1.5% cotton seed meal, 2.0% maltose, 0.5% (v/v) corn steep liquor and 1.5% agar.

Gustafsson (1969) also investigated the use of liquid media for mycelial growth, sporulation and resting spore production. However, he does not go into detail on the results with *E. culicis* other than to state that "preliminary experiments showed that several *Entomophthora* species grew and sporulated well" on a medium composed of 3% cotton seed meal, 2% maltose and 0.5% (v/v) corn steep liquor. In subsequent studies on the survival of different isolates of *Entomophthora*, in air-tight tins containing vermiculite as a substratum for growth, the tins were filled to about 50% capacity with growth medium and vermiculite (vermiculite was added at a proportion of 400 ml of medium per 100 g of vermiculite) and stored at room temperature (20–24 °C). When *E. culicis* was tested in this system, using a medium (V2) composed of 1.5% cotton seed meal, 1.5% soya bean meal, 0.5% milk powder, 0.5% yeast extract and 2% maltose, the cultures remained viable for at least five months.

Thus, several media are available for testing and modification under mass-fermentation conditions for resting spore and/or conidium production by *E. culicis* isolates. It should also be relatively easy to develop a medium for determining *E. culicis* resting spore viability (Nolan *et al.* 1976).

Entomophthora curvispora has been found on *Simulium latipes* (Lakon 1919); and Gustafsson (1965a) mentions that *E. culicis* often occurs with representatives of other species, including *E. curvispora*, which are found in similar humid to aquatic habitats and on the same insect host-species. Gustafsson (1965b) found that improved sporulation (conidia production) occurred when 0.2% corn meal extract was added to Sabouraud maltose agar. The optimum temperature for growth in a liquid medium (as based upon dry weight yield) was between 24 °C and 26 °C. Good growth was also obtained at 28 °C. While the optimum pH for growth was not determined, as in the case of *E. culicis*, pH 7.0 was used. When tested on seven potential nitrogen sources, the isolate of *E. curvispora* grew very well on neopeptone, casamino acids, ammonium nitrate and ammonium chloride (when glucose was the added carbon source), and best on *N*-acetyl-D-glucosamine, neopeptone, casamino acids and L-asparagine (when starch was the added carbon source). Inorganic nitrogen sources and glycine gave lesser dry-weight yields. Thirteen sole carbon sources were tested at a final level of 2%. The highest dry weight yields were obtained with glycogen and starch; whereas, yields reduced by 50% or more were obtained with the other carbon sources.

Additional growth stimulation studies similar to those previously discussed for *E. culicis* were not conducted. However, Gustafsson (1965b) suggested the use of 0.5–1.5% starch or glycogen as the carbon source, and 0.5–1.0% casein hydrolysate (casamino acids) or neopeptone with yeast extract as the nitrogen source for *E. curvispora*. As in the case of growth of *E. culicis*, Gustafsson (1965b) recommended a neutral or slightly acidic medium containing Mg, Fe and Zn. This author (1969) found, in initial experiments, that *E. curvispora* remained viable for up to one month in air-tight tins stored at 20 °C and filled approximately three-quarters full with a mixture of 30 g cotton seed meal, 20 g maltose and 5 ml corn steep liquor to 2 kg of soil with sufficient water to achieve a final moisture content of 55%. Studies on the use of vermiculite in the sealed tins were not conducted (Gustafsson 1969).

Gustafsson (1965a) refers to the overall similarity between *E. culicis* and *E. curvispora*, and he (1965b) also comments that media specifically used for *E. culicis* could be used for *E. curvispora*. Thus, the transition to the mass-fermentation level of production for *E. curvispora* should also be relatively easy if this author's assumptions are valid. The major problem with Gustafsson's studies (1965a,b; 1969) is the lack of quantitative information on the effects of various media on resting spore formation and, for the most part, on conidium production—because the chief emphasis for most species studied was on the mass (mycelial dry weight) of fungus produced. Nevertheless, these studies provide a solid foundation upon which to begin mass-fermentation investigations focusing on viable resting spore and conidium production for these two fungal species. Previous experience with small-scale (14 litres or less) fermentation production of resting spores of *Entomophthora virulenta*

(Soper *et al.* 1975; Latgé *et al.* 1977) indicates that the basic technology used in *B. thuringiensis* work is readily applicable to studies with isolates of *Entomophthora* which do not spontaneously form free-protoplasts and, thus, require additional morphogenetic development (Dunphy and Nolan 1977a,b; Nolan, unpublished data).

The aquatic phycomycete, *Coelomycidium simulii*, occurs in a wide variety of larval simuliid hosts in North America and Europe, including the USSR (Jenkins 1964; Strand *et al.* 1977). *In vitro* cultivation of *C. simulii* (L. S. Bauer, R. S. Soper and D. W. Roberts, personal communication) from surface-sterilized *Simulium venustum/verecundum* larvae has recently been accomplished using Grace's insect tissue culture medium (Grand Island Biological Company, Grand Island, NY). The fungus grows as spherical bodies each of which subsequently divides into two or more smaller bodies. Rapid growth occurs in a wide variety of tissue culture media, at different pH levels and at various concentrations (dilutions) of Grace's medium. Zoospores have not been observed in culture. The fungus did not grow following intra-haemocoelic injection of larvae of the lepidopteran, *Estigmene acrea*, but did grow in larvae of *Simulium pictipes*. Thus, if the fungus can be successfully maintained *in vitro* on a long-term basis, it should be possible to simplify the composition of the complex culture media following the procedures previously used with the aquatic phycomycete, *Catenaria anguillulae* (Nolan 1970a,b,c). This information on the physical and chemical growth requirements can then be used to design mass-fermentation media.

Metarhizium anisopliae possesses what is probably a very great, but untested, potential for biocontrol of both larval and adult blackflies. *Metarhizium anisopliae* isolates have been used against a wide variety of pests of economic importance and have been assessed by WHO as having reached stage III of their testing sequence (Laird 1978). This organism has also been grown at the pilot-plant level (Dulmage and Rhodes 1971). At least one mutant derived from a wild-type strain of *M. anisopliae* has been produced with increased virulence towards *Culex p. pipiens* (Al-Aidroos and Roberts 1978). Once the genetic basis of virulence is better understood, this facility in genetic manipulation may enhance the possibility of deriving mutants with hypervirulence towards simuliids. At least three media which have been used by Roberts (1966) are available for mass-fermentation growth of *M. anisopliae* isolates: (1) SDBY consisting of 1% neopeptone, 2% dextrose and 0.2% yeast extract; (2) CDBO composed of 1% neopeptone, 3% sucrose, 0.1% K_2HPO_4, 0.05% $MgSO_4$, 0.05% KCl and trace amounts of Cu (16 ppm), Mn (18 ppm), Zn (130 ppm), Mo (9.6 ppm) and Fe (5000 ppm); and (3) CDBI which was based upon the CDBO medium but contained 0.87% $NaNO_3$ instead of the neopeptone. The hydrogen ion content of the three media, after autoclaving, was approximately pH 7.2–7.3. The feasibility of using peanut oil as a conidial diluent, in order to obtain a liquid formulation and also to decrease the required conidial concentration, has been tested. The results indicate the possibility of reducing the required conidial dosage by approximately 80% (Roberts 1977).

Potential of Mermithids for Control and *In Vitro* Culture

J. R. Finney

The potential of mermithids (Nematoda:Mermithidae) for control of blackflies, lies in the fatal and sterilizing effects of mermithid parasitism on the host, and their ecological adaptation to the latter's aquatic larval and airborne adult stages. Their utilization in control programmes depends on man's innovative ability to manipulate pre-existing simuliid/mermithid populations and/or to mass-produce the worms so that they can be introduced into preimaginal blackfly habitats where they did not previously occur.

Life-cycle of Parasite and Effect on Host

The infective second-stage preparasitic larva hatches from the egg in the stream habitat and enters the host haemocoel by penetrating the cuticle (Anon. 1973b; Molloy and Jamnback 1975). Infection is never due to ingestion of the nematode egg or preparasite as claimed by Phelps and DeFoliart (1964). The several manifestations of parasitism in larval simuliids are well-documented by Strickland (1911) and Phelps and DeFoliart (1964). The effects are most obvious in the latter stages of nematode development, when the fat-body of the host is so depleted that the sizeable nematode(s) can be seen coiled in the host haemocoel. Simuliid larval development time is usually extended. Pupal and adult histoblasts of infected larvae do not develop to the extent of those in uninfected hosts. Condon and Gordon (1977) hypothesized that the presence of the mermithids in simuliids so stress the host that its endocrinology is disturbed. Whatever the disruptions to the host (physiologically or mechanically) during the parasitic phase of the mermithid, the ultimate emergence of the parasite(s) from the larval host is lethal. This serves to reduce, at the larval stage, a potential vector- or nuisance-population before it becomes a problem.

Both male and female adult simuliids may exhibit mermithids, too, the fat-body then being reduced or lacking. More importantly, the production of intersexes, induced sterility and mortality at emergence of mermithids, all serve to eliminate infected adults from a breeding population. The topic of intersexes in simuliids harbouring mermithids, has been reviewed by Welch (1964) and Gordon *et al.* (1973) and since noted by Fredeen (1970) and Hunter and Moorhouse (1976). Mermithid-induced sterility in adult blackflies has been associated with the reduction or total absence of ovaries (Peterson 1960; Shipitsina 1963; Phelps and DeFoliart 1964; Le Berre 1966; Hunter and Moorhouse 1976; Mondet *et al.* 1976; Anderson and Shemanchuk 1977; Mokry and Finney 1977). In extreme cases, the host abdomen is altogether empty (Hocking and Pickering 1954), although the presence of mature eggs in some infected females has been noted (Hunter and Moorhouse 1976).

The role of a mermithid in the induction of intersexuality and sterility in its host (in this instance, a chironomid) has been thoroughly investigated by Wülker (1964, 1975). Infected females normally make oviposition flights, depositing mermithids instead of eggs at the head of streams (Grunin 1949; Davies 1958; Peterson 1960; Colbo, personal communication). As in the case of larvae, adults die on emergence of the worm(s).

On emergence from either the larvae or adults, the mermithids enter a non-feeding, free-living stage during which they become fully adult and produce the next generation. During this time, they are sustained by the storage materials laid down at the expense of the host during the parasitic phase.

Natural Infection and Occurrence

The numerous and widespread records of mermithid infections in simuliids have been reviewed by Welch (1964), Gordon et al. (1973), and Poinar (1977). Some accounts of parasitism report infection rates of 5% and below (Ezenwa 1974a; Bailey and Gordon 1977; Brüder and Crans 1979), others of 90–99% (Anderson and Dicke 1960; Rubtsov 1963b). The latter figures have been used to indicate the potential of mermithids in blackfly control. However, in interpreting such data, it has to be borne in mind that although the incidence of parasitism, in the simuliid population sampled, automatically reflects the percentage mortality therein, it does not reflect the percentage of parasitism of the entire population. Samples of late instars will reflect only the numbers of slow-developing infected larvae which may bear no relevance to the entire population, the majority of which may have emerged already (Ebsary and Bennett 1975). Alternatively, as the nematodes are only clearly visible in the later instars, the infections in samples of early instars, or nematodes at an early stage of development in late instars, will go undetected without dissection of the host. Obviously, critical assessment of sampling techniques in such surveys (whether past or future) is needed.

With few exceptions, the data presented in the literature refer only to individual collections. They thus give little indication of the life-cycle patterns of mermithids in the context of stream ecology or of their possible manipulation for control. It is only by intensive study of entire streams that certain aspects of mermithid/simuliid ecology can be clarified. From those stream studies that have been carried out, the following factors emerge.

In Wisconsin, USA (Phelps and DeFoliart 1964), and the USSR (Welch and Rubtsov 1965), evidence has accumulated suggesting that mermithids may cause short-term elimination of individual simuliid species from localized areas. The localization of infection was emphasized by Rubtsov (1963b) and again by Welch (1964), who described mermithid distribution as haphazard—there being high parasitism in some localities, while nearby streams may totally lack mermithids. Even within an infested stream, the locations of parasitism may vary along its course (Glötzel 1973; Colbo 1974; Ebsary and Bennett 1975; Bailey and Gordon 1977). Colbo (1979) found that levels of infection of Prosimulium mixtum by Neomesomermis sp. in Newfoundland increased downstream, both in extent and intensity. Parasitism may also vary within blackfly sites in a stream, depending both on time and the developmental stage of the host (Bailey and Gordon 1977). The stage of the host, in turn, will be influenced by both environmental phenomena and the level of larval nutrition provided by the stream (Colbo and Porter 1979).

In addition, within a single stream and even within individual larval sites in that stream, only certain simuliid species may be infected by a particular mermithid; although several species of simuliids and mermithids may co-exist. It has been questioned whether host-specificity (as suggested by Welch 1964) in these nematodes, would be an advantage for their use as control agents, under ecologically favourable conditions. However, there is increasing evidence that the Mermithidae in question are not host-specific. Rather, as suggested by Mokry and Finney (1977), they simply appear to prefer certain hosts over others, ultimate selection depending on host-availability. Welch and Rubtsov (1965) recorded *Gastromermis boophthorae* larvae from four species of *Simulium* and one of *Eusimulium*, finding several other simuliid species uninfected. These authors also pointed out that a single simuliid species may serve as host for several mermithid species; e.g. *Isomermis* spp. and *Mesomermis* spp. were sometimes found together in a *G. boophthorae* population. Mondet *et al.* (1976) found that none of five local blackfly mermithids was specific for *S. damnosum s.l.* in West Africa. Other examples of this *genre* are cited by Welch (1964). The range of hosts for each of these worms has yet to be fully delineated. It is of fundamental importance that this be achieved for any species of mermithid under consideration as a biocontrol agent before actual biocontrol measures are proposed. Host-reaction or host-avoidance behaviour to certain invading larvae are possible reasons for the occurrence of uninfected species in the midst of several mermithid populations. However, timing (which is, in turn, affected by the environment) is probably the most important single factor in determining nematode infection levels and the maintenance of a mermithid population in a stream. Indeed, lack of synchrony may explain the restricted distribution of mermithids in certain simuliid streams.

In West Africa, Mondet *et al.* (1976) describe the sudden appearance of mermithids as suitable simuliid larval hosts become established in temporary flow areas. Ebsary and Bennett (1975) suggest that *N. flumenalis* is univoltine but has two isolated populations, one synchronous with *Prosimulium* hosts, the other with *Simulium*. These authors, together

with Phelps and DeFoliart (1964), Gordon *et al.* (1973) and Ezenwa (1974b), have indicated that in cold climates, mermithid and simuliid egg hatch are synchronous, both appearing to depend, to some extent, on the temperature rise in spring.

Ezwena (1974b), Molloy and Jamnback (1975) and Bailey and Gordon (1977) pointed out that invasion of *S. venustum* and *P. mixtum* by *N. flumenalis* only occurs in the 1st and 2nd-instars, later instars being refractory to infection. In the design of a control programme in such a situation, it would be essential that the microhabitat distribution of early instars be determined. The inundative application of preparasites to newly-emerged larvae, or the synchronous hatch of host and parasite, would result in multiply-infected 1st-instars. As sex in simuliid mermithids appears to be determined, in part, by their numbers harboured by the host (as in other mermithids), multiple infection should result in the emergence of predominantly male postparasites (Ezenwa 1974a; Ebsary and Bennett, 1975; Ezenwa and Carter 1975). In nature, as fewer hosts and parasites hatch, multiple infections would be replaced by single ones with subsequent emergence of female postparasites and maintenance of the nematode population in the stream. However, in a control situation, reapplication of a regulated dose of the mermithid would be needed to achieve the same result.

The factors leading to infection of adult simuliids are not clearly understood. *Isomermis lairdi* can regulate both larval and adult populations of *S. damnosum s.l.* depending on the time of infection (Mondet *et al.* 1977b), infections of late instars being carried over to the adult where development of the nematode is completed. Such a nematode could prove a useful tool for biocontrol of larvae and/or adults. In the latter case, the timing of application need not be as precise as that for a worm used inundatively, purely as a larviciding agent. Time of infection, however, cannot be the only factor governing mermithid carry-over to adult simuliids. *N. flumenalis*, which infects only early instars of *S. venustum* and *P. mixtum*, has been recovered from the adults of these species (Mokry and Finney 1977). The possibility of a mermithid species emerging from the larval stage of one host and (subject to availability) infecting yet another species in which it can carry-over to the adult, cannot be overlooked. In addition, Colbo (1979b) cites an example of infected young larvae in which development of the mermithid was never completed until after the host had matured. This further indicates the importance of evaluating the susceptibility of different hosts and instars to infection when assessing their potential for biocontrol. In addition to transporting mermithid infections to other locations, upstream flight of parasitized adults and deposition of nematodes by them, helps redress the balance of mermithid sex-ratios at the larval site (Colbo 1979b) so that the mermithid population remains established in the blackfly breeding area and is not swept away downstream (Wenk 1976). The manipulation of such nematode life-histories for biocontrol requires precise information on target simuliid ecology for the timing of regulated applications of host-specific nematodes.

Also, it has been suggested by Mondet *et al.* (1977) and Anderson and Shemanchuk (1977) that ingestion of a blood meal by the host is necessary for full development of certain mermithids occurring in adult blackflies. This has several implications for design of control programmes against nuisance or vector populations of simuliids. Thus, it would be futile to utilize a mermithid, incapable of completing its development, without its hosts having fed on blood for control of a serious cattle pest such as *S. arcticum*. In such a case, the population has to be regulated by other means, notably larvicides. Nevertheless, a mermithid with the same requirements could regulate a

vector population before actual transmission of a disease-causing agent occurred.

Simuliid infections by mermithids occur during relatively short seasons. There is still speculation as to how and at what stage mermithid populations survive dry (e.g. West African) or overwintering (e.g. North American) periods. Welch (1964), Ezenwa (1974b), and Ebsary and Bennett (1975) suggested that cold-climate mermithids tide over as eggs or as free-living adults in the substrate of the stream-bed during such periods. An alternative is exemplified by *N. flumenalis*, which has an extended parasitic phase during winter in its *Prosimulium* host (Bailey and Gordon 1977). The ecology of the free-living stages of mermithids, as well as their susceptibility (particularly at the egg stage) to desiccation and changes in temperature, are areas which have yet to be investigated thoroughly.

Mass-collection of Infected Simuliid Larvae and Maintenance of Free-living Nematode Stages

Strickland (1911) and Petersen (1924) suggested that one method of regulating simuliid numbers would be by moving mermithids from one stream to another. This can be achieved by transference of infected simuliid larvae. However, unless host-parasite relationships in the stream of origin are well-understood, it is quite conceivable that a poor sex-ratio of nematodes could be introduced. Moreover, the susceptibility of the proposed host to the introduced nematode must be pre-determined and the synchrony of host and parasite established. Gordon (1975) attempted such a transference, on a small scale. While he demonstrated that a mermithid, relocated in this way, could cause infections where they did not exist before, the nematode (*Isomermis* sp.) failed to become established. Colbo and Porter (1977) have further investigated the feasibility of this transfer method for enhancement of mermithid populations in Newfoundland streams.

Any other form of introduction or enhancement depends on mass-collection of infected simuliids and maintenance of emergent nematodes until they have reached a stage suitable for reintroduction. The collection of suitably large numbers of parasitized larvae is not always feasible due to sparse distribution, and unless the ecology of the particular stream is well-understood, a collection of a mixed population of hosts and parasites may cause confusion. Prior to collection, Bailey and Gordon (1977) recommended constant monitoring of streams, so that environmentally-induced changes in them could be recognized. Timing of a collection is important for two reasons. Firstly, it is the late instars which are usually collected, the larvae then not having to be maintained for long in the laboratory. Secondly, where necessitated by the host–parasite relationship, both early and late larval collections may be needed to ensure a viable sex-ratio of emergent nematodes.

Bailey *et al.* (1974) suggested a method for collecting infected simuliids in the field and transporting them back to the laboratory without undue stress to the host—such stress can stimulate premature emergence of incompletely-developed nematodes, which usually die. The emergent post-parasitic mermithids can be collected and maintained during the moult to the adult, copulation and oviposition (Bailey *et al.* 1977). Throughout this period, nematodes are highly susceptible to a variety of pathogens (Bailey *et al.* 1977), which may spread through an entire culture. High mortality in the laboratory due to fungal infection has been reported for *N. flumenalis* (Phelps and DeFoliart 1964; Ebsary and Bennett 1973). Two naturally-occurring fungal parasites of *N. flumenalis*—*Saprolegnia megasperma* and *Pleuropedium tricladioides*—have been described

and characterized by Nolan (1975a, b; 1977) and Murrin and Nolan (1977).

Maintenance of post-emergent mermithids in the laboratory has been achieved in a variety of containers designed in the belief that a substrate was essential to the postparasitic moult which releases the fully-formed adult. Clear, water-filled containers are, in fact, adequate. They have the advantage of allowing observation of the nematodes through all stages of development (Finney, unpublished). After the moult has been accomplished and the sexes fully differentiated, the adult sex-ratio can be manipulated for optimum egg-production. The rearing of the free-living stages of several mermithids has been accomplished; *Gastromermis viridis* and *I. wisconsinensis* (Phelps and DeFoliart 1964), these two species plus *N. flumenalis* (Ebsary and Bennett 1973), and *I. lairdi* (Mondet *et al.* 1977). The most relevant facts to emerge from these studies include the duration of the stages from emergence to egg-hatch of the commonly-found North American species (summarized in Ebsary and Bennett 1973), and the thermal tolerance of different species. The latter authors found that the optimum temperature for development of Newfoundland-collected *N. flumenalis* was 12 °C. Phelps and DeFoliart (1964) found *I. wisconsinensis* more cold-tolerant than *G. viridis*, correlating this with its relative abundance in colder streams; which indicates the importance of temperature in local distribution.

The stage at which these mermithids can best be stored for stockpiling and transporting to a test site, remains to be determined. So does their viability, particularly that of the infective stage in various habitats. The overriding problem still lies in the egg stage—without a mechanism for initiation of egg hatch at will, advances in several aspects of research will be severely restricted. These avenues can be explored using field-collected, laboratory-maintained mermithids.

Employing laboratory-hatched pre-parasites of *N. flumenalis*, Molloy and Jamnback (1975) achieved consistently high rates of parasitism in early instars of *S. vittatum* under standardized running-water laboratory conditions. Further-more, in 1977, they undertook a field trial and achieved reasonably high inci-dences of parasitism in *S. venustum* and *S. vittatum*. These authors, together with Bruder (1974), considered *N. flumenalis* of limited value as an efficient, inundative biocontrol agent. Basically, this is be-cause only early simuliid instars are sus-ceptible to this worm. Re-treatment, on a regular basis, thus being necessary (under prevailing circumstances of poor mermithid-availability, this would be prohibitively expensive).

As an alternative to inundative use of such mermithids, experiments are in progress in Newfoundland whereby attempts are being made to enhance the natural populations of these nematodes in certain streams by the addition of laboratory-maintained nematodes (Colbo and Porter 1977). The limited success of this venture, to date, has further demon-strated the importance of understand-ing the host–parasite system and the many facets of compatibility of one with the other. The feasibility and long-term benefits of "seeding" blackfly streams in this way have still to be determined.

Mermithids and Integrated Control

Field-use of mermithids should also be considered in the context of integrated programmes, utilizing other pathogens or chemical larvicides. Although there are reports of pathogens, especially Microsporida, parasitizing simuliids in the same streams as mermithids (Rubt-sov 1950; Abdelnur 1968; Colbo 1974; Ezenwa 1974a,b; Ebsary and Bennett 1975; Garris and Noblet 1975), little use

has been made of this information for feasibility studies of their combined ap-plication. Only Lichovoz (1975) investi-gated the effects of *Gastromermis booph-thorae* and *Pleistophora simulii* on larval *Boophthora erythrocephala* de Geer. The combined effects of these organisms re-sulted in earlier mortality than that of blackfly larvae parasitized by either alone.

Recently, the integrated use of the warm-water mosquito mermithid, *Romanomermis culicivorax*, and insecti-cides, has been demonstrated in the laboratory (Finney *et al.* 1977; Levy and Miller 1977; Winner *et al.* 1978). Similar procedures could be adopted for simuliid mermithids, primarily to determine com-patibility of agents and feasibility of ap-plication. Further study should reveal whether or not chemical applications adversely affect the natural regulatory potential of such mermithids on simuliid populations in the field. Abate® and methoxychlor are prime candidates for investigation. The former was, until 1980, the sole larvicidal chemical used against *Simulium damnosum s.l.* in WHO/OCP (see pp. ix, 89–90). Studies of its effect on mermithids have so far been limited to the work of Garris and Noblet (1975), North Carolina, who found that treatment of blackfly larvae with 2% Abate® had no significant effect on the incidence of mermithids within them. Methoxychlor has been used in ARBFRP (see p. 119) against *S. arcticum*, a known mermithid host (Anderson and Sheman-chuk 1977). Sufficient nematodes for laboratory trials, as suggested, could be made available through their field collec-tion and laboratory-maintenance.

Mermithid Culture

The eventual widespread introduction of mermithids into blackfly streams, either alone or more probably as part of integrated programmes, will only be accomplished when reliance on field collections can be replaced by the *in vivo/in vitro* mass-production of the free-living stages (Welch 1964; Gordon *et al.* 1973).

In Vivo

Even after initial field collection and maintenance in the laboratory to the egg/preparasite stage, the barriers to success in initiating an *in vivo* culture of a cold-climate blackfly mermithid are many. However, these obstacles could be overcome were there sufficient understanding of the host-specificity of the nematodes, combined with an efficient blackfly-rearing system. Phelps and DeFoliart (1964) used battery-jars with compressed air passed through air-stones to rear *S. vittatum*. In this system, they challenged the simuliid larvae with *G. viridis* and *I. wisconsinensis*, obtaining infections of up to 5%. These authors primarily investigated the mode of infection by preparasitic nematodes and their ensuing development during their 10–15 days inside the host. They neither achieved a continuous passage of the nematodes, nor gave any indication of host-survival in this system. Bailey and Gordon (1977) used a similar system to investigate the various facets of simuliid susceptibility to mermithids, but failed to obtain continuous culture of any of them.

The recently-developed, blackfly-rearing systems of Colbo and Thompson (1978) has greatly increased larval survival and further enhanced the chances of successful *in vivo* mermithid production. However, due to the absence of an adult blackfly-rearing system for most Simuliidae, a source of the host is not available throughout the year unless extensive egg collections are made seasonally and stored. With the advent of the continuous laboratory culture of at least one simuliid species, *S. decorum* (Simmons and Edman, 1978; and see p. 303 [Ed.]), it should be possible to investigate the bionomics of infection of this host by its natural mermithid parasites and, eventually, to culture them *in vivo*. However, there is not yet a method for obtaining synchronous hatch of mermithid eggs, especially of those in diapause. Small hatches over an extensive period preclude good infection rates in the host. As pointed out earlier, the sex of a nematode is determined, in part, by the number of worms harboured by an individual host, this, in turn, affecting the emergent male/female postparasitic ratios essential to good egg-production and subsequent continuation of the culture. Further, the natural life-cycle of cold-climate hosts is protracted. Therefore, any passages of nematodes through them will be subsequently affected, precluding rapid production of mermithids. The use of an alternative host for production of these mermithids deserves consideration. In the short-term, this might be feasible, but it would not circumvent the eventual high-cost problems associated with any large-scale *in vivo* culture.

In vivo culture of warm-climate simuliid mermithids recently showed some progress at IRO/OCCGE, Bouaké, Ivory Coast, although there are still difficulties to be overcome due to lack of a host colony. Mondet *et al.* (1977b), maintained *I. lairdi* in the laboratory and, using a magnetic stir-bar larval rearing method, successfully infected *S. damnosum s.l.* and passaged the nematodes over a 10–16-day period through the host. The emergent postparasites were viable, completing their free-living cycle in 15–23 days, but no attempt was made to continuously cycle the resultant preparasitic stage. However, a large-scale trough-type rearing system for *S. damnosum s.l.* has been constructed, with financial assistance from IDRC, under the recent protocol between that Organization and OCCGE (Berl and Prud'hom 1978, 1979). Perhaps this will help realize the *in vivo* mass-production of *S. damnosum s.l.* mermithids?

In Vitro

In vitro culture has the basic economic advantages over *in vivo* methods in that rearing of a host is not involved. The aim of this method is to simulate larval parasitic development in the haemocoel of the simuliid, by providing a substitute environment, where the nematode can abstract and store large quantities of nutrients for later utilization in its non-feeding, free-living stage. Towards the formulation of such an environment, the nature of the simuliid haemolymph has, in part, been elucidated (Rubtsov 1959; Gordon and Bailey 1974, 1976). Recent investigations concerning the nutrition of the parasitic stage of *N. flumenalis*, its storage products and their derivation from the host, have indicated that this mermithid severely depletes proteins and blood glucose in the haemolymph of *Simulium venustum* and *Prosimulium mixtum/fuscum* (Gordon *et al.* 1978). Condon and Gordon (1977) suggested that parasitism also resulted in depleted fat-body glycogen reserves in both host species. The accumulation of material within the developing trophosome, a limiting factor in previous *in vitro* culture attempts with mermithids (Finney 1976), was investigated by Gordon *et al.* (1979). They found storage lipid in the form of triacylglycerols, phospholipids, sterol esters and free sterols, in the trophosomes of two strains of *N. flumenalis*. This finding is of fundamental importance to the composition of culture media.

Notwithstanding our limited information on the physiological interrelationships between host and parasite, and the specific dietary requirements of the nematode, attempts at *in vitro* culture have been made using eggs and preparasites derived from laboratory-maintained nematodes. Myers (personal communication) initiated cultures of *N. flumenalis* from sterilized eggs and pre-parasites. He incubated them in a range of insect tissues culture media but obtained best growth of the nematodes in a formulation of Grace's medium. Finney (1976) further developed techniques for mass sterilization of the preparasitic stage, and capitalizing on Myers' preliminary work, achieved some success in the growth and development of *N. flumenalis* in a combination of Schneider's *Drosophila* medium and Grace's tissue culture medium, modified by the addition of foetal calf serum and kept at 10 °C. An alternative approach to mermithid culture lies in their propagation in cultured blackfly cells. In 1975, Shapiro was maintaining pupal cell cultures from *S. vittatum* and *S. venustum*, although no attempt was made to passage *N. flumenalis* through them. This initial work is being continued by Abercrombie (personal communication) who is in process of establishing an *S. vittatum* cell line for the purpose of studying the host–parasite relationship and the initiation of a mermithid culture. Rubtsov (1967b) postulated that the mermithid parasites of simuliids secrete proteolytic enzymes to predigest the fat-body and then absorb the hydrolysate through their cuticles. He suggested that a fat-body culture might provide the requisite materials for blackfly mermithid development. Finney (unpublished) adapted the *Galleria meleonella* fat-body culture method of Oberlander (1969) and Richman and Oberlander (1971), which utilizes Grace's tissue culture medium, for this purpose. To date, she has had some success in the culture of *N. flumenalis* using this system. This line of research is also being pursued in the USSR.

Progress in this field is slow, restricted as it is by lack of funding and of available material both for *in vitro* culture trials and intensive research into the host–parasite relationship. The future of *in vitro* culture depends on year-round availability of simuliid mermithids. This could be achieved by establishment of an efficient *in vivo* system or by utilization of stored mermithid eggs, preparasites or gravid females (Mondet *et al.* 1977), any of which would circumvent the problem.

Use of Other Mermithids for Control

In the absence of a natural simuliid mermithid which can be mass-produced, consideration was given to using another one of these worms which is not naturally a blackfly parasite, but which is available on a large scale. Finney (1975) found that the mosquito mermithid, *Romanomermis culicivorax*, penetrates early instars of three simuliid species under still-water conditions. Hansen and Hansen (1976) demonstrated, in Ivory Coast, that *S. damnosum s.l.* is sometimes attacked by this nematode in a moving-water system. After further tests had been carried out in a variety of moving-water systems, it became apparent that an uneconomic number of preparasites would have to be utilized to obtain even a low-percentage mortality (Mondet *et al.* 1978; Finney and Mokry 1980; Colbo *et al.* 1978). Although development of the parasite was initiated when infection occurred, it was never observed to completion (Poinar *et al.* 1979; Finney and Mokry 1980). It therefore seems unlikely that this nematode could become established even in warm-climate blackfly streams; while its inundative use would clearly prove uneconomical—only about 0.5% of exposed 1st-instar larval *S. damnosum s.l.* having been fully penetrated in three experiments (e.g. Colbo *et al.* 1978) in the IRO trough systems, using some 80 million preparasites. However, experimentation with *R. culicivorax*, in these systems, has given us a model for methodology of application of true blackfly mermithids when they become abundantly available through *in vivo* and *in vitro* culture.

Conclusion

Welch (1964) and Gordon *et al.* (1973), previously reviewed the potential of mermithids for the biocontrol of simuliids. These authors pointed out the necessity for further investigation into the many-faceted bionomics of the host–parasite system, emphasizing the importance of mass-culture of the worms in order to exploit their demonstrable potential as control agents. Further investigations, of both topics, are still needed—the onus being on field workers to locate and study all aspects of infected areas, and to collect mermithids for laboratory experimentation. Research into the market-potential of mosquito mermithids suggests that there is likely to be a demand for mermithids as blackfly control agents if they can be produced economically. This end would be best achieved by *in vitro* mass-production, the generation of which should be a major goal in future research on mermithids of simuliids, towards the eventual provision of adequate quantities of these worms for inoculative and inundative use against blackfly pests and vectors.

Factors Affecting Industrialization of Entomopathogens

T. L. Couch and A. L. Paul

Industrial interest in the development of entomopathogens of insects of economic (as opposed to medical and veterinary) importance has fluctuated over the past several decades. Historically, through the late 'forties, 'fifties and early 'sixties speculative research into, and limited commercialization of lepidopteran strains of *Bacillus thuringiensis* (*B.t.*) by as many as four different companies were undertaken. Generally, the investigative field research and development of this important entomopathogen lacked the coordination and direction needed for simultaneous determination of safety, specificity, biological and field stability, formulation effects, field efficacy and application parameters. Consequently, registration with federal agencies was delayed. This was the result of poor field data and the hesitancy of regulatory agencies to declare a naturally occurring pathogen as safe, specific, and exempt from residue tolerance requirements. Consequently, two of the four companies producing entomopathogens in the USA abandoned this promising new product area. It is a credit to Nutrilite Products

Inc. and International Minerals and Chemicals, Inc. that they persisted, and eventually obtained federal registration for *B.t.* Also, they expanded their research and development interests to include entomogenous fungi and baculoviruses.

Historical Background

An excellent review by Hall (1963) covered the early tests of the *B.t.* formulations. Inconsistency in field efficacy was a problem with these early products; lack of standardization and poor physical characteristics of the formulation prevented them from achieving a significant share of the insecticide market.

This situation persisted until 1969 when a new strain of *B.t.*, much more potent than previous ones, was made available to commercial interests by H. T. Dulmage. The results of this introduction are gratifying. Since the advent of the HD–1 strain, *B.t.* has become the standard control agent against which other (chemical and microbial control) agents are measured. While it is true that the new strain contributed to the final eventual success story of *B.t.*, other factors involved were proper strategic planning and positioning of this agent into a network which insured adequate research and development funding. It is the results of strategic planning studies which determine industrial involvement and investment in new research projects. These studies outline and define all areas which will influence cash flow and return on investment. Since return on investment dictates research thrusts within industry, this planning essentially defines the factors which ultimately affect industralization of entomopathogens.

All comments presented herein will be concerned with the type of decision process used within Abbott Laboratories to determine involvement with potential microbial products. Recognizing that other companies may use a different pathway, it is the purpose of this chapter to detail the components of Abbott's decision process.

Industrial Considerations

Couch (1976) reviewed some of the considerations involved in commercialization of entomogenous fungi. These were: technological background and needs; safety, specificity, and utility; reliability, predictability, and grower acceptance; aspects of production and cost; market impact and return on investment.

To better understand the impact of each of these considerations on eventual industrialization, they will be explained in some detail.

Technological Background and Needs

This area is particularly important since it includes development of a thorough knowledge of the organism's life-history and growth requirements. This is essential to determine whether a candidate biocontrol agent can be mass-produced. Allied with growth-cycle information is elucidation of the mode of action. This provides information to the fermentologist who can then use his skills to optimize insecticidal activity by altering fermentation conditions. This is important for eventual cost reduction in the manufacturing process.

Safety, Specificity, and Utility

Is the entomopathogen naturally occurring? Are by-products produced during fermentation non-toxic and non-infective? What is the insect spectrum? These are generally the first questions asked industrial scientists before a project with a new microbial is undertaken.

Closely associated with safety and specificity is utility. Is there a demonstrated market need for the potential product? Is it as effective or more effective than chemicals? Is a complex of insect pests susceptible or only one? Is this a disadvantage? *B.t.*, for example, has a relatively large host-spectrum in the insect order Lepidoptera and, consequently, the commercial potential has grown accordingly. The result has been that Abbott has invested considerable capital in the development of this pathogen.

Reliability, Predictability, and Grower Acceptance

Can the organism be grown and properly formulated to provide the crop-producer with a reliable and predictable insect control agent? Is extensive user education required for proper use of the pathogen? In the final analysis a farmer or control operative wants a material which is easy to use, inexpensive, and capable of being applied in a conventional manner. *B.t.* formulations and some baculoviruses fit these criteria, but nematodes and fungi could necessitate re-education of the end-user. The extent of this education will often have a negative impact on user acceptance. Without this acceptance, there would be no market; hence no industrial investment.

Aspects of Production and Cost

To be a viable candidate product, the entomopathogen must be capable of being mass-produced. Detailed studies must be undertaken to determine the feasibility of progressive scale-up from shaker flask to commercial fermentation conditions. Allied to this consideration is cost. Can the entomopathogen be produced, packaged and sold at an end-use cost comparable to materials currently available on the market?

Bacteria, baculoviruses, fungi and nematodes present some special and unique problems which affect production costs, formulation, packaging and inventory control. Since these are living organisms, product forms must be viable and virulent. Failure to achieve this produces obvious results. It is possible that some microbial agents formulated for sale will carry expiration dates, reflecting a time limit beyond which sale is prohibited. If this time is short (e.g. 6 months), production planning and inventory control become critical. This tends to increase cost, and cuts into the standard manufacturing margin.

Another factor of ultimate importance in an industrial decision to fund research projects, is the presence of patent or proprietary protection. Since most entomopathogens are commodities obtained from third parties, patent protection is usually not available. To make investment in these agents attractive, a company must possess, or be capable of developing, proprietary fermentation technology for the production of an economical, stable and efficacious product. Also, companies like Abbott already have fermentation capacity and, therefore, little additional capital investment is required to develop a research and production programme for a potential microbial agent.

Market Impact and Return on Investment

In the final analysis, any product developed by commercial interests must command a significant share of a defined market, and provide a reasonable return on the investment monies used to support new-product research and development. An entomopathogen is no exception; it must provide a reasonable profit pool to the company. This profit is then used for continued product development and new-product research.

Strategic Planning and Investment Considerations

The criteria described above are parts of the decision-making processes which occur daily in industry.

A typical pathway used for the strategic planning and development of entomopathogens will now be discussed.

"Strategic planning" is nothing more than a term which has been applied to the process used by most commercial companies to concentrate their resources into given areas of activity. The process begins with the selection of a long-range direction by asking and answering a series of questions: What is the desired configuration of the company in five years—10 years—perhaps even 20 years or more?; what are today's markets and how are they expected to change over time?; what forces are at work in the general external environment and how will they influence these markets? (economic, political, ecological, or competitive?); what is the present internal environment? (strengths and weaknesses relative to competition); production/manufacturing capabilities?; research and development capabilities?; marketing and sales capabilities?; financial capabilities?

The list of questions above, while not intended to be exhaustive, will serve to illustrate the overlapping grids of "Opportunity" and "Capability". These are drawn to establish a base-point, from which management can measure the impact of their alternatives in relation to their financial objectives.

The previous discussion in this chapter assumes that fermentation-produced microbial pesticides have been identified as an area of long-range opportunity for a company. We come now to a typical financial pathway for development of a specific entomopathogen within that broad category.

Obviously the process is dynamic and begins with a product-candidate with potential for activity—until lately, benefitting only some agronomic crop systems, but with vector control now entering into consideration as well. Generally, the only financial consideration at this point will be whether or not the potential market for this benefit is large enough to meet the criteria established by the company. In today's regulatory environment the development of a minor-use product is not significantly less expensive than that of a major-use one; and research-based companies are forced to reject many good product ideas simply because the potential sales would never be sufficient to recover the multi-million dollar development investment.

Assuming that our candidate meets the initial measurement criteria, relevant research will be undertaken to confirm its activity and to determine the use-rate required to ensure an acceptable return. When these parameters are established, the candidate is considered for full project status. This requires completion of a thorough financial analysis, which first considers the characteristics of the candidate in relation to competitive products already on the market. Is benefit superior—equal—inferior, and to what degree? This is the key parameter for establishing the price at which the company will introduce this product. Considering the price/benefit ratio at the user level, a sales forecast is entered. Additional financial data such as projected factory cost, incremental marketing and selling cost, research time/cost to complete development, and capital needs are also entered at this time.

The financial analysis provides management with a means of evaluating individual product-candidates relative to each other in terms of their financial impact. It does not tell management whether the project is good or bad. Generally, the analysis is expressed in one or more forms which consider the value of money over time. Discounted cash flow, present value index, and payback are commonly used measurements.

As previously stated, these criteria only express the relative value of projects, one to another. Nevertheless, they are useful tools for allocating *limited* resources. Few, if any, companies in today's environment enjoy the luxury of unlimited development resources, and an entomopathogen which meets all acceptance criteria may be killed or delayed because it has a relatively lower priority than other candidate opportunities.

Again, the financial pathway for development of an entomopathogen is dynamic. As the candidate-product progresses through research and development, and additional information is generated, it may have to survive additional financial evaluations, each of which requires a go/no-go decision by management. For each product that emerges for sale in the market there are literally thousands that did not survive this arduous process.

Conclusion

The preceeding discussion was presented to familiarize individuals involved in basic research areas with the type of decision-making processes affecting industrial research. An understanding of this process is essential since it underlines the necessity for cooperative ventures between federal and state research agencies and universities. These organizations must assume a primary role in the industrialization of entomopathogens, particularly where the market size is restrictive and insufficient to support industrial interest. From these cooperative approaches, the research expenses for commercialization can be defrayed to a degree sufficient to permit an industry, based in the growth of microorganisms, to continue interest in mass production—even if only on a contract-supply basis.

Not all entomopathogens will be profitably commercialized, but many of these which do not meet the criteria discussed in this chapter could be produced and used in their limited-impact areas if cooperative research with public agencies occurs. Results of these cooperative efforts between industry and public agencies could accelerate and simplify the often lengthy and costly pathway toward registration and ultimate use in integrated pest management systems in economic entomology, and eventually, integrated methodologies for vector control.

References

Abdelnur, O. M. (1968). The biology of some black flies (Diptera: Simuliidae) of Alberta. *Quaest. ent.* **4**, 113–174.

Ackermann, H.-W. and Smirnoff, W. A. (1978). Recherches sur la lysogénie chez *Bacillus thuringiensis* et *B. cereus. Can. J. Microbiol.* **24**, 818–826.

Adamicka, P. (1979). On feeding of *Cottus gobio*, Chap. 2.2 *Jber. Biol. Sta. Lunz* (1978) 2, 81–83. (*C. gobio* feeding on simuliid larvae).

Agnew, J. D. (1962). The distribution of *Centroptiloides bifasciata* (E.-P) (Baëtidae: Ephem.) in Southern Africa, with ecological observations on the nymphs. *Hydrobiologia* **20**, 367–372.

Akey, D. H., Potter, H. W. and Jones, R. H. (1978). Effects of rearing temperature and larval density on longevity, size and fecundity in the biting gnat *Culicoides variipennis. Ann. ent. Soc. Am.* **71**, 411–418.

Al-Aidroos, K. and Roberts, D. W. (1978). Mutants of *Metarhizium anisopliae* with increased virulence toward mosquito larvae. *Can. J. Genet. Cytol.* **20**, 211–219.

Allen, K. R. (1941). Studies on the biology of the early stages of the salmon (*Salmo salar*). 2. Feeding habits. *J. Anim. Ecol.* **10**, 47–76.

Allen, K. R. (1959). The distribution of stream bottom faunas. *Proc. N.Z. ecol. Soc.* **6**, 5–8.

Ambühl, H. (1959). Die Bedeutung der Strömung las ökologischer Faktor. *Schweiz. Z. Hydrol.* **21**, 133–264.

Ambühl, H. (1962). Die Besonderheiten der Wasserströmung in physikalischer, chemischer und biologischer Hinsicht. *Schweiz. Z. Hydrol.* **24**, 367–382.

Anderson, J. R. and DeFoliart, G. R. (1961). Feeding behaviour and host preferences of some black flies (Diptera: Simuliidae) in Wisconsin. *Ann. ent. Soc. Am.* **54**, 716–729.

Anderson, J. R. and DeFoliart, G. R. (1962). Nematode parasitism of black fly (Diptera: Simuliidae) larvae in Wisconsin. *Ann. ent. Soc. Am.* **55**, 542–546.

Anderson, J. R. and Dicke, R. J. (1960). Ecology of the immature stages of some Wisconsin black flies (Simuliidae: Diptera). *Ann. ent. Soc. Am.* **53**, 386–404.

Anderson, J. R. and Shemanchuk, J. A. (1977). Parity and mermithid parasitism of *Simulium arcticum* caught attacking cattle and flying over an Alberta River. *Proc. 1st. Inter-Regional Conf. on N. American Blackflies.* pp. 143–145.

Anderson, J. R., Trainer, D. O. and DeFoliart, G. R. (1962). Natural and experimental transmission of the waterfowl parasite, *Leucocytozoon simondi* M. & L., in Wisconsin. *Zoonoses Res.* **1**, 155–164.

Anon. (1932). Entomological investigations. *6th Ann. Rept. C.S.I.R.O. Australia 1931–32*, 20–23.

Anon. (1971). Water Quality Criteria Data Book. Vol. 3. Effects of chemicals on aquatic life. Washington. E.P.A.

Anon. (1973a). Capsulation techniques, development of diets for larval and post-larval aquatic animals reported. *Feedstuffs* **45**, 35.

Anon. (1973b). Third Biannual Report. Res. Unit Vector Path. Memorial Univ. Newfoundl.

Arnason, A. P., Brown, A. W. A., Fredeen, F. J. H., Hopewell, W. W. and Rempel, J. G. (1949). Experiments in the control of *Simulium arcticum* by means of DDT in the Saskatchewan River. *Scient. Agric.* **29**, 527–537.

Asibey, E. O. A. (1975). Black fly and the environment. *Envir Conserv.* **2**, 25–28.

Badcock, R. M. (1949). Studies in stream life in tributaries of the Welsh Dee. *J. Anim. Ecol.* **18**, 193–208.

Bailey, C. H. (1977). Field and laboratory observations on a cytoplasmic polyhedrosis virus of blackflies (Diptera: Simuliidae). *J. Invert. Pathol.* **29**, 69–73.

Bailey, C. H. and Gordon, R. (1977). Observations on the occurrence and collection of mermithid nematodes from blackflies (Diptera: Simuliidae). *Can. J. Zool.* **55**, 148–154.

Bailey, C. H., Gordon, R. and Mills, C. (1977). Laboratory culture of the free-living stages of *Neomesomermis flumenalis*, a mermithid nematode parasite of Newfoundland blackflies (Diptera: Simuliidae). *Can J. Zool.* **55**, 391–397.

Bailey, C. H., Gordon, R. and Mokry, J. (1974). Procedure for mass collection of mermithid postparasites (Nematoda: Mermithidae) from larval blackflies (Diptera: Simuliidae). *Can. J. Zool.* **52**, 660–661.

Bailey, C. H., Shapiro, M. and Granados, R. R. (1975). A cytoplasmic polyhedrosis virus from the larval blackflies *Cnephia mutata* and *Prosimulium mixtum* (Diptera: Simuliidae). *J. Invert. Pathol.* **25**, 273–274.

Bain, O. (1971). Transmission des filarioses Limitation des passages des microfilaires ingénées vers l'hémocèle du vecteur; interpretation. *Annls Parasit. hum. comp.* **40**, 613–637.

Baker, J. H. and Bradnam, L. A. (1976). The role of bacteria in the nutrition of aquatic detritivores. *Oecologia* **24**, 95–104.

Baker, J. R. (1970). Transmission of *Leucocytosoon sakharoffi* in England by *Simulium angustitarse. Parasitology* **60**, 417–423.

Balay, G. (1964). Observations sur l'oviposition de *Simulium damnosum* Theobald et *Simulium adersi* Pomeroy (Diptera, Simuliidae) dans l'est de la Haute-Volta. *Bull. Soc. Path. exot.* **57**, 588–611.

Balay, G. and Grenier, P. (1964). *Lispe nivalis* Wiedemann (Muscidae, Lispinae) et *Ochthera* sp. (Ephydridae), Diptera prédateurs de *Simulium damnosum* Theobald et *S. adersi* Pomeroy en Haute-Volta. *Bull. Soc. Path. exot.* **57**, 611–619.

Baldwin, W. F., Allen, J. R. and Slater, N. S. (1966). A practical field method for the recovery of black flies labelled with phosphorus-32. *Nature, Lond.* **212**, 959–960.

Baldwin, W. F. and Gross, H. P. (1972). Fluctuations in numbers of adult black flies (Diptera: Simuliidae) in Deep River, Ontario. *Can. Ent.* **104**, 1465–1470.

Baldwin, W. F., West, A. S. and Gomery, J. (1975). Dispersal pattern of black flies (Diptera: Simuliidae) tagged with ^{32}P. *Can. Ent.* **107**, 113–118.

Baranov, N. (1934). Golubačka mušica u godini 1934. *Vet. Arh.* **4**, 346–393. (*Rev. appl. Ent.* B22, 203).

Baranov, N. (1936). Studien an pathogenen und parasitischen Insekten. IV. *Simulium (Danubiosimulium) columbaczense* Schönb. in Yougoslavie. *Arb. parasit. Abt. Inst. Hyg. Zagreb.* **4**, 1–36.

Baranov, N. (1937). Die Kolumbatscher Mücke in Jugoslawien im Jahre 1937. *Arch.* wiss. prakt. Tierheilk. **72**, 158–164. (*Rev. appl. Ent.* B26, 33).

Baranov, N. (1938a). Contribution to the knowledge of the Golubatz fly. VI. Study of the fly and its synbiocenonts. *Vet. Arhiv.* **8**, 313–328. (*Rev. appl. Ent.* B26, 214).

Baranov, N. (1938b). Contribution to the knowledge of natural enemies of the Golubatz fly of the class of insects. *Arh. Minist. Poljopr* **5**, 106–116. (*Rev. appl. Ent.* B27, 15–16).

Barnley, G. R. and Prentice, M. A. (1958). *Simulium neavei* in Uganda. *E. Afr. med. J.* **35**, 475–485.

Basrur, P. K. (1959). The salivary gland chromosomes of seven segregates of *Prosimulium* (Diptera: Simuliidae) with a transformed centromere. *Can. J. Zool.* **37**, 527–570.

Basrur, P. K. (1962). The salivary gland chromosomes of seven species of *Prosimulium* from Alaska and British Columbia. *Can. J. Zool.* **40**, 1019–1033.

Basrur, V. R. and Rothfels, K. H. (1959). Triploidy in natural populations of the blackfly *Cnephia mutata* (Malloch). *Can. J. Zool.* **37**, 571–589.

Batson, B. S., Johnson, M. R. L., Arnold, M. K. and Kelly, D. C. (1976). An iridescent virus from *Simulium* sp. (Diptera: Simuliidae) in Wales. *J. Invert. Pathol.* **27**, 133–135.

Beaudoin, R. and Wills, W. (1965). A description of *Caudospora pennsylvanica* sp.n. (Caudosporidae, Microsporidia), a parasite of the larvae of the blackfly *Prosimulium magnum* Dyar and Shannon. *J. Invert. Pathol.* **7**, 152–155.

Beaudoin, R. L. and Wills, W. (1968). *Haplosporidium simulii* sp.n. (Haplosporida: Haplosporidiidae) parasitic in larvae of *simulium venustum* Say. *J. Invert. Pathol.* **10**, 374–378.

Beaumont, P. (1975). Hydrology. *In* River Ecology. Studies in Ecology, Vol. 2, pp. 1–38. (B. A. Whittin, Ed.). Oxford, Blackwell Scientific.

Bedo, D. G. (1975). Polytene chromosomes of three species of black flies in the *Simulium pictipes* group. *Can. J. Zool.* **53**, 1147–1164.

Bedo, D. G. (1976). Polytene chromosomes in pupal and adult blackflies (Diptera: Simuliidae). *Chromosoma* **57**, 387–396.

Bedo, D. G. (1977). Cytogenetics and evolution of *Simulium ornatipes* Skuse (Diptera:

Simuliidae). 1. Sibling speciation. *Chromosoma* **64**, 37–65.

Bedo, D. G. (1979). Cytogenetics and evolution of *Simulium ornatipes* Skuse. 11. Temporal variation in chromosomal polymorphism and homosequential sibling species. *Evolution* **33**, 296–308.

Bellec, C. (1974). Les méthodes d'échantillonage des populations adultes de *Simulium damnosum* Theobald, 1903 (Diptera, Simuliidae) en Afrique de l'Ouest. Thèse (Univ. Paris-Sud Centre d'Orsay) ORSTOM, Paris.

Bellec, C. (1976). Captures d'adultes de *Simulium damnosum* Theobald, 1903 (Diptera: Simuliidae) à l'aide de plaques d'aluminium, en Afrique de l'Ouest. *Cah. ORSTOM sér. Ent. med. Parasit.* **14**, 209–217.

Bellec, C. and Hébrard, G. (1977). Captures d'adultes de Simuliidae, en particulier de *Simulium damnosum* Theobald 1903, à l'aide de pièges d'interception: les pièges vitres. *Cah. ORSTOM sér. Ent. méd. Parasit.* **15**, 41–54.

Beltaos, S. (1977). Evaluation of insecticide mixing in the Athabasca River downstream of Athabasca. Ms. Rept., Alta. Res. Council, Edmonton. University of Alberta.

Beltaos, S. (1980). Mixing and effects of insecticides: A working hypothesis for an analytical model. *In* Control of black flies in the Athabasca River: Technical Report, pp. 97–122. (W. O. Haufe and G. C. R. Croome, Eds.). Edmonton, Alta. Environ.

Beltaos, S. and Charnetski, W. A. (1980). Mixing of insecticide: One-dimensional analysis of methoxychlor concentration data. *In* Control of black flies in the Athabasca River: Technical Report, pp. 123–130. (W. O. Haufe and G. C. R. Croome, Eds.). Edmonton. Alta. Environ.

Bengtson, S.-A. (1972). Breeding ecology of the harlequin duck *Histrionicus histrionicus* (L.) in Iceland. *Ornis Scand.* **3**, 1–19.

Bening, A. L. (1924). About blackflies from the lower Volga. *Russk. gidrobiol. Zh.* (reference incomplete). (In Russian).

Bennett, G. F. (1960). On some ornithophilic blood-sucking Diptera in Algonquin Park, Ontario, Canada. *Can. J. Zool.* **38**, 377–389.

Bennett, G. F. (1963). The salivary gland as an aid in the identification of some simuliids. *Can. J. Zool.* **41**, 947–952.

Bennett, G. F. and Fallis, A. M. (1971). Flight range, longevity and habitat preference of female *Simulium euryadminiculum* Davies (Diptera: Simuliidae). *Can J. Zool.* **49**, 1203–1207.

Bennett, G. F., Fallis, A. M. and Campbell, A. G. (1972). The response of *Simulium (Eusimulium) euryadminiculum* Davies (Diptera: Simuliidae) to some olfactory and visual stimuli. *Can. J. Zool.* **50**, 793–800.

Bequaert, J. C. (1930). Medical and economic entomology, p. 849. *In* The African Republic of Liberia and the Belgian Congo, No. 2, pp. 797–1001 (R. P. Strong *et al.* Eds.).

Bequaert, J. C. (1934). Notes on the black flies or Simuliidae with special reference to those of the *Onchocerca* region of Guatemala. Part 111 (pp. 175–224). *In* Onchocerciasis with special reference to the Central American form of the disease. *Contrib. Dept. Trop. Med. and Inst. Trop. Biol. Med.* Harvard Univ. **6**, 1–234.

Berl, D. and Prud'hom, J. M. (1978). Un nouveau système d'élevage de masse de *Simulium damnosum s.l.* 1. Description et premières expériences. *ORSTOM/OCCGE. No. 15/ONCHO/Rap./78.* Mimeogr. Doc.

Berl, D. and Prud'hom J. M. (1979). Un nouveau système d'élevage de masse de *Simulium damnosum s.l.* 11. Modification et nouvelles expériences. *ORSTOM/OCCGE No. 9/ONCHO/RAP. 79* Mimeogr. Doc.

Berrie, A. D. (1976). Detritus, microorganisms and animals in fresh water. *In* The role of aquatic and terrestrial organisms in decomposition processes, pp. 323–338 (J. M. Anderson and A. Macfadgen, Eds.). Oxford. Blackwell Scientific.

Berzina, A. I. (1953). Attack by simuliids on man in nature. (In Russian). *Parazit. Sb.* **15**, 353–385.

Bishop, J. E. (1973a). Observations on the vertical distribution of the benthos in a Malaysian stream. *Freshwat. Biol.* **3**, 147–156.

Bishop, J. E. (1973b). Limnology of a small Malayan river Sungai Gombak. *Monographiae biol.* **22**, 1–485.

Biswas, S. and Mokry, J. (1978). Benthobservatory studies of microflora in a Newfoundland stream. *Hydrobiologia* **60**, 213–219.

Blacklock, D. B. (1926). The development of *Onchocerca volvulus* in *Simulium damnosum* *Ann. Trop. Med. Parasit.* **20**, 1–48.

Bobrova, S. E. (1971). Concerning the parasite and predators of blackflies. *Izv. sib. Otdel. Acad. Nauk SSSR (Biol.)* **10**. 172–173. (In Russian).

Boemare, N. and Maurand, J. (1976). Recherches sur le metabolisme respiratoire des larves de simulies saines et atteintes de microsporidioses. *Bull. Soc. Zool. Fr.* **101**, 377–385.

Bond, W. A. and Berry, D. (1980). Ms Rept. Freshwater Institute, AOSER Program, Winnipeg, Manitoba.

Bradbury, W. C. and Bennett, G. F. (1974a). Behavior of adult Simuliidae (Diptera). 1. Response to color and shape. *Can. J. Zool.* **52**, 251–259.

Bradbury, W. C. and Bennett, G. F. (1974b). Behavior of adult Simuliidae (Diptera). 11. Vision and olfaction in near-orientation and landing. *Can. J. Zool.* **52**, 1355–1364.

Bradley, G. H. (1935a). Notes on the southern buffalo gnat, *Eusimulium pecuarum* (Riley) (Diptera: Simuliidae). *Proc. ent. Soc. Wash.* **37**, 60–64.

Bradley, G. H. (1935b). The hatching of eggs of the southern buffalo gnat. *Science* **82**, 277–278.

Bradt, S. (1932). Notes on Puerto Rican black flies. *Puerto Rico J. publ. Hlth trop Med.* **8**, 69–81. (English and Spanish).

Breyev, K. A. (1950). The behaviour of bloodsucking Diptera and bot-flies when attacking reindeer and the responsive reactions of the reindeer. 1. The behaviour of bloodsucking Diptera and bot-flies when attacking reindeer. (In Russian). *Parazit. Sb.* **12**, 167–198.

Briggs, J. D. (1963). Commercial production of insect pathogens. *In* Insect Pathology. An Advanced Treatise. Vol. 2, pp. 519–548. (E. A. Steinhaus, Ed.). New York. Academic Press.

Brooks, G. T. (1976). Selective toxicity of insecticides. *In* The Future for Insecticides—Needs and Prospects, pp. 97–143 (R. Metcalf and J. McKelvey, Eds.). New York. Wiley.

Bruder, K. W. (1974). The blackflies (Simuliidae: Diptera) of the Stony Brook Watershed of New Jersey with emphasis on parasitism by mermithid nematodes (Mermithidae: Nematoda). PhD thesis, Rutgers University, The State University of New Jersey.

Bruder, K. W. and Crans, W. J. (1979). The blackflies (Simuliidae: Diptera) of the Stony Brook Watershed of New Jersey, with emphasis on parasitism by mermithid nematodes (Mermithidae: Nematoda). *Bull. No. 851, N.J. Agric. Expt. Stat. State Univ., N.J.*

Brumpt, E. (1936). Précis de parasitologie. 5e éd., Masson. Paris, 1572–1686.

Burdick, G., Dean, H., Skea, J. and Frisa, C. (1974). Effect of blackfly larviciding in some Adirondack streams. *N.Y. Fish Game J.* **21**, 1–17.

Burdick, G., Dean, H., Harris, E. J., Skea, J., Frisca, C. and Sweeney, C. (1968). Methoxychlor as a blackfly larvicide, persistence of its residues in fish and its effect on stream arthropods. *N.Y. Fish Game J.* **15**, 120–142.

Burges, H. D. (1967). The standardization of products based on *Bacillus thuringiensis*. *In* Insect Pathology and Microbial Control, pp. 306–314 (P. A. van der Lann, Ed.). Amsterdam. North-Holland Publishing.

Burges, H. D. and Thomson, E. M. (1971). Standardization and assay of microbial insecticides. *In* Microbial Control of Insects and Mites, pp. 591–622 (H. D. Burges and N. W. Hussey, Eds.). New York. Academic Press.

Burton, G. J. (1964). An exposure-tube for determining the mortality of *Simulium* larvae in rivers following larvicidal operations. *Ann. trop. Med. Parasit.* **58**, 339–342.

Burton, G. J. (1966). Observations on cocoon formation, the pupal stage, and emergence of the adult of *Simulium damnosum* Theobald in Ghana. Ann. trop. Med. Parasit. **60**, 48–56.

Burton, G. J. (1971). Cannibalism among *Simulium damnosum* (Simulidae: [sic]) larvae. *Mosquito News* **31**, 602–603.

Burton, G. J. (1973). Feeding of *Simulium hargreavesi* Gibbons larvae on *Oedogonium* algal filaments in Ghana. *J. med. Ent.* **10**, 101–106.

Burton, G. J. and McRae, T. M. (1972a). Observations on trichopteran predators of aquatic stages of *Simulium damnosum* and other *Simulium* species in Ghana. *J. med. Ent.* **9**, 289–294.

Burton, G. J. and McRae, T. M. (1972b). Phoretic attachment of *Simulium* larvae and pupae to mayfly and dragonfly nymphs. *Mosquito News* **32**, 436–443.

Cabejszek, I., Luczac, J., Maleszewska, J. and Stanislawska, J. (1966). Der Einfluss von Insektiziden (Aldrin und Methoxychlor) auf physikalisch-chemische Eigenschaften des Wassers und auf Wasser-Organismen. *Verh. Int. Verein. theor. angew. Limnol.* **16**, 963–968.

Cadwallader, P. L. (1975). The food of the

New Zealand common river galaxies, *Galaxias vulgaris* Stockell (Pisces: Salmoniformes). *Aust. J. Mar. Freshw. Res.* **26**, 15–30.

Cameron, A. E. (1922). The morphology and biology of a Canadian cattle-infesting black fly, *Simulium simile* Mall. (Diptera: Simuliidae). *Bull. Dep. Agric. Dom. Can.* (n.s.) **5**, 1–26.

Carlsson, G. (1962). Studies on Scandinavian black flies. *Opusc. ent. Suppl.* **21**, 1–280.

Carlsson, G. (1967). Environmental factors influencing blackfly populations. *Bull. Wld Hlth Org.* **37**, 139–150.

Carlsson, G. (1968). Benthonic fauna in African watercourses with special reference to blackfly populations. *Scand. Inst. Afr. Studies, Res. rep.* **3**, 1–13.

Carlsson, G. (1970). Biology of immature stages of blackflies in parts of East Africa with special reference to *Simulium damnosum*. *WHO/ONCHO 70.81, WHO/VBC/ 70.232. Mimeogr. Doc.*

Carlsson, G. and Müller, H. (1978). Studie zum Lebenszyklus von *Simulium morsitans* (Diptera: Simuliidae) in Nordschweden. *Ent. Z. Stuttgart* **88**, 6–10.

Carlsson, M., Nilsson, L. M., Svensson, Bj., Ulfstrand, S. and Wotton, R. S. (1977). Lacustrine seston and other factors influencing the blackflies (Diptera: Simuliidae) inhabiting lake outlets in Swedish Lapland. *Oikos* **29**, 229–238.

Cazal, M. and Maurand, J. (1966). Système stomatogastrique et glands endocrines de la larve de *Simulium ornatum* Meig. *Bull. Soc. Zool. Fr.* **91**, 321–325.

Challier, A. (1977). Trapping technology. *In* Tsetse: The future for biological methods in integrated control, pp. 109–123. (M. Laird, Ed.). Ottawa, IDRC-077e.

Chance, M. M. (1970a). A review of chemical control methods for blackfly larvae (Diptera: Simuliidae). *Quaest. ent.* **6**, 287–292.

Chance, M. M. (1970b). The functional morphology of the mouthparts of blackfly larvae (Diptera: Simuliidae). *Quaest. ent.* **6**, 245–284.

Charnetski, W. A. and Currie, R. A. (1980). Pretreatment background insecticide and PCB residues and post-treatment methoxychlor insecticide residues in fish from the Athabasca River. *In* Control of black flies in the Athabasca River: Technical Report, pp. 75–87 (W. O. Haufe and G. C. R. Croome, Eds.). Edmonton. Alta. Environ.

Charnetski, W. A., Currie, R. A. and Calder, L. (1980a). Methoxychlor organochlorine, and organophosphorus insecticides and unidentified hydrocarbon residues in bed material of Lake Athabasca and the Athabasca Delta. *In* Control of black flies in the Athabasca River: Technical Report, pp. 89–92 (W. O. Haufe and G. C. R. Croome, Eds.). Edmonton. Alta. Environ.

Charnetski, W. A. and Depner, K. R. (1980). Distribution and persistence of methoxychlor in Athabasca River mud and bedload. *In* Control of black flies in the Athabasca River: Technical Report, pp. 63–73 (W. O. Haufe and G. C. R. Croome, Eds.). Edmonton. Alta. Environ.

Charnetski, W. A., Depner, K. R. and Beltaos, S. (1980b). Distribution and persistence of methoxychlor in Athabasca River water. *In* Control of black flies in the Athabasca River: Technical Report, pp. 39–61 (W. O. Haufe and G. C. R. Croome, Eds.). Edmonton. Alta. Environ.

Chaston, I. (1968). A study on the exploitation of invertebrate drift by brown trout (*Salmo trutta* L.) in a Dartmoor stream. *J. appl. Ecol.* **5**, 721–729.

Chaston, I (1972). Non-catastrophic invertebrate drift in lotic systems. *In* Essays in Hydrobiology, pp. 33–51. Exeter.

Chutter, F. M. (1968). On the ecology of the fauna of stones in the current in a South African river supporting a very large *Simulium* (Diptera) population. *J. appl. Ecol.* **5**, 531–561.

Chutter, F. M. (1970). A preliminary study of factors influencing the number of oocytes present in newly emerged blackflies (Diptera: Simuliidae) in Ontario. *Can. J. Zool.* **48**, 1389–1400.

Chutter, F. M. (1971). Hydrobiological studies in the catchment of Vaal Dam, South Africa. Part 2. The effects of stream contamination on the fauna of stones-in-current and marginal vegetation biotopes. *Int. Revue ges. Hydrobiol.* **56**, 227–240.

Chutter, F. M. (1972a). A reappraisal of Needham and Usinger's data on the variability of a stream fauna when sampled with a Surber sampler. *Limnol. Oceanogr.* **17**, 139–141.

Chutter, F. M. (1972b). Notes on the biology of South African Simuliidae particularly *Simulium (Eusimulium) nigritarse* Coquillet[t]. *News Lett. Limnol. Soc. Afr.* **18**, 10–18.

Chutter, F. M. and Noble, R. G. (1966). The

reliability of a method of sampling stream invertebrates. *Arch. Hydrobiol.* **62**, 95–103.

Claus, A. (1937). Vergleichend-physiologische Untersuchungen zur Ökologie der Wasserwanzen mit besonderer Berücksichtigung der Brackwasserwanze *Sigara lugubris* Fieb. *Zool. Jb.* (Allg. Zool.) **58**, 365–432.

Clifford, H. F. (1972). Drift of invertebrates in an intermittent stream draining marshy terrain of west-central Alberta. *Can. J. Zool.* **50**, 985–991.

Coffman, W. P. (1967). Community structure and trophic relations in a small woodland stream, Linesville Creek, Crawford County, Pennsylvania. PhD thesis. Pittsburg, Pennsylvania, USA. Univ. Pittsburg.

Colbo, M. H. (1974). Studies on the biology of the Simuliidae in North-eastern Australia with reference to their potential as vectors of pathogens. PhD thesis Australia. Univ. of Queensland.

Colbo, M. H. (1976). Four new species of *Simulium* Latreille (Diptera: Simuliidae) from Australia. *J. Aust. ent. Soc.* **15**, 253–269.

Colbo, M. H. (1977). Diurnal emergence patterns of two species of Simuliidae (Diptera) near Brisbane, Australia. *J. med. Ent.* **13**, 514–515.

Colbo, M. H. (1979a). Distribution of winter-developing Simuliidae (Diptera) in eastern Newfoundland. *Can. J. Zool.* **57**, 2143–2152.

Colbo, M. H. (1979b). Simuliid Mermithidae: Effect of host stage parasitized on the distribution of infected simuliids in a stream with consideration for biological control. *Proc. Int. Colloqu. Invert. Path. Prague.* **??**, 37–38.

Colbo, M. H., Fallis, A. M. and Reye, E. J. (1977). The distribution and biology of *Austrosimulium pestilens*, a serious biting-fly pest following flooding. *Aust. Vet. J.* **53**, 135–138.

Colbo, M. H., Laird, M. and Peterson, J. (1978). *Ann. Rpt Res. Unit Vec. Path.*, Memorial Univ. Newfoundland.

Colbo, M. H. and Moorhouse, D. E. (1974). The survival of the eggs of *Austrosimulium pestilens* Mack. and Mack. (Diptera: Simuliidae). *Bull. ent. Res.* **64**, 629–632.

Colbo, M. H. and Moorhouse, D. E. (1979). The ecology of pre-imaginal Simuliidae (Diptera) in South-Eastern Queensland, Australia. *Hydrobiologia* **63**, 63–79.

Colbo, M. H. and Porter, G. N. (1977). *Ann Rpt Res. Unit Vec. Path.*, Memorial Univ. Newfoundland.

Colbo, M. H. and Porter, G. N. (1979). Effects of the food supply on the life history of Simuliidae (Diptera). *Can. J. Zool.* **57**, 301–306.

Colbo, M. H. and Thompson, B. H. (1978). An efficient technique for laboratory rearing of *Simulium verecundum* S. & J. (Diptera: Simuliidae). *Can. J. Zool.* **56**, 507–510.

Colbo, M. H. and Undeen, A. H. (1980). Effect of *Bacillus thuringiensis* var. *israelensis* on non-target insects in stream trials for control of Simuliidae. *Mosquito News* **40**, 368–371.

Coleman, M. J. and Hynes, H. B. N. (1970). The vertical distribution of the invertebrate fauna in the bed of a stream. *Limnol. Oceanogr.* **15**, 31–40.

Collins, R. C. (1977). Reunion Internacional Sobreoncho (Informe Preliminar), **31**, 53.

Collins, R. C., Campbell, C. C., Wilton, D. P. and Newton, L. (1977). Quantitive aspects of the infection of *Simulium ochraceum* by *Onchocerca volvulus. Tropenmed. Parasit.* **28**, 235–243.

Collins, R. C., Wilton, D. P., Figueroa, H. and Campbell, C. C. (1976). An improved methodology for collecting, transporting and maintaining *S. ochraceum* to the infective stage. *WHO/VBC/70.232.* Mimeogr. Doc.

Conway, C. H. (1952). The life history of the water shrew (*Sorex palustris navigator*). *Am. Midl. Nat.* **48**, 219–248.

Condon, W. J. and Gordon, R. (1977). Some effects of mermithid parasitism on the larval blackflies *Prosimulium mixtum fuscum* and *Simulium venustum. J. Invert Pathol.* **29**, 56–62.

Condon, W. J., Gordon, R. and Bailey, C. H. (1976). Morphology of the neuroendocrine systems of two larval blackflies, *Prosimulium mixtum/fuscum* and *Simulium venustum. Can. J. Zool.* **54**, 1579–1584.

Convit, J. (1974). Onchocerciasis in Venezuela. Int. Symp. Research and Control of Onchocerciasis in the Western Hemisphere. *PAHO Scient. Publ.* **398**, 105–109.

Corbet, P. S. (1958a). Some effects of DDT on the fauna of the Victoria Nile. *Revue Zool. Bot. afr.* **57**, 73–95.

Corbet, P. S. (1958b). Effects of *Simulium* control on insectivorous fishes. *Nature, Lond.* **181**, 570–571.

Corbet, P. S. ((1961a). The food of non-cichlid fishes in the Lake Victoria Basin, with remarks on their evolution and adaptation to lacustrine conditions. *Proc. zool. Soc. Lond.* **136,** 1–101.

Corbet, P. S. (1961b). The biological significance of the attachment of immature stages of *Simulium* to mayflies and crabs. *Bull. ent. Res.* **52,** 695–699.

Corbet, P. S. (1962). Observations on the attachment of *Simulium* pupae to larvae of Odonata. *Ann. trop. Med. Parasit.* **56,** 136–140.

Corbet, P. S. (1967). The diel oviposition periodicity of the black fly, *Simulium vittatum. Can. J. Zool.* **45,** 583–584.

Coscarón, S. and Wygodzinsky, P. (1972). Taxonomy and distribution of the blackfly subgenus *Simulium (Pternaspatha)* Enderlein Simuliidae, (Diptera, Insecta). *Bull. Am. Mus. nat. Hist.* **147,** 199–240.

Couch, T. L. (1976). *In* Proceedings of the First International Colloquium on Invertebrate Pathology and IXth Annual Meeting. Society for Invertebrate Pathology, pp. 305–308. Kingston, Canada. Queen's University.

Coutant, C. C. (1964). Insecticide Sevin: Effect of aerial spraying on drift on stream insects. *Science* (n.s.) **146,** 420–421.

Couvert, L. (1970). Studio morfologico delle capsule cefaliche delle larve di *Prosimulium conistylum* Rubtzov e *Liponeura cinerascens* Loew (Diptera: Nematocera). *Memorie Soc. ent. ital.* **49,** 159–188.

Craig, D. A. (1969). The embryogenesis of the larval head of *Simulium venustum* Say (Diptera: Nematocera). *Can. J. Zool.* **47,** 495–503.

Craig, D. A. (1972). Blackflies. Univ. Alberta, Dept. Ent. Ext. Leaflet 921–a.

Craig, D. A. (1974). The labrum and cephalic fans of larval Simuliidae (Diptera: Nematocera). *Can. J. Zool.* **52,** 133–159.

Craig, D. A. (1975). The larvae of Tahitian Simuliidae (Diptera: Nematocera). *J. med. Ent.* **12,** 463–476.

Craig, D. A. (1977a). A re-assessment of the systematic position of *Pseudosimulium humidum* (Westwood), an Upper Jurassic fossil dipteran. *Entomologist's Gaz.* **28,** 175–179.

Craig, D. A. (1977b). Mouthparts and feeding behaviour of Tahitian larval Simuliidae (Diptera: Nematocera). *Quaest. ent.* **13,** 195–218.

Crisp, G. (1956a). *Simulium* and Onchocerciasis in the Northern Territories of the Gold Coast. London, Lewis.

Crisp, G. (1956b). An ephemeral fauna of torrents in the northern territories of the Gold Coast, with special reference to the enemies of *Simulium. Ann. trop. Med. Parasit.* **50,** 260–267.

Croft, B. A. and Brown, A. W. A. (1975). Responses of arthropod natural enemies to insecticides. *A. Rev. Ent.* **20,** 285–335.

Crosby, T. K. (1974). Life history stages and taxonomy of *Austrosimulium (Austrosimulium) tillyardianum* (Diptera: Simuliidae). *N.Z. Jl Zool.* **1,** 5–28.

Crosby, T. K. (1975). Food of the New Zealand trichopterans *Hydrobiosis parumbripennis* McFarlane and *Hydropsyche colonica* McLachlan. *Freshwat. Biol.* **5,** 105–114.

Crosskey, R. W. (1954). Infection of *Simulium damnosum* with *Onchocerca volvulus* during the wet season in Northern Nigeria. *Ann. trop. Med. Parasit.* **48,** 152–159.

Crosskey, R. W. (1955). Observations on the bionomics of adult *Simulium damnosum* Theobald (Diptera: Simuliidae) in Northern Nigeria. *Ann. trop. Med. Parasit.* **49,** 142–153.

Crosskey, R. W. (1956). The distribution of *Simulium damnosum* Theobald in Northern Nigeria. *Trans. R. Soc. trop. Med. Hyg.* **50,** 379–392.

Crosskey, R. W. (1957). Man-biting behaviour in *Simulium bovis* De Meillon in Northern Nigeria, and infection with developing filariae. *Ann. trop. Med. Parasit.* **51,** 80–86.

Crosskey, R. W. (1958a). First results in the control of *Simulium damnosum* Theobald (Diptera: Simuliidae) in Northern Nigeria. *Bull. ent. Res.* **49,** 715–735.

Crosskey, R. W. (1958b). The body weight in unfed *Simulium damnosum* Theobald, and its relation to the time of biting, the fat-body and age. *Ann. trop. Med. Parasit.* **52,** 149–157.

Crosskey, R. W. (1960). A taxonomic study of the larvae of West African Simuliidae (Diptera: Nematocera) with comments on the morphology of the larval blackfly head. *Bull. Br. Mus. nat. Hist.* (Ent.) **10,** 1–74.

Crosskey, R. W. (1962). Observations on the uptake of human blood by *Simulium damnosum*: The engorgement time and size of the blood-meal. *Ann. trop. Med. Parasit.* **56,** 141–148.

Crosskey, R. W. (1967a). The classification of *Simulium* Latreille (Diptera: Simuliidae) from Australia, New Guinea and the Western Pacific. *J. nat. Hist.* **1,** 23–51.

Crosskey, R. W. (1967b). A preliminary revision of the black-flies (Diptera: Simuliidae)

of the Middle East. *Trans. R. ent. Soc. Lond.*
119, 1–45.

Crosskey, R. W. (1969). A re-classification of
the Simuliidae (Diptera) of Africa and its
islands. *Bull. Br. Mus. nat. Hist.* (Ent.)
Suppl. **14**, 1–195.

Crosskey, R. W. (1973a). Family Simuliidae.
In A Catalog of the Diptera of the Oriental
Region, pp. 423–430 (M. D. Delfinado and
D. E. Hardy, Eds.). Honolulu. University
Press of Hawaii.

Crosskey, R. W. (1973b). Simuliidae. *In* Insects and other Arthropods of Medical Importance, pp. 109–153 (K. G. V. Smith, Ed.).
London. British Museum (Natural History).

Crosskey, R. W. (1977). La faune terrestre de
l'île de Sainte-Hélène. Troisième partie. 6.
Fam. Simuliidae. *Annls Mus. r. Afr. cent.
Sér. 8ᵛᵒ* (Sci. Zool.) **215** (1976), 31–43.
(Volume dated 1976 but not issued until
mid 1977).

Crosskey, R. W. (1980). Family Simuliidae. *In*
Catalogue of the Diptera of the Afrotropical
Region, pp. 203–210 (R. W. Crosskey, Ed.).
London. British Museum (Natural History).

Crosskey, R. W. and Davies, J. B. (1962).
Xenomyia oxycera Emden, a muscid predator
on *Simulium damnosum* Theobald in Northern Nigeria. *Proc. R. ent. Soc. Lond.* (A) **37**,
22–26.

Cummins, K. W. (1973). Trophic relations of
aquatic insects. *A. Rev. Ent.* **18**, 183–206.

Cummins, K. W. (1975). Macroinvertebrates
in River Ecology. Studies in Ecology 2 (B. A.
Whitton, Ed.). Oxford. Blackwell Scientific.

Cupp, E. W. and Collins, R. C. (1979). The
gonotrophic cycle in *Simulium ochraceum*.
Am. J. trop. Med. Hyg. **28**, 422–426.

Cupp. E. W., Lok, J. B., Bernardo, J., Bren-
ner, R. J., Pollack, R. J. and Scoles, G. A.
(1981). Complete generation rearing of
Simulium damnosum s.l. (Diptera: Simuliidae) in the laboratory. *Ann. trop. Med.
Parasit.* (in press).

Curry, R. R. (1972). *In* Rivers—a Geomorphic
and Chemical Overview on River Ecology
and Man, pp. 9–31 (R. T. Oglesby, C. A.
Carlson and J. A. McCann, Eds.). New
York. Academic Press.

Dale, W., Miles, J. and Guerrant, G. (1975).
Monitoring residues of abate in streams
treated for *Simulium* control. *Envir. Q. Safety*
3, 780–783.

Dalmat, H. T. (1950a). Induced oviposition of
Simulium flies by exposure to CO_2. *Publ.
Hlth Rep. Wash.* **65**, 545–546.

Dalmat, H. T. (1950b). Studies on the flight
range of certain Simuliidae, with the use of
aniline dye marker. *Ann. ent. Soc. Am.* **43**,
537–545.

Dalmat, H. T. (1952). Longevity and further
flight range studies on the blackflies (Diptera: Simuliidae), with the use of dye markers. *Ann. ent. Soc. Am.* **45**, 23–37.

Dalmat, H. T. (1954). Ecology of simuliid
vectors of onchocerciasis in Guatemala. *Am.
Midl. Nat.* **52**, 175–196.

Dalmat, H. T. (1955). The black flies (Diptera:
Simuliidae) of Guatemala and their role as
vectors of onchocerciasis. *Smithson. misc.
Collns* **125**, 1–425.

Dalmat, H. T. (1958). Biology and control of
Simuliidae (Diptera) vectors of on-
chocerciasis in Central America. *Proc. Xth
Int. Congr. Ent.* (1956) **3**, 517–533.

Dalmat, H. T. and Gibson, C. L. (1952). A
study of flight range and longevity of blackflies (Diptera: Simuliidae) infected with
Onchocerca volvulus. Ann. ent. Soc. Am. **45**,
605–612.

Davies, D. M. (1949a). The ecology and life
history of black-flies (Simuliidae: Diptera) in
Ontario with a description of a new species.
PhD thesis, Univ. Toronto, Canada.

Davies, D. M. (1949b). Variation in taxonomic
characters of some Simuliidae (Diptera).
Can. Ent. **81**, 18–21.

Davies, D. M. (1950). A study of the black fly
population of a stream in Algonquin Park,
Ontario. *Trans. R. Can. Instit.* **28**, 121–159.

Davies, D. M. (1952). The population and
activity of adult female black flies in the
vicinity of a stream in Algonquin Park,
Ontario. *Can. J. Zool.* **30**, 287–321.

Davies, D. M. (1953). Longevity of black flies
in captivity. *Can. J. Zool.* **31**, 304–312.

Davies, D. M. (1958). Some parasites of Canadian black flies (Diptera, Simuliidae). *Proc.
XVth Int. Congr. Zool.*, 660–661.

Davies, D. M. (1959). The parasitism of black
flies (Diptera, Simuliidae) by larval water
mites mainly of the genus *Sperchon. Can. J.
Zool.* **37**, 353–369.

Davies, D. M. (1960). Microsporidia in a sper-
chonid mite, and further notes on Hydracarina and simuliids (Diptera). *Proc. ent.
Soc. Ont.* **90**, 53.

Davies, D. M. (1972). The landing of blood-
seeking female black-flies (Simuliidae: Diptera) on coloured materials. *Proc. ent. Soc.
Ont.* **102**, (1971), 124–155.

Davies, D. M. (1978). Ecology and behaviour

of adult black flies (Simuliidae): a review. *Quaest. ent.* **14**, 3–12.

Davies, D. M., Györkös, H. and Raastad, J. E. (1977). Simuliidae (Diptera) of Rendalen, Norway. IV. Autogeny and anautogeny. *Norw. J. Ent.* **24**, 19–23.

Davies, D. M. and Peterson, B. V. (1956). Observations on the mating, feeding, ovarian development and oviposition of adult black flies (Simuliidae, Diptera). *Can. J. Zool.* **34**, 615–655.

Davies, D. M., Peterson, B. V. and Wood, D. M. (1962). The black flies (Diptera, Simuliidae) of Ontario. Part 1. Adult identification and distribution with description of six new species. *Proc. ent. Soc. Ont.* **92**, 70–154.

Davies, D. M. and Syme, P. D. (1958). Three new Ontario black flies of the genus *Prosimulium* (Diptera, Simuliidae). Part II. Ecological observations and experiments. *Can. Ent.* **90**, 744–759.

Davies, J. B. (1962). Egg-laying habits of *Simulium damnosum* Theobald and *Simulium medusaeforme* form *hargrevesi* [sic] Gibbins in Northern Nigeria. *Nature, Lond.* **196**, 149–150.

Davies, J. B., Crosskey, R. W., Johnston, M. R. L. and Crosskey, M. E. (1962). The control of *Simulium damnosum* at Abuja, Northern Nigeria, 1955–60. *Bull. Wld Hlth Org.* **27**, 491–510.

Davies, L. (1957). A study of the blackfly, *Simulium ornatum* Mg. (Diptera), with particular reference to its activity on grazing cattle. *Bull. ent. Res.* **48**, 407–424.

Davies, L. (1961). Ecology of two *Prosimulium* species (Diptera) with reference to their ovarian cycles. *Can. Ent.* **93**, 1113–1140.

Davies, L. (1963). Seasonal and diurnal changes in the age-composition of adult *Simulium venustum* Say (Diptera) populations near Ottawa. *Can. ent.* **95**, 654–667.

Davies, L. (1965). On spermatophores in Simuliidae (Diptera). *Proc. R. ent. Soc. Lond.* (A) **40**, 30–34.

Davies, L. (1968). A key to the British species of Simuliidae (Diptera) in the larval, pupal and adult stages. *Scient. Publs. Freshwat biol. Ass.* **24**, 1–126.

Davies, L. (1974). Evolution of larval headfans in Simuliidae (Diptera) as inferred from the structure and biology of *Crozetia crozetensis* (Womersley) compared with other genera. *Zool. J. Linn. Soc.* **55**, 193–224.

Davies, L. (1975). Book Review of Rubtsov, 1974c. *J. nat. Hist.* **9**, 720.

Davies, L., Downe, A. E. R., Weitz, B. and Williams, C. B. (1962). Studies on black flies (Diptera, Simuliidae) taken in a light trap in Scotland. II. Blood-meal identification by precipitin tests. *Trans. R. ent. Soc. Lond.* **114**, 21–27.

Davies, L. and Roberts, D. M. (1973). A net and a catch-segregating apparatus mounted in a motor vehicle for field studies on flight activity of Simuliidae and other insects. *Bull. ent. Res.* **63**, 103–112.

Davies, L. and Smith, C. D. (1958). The distribution and growth of *Prosimulium* larvae (Diptera, Simuliidae) in hill streams in Northern England. *J. Anim. Ecol.* **27**, 335–348.

Davies, L. and Williams, C. B. (1962). Studies on black flies (Diptera, Simuliidae) taken in a light trap in Scotland. I. Seasonal distribution, sex ratio and internal condition of catches. *Trans. R. ent. Soc. Lond.* **114**, 1–20.

Davis, C. C. (1971). A study of the hatching process in aquatic invertebrates. XXV. Hatching in the blackfly, *Simulium* (probably *venustum*) (Diptera, Simuliidae). *Can. J. Zool.* **49**, 333–336.

Davis, M. B. (1934). Habits of the Trichoptera. *In* The Caddis Flies or Trichoptera of New York State, (C. Betten, Ed.). *Bull. N.Y. St. Mus.* **292**, 1–576.

Debaisieux, P. (1919). Une chytridinée nouvelle: *Coelomycidium simulii*, nov. gen. nov. spec. *C.r. Séanc. Soc. Biol.* **82**, 899–900.

Debaisieux, P. (1926). A propos d'une microsporidie nouvelle: *Octosporea simulii*. *Annls Soc. scient. Brux.* **46**, 594–601.

De Barjac, H. (1978). Une nouvelle variété de *Bacillus thuringiensis* très toxique pour les moustiques: *B. thuringiensis* var. *israelensis* sérotype 14. *C.r. hebd. Séanc. Acad. Sci. Paris* (D) **286**, 797–800.

Décamps, H., Larrouy, G. and Trivellato, D. (1975). Approche hydrodynamique de la microdistribution d'invertébrés benthiques en eau courante. *Annls Limnologie* **11**, 79–100.

DeFoliart, G. R. and Rao, M. R. (1965). The ornithophilic black fly *Simulium meridionale* Riley (Diptera, Simuliidae) feeding on man during autumn. *J. med. Ent.* **2**, 84–85.

Dejoux, C. (1975). Nouvelle technique pour tester *in situ* l'impact de pesticides sur la faune aquatique non cible. *Cah. ORSTOM Sér Ent. méd. Parasit.* **13**, 75–80.

Dejoux, C. (1977a). Action de l'abate sur les invertébrés aquatiques. III. Effets de premiers traitements de la Bagoué. *ORSTOM, Bouaké Rap. 14*. Mimeogr. Doc.

Dejoux, C. (1977b). Action de l'abate sur les invertébrés aquatiques. IV. Devenir des organismes dérivants à la suite des traitments. *ORSTOM, Bouaké Rap. 15*. Mimeogr. Doc.

Dejoux, C. (1978). Impact of insecticide treatments against *Simulium* larvae on non-target fauna. *OCP/SWG/78.10*. Mimeogr. Doc.

Dejoux, C. and Elouard, J. M. (1975). Étude quantitative de l'entomofaune aquatique. Méthodologie et normalisation. *ORSTOM, Bouaké Rap. 191*. Mimeogr. Doc.

Dejoux, C. and Elouard, J. M (1977). Action de l'abate sur les invertébrés aquatiques cinétique de decrochement à court et moyen terme. *Cah. ORSTOM, Sér Hydrobiol.* **11**, 217–230.

Dejoux, C. and Troubat, J. J. (1976). Toxicité comparée de deux insecticides organophosphores sur la faune aquatique non cible en milieu tropical. *ORSTOM, Bouaké Rap. 1*. Mimeogr. Doc.

De León, J. R. (1947). Entomologia de la onchocercosis. *In* "Onchocercosis", pp. 147–172. Guayaquil, Guatemala. Univ. San Carlos.

De León, J. R. and Duke, B. O. L. (1966). Experimental studies on the transmission of Guatemalan and West African strains of *Onchocerca volvulus* by *Simulium ochraceum, S. metallicum* and *S. callidum*. *Trans. R. Soc. trop. Med. Hyg.* **60**, 735–752.

Depner, K. R. and Charnetski, W. A. (1978). Divers and television for examining riverbed material and populations of blackfly larvae in the Athabasca River. *Quaest. ent.* **14**, 441–444.

Depner, K. R., Charnetski, W. A. and Haufe, W. O. (1980a). Population reduction of the blackfly *Simulium arcticum* at breeding sites in the Athabasca River. *In* Control of black flies in the Athabasca River: Technical Report, pp. 21–37 (W. O. Haufe and G. C. R. Croome, Eds.). Edmonton. Alta. Environ.

Depner, K. R., Charnetski, W. A. and Haufe, W. O. (1980b). Effect of methoxychlor on resident populations of the invertebrates of the Athabasca River. *In* Control of black flies in the Athabasca River: Technical Report, pp. 141–150 (W. O. Haufe and G. C. R. Croome, Eds.). Edmonton. Alta. Environ.

Derr, S. K. and Zabik, M. J. (1972). Biological-

ly active compounds in the aquatic environment. The uptake and distribution of DDE by *Chironomus tentans* Fabricius (Diptera, Chironomidae). *Trans. Am. Fish. Soc.* **101**, 323–329.

Dietrich, W. (1909). Die Facettenaugen der Dipteren. *Z. wiss. Zool.* **92**, 465–539.

Dimick, R. E. and Mote, D. C. (1934). A preliminary survey of the food of Oregon trout. *Bull. Ore. agric. Exp. Stn.* **323**, 1–23.

Dimond, J. (1967). Pesticides and stream insects. *Me Forest Serv. Conserv. Edn. Bull.* **23**, 17 pp.

Dinulescu, G. (1966). Diptera Fam. Simuliidae (Mustele columbace). *Fauna Repub. pop. rom.* (Insecta) **11**(8), 1–600.

Disney, R. H. L. (1969). The timing of adult eclosion in blackflies (Diptera, Simuliidae), in West Cameroon. *Bull. ent. Res.* **59** (1968), 485–503.

Disney, R. H. L. (1970a). A note on the uptake of sugar by some blood sucking flies. *Proc. R. ent. Soc. Lond.* (A) **45**, 51–54.

Disney, R. H. L. (1970b). The timing of the first blood meal in *Simulium damnosum* Theobald. *Ann. trop. Med. Parasit.* **64**, 123–128.

Disney, R. H. L. (1971a). Notes on *Simulium ovazzae* Grenier ànd Mouchet (Diptera: Simuliidae) and river crabs (Malacostraca: Potamidae) and their association. *J. nat. Hist.* **5**, 677–689.

Disney, R. H. L. (1971b). Associations between blackflies (Simuliidae) and prawns (Atyidae), with a discussion of the phoretic habit in simuliids. *J. Anim. Ecol.* **40**, 83–92.

Disney, R. H. L. (1971c). *Simulium berneri* Freeman (Diptera: Simuliidae) and its mayfly host (Ephemeroptera: Oligoneuridae). *J. Ent.* (A) **46**, 39–51.

Disney, R. H. L. (1971d). Two phoretic blackflies (Diptera: Simuliidae) and their associated mayfly host (Ephemeroptera: Heptageniidae) in Cameroon. *J. Ent.* (A) **46**, 53–61.

Disney, R. H. L. (1972a). Observations on chicken-biting blackflies in Cameroon with a discussion of parous rates of *Simulium damnosum*. *Ann. trop. Med. Parasit.* **66**, 149–158.

Disney, R. H. L. (1972b). Observations on sampling pre-imaginal populations of blackflies (Dipt., Simuliidae) in West Cameroon. *Bull. ent. Res.* **61**, 485–503.

Disney, R. H. L. (1973). Larval Hydroptilidae (Trichoptera) that prey upon Simuliidae

(Diptera) in Cameroon. *Entomologist's mon. Mag.* **108** (1972). 84–85.

Disney, R. H. L. (1975a). *Drosophila gibbinsi* larvae also eat *Simulium. Trans. R. Soc. trop. Med. Hyg.* **69**, 365–366.

Disney, R. H. L. (1975b). A survey of blackfly populations (Dipt., Simuliidae) in West Cameroon. *Entomologist's mon. Mag.* **111**, 211–227.

Disney, R. H. L. and Boreham, P. F. L. (1969). Blood gorged resting blackflies in Cameroon and evidence of zoophily in *Simulium damnosum. Trans. R. Soc. trop. Med. Hyg.* **63**, 286–287.

Doby, J. M., David, F. and Rault, B. (1959). L'élevage en laboratoire, de l'oeuf à l'adulte de *Simulium ornatum* Meigen, 1818, *S. aureum* Fries, 1824, *S. erythrocephalum* De Geer, 1776 et *S. decorum* Walker, 1848 (Diptères Nématocères Simuliidés). Observations biologiques concernant ces espèces. *Annls Parasit. hum. comp.* **34**, 676–693.

Doby, J. M. and Laurent, P. (1953). Mermithidés parasites de larves de simulies en provenance de l'Avre et de la Semoy. *Annls Parasit. hum. comp.* **28**, 330–332.

Doby, J. M. and Saguez, F. (1964). *Weiseria*, genre nouveau de microsporidies et *Weiseria laurenti* n.sp., parasite de larves de *Prosimulium inflatum* Davies, 1957 (Diptères Paranématocères). *C.r. hebd. Séanc. Acad. Sci., Paris* (D) **259**, 3614–3617.

Downe, A. E. R. and Caspary, V. G. (1973). The swarming behaviour of *Chironomus riparius* (Diptera: Chironomidae) in the laboratory. *Can. Ent.* **105**, 165–171.

Downes, J. A. (1958). Assembly and mating in the biting Nematocera. *Proc. Xth Int. Congr. Ent.* (1956) **2**, 425–434.

Downes, J. A. (1965). Adaptations of insects in the arctic. *A. Rev. Ent.* **10**, 257–274.

Downes, J. A. (1969). The swarming and mating flight of Diptera. *A. Rev. Ent.* **14**, 271–298.

Downes, J. A. (1971). The ecology of bloodsucking Diptera: an evolutionary perspective. *In* Ecology and Physiology of Parasites, pp. 232–258 (A. M. Fallis, Ed.). Toronto. Univ. Toronto Press.

Drummond, F. H. N. (1933). West Australian Simuliidae. *J. Proc. R. Soc. West. Aust.* **18** (1931–32), 1–12.

Dubitskii, A. M. (1964). *In* Blood-feeding insects in Kazakhstan, pp. 1–35. [In Russian]. Alma-Ata. Izd-vo ANKazSSR.

Dubitskii, A. M. (1978). Biological control of bloodsucking flies in the USSR. *Inst. Zool. Acad. Sci. Alma-Ata. KazSSR* [In Russian].

Duke, B. O. L. (1962a). Studies on factors influencing the transmission of onchocerciasis. I. The survival rate of *Simulium damnosum* under laboratory conditions and the effect upon it of *Onchocerca volvulus. Ann. trop. Med. Parasit.* **56**, 130–135.

Duke, B. O. L. (1962b). Studies on factors influencing the transmission of onchocerciasis. II. The intake of *Onchocerca volvulus* microfilariae by *Simulium damnosum* and the survival of the parasites in the fly under laboratory conditions. *Ann. trop. Med. Parasit.* **56**, 255–263.

Duke, B. O. L. (1966). *Onchocerca-Simulium* complexes. III. The survival of *Simulium damnosum* after high intakes of microfilariae of incompatible strains of *Onchocerca volvulus* and the survival of the parasites in the fly. *Ann. trop. Med. Parasit.* **60**, 495–500.

Duke, B. O. L. (1967). Infective filaria larvae, other than *Onchocerca volvulus*, in *Simulium damnosum. Ann. trop. Med. Parasit.* **61**. 200–205.

Duke, B. O. L. (1968a). Studies on factors influencing the transmission of onchocerciasis. IV. The biting cycles, infective biting density and transmission potential of "forest" *Simulium damnosum. Ann. trop. Med. Parasit.* **62**, 95–106.

Duke, B. O. L. (1968b). Studies on factors influencing the transmission of onchocerciasis. V. The stages of *Onchocerca volvulus* in the wild "forest" *Simulium damnosum*, the fate of the parasites in the fly, and the age-distribution of the biting population. *Ann. trop. Med. Parasit.* **62**, 107–116.

Duke, B. O. L. (1968c). Studies on factors influencing the transmission of onchocerciasis. VI. The infective biting potential of *Simulium damnosum* in different bioclimatic zones and its influence on the transmission potential. *Ann. trop. Med. Parasit.* **62**, 164–170.

Duke, B. O. L. (1970). Quantitative approach to the transmission of *O. vulvulus. Trans. R. Soc. trop. Med. Hyg.* **64**, 311–312.

Duke, B. O. L. (1973). Studies on factors influencing the transmission of onchocerciasis. VIII: The escape of infective *Onchocerca volvulus* larvae from feeding "forest" *Simulium damnosum. Ann. trop. Med. Parasit.* **67**, 95–99.

Duke, B. O. L. (1975). The differential dispersal of nulliparous and parous *Simulium dam-nosum*. *Tropenmed. Parasit.* **26**, 88–97.

Duke, B. O. L. and Beesley, W. N. (1958). The vertical distribution of *Simulium damnosum* bites on the human body. *Ann. trop. Med. Parasit.* **52**, 274–281.

Duke, B. O. L., Lewis, D. J. and Moore, P. J. (1966). *Onchocerca-Simulium* complexes. I. Transmission of forest and Sudan-savanna strains of *Onchocerca volvulus*, from Came-roon, by *Simulium damnosum* from various West African bioclimatic zones. *Ann. trop. Med. Parasit.* **60**, 318–336.

Duke, B. O. L. and Moore, P. J. (1968). The contributions of different age groups to the transmission of onchocerciasis in a Came-roon forest village. *Trans. R. Soc. trop. Med. Hyg.* **62**, 22–28.

Duke, B. O. L., Moore, P. J. and De León, J. R. (1967a). *Onchocerca-Simulium* complexes. IV. Transmission of a variant of the forest strain of *Onchocerca volvulus*. *Ann. trop. Med. Parasit.* **61**, 326–331.

Duke, B. O. L., Moore, P. J. and De León, J. R. (1967b). *Onchocerca-Simulium* complexes. V. The intake and subsequent fate of micro-filariae of a Guatemalan strain of *Onchocerca volvulus* in forest and Sudan-savanna forms of West African *Simulium damnosum*. *Ann. trop. Med. Parasit.* **61**, 332–337.

Duke, B. O. L., Scheffel, P. D., Guyon, J. and Moore, P. J. (1967). The concentration of *Onchocerca volvulus* microfilariae in skin snips taken over twenty-four hours. *Ann. trop. Med. Parasit.* **61**, 206–219.

Dulmage, H. T. (1970). Production of the spore-δ-endotoxin complex by variants of *Bacillus thuringiensis* in two fermentation media. *J. Invert. Pathol.* **16**, 385–389.

Dulmage, H. T. (1971). Production of δ-endotoxin by eighteen isolates of *Bacillus thuringiensis*, serotype 3, in 3 fermentation media. *J. Invert. Pathol.* **18**, 353-358.

Dulmage, H. T. (1973). Assay and standar-dization of microbial insecticides. *Ann. N.Y. Acad. Sci.* **217**, 187–199.

Dulmage, H. T. and De Barjac, H. (1973). HD-187, A new isolate of *Bacillus thuringien-sis* that produces high yields of δ-endotoxin. *J. Invert. Pathol.* **22**, 273–277.

Dulmage. H. T. and Rhodes, R. A. (1971). Production of pathogens in artificial media. *In* Microbial Control of Insects and Mites, pp. 507–540 (H. D. Burges and N. W. Hus-sey, Eds.). New York. Academic Press.

Dumbleton, L. J. (1973). The genus *Austrosi-*

mulium Tonnoir (Diptera: Simuliidae) with particular reference to the New Zealand fauna. *N.Z. J. Sci.* **15** (1972), 480–584.

Dunbar, R. W. (1959). The salivary gland chromosomes of seven forms of blackflies included in *Eusimulium aureum* Fries. *Can. J. Zool.* **37**, 495–525.

Dunbar, R. W. (1966). Four sibling species included in *Simulium damnosum* Theobald (Diptera: Simuliidae) from Uganda. *Nature, Lond.* **209**, 597–599.

Dunbar, R. W. (1969). Nine cytological segre-gates in the *Simulium damnosum* complex (Diptera: Simuliidae). *Bull. Wld Hlth Org.* **40**, 974–979.

Dunbar, R. W. (1972). Polytene chromosome preparations from tropical Simuliidae. *WHO/ONCHO/72.95.* Mimeogr. Doc.

Dunbar, R. W. (1976). The East African situa-tion and a review of the *Simulium damnosum* complex as a whole. *WHO/VBC/SC/76.20.* Mimeogr. Doc.

Dunbar, R. W. and Grunewald, J. (1974). Distribution of four species near *Simulium damnosum* along a mountain river. *Proc. III Int. Congr. Parasit.* **2**, 922–923.

Dunbar, R. W. and Vajime, C. G. (1971). Cytotaxonomic analysis of the *Simulium damnosum* complex. *WHO/ONCHO/71.87, WHO/VBC/71.320.* Mimeogr. Doc.

Dunbar, R. W. and Vajime, C. G. (1972). The *Simulium (Edwardsellum) damnosum* com-plex. A report on cytotaxonomic studies to April 1972. *WHO/ONCHO/72.100.* Mimeogr. Doc.

Dunphy, G. B. and Nolan, R. A. (1977a). Regeneration of protoplasts of *Entomoph-thora egressa*, a fungal pathogen of the east-ern hemlock looper. *Can. J. Bot.* **55**, 107–116.

Dunphy, G. B. and Nolan, R. A. (1977b). Morphogenesis of protoplasts of *Entomoph-thora egressa* in simplified culture media. *Can. J. Bot.* **55**, 3046–3053.

Ebsary, B. A. and Bennett, G. F. (1973). Molting and oviposition of *Neomesomermis flumenalis* (Welch, 1962) Nickle, 1972, a mer-mithid parasite of blackflies. *Can. J. Zool.* **51**, 637–639.

Ebsary, B. A. and Bennett, G. F. (1975). Studies on the bionomics of mermithid nematode parasites of blackflies in New-foundland. *Can. J. Zool.* **53**, 1324–1331.

Eder, S. and Carlson, C. A. (1977). Food habits of carp and white suckers in the South Platte and St. Vrain Rivers and

Goosequill Pond, Weld County, Colorado. *Trans. Am. Fish. Soc.* **106**, 339–346.

Edwards, A. J. and Trenholmè, A. A. G. (1976). Diel periodicity in the adult eclosion of the blackfly *Simulium damnosum* Theobald, in the Ivory Coast. *Ecol. Ent.* **1**, 279–282.

Edwards, E. E. (1956). Human onchocerciasis in West Africa with special reference to the Gold Coast. *Jl W. Afr. Sci. Ass.* **2**, 1–35.

Edwards, F. W. (1920). On the British species of *Simulium*. II. The early stages; with corrections and additions to part I. *Bull. ent. Res.* **11**, 211–246.

El Bashir, S., El Jack, M. H. and El Hadi, H. M. (1976). The diurnal activity of the chicken-biting blackfly, *Simulium griseicolle* Becker (Diptera: Simuliidae) in Northern Sudan. *Bull. ent. Res.* **66**, 481–487.

Eligh, G. S. (1952). Factors influencing the performance of the precipitin test in the determination of blood meals of insects. *Can. J. Zool.* **30**, 213–218.

Elizarov, Yu. and Chaika, S. Yu. (1975). Electron microscopic investigation of the taste and olfactory sensilla of the blackfly *Boophthora erythrocephala* (De Geer) (Simuliidae: Diptera). *Vest. Mosk. Univ.* (Biol.) **5**, 3–11.

Elliott, J. M. (1967). The food of trout (*Salmo trutta*) in a Dartmoor stream. *J. appl. Ecol.* **4**, 59–71.

Elliott, J. M. (1971). Some methods for the statistical analysis of samples of benthic invertebrates. *Scient. Publs Freshwat. biol. Ass.* **25**, 1–144.

Elliott, J. M. (1973). The food of brown and rainbow trout (*Salmo trutta* and *S. gairdneri*) in relation to the abundance of drifting invertebrates in a mountain stream. *Oecologia* **12**, 329–347.

Elouard, J. M. (1975). Effets toxiques sur la faune noncible de l'Abate Standard Procida épandu lors d'un surdosage accidentel. *ORSTOM, Bouaké Rap.* 147, Mimeogr. Doc.

Elouard, J. M. (1978). Identification biométrique des stades larvaires de *S. damnosum* s.l. et de *S. adersi* (Diptera: Simuliidae). *Tropenmed. Parasit.* **29**, 183–187.

Elouard, J. M. and Elsen, P. (1975). Variations de l'absorption des particules alimentaires et de la vitesse de transit digestif au cours d'un nycthémère chez les larves de *Simulium damnosum* Theobald. *ORSTOM, Bouaké Rap.* 623. Mimeogr. Doc.

Elouard, J. M. and Elsen, P. (1977). Variations de l'absorption des particules alimentaires et de la vitesse de transit digestif en fonction de certains paràmetres du milieu chez les larves de *Simulium damnosum* Theobald 1903 (Diptera: Simuliidae). *Cah. ORSTOM, Sér Ent. méd. Parasit.* **15**, 29–39.

Elouard, J. M. and Forge, P. (1977). Action de l'Abate sur les invertébrés aquatiques. II. Effets d'un mois de suspension des traitements sur la faune aquatiques de gîte Grechan (Leraba). *ORSTOM, Bouaké Rap.* 13. Mimeogr. Doc.]

Elouard, J. M., Lebtahi, F., Lévêque, C. and Vernard, P. (1974). Effets de deux formulations d'Abate sur l'entomofaune associée aux larves de *Simulies* et sur l'ichtyofaune. *ORSTOM*, 30 pp. [Reference incomplete and original unavailable; probably *ORSTOM, Bouaké Rap.* Mimeogr. Doc.]

Elouard, J. M. and Lévêque, C. (1977). Rythme nycthéméral de dérive des insectes et des poissons dans les rivières de Côte d'Ivoire. *Cah. ORSTOM, sér. Hydrobiol.* **11**, 179–183.

Elsen, P. (1977a). Note biologique sur *Xenomyia oxycera* Emden (Muscidae, Limnophorinae) et *Ochthera insularis* Becker (Ephydridae), deux Diptères prédateurs de *Simulium damnosum* Theobald (Diptera: Simuliidae) en Côte d'Ivoire. *Revue Zool. afr.* **91**, 732–736.

Elsen, P. (1977b). Méthodes d'échantillonnage des populations préimaginales de *Simulium damnosum* Theobald, 1903 (Diptera: Simuliidae) en Afrique de l'Ouest. I. Distribution verticale des larves et des nymphes, observations préliminaires. *Tropenmed. Parasit.* **28**, 91–96.

Elsen, P. (1978). Some notes on *Xenomyia oxycera* Emden (Muscidae, Limnophorinae), a dipteran predator of *Simulium damnosum* Theobald (Diptera: Simuliidae) in the Ivory Coast. *Proc. British Soc. Parasit.* Spring Meet. held at Univ. of Kent, Canterbury, 5–7 April 1978.

Elsen, P. and Hébrard, G. (1977a). A new rearing technique for studying individually the time of development of preimaginal instars of *Simulium damnosum* Theobald (Diptera: Simuliidae). *Trans. R. Soc. trop. Med. Hyg.* **71**, 269–270.

Elsen, P. and Hébrard, G. (1977b). Méthodes d'échantillonage des populations préimaginales de *Simulium damnosum* Theobald, 1903 (Diptera: Simuliidae) en Afrique de l'Ouest. II. Observations sur le choix des couleurs, l'évolution du peuplement et la réparation horizontale au moyen de rubans en plastique. *Tropenmed. Parasit.* **28,** 471–477.

Elsen, P., Quillévéré, D. and Hébrard, G. (1978). Le transit intestinal chez les larves du complexe *Simulium damnosum* (Diptera: Simuliidae) en Afrique de l'Ouest. I. Influence du sexe et de l'espèce. *Annls. Soc. belge. Méd. trop.* **58,** 209–217.

Elson, P. F. (1967). Effects on wild young salmon of spraying DDT over New Brunswick forests. *J. Fish. Res. Bd Can.* **24,** 731–767.

Elton, C. (1927). Animal Ecology. Sidgwick and Jackson Ltd.

Emery, W. T. (1914). The morphology and biology of *Simulium vittatum* and its distribution in Kansas. *Kans. Univ. Sci. Bull.* **8,** (1913) 323–362.

Enderlein, G. (1931). Zur Beurteilung und Bekämpfung der Kriebelmücken-Schäden des Leinegebietes. *Arch. wiss. Prakt. Tierheilk.* **63,** 475–528.

Eriksen, C. H. (1968). Ecological significance of respiration and substrate for burrowing Ephemeroptera. *Can. J. Zool.* **46,** 93–103.

Escaffre, H., Brunhes, J., Sechan, Y., Carlsson, G., Stiles, A., Kulzer, H., Pawlick, O. and Mertens, J. (1976). Control of *Simulium damnosum*, vector of human onchocerciasis in West Africa. VII. Helicopter applications of new insecticides and formulations in the Korhogo Region of Ivory Coast (1974). *WHO/VBC/76.620.* Mimeogr. Doc.

Estrada Sandoral, C. *et al.* (1963). Epidemiologia, Enfermedad de Robles. Editorial Universitaria, 83–121.

Ezenwa, A. O. (1973). Mermithid and microsporidan parasitism of blackflies (Diptera: Simuliidae) in the vicinity of Churchill Falls, Labrador. *Can. J. Zool.* **51,** 1109–1111.

Ezenwa, A. O. (1974a). Ecology of Simuliidae, Mermithidae and Microsporida in Newfoundland freshwaters. *Can. J. Zool.* **52,** 557–565.

Ezenwa, A. O. (1974b). Studies on host-parasite relationships of Simuliidae with mermithids and microsporidans. *J. Parasit.* **60,** 809–813.

Ezenwa, A. O. and Carter, N. E. (1975). Influence of multiple infections on sex ratios of mermithid parasites of blackflies. *Environ. Ent.* **4,** 142–144.

Ezenwa, A. O., Howlett, T. and Hedge, K. (1974). Microsporidan infection of *Simulium damnosum* in West Africa. *J. Parasit.* **60,** 975.

Fairchild, G. B. and Barreda, E. A. (1945).

DDT as a larvicide against *Simulium. J. econ. Ent.* **38,** 694–699.

Fallis, A. M. (1964). Feeding and related behavior of female Simuliidae (Diptera). *Expl. Parasit.* **15,** 439–470.

Fallis, A. M., Bennett, G. F., Griggs, G. and Allen, T. (1967). Collecting *Simulium venustum* females in fan traps and on silhouettes with the aid of carbon dioxide. *Can. J. Zool.* **45,** 1011–1017.

Fallis, A. M., Desser, S. S. and Khan, R. A. (1974). On species of *Leucocytozoon. In* "Advances in Parasitology" No. 12 (B. Dawes, Ed.), pp. 1–67. New York. Academic Press.

Fallis, A. M., Jacobson, R. L. and Raybould, J. N. (1973a). Experimental transmission of *Trypanosoma numidae* Wenyon to guinea fowl and chickens in Tanzania. *J. Protozool.* **20,** 436–437.

Fallis, A. M., Jacobson, R. L. and Raybould, J. N. (1973b). Haematozoa in domestic chickens and guinea fowl in Tanzania and transmission of *Leucocytozoon neavei* and *Leucocytozoon schoutedeni. J. Protozool.* **20,** 438–442.

Fallis, A. M. and Raybould, J. N. (1975). Response of two African simuliids to silhouettes and carbon dioxide. *J. med. Ent.* **12,** 349–351.

Fallis, A. M. and Smith, S. M. (1964). Ether extracts from birds and CO_2 as attractants for some ornithophilic simuliids. *Can. J. Zool.* **42,** 723–730.

Faragher, R. A., Grant, T. R. and Carrick, F. N. (1979). Food of the platypus (*Ornithorhynchus anatinus*) with notes on the food of the brown trout (*Salmo trutta*) in the Shoalhaven River, N.S.W. *Aust. J. Ecol.* **4,** 171–179.

Farquharson, I. D. (1976). Pesticides: A guide to terminology. *In* Pesticides and Human Welfare, pp. 256–273 (D. Gunn and J. Stevens, Eds.). Oxford. Oxford University Press.

Federici, B. A. (1976). Pathology and histochemistry of a densonucleosis virus in larvae of the blackfly *Simulium vittatum. Proc. 1st Int. Colloq. Invert. Path.*, Kingston, Ontario, 341–342.

Federici, B. A. and Lacey, L. A. (1976). Densonucleosis virus and cytoplasmic polyhedrosis virus diseases in larvae of the blackfly, *Simulium vittatum. Proc. Pap. a. Conf. Calif. Mosquito Control Ass.* **44,** 124.

Ferguson, D. E. (1969). The compatible existence of non-target species to pesticides. *Bull. ent. Soc. Am.* **15**, 363–366.

Field, G., Duplessis, R. J. and Breton, A. P. (1967). Progress report on laboratory rearing of black flies (Diptera: Simuliidae). *J. med. Ent.* **4**, 304–305.

Figueroa Marroquin, H. (1974). Robles' disease (American onchocerciasis) in Guatemala. Int. Symp. Research and Control of onchocerciasis in Western Hemisphere, *PAHO Scient. Publ.* **298**, 100–104.

Filipjev, I. N. (1934). Harmful and useful nematodes in forestry. *Selkozgiz* **1**, 440.

Filteau, G. (1959). Effets des vaporisations aériennes du DDT sur les insectes aquatiques. *Nat. Can.* **86**, 113–128.

Finney, J. R. (1975). The penetration of three simuliid species by the nematode *Reesimermis nielseni*. *Bull. Wld Hlth Org.* **52**, 235.

Finney, J. R. (1976). The *in vitro* culture of mermithid parasites of blackflies and mosquitoes. *J. Nematology* **8**, 284.

Finney, J. R. (1978). The development of *Romanomermis culicivorax* in *in vitro* culture. *Nematologica* **23**, 479–480.

Finney, J. R., Gordon, R., Condon, W. J. and Rusted, T. N. (1977). Laboratory studies on the feasibility of integrated mosquito control using an insect growth regulator and a mermithid nematode. *Mosquito News* **37**, 6–11.

Finney, J. R. and Mokry, J. E. (1980) *Romanomermis culicivorax* and simuliids. *J. Invert. Pathol.* **35**, 211–213.

Flannagan, J. F. (1976). Preliminary report on studies to determine the effect of methoxychlor treatment (1975) on the aquatic invertebrates of the Athabasca River, Alberta. *Ms. Rpt. Freshwat. Inst., Winnipeg, Man.*, 15 pp.

Flannagan, J. F., Townsend, B. E. and de March, B. G. E. (1980a). Acute and long-term effects of methoxychlor larviciding on the aquatic invertebrates of the Athabasca River, Alberta. *In* Control of black flies in the Athabasca River: Technical Report, pp. 151–158 (W. O. Haufe and G. C. R. Croome, Eds.). Edmonton. Alta. Environ.

Flannagan, J. F., Townsend, B. E., de March, B. G. E., Friessen, M. and Leonard, S. L. (1979). *Ms. Rpt. Freshwat. Inst., Winnipeg, Man.*

Flannagan, J. F., Townsend, B. F., de March, B. G. E., Friessen, M. K. and Leonard, S. L. (1980b). Effects of an experimental injection

of methoxychlor in 1974 on aquatic invertebrates: Accumulation, standing crop, and drift. *In* Control of black flies in the Athabasca River: Technical Report, pp. 131–140 (W. O. Haufe and G. C. R. Croome, Eds.). Edmonton. Alta. Environ.

Fredeen, F. J. H. (1959). Collection, extraction, sterilization and low-temperature storage of black-fly eggs (Diptera: Simuliidae). *Can. Ent.* **91**, 450–453.

Fredeen, F. J. H. (1960). Bacteria as a source of food for black-fly larvae. *Nature, Lond.* **187**, 963.

Fredeen, F. J. H. (1961). A trap for studying the attacking behaviour of black flies, *Simulium arcticum* Mall. *Can. Ent.* **93**, 73–78.

Fredeen, F. J. H. (1962a). Proc. 3rd Conf. on Black Flies (Diptera: Simuliidae), p. 77. Algonquin Park, Ontario, Canada (B. V. Peterson, Ed.).

Fredeen, F. J. H. (1962b). DDT and Heptachlor as black-fly larvicides in clear and turbid water. *Can. Ent.* **94**, 875–880.

Fredeen, F. J. H. (1964). Bacteria as food for blackfly larvae (Diptera: Simuliidae) in laboratory cultures and in natural streams. *Can. J. Zool.* **42**, 527–548.

Fredeen, F. J. H. (1969). A new procedure allowing replicated miniature larvicide tests in a large river. *Can. Ent.* **101**, 713–725.

Fredeen, F. J. H. (1970). Sexual mosaics in the black fly *Simulium arcticum* (Diptera: Simuliidae). *Can. Ent.* **102**, 1585–1592.

Fredeen, F. J. H. (1974). Tests with single injections of methoxychlor black fly (Diptera: Simuliidae) larvicides in large rivers. *Can. Ent.* **106**, 285–305.

Fredeen, F. J. H. (1975). Effects of a single injection of methoxychlor black-fly larvicide on insect larvae in a 161-km (100 mile) section of the North Saskatchewan River. *Can. Ent.* **107**, 807–817.

Fredeen, F. J. H. (1976). The seven larval instars of *Simulium arcticum* (Diptera: Simuliidae). *Can. Ent.* **108**, 591–600.

Fredeen, F. J. H. (1977a). A review of the economic importance of black flies (Simuliidae) in Canada. *Quaest. ent.* **13**, 219–229.

Fredeen, F. J. H. (1977b). Black fly control and environmental quality with reference to chemical larviciding in Western Canada. *Quaest. ent.* **13**, 321–325.

Fredeen, F. J. H., Arnason, A. P. and Berck, B. (1953). Adsorption of DDT on suspended solids in river water and its role in black fly control. *Nature, Lond.* **171**, 700–701.

Fredeen, F. J. H., Arnason, A. P., Berck, B. and Rempel, J. G. (1953b). Further experiments with DDT in the control of *Simulium arcticum* Mall. in the North and South Saskatchewan Rivers. *Can. J. Agric. Sci.* **33**, 379–393.

Fredeen, F. J. H., Rempel, J. G. and Arnason, A. P. (1951). Egg-laying habits, overwintering stages, and life-cycle of *Simulium arcticum* Mall. (Diptera: Simuliidae). *Can. Ent.* **83**, 73–76.

Fredeen, F. J. H., Saha, J. G. and Balba, M. H. (1975). Residues of methoxychlor and other chlorinated hydrocarbons in water, sand, and selected fauna following injections of methoxychlor black fly larvicide into the Saskatchewan River, 1972. *Pestic. Monit. J.* **8**, 241–246.

Fredeen, F. J. H., Saha, J. G. and Royer, L. M. (1971). Residues of DDT, DDE, and DDD in fish in the Saskatchewan River after using DDT as a blackfly larvicide for twenty years. *J. Fish. Res. Bd Can.* **28**, 105–109.

Fredeen, F. J. H. and Shemanchuk, J. A. (1960). Black flies (Diptera: Simuliidae) of irrigation systems in Saskatchewan and Alberta. *Can. J. Zool.* **38**, 723–735.

Fredeen, F. J. H., Spinks, J. W. T., Anderson, J. R., Arnason, A. P. and Rempel, J. G. (1953). Mass tagging on black flies (Diptera: Simuliidae) with radiophosphorus. *Can, J. Zool.* **31**, 1–15.

Fredeen, F. J. H. and Spurr, D. T. (1978). Collecting semi-quantitative samples of blackfly larvae (Diptera: Simuliidae) and other aquatic insects from large rivers with the aid of artificial substrates. *Quaest. ent.* **14**, 411–431.

Freeman, P. and de Meillon, B. (1953). Simuliidae of the Ethiopian Region. London, British Museum (Natural History).

Friederichs, K. (1919). Untersuchungen über Simuliiden. *Z. angew. Ent.* **6**, 61–83.

Friederichs, K. (1922). Untersuchungen über Simuliiden (Teil II). *Z. angew. Ent.* **8**, 31–92.

Friend, W. G. (1965). The gorging response in *Rhodnius prolixus* Stahl. *Can. J. Zool.* **43**, 125–132.

Frommer, R. L., Carestia, R. R. and Vavra, R. W. (1974). A modified CDC trap using carbon dioxide for trapping blackflies (Simuliidae: Diptera). *Mosquito News* **34**, 468–469.

Frommer, R. L., Schiefer, B. A. and Vavra, R. W. (1976). Comparative effects of CO_2 flow rates using modified CDC light traps on trapping adult black flies (Simuliidae: Diptera). *Mosquito News* **36**, 355–358.

Frost, S. (1970). Microsporidia (Protozoa: Microsporidia) in Newfoundland blackfly larvae (Diptera: Simuliidae). *Can. J. Zool.* **48**, 890–891.

Frost, S. and Nolan, R. A. (1972). The occurrence and morphology of *Caudospora* spp. (Protozoa: Microsporida) in Newfoundland and Labrador blackfly larvae (Diptera: Simuliidae). *Can. J. Zool.* **50**, 1363–1366.

Frost, W. E. (1943). The natural history of the minnow (*Phoxinus phoxinus*). *J. Anim. Ecol.* **12**, 139–162.

Frost, W. E. (1945). River Liffey survey. VI. Discussion on the results obtained from investigations on the food and growth of brown trout (*Salmo trutta* L.) in alkaline and acid waters. *Proc. R. Ir. Acad.* **50**, (B19), 321–342.

Frost, W. E. (1946). Observations on the food of eels (*Anguilla anguilla*) from the Windermere catchment area. *J. Anim. Ecol.* **15**, 43–53.

Frost, W. E. (1950). The growth and food of young salmon (*Salmon salar*) and trout (*S. trutta*) in the river Forss, Caithness. *J. Anim. Ecol.* **19**, 147–158.

Fuller, R. L. and Stewart, K. W. (1977). The food habits of stoneflies (Plecoptera) in Upper Gunnison River, Colorado. *Envir. Ent.* **6**, 293–302.

Galun, R. and Margalit, J. (1969). Adenine nucleotides as feeding stimulants of the tsetse fly *Glossina austeni* Newst., *Nature, Lond.* **222**, 583–584.

Gambrell, F. L. (1933). The embryology of the black fly, *Simulium pictipes* Hagen. *Ann. ent. Soc. Am.* **26**, 641–671.

Garcia Manzo, G. A. (1965). Enfermedad de Robles, su morbilidad y proyecciones socioeconomicas. *Boletin Sanitario* **61**, 121–130.

Garcia Manzo, G. A. (1971). La onchocercosis en Guatemala. Breve resumen preparado para informacion de los asistentes al congreso nacional de salud. Unpubl. Doc.

Gardner, D. R. and Bailey, J. R. (1975). Methoxychlor: its effects on environmental quality. *Natn. Res. Coun. Canada*, N.R.C.C. 14102.

Garms, R. (1973). Quantitative studies on the transmission of *Onchocerca volvulus* by *Simulium damnosum* in the Bong Range, Liberia. *Z. Tropenmed. Parasit.* **24**, 358–372.

Garms, R. (1975). Observations on filarial

infections and parous rates of anthropophilic blackflies in Guatemala with reference to the transmission of *Onchocerca volvulus*. *Tropenmed. Parasit.* **26**, 169–182.

Garms, R. (1978). Use of morphological characters in the study of *Simulium damnosum* s.l. populations in West Africa. *Tropenmed. Parasit.* **29**, 483–491.

Garms, R. and Post, A. (1967a). Die Simulien der Republik Guinea, Westafrika. *Int. Revue ges. Hydrobiologia* **52**, 1–36.

Garms, R. and Post, A. (1967b). Freilandversuche zur Wirksamkeit von DDT und Baytex gegen Larven von *Simulium damnosum* in Guinea, Westafrika. *Anz. Schädlingsk.* **40**, 49–56.

Garms, R. and Vajime, C. G. (1975). On the ecology and distribution of the *Simulium damnosum* complex in different bioclimatic zones of Liberia and Guinea. *Tropenmed. Parasit.* **26**, 375–380.

Garms, R. and Voelker, J. (1969). Unknown filarial larvae and zoophily in *Simulium damnosum* in Liberia. *Trans. R. Soc. trop. Med. Hyg.* **63**, 676–677.

Garms, R., Walsh, J. F. and Davies, J. B. (1979). Studies on the reinvasion of the Onchocerciasis Control Programme in the Volta River Basin by *Simulium damnosum s.l.* with emphasis on the south-western areas. *Tropenmed. Parasit.* **30**, 345–362.

Garms, R. and Weyer, F. (1968). Natürliche Infektion von *Simulium damnosum* mit *Onchocerca volvulus* in Savannengebieten Guineas. *Z. Tropenmed. Parasit.* **19**, 289–296.

Garnham, P. C. C. and Lewis, D. J. (1959). Parasites of British Honduras with special reference to leishmaniasis. *Trans. R. Soc. trop. Med. Hyg.* **53**, 12–40.

Garnham, P. C. C. and McMahon, J. P. (1947). The eradication of *Simulium neavei*, Roubaud, from an onchocerciasis area in Kenya Colony. *Bull. ent. Res.* **37**, 619–627.

Garris, G. I. and Noblet, R. (1975). Notes on parasitism of black flies (Diptera: Simuliidae) in streams treated with Abate®. *J. med. Ent.* **12**, 481–482.

Gassouma, M. S. S. (1972). Microsporidan parasites of *Simulium ornatum* Mg. in South England. *Parasitology, Cambridge* **65**, 27–45.

Gaufin, A. R., Harris, E. K., and Walter, H. J. (1956). A statistical evaluation of stream bottom sampling data obtained from three standard samplers. *Ecology*, **37**, 643–648.

Gelbič, I. and Knoz, J. (1972). Differences in the structure of the labrofrontal nerve in

larvae of various groups of the family Simuliidae (Diptera: Nematocera). *Acta. ent. bohemoslovaca* **69**, 305–311.

Germain, M. and Grenier, P. (1967). Observations biologiques et écologiques sur l'association de *Simulium berneri kumboense* Grenier, Germain et Mouchet, 1965, avec *Elassoneuria* sp. (Ephemeroptera, Oligoneuriidae). *Cah. ORSTOM sér. Ent. Méd.* **5**, 71–92.

Germain, M., Grenier, P. and Mouchet, J. (1968). Les simulies du Cameroun occidental. Influence du milieu physique sur leur répartition. *Cah. ORSTOM sér. Ent. Méd.* **6**, 167–190.

Gessner, F. (1959). Hydrobotanik. Die physiologischen Grundlagen der Pflanzenverbreitung im Wasser. II. Stoffhaushalt. Berlin, VEB Deutscher Verlag der Wissenschaften.

Giaquinto, A. (1937). Contributo agli studi sul problema della trasmissione della onchocercosi nel Guatemala. *Annali Ig. sper.* **47**, 108–125.

Gibbins, E. G. (1933). Studies on Ethopian Simuliidae. *Simulium damnosum*. Theo. *Trans. R. ent. Soc. Lond.* **81**, 37–51.

Gibson, C. L. (1951). Parasitological studies on onchocerciasis in Guatemala. PhD thesis, Univ. of Michigan.

Gibson, C. L. and Dalmat, H. T. (1952). Three new potential intermediate hosts of human onchocerciasis in Guatemala. *Am. J. trop. Med. Hyg.* **1**, 848–851.

Gibson, R. J. and Galbraith, D. (1975). The relationships between invertebrate drift and salmonid populations in the Matamek river, Quebec, below a lake. *Trans. Am. Fish. Soc.* **104**, 529–535.

Gillies, M. T. (1958). A simple autoradiographic method for distinguishing insects labelled with Phosphorus-32 and Sulphur-35. *Nature, Lond.* **182**, 1683–1684.

Giudicelli, J. (1962). Recoltés de Simuliidae (Diptera) en Corse. Étude faunistique et écologique. *C. r. 86e Congr. Soc. sav., Montpellier* **1961**, 715–723.

Giudicelli, J. (1966). Récoltes de Simulies en Côte d'Ivoire. Étude de l'activité diurne des femelles de *Simulium damnosum* Theobald. *Annls Soc. ent. Fr.* (n.s.) **2**, 325–342.

Gjullin, C. M., Cross, H. F. and Applewhite, K. H. (1950). Tests with DDT to control blackfly larvae in Alaskan streams. *J. econ. Ent.* **43**, 696–697.

Glatthaar, R. (1978). Verbreitung und Okologie der Kriebelmücken (Diptera: Simuliidae) in der Schweiz. *Vjschr. naturf. Ges. Zürich* **123**, 71–124.

Glötzel, R. (1973). Populationsdynamik und Ernährungsbiologie von Simuliiden-larven in einem mit organischen Abwassern verunreinigten Gebirgsbach. *Arch. Hydrobiol. suppl.* **42**, 406–451.

Goldberg, L. J. and Margalit, J. (1977). A bacterial spore demonstrating rapid larvicidal activity against *Anopheles sergentii, Uranotaenia unguiculata, Culex univitattus, Aedes aegypti* and *Culex pipiens. Mosquito News* **37**, 355–358.

Golini, V. I. (1974). Relative response to coloured substrates by ovipositing blackflies (Diptera: Simuliidae). III. Oviposition by *Simulium (Psilozia) vittatum* Zetterstedt. *Proc. ent. Soc. Ont.* **105**, 48–55.

Golini, V. I. (1975). Simuliidae (Diptera) of Rendalen, Norway. I. *Eusimulium rendalense* n.sp. and *E. fallisi* n.sp. feeding on ducks. *Ent. Scand.* **6**, 229–239.

Golini, V. I. and Davies, D. M. (1975a). Relative response to colored substrates by ovipositing blackflies (Diptera: Simuliidae). I. Oviposition by *Simulium (Simulium) verecundum* Stone and Jamnback. *Can. J. Zool.* **53**, 521–535.

Golini, V. I. and Davies, D. M. (1975b). Relative response to colored substrates by ovipositing blackflies (Diptera: Simuliidae). II. Oviposition by *Simulium (Odagmia) ornatum* Meigen. *Norw. J. Ent.* **22**, 89–94.

Golini, V. I., Davies, D. M. and Raastad, J. E. (1976). Simuliidae (Diptera) of Rendalen, Norway. II. Adult females attacking cows and humans. *Norw. J. Ent.* **23**, 79–86.

Golterman, H. L. (1975). Physiological Limnology. An approach to the physiology of lake ecosystems. Oxford, New York. Elsevier, Amsterdam.

Gorayeb, I. S. and Mok, W. Y. (1978). Comparison of precipitin and immunodiffusion tests in the detection of *Simulium fulvinotum* larval predator. Unpubl. ms.

Gorayeb, I. S. and Pinger, R. R. (1978). Deleccao de predadores naturals das larvas de *Simulium fulvinotum* Cerq. e Mello, 1968 (Diptera: Nematocera). *Acta Amazonica* **8**, 629–637.

Gordon, R. (1975). *Ann. Rpt, Res. Unit Vec. Path.*, Memorial Univ. Newfoundland.

Gordon, R. and Bailey, C. H. (1974). Free amino acid composition of the hemolymph of the larval blackfly *Simulium venustum* (Diptera: Simuliidae). *Experientia* **30**, 902–903.

Gordon, R. and Bailey, C. H. (1976). Free amino acids, ions, and osmotic pressure of the hemolymph of three species of blackflies. *Can. J. Zool.* **54**, 399–404.

Gordon, R., Condon, W. J., Edgar, W. J. and Babie, S. J. (1978). Effects of mermithid parasitism on the haemolymph composition of the larval blackflies *Prosimulium mixtum/fuscum* and *Simulium venustum. Parasitology* **77**, 367–374.

Gordon, R., Ebsary, B. A. and Bennett, G. F. (1973). Potentialities of mermithid nematodes for the biocontrol of blackflies (Diptera: Simuliidae)—A review. *Expl. Parasit.* **33**, 226–238.

Gordon, R., Finney, J. R., Condon, W. J. and Rusted, T. N. (1979). Lipids in the storage organs of three mermithid nematodes and in the hemolymph of their hosts. *Comp. Biochem. Physiol.* **64B**, 369–374.

Gorham, J. R. (1961). Aquatic insects and DDT forest sprayings in Maine. *Me. Forest Serv. Conserv. Fdn. Bull.* No. 19.

Gosbee, J., Allen, J. R. and West, A. S. (1969). The salivary glands of adult blackflies. *Can. J. Zool.* **47**, 1341–1344.

Gouteux, J.-P. (1975). Larves de simulies (Diptera: Simuliidae) du Kivu rattachables au complexe "damnosum". Description de quatre types distincts morphologiquement. *Cah. ORSTOM, Ent. méd. Parasit.* **13**, 237–243.

Gouteux, J.-P. (1976). Nouveau gîte, données éthologiques et morphologiques pour une drosophile aquatique du groupe *simulivora* Tsacas and Disney 1974: *Drosophila gibbinsi* Aubertin 1937 (Diptera: Drosophilidae). *C.R. Acad. Sc. Paris, Sér. D.* **282**, 2191–2194.

Gouteux, J.-P. (1977). Micromorphologie du revêtement cuticulaire, larvaire et nymphal, de simulies du Zaire (Diptera: Simuliidae) rattachables au complexe *Simulium damnosum. Tropenmed. Parasit.* **28**, 97–99.

Graham, R. J. and Scott, D. O. (1959). Effects of an aerial application of DDT on fish and aquatic insects in Montana. *Montana State Fish and Game Dept., US Fish and Wildl. Serv., and US Forest Serv. Final rep.*

Grant, C. D. and Brown, A. W. A. (1967). Development of DDT resistance in certain mayflies in New Brunswick. *Can. Ent.* **99**, 1040–1050.

Grebel'skii, C. G. (1963). *In* Control of blood-sucking insects in middle Angara basin, pp. 1–63. Irkutsk Irkutskoe knizhnoe izdatel'stvo. [In Russian].

Grenier, P. (1943). Observations sur quelques stations de simulies. Parasites et prédateurs des larves et nymphes. *Bull. Soc. Path. exot.* **36**, 105–110.

Grenier, P. (1945a). Remarques sur la biologie de quelques ennemis des simulies. *Bull. Soc. ent. Fr.* **49** (1944), 130–133.

Grenier, P. (1945b). Quelques remarques sur la biologie larvaire d'une simulie *S. (Eusimulium) costatum* Fried., peuplant certain ruisseaux du bois de Meudon. *Cahiers Comm. Seine, Trav. biol.* **4**, 11–15.

Grenier, P. (1949). Contribution à l'étude biologique des simuliides de France. *Physiol. comp. Oecol.* **1**, 165–330.

Grenier, P. (1953). Simuliidae de France et d'Afrique de Nord. *Encycl. ent. Sér. A.* **29**, 7–170.

Grenier, P., Ovazza, M. and Valade, M. (1960). Notes biologiques et faunistiques sur *S. damnosum* et les Simuliidae d'Afrique occidentale (Haute-Volta, Côte d'Ivoire, Dahomey, Soudan). *Bull. Inst. fr. Afr. noire (A)* **22**, 892–918.

Grunewald, J. (1972). Die hydrochemischen Lebensbedingungen der präimaginalen Stadien von *Boophthora erythrocephala* De Geer (Diptera: Simuliidae). 1. Freilanduntersuchungen. *Z. Tropenmed. Parasit.* **23**, 432–445.

Grunewald, J. (1973). Die hydrochemischen Lebensbedingungen der präimaginalen Stadien von *Boophthora erythrocephala* De Geer (Diptera: Simuliidae). 2. Die Entwicklung einer Zucht unter experimentellen Bedingungen. *Z. Tropenmed. Parasit.* **24**, 232–249.

Grunewald, J. (1974). The hydro-chemical living conditions of the immature stages of some forms of the *Simulium damnosum* complex with regard to their laboratory colonization. *Proc. III. Int. Congr. Parasit.* **2**, 914–915.

Grunewald, J. (1976a). The hydro-chemical and physical conditions of the environment of the immature stages of some species of the *Simulium (Edwardsellum) damnosum* complex (Diptera). *Tropenmed. Parasit.* **27**, 438–454.

Grunewald, J. (1976b). The hydro-chemical and physical conditions of the environment of the aquatic stages of some West African cytotypes of the *Simulium damnosum* complex, and studies on the water quality at some potential locations for a rearing laboratory. *Wld Hlth Org. Rpt.* 1–27.

Grunewald, J. (1978). Die Bedeutung der Stickstoff-Exkretion und Ammoniak-Empfindlichkeit von Simuliiden-Larven (Diptera) für den Aufbau von Laboratoriumskulturen. *Z. angew. Ent.* **85**, 52–60.

Grunewald, J. and Grunewald, E. B. (1978). Der Einfluss der Wasserstoffionen- und Gesamtionenkonzentration sowie der Ionenkomposition auf die aquatischen Stadien zweier Zytoarten des *Simulium damnosum*-Komplexes (Diptera: Simuliidae) Ostafrikas. *Arch. Hydrobiol.* **82**, 419–431.

Grunewald, J., Grunewald, E. B., Raybould, J. N. and Mhiddin, H. K. (1979). The hydrochemical and physical characteristics of the breeding sites of the *Simulium neavei* Roubaud group and their associated crabs in the Eastern Usambara Mountains in Tanzania. *Int. Revue ges. Hydrobiol.* **64**, 71–88.

Grunewald, J. and Wirtz, H. P. (1978). Künstliche Blutfütterung einiger afrikanischer und paläarktischer Simuliiden (Diptera). *Z. angew. Ent.* **85**, 425–435.

Grunin, K. Ya. (1949). A mistake in instinct resulting from parasitic castration in *Prosimulium hirtipes* Fries (Diptera: Simuliidae). *Dokl. Akad. Nauk. SSSR* **66**, 305–307 [In Russian].

Guillet, P., Escaffre, M., Ouedraogo, M. and Quillévéré, D. (1981) Mise en évidence d'une résistance au téméphos dans le complexe *Simulium damnosum* [*S. sanctipauli* et *S. soubrense*] en Côte d'Ivoire. (Zone du programme de lutte contre l'onchocercose dans la région du bassin de la Volta). *Cah. OR STOM, Ent. méd. Parasit.* **18** (1980), 291–299.

Gustafsson, M. (1965a). On species of the genus *Entomophthora* Fres. in Sweden. I. Classification and distribution. *Lantbr. Högsk. Annlr.* **31**, 103–212.

Gustafsson, M. (1965b). On species of the genus *Entomophthora* Fres. in Sweden. II. Cultivation and physiology. *Lantbr. Högsk. Annlr.* **31**, 405–457.

Gustafsson, M. (1969). On species of the genus *Entomophthora* Fres. in Sweden. III. Possibility of usage in biological control. *Lantbr. Högsk. Annlr.* **35**, 235–274.

Guttman, D. (1972). The biting activity of black flies (Diptera: Simuliidae) in three types of habitats in Western Colombia. *J. med. Ent.* **9**, 269–276.

Hall, I. M. (1963). Microbial Control. *In* Insect Pathology. An Advanced Treatise, pp. 477–517 (E. A. Steinhaus, Ed.), Vol. 2. New York. Academic Press.

Hamelink, J. L., Waybrant, R. C. and Ball, R. C. (1971). A proposal: exchange equilibria control the degree chlorinated hydrocarbons are biologically magnified in lentic environments. *Trans. Am. Fish. Soc.* **100**, 207–214.

Hamilton, A. L. (1969). On estimating annual production. *Limnol. Oceanogr.* **14**, 771–782.

Hamm, A. H. and Richards, O. W. (1926). The biology of British Crabronidae. *Trans. ent. Soc. Lond.* **74**, 297–331.

Hamon, J. (1974a). Onchocerciasis vectors in the Western Hemisphere. A. Vector biology and vector-parasite relationship. Int. Symp. Research and Control of Onchocerciasis in Western Hemisphere. *PAHO Scient. Publ.* **298**, 58–68.

Hamon, J. (1974b). Onchocerciasis vectors in the Western Hemisphere. B. Vector control aspects. Int. Symp. Research and control of onchocerciasis in Western Hemisphere. *PAHO Scient. Publ.* **298**, 69–84.

Hansen, E. L. and Hansen, J. W. (1976). Parasitism of *Simulium damnosum* by *Romanomermis culicivorax*. *I.R.C.S. Med. Sci.* **4**, 508.

Hargrove, J. W. (1977). Some advances in the trapping of tsetse (*Glossina* spp.) and other flies. *Ecol. Ent.* **2**, 123–137.

Harrison, A. D. (1958). Hydrobiological studies on the Great Berg River, Western Cape Province. Part 2. Quantitative studies on sandy bottoms, notes on tributaries and further information on the fauna arranged systematically. *Trans. R. Soc. S. Afr.* **35**, 227–276.

Harrod, J. J. (1964a). The distribution of invertebrates on submerged aquatic plants in a chalk stream. *J. Anim. Ecol.* **33**, 335–348.

Harrod, J. J. (1964b). The instars of *Simulium ornatum* var. *nitidifrons* Edwards (Dipt., Simuliidae). *Entomologist's mon. Mag.* **100**, 34–35.

Harrod, J. J. (1965). Effect of current speed on the cephalic fans of the larva of *Simulium ornatum* var. *nitidifrons* Edwards (Diptera: Simuliidae). *Hydrobiologia* **26**, 8–12.

Hart, C. W. and Fuller, S. L. (1974). Pollution Ecology of Freshwater Invertebrates. New York. Academic Press.

Hartley, P. H. T. (1948). Food and feeding relationships in a community of fresh-water fishes. *J. Anim. Ecol.* **17**, 1–14.

Hastings, E., Kittams, W. H. and Pepper, J. H. (1961). Repopulation by aquatic insects in streams sprayed with DDT. *Ann. ent. Soc. Am.* **54**, 436–437.

Hatfield, C. H. (1969). Effects of DDT larvicid-ing on aquatic fauna of Bobby's Brook, Labrador. *Can. Fish. Cult.* **40**, 61–72.

Haufe, W. O. (1980). Control of black flies in the Athabasca River: Evaluation and recommendations for chemical control of *Simulium arcticum* Malloch. Alta. Environ. Publ. Edmonton.

Haufe, W. O. and Croome, G. C. R. (Eds.). (1980). Control of black flies in the Athabasca River: Technical Report. An interdisciplinary study for the chemical control of *Simulium arcticum* Malloch in relation to the bionomics of biting flies in the protection of human, animal, and industrial resources and its impact on the aquatic environment. Edmonton. Alta. Environ.

Haufe, W. O., Depner, K. R. and Charnetski, W. A. (1980a). Impact of methoxychlor on drifting aquatic invertebrates. *In* Control of black flies in the Athabasca River: Technical Report, pp. 159–168 (W. O. Haufe and G. C. R. Croome, Eds.). Edmonton. Alta. Environ.

Haufe, W. O., Depner, K. R. and Kozub, G. C. (1980b). Parameters for monitoring displacement of drifting aquatic invertebrates. *In* Control of black flies in the Athabasca River: Technical Report, pp. 169–182 (W. O. Haufe and G. C. R. Croome, Eds.). Edmonton. Alta. Environ.

Häusermann, W. (1969). On the biology of *Simulium damnosum* Theobald, 1903, the main vector of onchocerciasis in the Mahenge Mountains, Ulanga, Tanzania. *Acta Trop.* **26**, 29–69.

Häusermann, W. (1971). *In* A bibliography on diseases and enemies of medically important arthropods 1963–1967. (M. Laird). *In* Microbial Control of Insects and Mites. (H. D. Burges and N. W. Hussey, Eds.). London. Academic Press.

Hawkes, H. A. (1975). River zonation and classification in river ecology. Studies in Ecology 2 (B. A. Whitton, Ed.). Oxford. Blackwell Scientific.

Hazard, E. I. and Oldacre, S. W. (1975). Revision of Microsporidia (Protozoa) close to *Thelohania*, with descriptions of one new family, eight new genera, and thirteen new species. *Agric. Res. Serv., U.S. Dept. Agric. Tech. Bull. No. 1530.*

Hechler, J. (1978). Das "Lutz'sche Organ" der Simuliiden (Diptera). Diss. (Biol.) Hamburg.

Hechler, J. and Rühm, W. (1976). Ergänzende Untersuchungen zur potentiellen Natalität verschiedener Kriebelmückenarten (Simuliidae: Diptera). *Z. angew. Ent.* **81**, 208–214.

Hellawell, J. M. (1971). The autecology of the chub, *Squalius cephalus* (L.), of the River Lugg and Afon Llynfi. III. Diet and feeding habits. *Freshwat. Biol.* **1**, 369–387.

Helson, B. V. (1972). The selective effects of particulate formulation of insecticides on stream fauna when applied as blackfly larvicides. MSc Thesis, Queen's Univ., Kingston, Ontario.

Hildrew, A. G. and Townsend, C. R. (1976). The distribution of two predators and their prey in an iron rich stream. *J. Anim. Ecol.* **45**, 41–57.

Hinton, H. E. (1958). The pupa of the fly *Simulium* feeds and spins its own cocoon. *Entomologist's mon. Mag.* **94**, 14–16.

Hinton, H. E. (1964). The respiratory efficiency of the spiracular gill of *Simulium*. *J. Insect Physiol.* **10**, 73–80.

Hinton, H. E. (1968). Spiracular gills. *Adv. Insect Physiol.* **5**, 65–162.

Hinton, H. E. (1971). Some neglected phases in metamorphosis. *Proc. R. ent. Soc. Lond.* (C) **35**, 55–64.

Hinton, H. E. (1976). The fine structure of the pupal plastron of simuliid flies. *J. Insect Physiol.* **22**, 1061–1070.

Hitchcock, S. W. (1965). Effects of an aerial DDT spray on aquatic insects in Connecticut. *J. econ. Ent.* **53**, 608–611.

Hitchen, C. S. and Goiny, H. H. (1966). Note on the control of *Simulium damnosum* in the region of the Kainji Dam Project, Northern Nigeria. *WHO/ONCHO/66.43*, Mimeogr. Doc.

Hobby, B. M. (1931a). The prey of dung-flies (Diptera: Cordyluridae). *Proc. ent. Soc. Lond.* (A) **6**, 47–49.

Hobby, B. M. (1931b). A list of prey of dung-flies (Diptera: Cordyluridae). *Trans. ent. Soc. S. Engl.* **7**, 35–39.

Hobby, B. M. (1931c). The British species of Asilidae (Diptera) and their prey. *Trans. ent. Soc. S. Engl.* **6**, (1930), 1–42.

Hocking, B. (1950). Further tests of insecticides against black flies (Diptera: Simuliidae) and a control procedure. *Scient. Agric.* **30**, 489–508.

Hocking, B. (1952). Two predators as prey. *Can. Fld Nat.* **66**, 107.

Hocking, B. (1953). The intrinsic range and speed of flight of insects. *Trans. R. ent. Soc. Lond.* **104**, 223–345.

Hocking, B. and Hocking, J. M. (1962). Entomological aspects of African onchocerciasis and observations on *Simulium* in the Sudan. *Bull. Wld Hlth Org.* **27**, 465–472.

Hocking, B. and Pickering, L. R. (1954). Observations on the bionomics of some northern species of Simuliidae (Diptera). *Can. J. Zool.* **32**, 99–119.

Hocking, B. and Richards, W. R. (1952). Biology and control of Labrador black flies (Diptera: Simuliidae). *Bull. ent. Res.* **43**, 237–257.

Hocking, B., Richards, W. R. and Twinn, C. R. (1950). Observations on the bionomics of some northern species of mosquitoes (Culicidae: Diptera). *Can. J. Res.* (D). **28**, 58–80.

Hoffmann, C. C. (1930). Nuevas investigaciones acerca de la transmisión de la oncocercosis de Chiapas. *Revta. méx. Biol.* **10**, 131–140.

Hoffmann, C. C. (1931). Estudios entomológicos y parasitológicos acerca de la oncocercosis en Chiapas. *Salubridad* **3**, 669–697.

Hoffmann, C. H. and Drooz, A. T. (1953). Effect of a C–47 airplane application of DDT on fish-food organisms in two Pennsylvania watersheds. *Am. Midl. Nat.* **50**, 172–188.

Hoffmann, C. H. and Surber, E. W. (1948). Effects of an aerial application of wettable DDT on fish and fish-food organisms in Back Creek, W..Virginia. *Trans. Am. Fish. Soc.* **75**, 48–57.

Hollingworth, R. M. (1971). Comparative metabolism and selectivity of organophosphate and carbamate insecticides. *Bull. Wld Hlth Org.* **44**, 155–170.

Hosoi, T. (1958). Adenosine-5-phosphates as the stimulating agent in blood for inducing gorging of the mosquito. *Nature, Lond.* **181**, 1664–1665.

Howard, L. O. (1887). *In* Riley, C. V. Report of the entomologist. *Rep. U.S. Dept. Agric.* 1886, 459–592.

Howard, L. O. (1888). Note on a *Simulium* common at Ithaca, *Insect Life* **1**, 99–101.

Howard, L. O. (1901). The Insect Book. New York, Doubleday Page and Co.

Hudson, D. K. M. and Hays, K. I.. (1975). Some factors affecting the distribution and abundance of black fly larvae in Alabama. *J. Georgia ent. Soc.* **10**, 110–122.

Hughes, M. H. (1952). Some observations on the bionomics of *Simulium damnosum* Theobald in the southern Gold Coast. *W. Afr. med. J.* (n.s.) **1**, 16–20.

Hughes, M. H. and Daly, P. F. (1951). Onchocerciasis in southern Gold Coast. *Trans. R. Soc. trop. Med. Hyg.* **45**, 243–252.

Hunter, D. M. (1977a). Eclosion and oviposition rhythms in *Simulium ornatipes* (Diptera: Simuliidae). *J. Aust. ent. Soc.* **16**, 215–220.

Hunter, D. M. (1977b). Sugar-feeding in some Queensland black flies (Diptera: Simuliidae). *J. med. Ent.* **14**, 229–232.

Hunter, D. M. (1978). The sequence of events in outbreaks of *Austrosimulium pestilens* Mackerras and Mackerras (Diptera: Simuliidae). *Bull. ent. Res.* **68**, 307–312.

Hunter, D. M. and Moorhouse, D. E. (1976). Sexual mosaics and mermithid parasitism in the *Austrosimulium bancrofti* (Tayl.) (Diptera: Simuliidae). *Bull. ent. Res.* **65**, 549–553.

Hynes, H. B. N. (1941). The taxonomy and ecology of the nymphs of British Plecoptera with notes on the adults and eggs. *Trans. R. ent. Soc. Lond.* **91**, 459–557.

Hynes, H. B. N. (1950). The food of freshwater sticklebacks (*Gasterosteus aculeatus* and *Pygosteus pungitius*), with a review of methods used in studies of the food of fishes. *J. Anim. Ecol.* **19**, 36–58.

Hynes, H. B. N. (1960). A plea for caution in the use of DDT in the control of aquatic insects in Africa. *Ann. trop. Med. Parasit.* **54**, 331–332.

Hynes, H. B. N. (1970a). The Ecology of Running Waters. Toronto, University of Toronto Press.

Hynes, H. B. N. (1970b). The ecology of stream insects. *A. Rev. Ent.* **15**, 25–42.

Hynes, H. B. N. (1974). Further studies on the distribution of stream animals within the substratum. *Limnol. Oceanogr.* **19**, 92–99.

Hynes, H. B. N. (1975). Edgardo Baldi Memorial Lecture. The stream and its valley. *Verh. Int. Verein. Limnol.* **19**, 1–15.

Hynes, H. B. N. and Coleman, M. J. (1968). A simple method of assessing the annual production of stream benthos. *Limnol. Oceanogr.* **13**, 569–573.

Hynes, H. B. N., Williams, D. D. and Williams, N. E. (1976). Distribution of the benthos within the substratum of a Welsh mountain stream. *Oikos* **27**, 307–310.

Hynes, H. B. N. and Williams, T. R. (1962). The effect of DDT on the fauna of a central African stream. *Ann. trop. Med. Parasit.* **56**, 78–91.

Ide, F. P. (1942). Availability of aquatic insects as food of the speckled trout, *Salvelinus fontinalis*. *Trans. 7th N. Amer. Wildl. Conf. Wildl. Inst.* 442–450.

Ide, F. P. (1956). Effect of forest spraying with DDT on aquatic insects of salmon streams. *Trans. Am. Fish. Soc.* **86**, 208–219.

Ide, F. P. (1967). Effects of forest spraying with DDT on aquatic insects of salmon streams in New Brunswick. *J. Fish. Res. Bd Can.* **24**, 769–805.

Idyll, C. (1942). Food of rainbow, cutthroat and brown trout in the Cowichan River system, B.C. *J. Fish. Res. Bd. Can.* **5**, 448–458.

Issi, I. V. (1968). *Stempellia rubtsovi* sp.n. (Microsporidia: Nosematidae), a microsporidian parasite of *Odagmia caucasica* larvae (Diptera: Simuliidae). *Acta Protozool.* **6**, 345–352.

Ivashchenko, L. A. (1977). The effect of oxygen and light on embryonic development and time of hatching of simuliid larvae (Diptera: Simuliidae). *Medskaya Parazit.* **46**, 37–41.

James, H. G. (1951). Notes on a preliminary survey of natural control of biting flies (Diptera: Culicidae, Tabanidae) at Churchill, Manitoba. *Rep. Quebec Soc. Protect. Plants* **32–33**, 119–121.

Jamnback, H. (1973). Recent developments in control of blackflies. *A. Rev. Ent.* **18**, 281–304.

Jamnback, H. (1976). Simuliidae (Blackflies) and their control. IV. *WHO/VBC/76.653.* Mimeogr. Doc.

Jamnback, H., Duflo, T. and Marr, D. (1970). Aerial application of larvicides for control of *Simulium damnosum* in Ghana: a preliminary trial. *Bull. Wld Hlth Org.* **42**, 826–828.

Jamnback, H. and Frempong-Boadu, J. (1966). Testing blackfly larvicides in the laboratory and in streams. *Bull. Wld Hlth Org.* **34**, 405–421.

Jamnback, H. and West, A. S. (1970). Decreased susceptibility of blackfly larvae to p, p'-DDT in New York State and eastern Canada. *J. econ. Ent.* **63**, 218–221.

Jedlička, L. (1978a). Influence of ecological variability on correlation of larval characters in *Odagmia ornata* (Meigen, 1818) and *Odagmia spinosa* (Doby et Deblock, 1957) (Diptera: Simuliidae), pp. 111–117 *in* I. Orszagh (Ed.). Dipterologica Bohemoslovaca 1. Veda, Bratislava.

Jedlička, L. (1978b). Variability of some characters in *Odagmia ornata* (Meigen, 1818) and *Odagmia spinosa* (Doby et Deblock, 1957) (Diptera: Simuliidae). *Acta Fac. Rerum. nat. Univ. comen.* **23**, 23–76.

Jenkins, D. W. (1964). Pathogens, parasites and predators of medically important arthropods. (Annotated list and bibliography). *Bull. Wld Hlth Org.* **30** (Suppl.), 1–150.

Jenkins, T. M. Jr., Feldmeth, C. R. and Elliott, G. V. (1970). Feeding of rainbow trout (*Salmo gairdneri*) in relation to abundance of drifting invertebrates in a mountain stream. *J. Fish. Res. Bd Can.* **27**, 2356–2361.

Jensen, L. D. and Gaufin, A. R. (1964a). Effects of ten organic insecticides on two species of stonefly naiads. *Trans. Am. Fish. Soc.* **93**, 27–34.

Jensen, L. D. and Gaufin, A. R. (1964b). Long-term effect of organic insecticides on two species of stonefly naiads. *Trans. Am. Fish. Soc.* **93**, 357–363.

JICA (1975a). Report of the first mission (Preliminary survey team) for the cooperation of onchocerciasis control pilot project in Guatemala. *JICA*.

JICA (1975b). Report of the second mission (Implementation survey team) for the cooperation of onchocerciasis control pilot project in Guatemala. *JICA*.

JICA (1977). Report of the third mission (Operation and review survey team) for the cooperation of onchocerciasis control pilot project in Guatemala. *JICA*.

JICA (1978). Onchocerciasis control project in Guatemala, First report. *JICA*.

Jírovec, O. (1943). Revision der in *Simulium*-Larven parasitierenden Mikrosporidien. *Zool. Anz.* **142**, 173–179.

Johannsen, O. A. (1903). Aquatic nematocerous Diptera. Part 6, p. 343. *In* Aquatic Insects of New York State (J. S. Needham, A. D. MacGillivray, O. A. Johannsen, and K. C. Davis, Eds.). *Bull. N.Y. St. Mus.* **68** (Entomol. 18), 199–517.

Johnson, A. F. (1969). Studies on the biology of black-fly larvae (Diptera: Simuliidae) with reference to the structure and function of the feeding organs. Unpubl. ms.

Johnson, B. T., Saunders, C. R., Sanders, H. O. and Campbell, R. S. (1971). Biological magnification and degradation of DDT and Aldrin by freshwater invertebrates. *J. Fish. Res. Bd Can.* **28**, 705–709.

Jones, J. R. E. (1949). A further ecological study of calcareous streams in the "Black Mountain" district of South Wales. *J. Anim. Ecol.* **18**, 142–159.

Jones, J. R. E. (1950). A further ecological study of the River Rheidol: the food of the common insects of the main-stream. *J. Anim. Ecol.* **19**, 159–174.

Jordan, A. M. (1974). Recent developments in the ecology and methods of control of tsetse flies (*Glossina* spp.) (Dipt., Glossinidae). A Review. *Bull. ent. Res.* **63**, 361–399.

Jost, O. (1975). Zur Ökologie der Wasseramsel (*Cinclus cinclus*) mit besonderer Berücksichtigung ihrer Ernährung. *Bonn. zool. Monogr.* **6**, 1–183.

Kaneko, K., Saito, K. and Wonde, T. (1973). Observations on the diurnal rhythm of the biting activity of *Simulium damnosum* in Omo-Gibe and Gojjeb Rivers, south-west Ethiopia. *Jap. J. sanit. Zool.* **24**, 175–180.

Kapoor, I. P., Metcalf, R. L., Nystrom, R. F. and Sangha, G. K. (1970). Comparative metabolism of methoxychlor, methiochlor and DDT in mouse, insects and in a model ecosystem. *J. Agric. Food Chem.* **18**, 1145–1152.

Kaushik, N. K. and Hynes, H. B. N. (1971). The fate of the dead leaves that fall into streams. *Arch. Hydrobiol.* **68**, 465–515.

Keast, A. (1966). Trophic interrelationships in the fish fauna of a small stream. *Great Lakes Res. Div., Univ. Michigan Publ.* **15**, 51–79.

Keenleyside, M. H. A. (1967). Effects of forest spraying with DDT in New Brunswick on food of young Atlantic salmon. *J. Fish. Res. Bd Can.* **24**, 807–822.

Kenaga, E. E. (1974). Partitioning and uptake of pesticides in biological systems. *Proc. Int. Conf. Transport of Persistent Chemicals in Aquatic Ecosystems, Ottawa, Canada II*, 19–22.

Kershaw, W. E., Williams, T. R., Frost, S. and Hynes, H. B. (1965). Selective effect of particulate insecticides on *Simulium* among stream fauna. *Nature, Lond.* **208**, 199.

Kershaw, W. E., Williams, T. R., Frost, S., Matchett, R. L., Mills, M. E. and Johnson, R. D. (1968). The selective control of *Simulium* larvae by particulate insecticides and its significance in river management. *Trans. R. Soc. trop. Med. Hyg.* **62**, 35–40.

Kerswill, C. J. (1967). Studies on effects of forest sprayings with insecticides, 1952–63, on fish and aquatic invertebrates in New Brunswick streams: Introduction and summary. *J. Fish. Res. Bd Can.* **24**, 701–708.

Khan, M. A. (1980). Protection of cattle from black flies. *In* Control of black flies in the Athabasca River: Technical Report, pp. 217–232 (W. O. Haufe and G. C. R. Croome, Eds.). Edmonton. Alta. Environ.

Khan, R. A. and Fallis, A. M. (1970). Life cycles of *Leucocytozoon dubreuili* Mathis and Leger, 1911 and *L. fringillinarum* Woodstock, 1910 (Haemosporidia: Leucocytozoidae). *J. Protozool.* **17**, 642–658.

Kirschfeld, K. and Wenk, P. (1976). The dorsal compound eye of simuliid flies: An eye specialized for the detection of small, rapidly moving objects. *Z. Naturf.* **31**(C), 764–765.

Knoll, F. (1926). Insekten und Blumen. Experimentelle Arbeiten zur Vertiefung unserer Kenntnisse über die Wechselbeziehungen zwischen Pflanzen und Tieren. IV. Die Arum-Blütenstände und ihre Besucher. *Abh. zool.-bot. Ges. Wien* **12**, 379–481.

Knoz, J. (1965). To identification of Czechoslovakian black-flies (Diptera: Simuliidae). *Folia přírod. Fak. Univ. Purkyně* **6**(5), 1–52.

Kolesnikov, V. A. (1977). The sphecid wasps (Hymenoptera, Sphecidae) of Bryansk Province as entomophages. *Ent Obozr.* **56**, 315–325. (In Russian: translation *Ent. Rev., Wash.* **56**(2), 57–65).

Konurbayev, E. O. (1978). The ecological classification of running waters in Soviet Central Asia and the distribution pattern of black flies (Diptera: Simuliidae) in water courses of different types. *Ent. obozr.* **56**, 736–750 (In Russian: translation *Ent. Rev., Wash.* **56**(4), 17–27).

Krombein, K. V. (1960). Biological notes on some Hymenoptera that nest in sumach pith. *Ent. News* **71**, 63–69.

Kureck, A. (1969). Tagesrhythmen lappländischer Simuliiden (Diptera). *Oecologia* **2**, 385–410.

Kurtak, D. C. (1974). Overwintering of *Simulium pictipes* Hagen (Diptera: Simuliidae) as eggs. *J. med. Ent.* **11**, 383–384.

Kurtak, D. C. (1978). Efficiency of filter-feeding of blackfly larvae (Diptera: Simuliidae). *Can. J. Zool.* **56**, 1608–1623.

Lacey, L. A. and Mulla, M. S. (1977). Evaluation of *Bacillus thuringiensis* as a biocide of blackfly larvae (Diptera: Simuliidae). *J. Invert. Pathol.* **30**, 46–49.

Lacey, L. A. and Mulla, M. S. (1978). Factors affecting the activity of diflubenzuron against *Simulium* larvae (Diptera: Simuliidae). *Mosquito News* **38**, 264–268.

Ladle, M., Bass, J. A. B. and Jenkins, W. R. (1972). Studies on production and food consumption by the larval Simuliidae (Diptera) of a chalk stream. *Hydrobiologia* **39**, 429–448.

Ladle, M., Bass, J. A. B., Philpott, F. R. and Jeffery, A. (1977). Observations on the ecology of Simuliidae from the River Frome, Dorset. *Ecol. Ent.* **2**, 197–204.

Laird, M. (1977a). Osiris, Asklepios, and the Harpies. The development of an African River basin. *In* A Time to Hear and Answer. Essays for the Bicentennial Season, pp. 103–140. The Franklin Lectures in the Sciences and Humanities, 4th Series. Auburn, Univ. of Alabama Press.

Laird, M. (1977b). Review of Weiser, J. "An Atlas of Insect Diseases". (2nd revised ed.). *J. Parasit.* **63**, 1114–1115.

Laird, M. (1978). The status of biocontrol investigations concerning Simuliidae. *Envir. Conserv.* **5**, 133–142.

Laird, M., Hansen, E. L. and Hansen, J. W. (1978). Recent developments concerning the potentialities of biological control against *Simulium damnosum*. *Abst. Fourth Int. Cong. Parasit., Warsaw*.

Laird, M., Mokry, J. and Noah, R. (1974). A benthobservatory for studies of the biology of larval Simuliidae. *Proc. III Int. Congr. Parasit.* **2**, 912–913.

Lakon, G. (1919). Die Insektenfeinde aus der Familie der Entomophthoreen. Beiträge zu einer Monographie der insektentötenden Pilze. *Z. angew. Ent.* **5**, 161–216.

Lamontellerie, M. (1963). Observations sur *Simulium adersi* Pomeroy en zone de savane sèche (Région de Garango, Haute-Volta). *Bull. Inst. fr. Afr. noire* (A)**25**, 467–484.

Lamontellerie, M. (1964). *Simulium damnosum* Theobald (Diptera: Simuliidae) en zone de savane sèche (Région de Garango, Haute-Volta). I. Réinfestation de la Volta-Blanche en début de saison des pluies. *Bull. Inst. fr. Afr. noire* (A)**26**, 1298–1312.

Lamontellerie, M. (1965). *Simulium damnosum* Théobald en zone de savane sèche (Région de Garango, Haute-Volta). II. Infestation par *Onchocerca volvulus* Leuckart. *Bull. Inst. fr. Afr. noire* (A)**27**, 219–228.

Lamontellerie, M. (1967). Captures de diptères Simuliidae de nuit en zone de savane sèche. *Bull. Inst. fr. Afr. noire* (A)**29**, 1812–1832.

Lamotte, M. and Bertrand, J.-Y. (1975). Rapport sur l'étude de quelques effets de l'abate—larvicide anti-simulies—sur la faune non cible du Bandama (Lamto, Côte d'Ivoire). Rapport à l'OMS, réf, OMS V2–181–80. Mimeogr. Doc.

Landau, R. (1962). Four forms of *Simulium tuberosum* (Lundstr.) in southern Ontario: a salivary gland chromosome study. *Can. J. Zool.* **40**, 921–939.

Latgé, J.-P., Soper, R. S. and Madore, C. D.

(1977). Media suitable for industrial production of *Entomophthora virulenta* zygospores. *Biotech. Bioeng.* **19**, 1269–1284.

Laurence, B. R. (1966). Intake and migration of the microfilariae of *Onchocerca volvulus* (Leuckart) in *Simulium damnosum* Theobald. *J. Helminth.* **40**, 337–342.

Lauzanne, L. and DeJoux, C. (1973). Étude de terrain de la toxicité sur la faune aquatique non cible de nouveaux insecticides employés en lutte anti-simulies. *ORSTOM, Bouaké Rap.* 15. Mimeogr. Doc.

Laws, E. R., Sedlak, V. A., Miles, J. W., Joseph, C. R., Lacomba, J. R. and Rivera, A. D. (1968). Field study of the safety of Abate for treating potable water and observations on the effectiveness of a control programme involving both Abate and Malathion. *Bull. Wld Hlth Org.* **38**, 439–445.

Lea, A. O. and Dalmat, H. T. (1955a). A pilot study of area larval control of black flies in Guatemala. *J. econ. Ent.* **48**, 378–383.

Lea, A. O. and Dalmat, H. T. (1955b). Field studies on larval control of black flies in Guatemala. *J. econ. Ent.* **48**, 274–278.

Le Berre, R. (1966). Contribution à l'étude biologique et écologique de *Simulium damnosum* Theobald, 1903 (Diptera: Simuliidae). *Mem. ORSTOM* **17**, 1–204.

Le Berre, R. (1968). Bilan sommaire pour 1967 de lutte contre le vecteur de l'onchocercose. *Méd. Afr. Noire* **15**, 71–72.

Le Berre, R., Balay, G., Brengues, J. and Coz, J. (1964). Biologie et écologie des femelles de *Simulium damnosum* Theobald 1903, en fonction des zones bioclimatiques d'Afrique occidentale. Influence sur l'épidemiologie de l'onchocercose. *Bull. Wld Hlth Org.* **31**, 843–855.

Le Berre, R., Philippon, B., Grébaut, S., Séchan, Y., Lenormand, J., Etienne, J. and Garreta, P. (1976). Control of *Simulium damnosum*, the vector of human onchocerciasis in West Africa. I. Supplementary trials of new insecticides. *WHO/VBC/76.614.* Mimeogr. Doc.

Le Berre, R. and Wenk, P. (1966). Beobachtungen über das Schwarmverhalten bei *Simulium damnosum* (Theobald) in Obervolta und Kamerun. *Verh. dt. zool. Ges.* **24**, 367–372.

Leclercq, J. (1954). Monographie systématique, phylogénétique et zoogéographique des Hyménoptères Crabroniens. Thèse, Fac. Sci., Univ. Liège, Belgique.

Lee, C. W. (1973). Aerial spraying trials in West Africa for blackfly control. *PANS* **19**, 190–192.

Lee, C. W., Parker, J. D., Philippon, B. and Baldry, D. A. T. (1973). Experiments in Ivory Coast with a prototype rapid-release system fitted to a Pilatus Porter aircraft to apply larvicide for the control of *Simulium damnosum*. *WHO/VBC/73.18.* Mimeogr. Doc.

Léger, L. (1897). Sur une nouvelle myxosporidie de la famille des glugeïdées. *C.r. hebd. Séanc. Acad. Sci., Paris.* **125**, 260–262.

Levander, K. M. (1923). Simulier som forellföda. *Notul. ent.* **3**, 57.

Levchenko, N. G. and Issi, I. V. (1973). Microsporida in bloodsucking arthropods. *In* Regulators of Numbers of Bloodsucking Flies in the South-east of Kazakhstan. (A. M. Dubitskii, Ed.). (English version *in press*, 1981, *Medical Library, Natl Insts Hlth, Bethesda, MD, USA*).

Lévêque, C., Odei, M. and Pugh Thomas, M. (1977). The Onchocerciasis Control Programme and the monitoring of its effect on the riverine biology of the Volta River Basin. *In* Linnean Society Symposium Series No. 5, Ecological Effects of Pesticides, pp. 133–143 (F. H. Perring and K. Mellanby, Eds.). London and New York. Academic Press.

Levy, R. and Miller, T. W., Jr. (1977). Susceptibility of the mosquito nematode *Romanomermis culicivorax* (Mermithidae) to pesticides and growth regulators. *Envir. Ent.* **6**, 447–448.

Lewis, D. J. (1953). *Simulium damnosum* and its relation to onchocerciasis in the Anglo-Egyptian Sudan. *Bull. ent. Res.* **43**, 597–644.

Lewis, D. J. (1958). Observations on *Simulium damnosum* Theobald at Lokoja in Northern Nigeria. *Ann. trop. Med. Parasit.* **52**, 216–231.

Lewis, D. J. (1960). Observations on *Simulium damnosum* in the southern Cameroons and Liberia. *Ann. trop. Med. Parasit.* **54**, 208–223.

Lewis, D. J. (1965). Features of the *Simulium damnosum* population of the Kumba area in West Cameroon. *Ann. trop. Med. Parasit.* **59**, 365–374.

Lewis, D. J. (1973). The Simuliidae (Diptera) of Pakistan. *Bull. ent. Res.* **62**, 453–470.

Lewis, D. J. (1974). A review of aerial control of blackflies with reference to tropical Africa. *WHO/VBS/74.471.* Mimeogr. Doc.

Lewis, D. J. and Bennett, G. F. (1974a). An artificial substrate for the quantitative comparison of the densities of larval simuliid (Diptera) populations. *Can. J. Zool.* **52,** 773–775.

Lewis, D. J. and Bennett, G. F. (1974b). The blackflies (Diptera: Simuliidae) of insular Newfoundland. II. Seasonal succession and abundance in a complex of small streams on the Avalon Peninsula. *Can. J. Zool.* **52,** 1107–1113.

Lewis, D. J. and Bennett, G. F. (1975). The blackflies (Diptera: Simuliidae) of insular Newfoundland. III. Factors affecting the distribution and migration of larval simuliids in small streams on the Avalon Peninsula. *Can. J. Zool.* **53,** 114–123.

Lewis, D. J. and Ibañez de Aldecoa, R. (1962). Simuliidae and their relation to human onchocerciasis in northern Venezuela. *Bull. Wld Hlth Org.* **27,** 449–464.

Lewis, D. J. and Duke, B. O. L. (1966). *Onchocerca-Simulium* complexes. II-Variation in West African female *Simulium damnosum. Ann. trop. Med. Parasit.* **60,** 337–346.

Lewis, D. J., Lyons, G. R. L. and Marr. J. D. M. (1961). Observations on *Simulium damnosum* from the Red Volta in Ghana. *Ann. trop. Med. Parasit.* **55,** 202–210.

Lewis, D. J. and Raybould, J. N. (1974). The subgenus *Lewisellum* of *Simulium* in Tanzania (Diptera: Simuliidae). *Revue zool. Afr.* **88,** 225–240.

Lewis, D. J. and Stam, A. B. (1968). Simuliidae (Diptera) form Mbandaka, Congo (Kinshasa). *Entomologist's mon. Mag.* **104,** 91–95.

Lichovoz, L. K. (1975). The combined parasitism of larvae of simuliids by mermithids and Microsporidia. *Medskya Parazit.* **44,** 230–233. [In Russian].

Lichovoz, L. K. (1978). Mermithids (Nematoda: Mermithidae) of blackflies (Diptera: Simuliidae) in the western regions of the Ukraine SSR. *XI Ann. Mtg. Soc. Invert. Pathol. (Abst.),* Prague, p. 71.

Linduska, J. P. and Surber, E. W. (1947). Effects of DDT and other insecticides on fish and wildlife, summary of investigations during 1947. *Fishery Leafl. Fish. Wildl. Serv. US.* **15,** 1–19.

Linley, J. R. and Carlson, D. A. (1978). A contact mating pheromone in the biting midge, *Culicoides melleus. Insect. Physiol.* **24,** 423–427.

Liu, T. P. (1972). Ultrastructural changes in the nuclear envelope of larval fat body cells

of *Simulium vittatum* (Diptera) induced by microsporidian infection of *Thelohania bracteata. Tissue Cell* **4,** 493–501.

Liu, T. P. (1973). Ultrastructure of the yolk protein granules in the frozen-etched oocyte of an insect. *Cytobiologie* **7,** 33–41.

Liu, T. P. (1974). Ultrastructure of the lipid inclusions of the yolk in the freeze-etched oocyte of an insect. *Cytobiologie* **8,** 412–420.

Liu, T. P., Darley, J. J. and Davies, D. M. (1975). Differentiation of ovariolar follicular cells and formation of previtelline-membrane substance in *Simulium vittatum* Zetterstedt (Diptera: Simuliidae). *Inst. J. Insect Morph. Embryol.* **4,** 331–340.

Liu, T. P. and Davies, D. M. (1971). Ultrastructural localization of glycogen in the flight muscle of the blackfly, *Simulium vittatum* Zett. *Can. J. Zool.* **49,** 219–221.

Liu, T. P. and Davies, D. M. (1972a). Ultrastructure of the nuclear envelope from blackfly fat-body cells with or without microsporidian infection. *J. Invert. Pathol.* **20,** 176–182.

Liu, T. P. and Davies, D. M. (1972b). Ultrastructure of the cytoplasm in fat-body cells of the blackfly, *Simulium vittatum,* with microsporidian infection; a freeze-etching study. *J. Invert. Pathol.* **19,** 208–214.

Liu, T. P. and Davies, D. M. (1972c). Ultrastructural localization of glutamic oxaloacetic transaminase in mitochondria of the flight muscle of Simuliidae. *J. Insect. Physiol.* **18,** 1665–1671.

Liu, T. P. and Davies, D. M. (1972d). An autoradiographic and ultrastructural study of glycogen metabolism and function in the adult fat body of a black-fly during oogenesis. *Entomologia exp. appl.* **15,** 265–273.

Liu, T. P. and Davies, D. M. (1972e). Ultrastructure of protein and lipid inclusions in frozen-etched black-fly oöcytes (Simuliidae: Diptera). *Can. J. Zool.* **50,** 59–62.

Liu, T. P. and Davies, D. M. (1972f). Fine structure of frozen-etched lipid granules in the fat body of an insect. *J. Lipid Res.* **13,** 115–118.

Liu, T. P. and Davies, D. M. (1973). Intramitochondrial transformations during lipid vitellogenesis in oocytes of a blackfly, *Simulium vittatum* Zetterstedt (Diptera: Simuliidae). *Int. J. Insect Morph. Embryol.* **2,** 233–245.

Lockhart, W. L. (1980). Methoxychlor studies

with fish: Athabasca River exposures and experimental exposures. *In* Chemical control of the blackfly *Simulium arcticum* Malloch in the Athabasca River, pp. 183–196 (W. O. Haufe and G. C. R. Croome, Eds.). Edmonton. Alta. Environ.

Lockhart, W. L., Metner, D. A. and Solomon, J. (1977). Methoxychlor residue studies in caged and wild fish from the Athabasca River, Alberta, following a single application of a blackfly larvicide. *J. Fish. Res. Bd Can.* **34**, 626–632.

Longstaff, T. G. (1932). An ecological reconnaissance in West Greenland. *J. Anim. Ecol.* **1**, 119–142.

Lowther, J. K. and Wood, D. M. (1964). Specificity of a black fly, *Simulium euryadminiculum* Davies, towards its host, the common loon. *Can. Ent.* **96**, 911–913.

Luckey, T. D. (1968). Insect hormoligosis. *J. econ. Ent.* **61**, 7–12.

Lugger, O. (1896). Insects injurious in 1896. *Bull. Minn. agric. Exp. Stn. Ent. Div.* **48**.

Lutz, A. (1910). Segunda contribuição para o conhecimento das especies brazileiras do genero "*Simulium*". *Mems. Inst. Oswaldo Cruz* **2**, 213–267.

Macan, T. T. (1962). Ecology of aquatic insects. *A. Rev. Ent.* **7**, 261–288.

MacArthur, R. H. and Wilson, E. O. (1967). The Theory of Island Biogeography. Princeton, N.H. Princeton Univ. Press.

Maciolek, J. A., Tunzi, M. G. (1968). Microseston dynamics in a simple Sierra Nevada lake-stream system. *Ecology* **49**, 60–75.

Mackereth, J. C. (1957). Notes on the Plecoptera from a stony stream. *J. Anim. Ecol.* **26**, 343–351.

Mackerras, M. J. and Mackerras, I. M. (1948). Simuliidae (Diptera) from Queensland. *Aust. J. scient. Res.* (B)**1**, 231–270.

Magor, I. J., Rosenberg, L. J. and Pedgley, D. E. (1975). Windborne movement of *Simulium damnosum*. Final report of the WHO-COPR studies in 1975. *COPR 39/3/3.* Mimeogr. Doc.

Maitland, P. S. (1965). The feeding relationships of salmon, trout, minnows, stone loach and three-spined sticklebacks in the River Endrick, Scotland. *J. Anim. Ecol.* **34**, 109–133.

Maitland, P. S. and Penney, M. M. (1967). The ecology of the Simuliidae in a Scottish river. *J. Anim. Ecol.* **36**, 179–206.

Mallén, M. S. (1974). Onchocerciasis in Mexico. Int. Symp. Research and Control of Onchocerciasis in the Western Hemisphere. *PAHO Scient. Publ.* **298**, 112–115.

Malone, K. M. and Nolan, R. A. (1978). Aerobic bacterial flora of the larval gut of the black fly *Prosimulium mixtum* (Diptera: Simuliidae) from Newfoundland, Canada. *J. med. Ent.* **14**, 641–645.

Mann, R. H. K. (1976). Observations on the age, growth, reproduction and food of the chub, *Squalius cephalus* (L.) in the River Stour, Dorset. *J. Fish. Biol.* **8**, 265–288.

Mann, R. H. K. and Orr, D. R. O. (1969). A preliminary study of the feeding relationships of fish in a hard-water and a soft-water stream in southern England. *J. Fish. Biol.* **1**, 31–44.

Mansingh, A. (1971). Physiological classification of dormancies in insects. *Can. Ent.* **103**, 983–1009.

Mansingh, A. and Steele, R. W. (1973). Studies on insect dormancy. 1. Physiology of hibernation in larvae of the blackfly *Prosimulium mysticum* Peterson. *Can. J. Zool.* **51**, 611–618.

Mansingh, A., Steele, R. W. and Helson, B. V. (1972). Hibernation in the blackfly *Prosimulium mysticum*: quiescence or oligopause? *Can. J. Zool.* **50**, 31–34.

Marlier, C. (1952). Fish feeding on *Simulium* larvae. *Nature, Lond.* **170**, 496.

Marr, J. D. M. (1962). The use of an artificial breeding site and cage in the study of *Simulium damnosum* Theobald. *Bull. Wld Hlth Org.* **27**, 622–629.

Marr, J. D. M. (1971). Observations on resting *Simulium damnosum* (Theobald) at a dam site in northern Ghana. *WHO/ONCHO/71.85, WHO/VBC/71.298.* Mimeogr. Doc.

Marr, J. D. M. and Lewis, D. J. (1963). Colour variation in *Simulium damnosum* (description of exhibit). *Trans. R. Soc. trop. Med. Hyg.* **57**, 7.

Marr, J. D. M. and Lewis, D. J. (1964). Observations on the dry-season survival of *Simulium damnosum* Theo. in Ghana. *Bull. ent. Res.* **55**, 547–564.

Martof, B. S. (1962). Some aspects of the life history and ecology of the salamander *Leurognathus*. *Am. Midl. Nat.* **67**, 1–35.

Martof, B. S. and Scott. D. C. (1957). The food of the salamander *Leurognathus*. *Ecology* **38**, 494–501.

Matsuo, K. *et al.* (1978). Taxonomic and ecological investigations of onchocerciasis vector in Guatemala. *Onchocerciasis control project in Guatemala*, First report. *JICA* 13–54

Maurand, J. (1975). Les microsporidies des larves de simulies: systématique, données cytochimiques, pathologiques et écologiques. *Annls. Parasit. hum. comp.* **50,** 371–396.

Maurand, J. and Manier, J. F. (1968). Actions histopathologiques comparées de parasites coelomiques des larves de simulies (Chytridiales: Microsporidies). *Annls. Parasit. hum. comp.* **43,** 79–85.

Mayr, E. (1969). Principles of Systematic Zoology. New York. McGraw-Hill.

McAtee, W. L. (1932). Effectiveness in nature of the so-called protective adaptations in the animal kingdom, chiefly as illustrated by the food habits of nearctic birds. *Smithson. misc. Collns.* **85**(7), 1–201.

McAtee, W. L. (1934). *In* Bequaert, J. C. (1934).

McCormack, J. C. (1962). The food of young trout (*Salmo trutta*) in two different becks. *J. Anim. Ecol.* **31,** 305–316.

McCrae, A. W. R. (1966). Hertig [sic, error for Herting] "window traps" trials. *Rep. E. Afr. Virus Res. Inst.* **16,** 36.

McCrae, A. W. R. (1967). The *Simulium damnosum* species complex. *Rep. E. Afr. Virus Res. Inst.* **17,** 67–70.

McCrae, A. W. R. (1968). Considerations of *Simulium damnosum* Theo. as a species complex and its relevance to control. *Proc. VIII Int. Congr. trop. Med. Malar., Teheran,* 135–136.

McCrae, A. W. R. (1969). Ecology and speciation in African blackflies (Diptera: Simuliidae). *Biol. J. Linn. Soc.* **1,** 43–49.

McCrae, A. W. R. and Manuma, P. (1967). Trials of modified Fredeen traps baited with dry ice. *Rep. E. Afr. Virus Res. Inst.* **17,** 65–66.

McMahon, J. P. (1967). A review of the control of *Simulium* vectors of onchocerciasis. *Bull. Wld Hlth Org.* **37,** 415–430.

McMahon, J. P. (1968). Artificial feeding of *Simulium* vectors of human and bovine onchocerciasis. *Bull. Wld Hlth Org.* **38,** 957–966.

McMahon, J. P. (1971). Report on the visit to the endemic areas of Robles disease (Onchocerciasis) in Guatemala, 8–25 September 1971. *PAHO.* Mimeogr. Doc.

McMahon, J. P., Highton, R. B. and Goiny, H. (1958). The eradication of *Simulium neavei* from Kenya. *Bull. Wld Hlth Org.* **19,** 75–107.

Mecom, J. O. (1972). Feeding habits of Trichoptera in a mountain stream. *Oikos* **23,** 401–407.

Mellanby, K. (1976). Pesticides, the environment, and the balance of nature. *In* Pesticides and human welfare, pp. 217–227 (D. Gunn and J. Stevens, Eds.). Oxford. Oxford Univ. Press.

Mercer, K. L. and McIver, S. B. (1973a). Studies on the antennal sensilla of selected blackflies (Diptera: Simuliidae). *Can. J. Zool.* **51,** 729–734.

Mercer, K. L. and McIver, S. B. (1973b). Sensilla on the palps of selected blackflies (Diptera: Simuliidae). *J. med. Ent.* **10,** 236–239.

Merna, J. W., Bender, M. E. and Novy, J. R. (1972). The effects of methoxychlor on fishes. I. Acute toxicity and breakdown studies. *Trans. Am. Fish. Soc.* **101,** 298–301.

Merritt, R. W., Mortland, M. M., Gersabeck, E. F. and Ross, D. H. (1978). X-ray diffraction of particles ingested by filter-feeding animals. *Entomologia exp. appl.* **24,** 27–34.

Merritt, R. W., Ross, D. H. and Peterson, B. V. (1978). Larval ecology of some Lower Michigan black flies (Diptera: Simuliidae) with keys to the immature stages. *Gt. Lakes Ent.* **11,** 177–208.

Metcalf, R. L. (1977). Model ecosystem approach to insecticide degradation: A critique. *A. Rev. Ent.* **22,** 241–261.

Metcalf, R. L. and McKelvey, J. (Eds.). (1976). The Future for Insecticides—Needs and Prospects. New York. Wiley.

Metcalf, R. L., Sangha, G. and Kapoor, I. (1971). Model ecosystem for the evaluation of pesticide biodegradability and ecological magnification. *Envir. Sci. Technol.* **5,** 709–713.

Miall, L. C. (1895). The Natural History of Aquatic Insects. London. MacMillan.

Miall, L. C. (1912). The Natural History of Aquatic Insects. London. MacMillan.

Millais, J. G. (1902). The natural history of British surface-feeding ducks. *In* The Ducks, Geese and Swans of North America by Kortright, F. H. (1942). *Amer. Wildl. Inst. Wash.*

Miller, J. M. (1974). The food of brook trout *Salvelinus fontinalis* (Mitchill) fry from different subsections of Lawrence Creek, Wisconsin. *Trans. Am. Fish. Soc.* **103,** 130–134.

Miller, R. C. and Kurczewski, F. E. (1975). Comparative behavior of wasps in the genus *Lindenius* (Hymenoptera: Sphecidae, Carboninae). *Jl. N.Y. ent. Soc.* **83,** 82–120.

Mills, A. R. (1969). A quantitative approach to the epidemiology of onchocerciasis in West Africa. *Trans. R. Soc. trop. Med. Hyg.* **63**, 591–602.

Mills, D. H. (1964). Ecology of the young stages of Atlantic salmon (*Salmo salar* L.) in the River Bran, Ross-shire. PhD Thesis. Univ. London.

Milne, M. D., Schriber, M. A. and Crawford, M. A. (1958). Non-ionic diffusion and the excretion of weak acids and bases. *Am. J. Med.* **24**, 709–729.

Minář, J. (1962). Observations on the effect of meteorological factors on some parasitic Diptera (Ceratopogonidae: Simuliidae: Tabanidae). *Čslká Parasit.* **9**, 331–342.

Minkiewicz, R. (1932). Nids et proies de sphégiens de Pologne. *Polskie Pismo ent.* **11**, 98–112.

Minshall, G. W. (1978). Autotrophy in stream ecosystems. *Bioscience* **28**, 767–771.

Mokry, J. E. (1975). Studies on the ecology and biology of blackfly larvae utilizing an *in situ* benthobservatory. *Verh. Internat. Verein. Limnol.* **19**, 1546–1549.

Mokry, J. E. (1976a). Laboratory studies on the larval biology of *Simulium venustum* Say (Diptera: Simuliidae). *Can. J. Zool.* **54**, 1657–1663.

Mokry, J. E. (1976b). A simplified membrane technique for feeding blackflies (Diptera: Simuliidae) on blood in the laboratory. *Bull. Wld Hlth Org.* **53**, 127–129.

Mokry, J. E. (1978). Progress towards the colonization of *Cnephia mutata* (Diptera: Simuliidae) *Bull. Wld Hlth Org.* **56**, 455–456.

Mokry, J. E. (1980a). Laboratory studies on blood-feeding of blackflies (Diptera: Simuliidae) I. Factors affecting the feeding rate. *Tropenmed. Parasit.* **31** 367–373.

Mokry, J. E. (1980b). Laboratory studies on blood-feeding of blackflies (Diptera: Simuliidae). II. Factors affecting the fecundity rate. *Tropenmed. Parasit.* **31** 374–380.

Mokry, J. E. and Finney, J. R. (1977). Notes on mermithid parasitism of Newfoundland blackflies, with the first record of *Neomesomermis flumenalis* from adult hosts. *Can. J. Zool.* **55**, 1370–1372.

Molloy, D. (1979). Description and bionomics of *Neomesomermis camdenensis* sp.n. (Nematoda: Mermithidae), a parasite of blackflies (Diptera: Simuliidae). *J. Nematology* **11**, 321–328.

Molloy, D. and Jamnback, H. (1975). Laboratory transmission of mermithids parasitic in blackflies. *Mosquito News.* **35**, 337–342.

Molloy, D. and Jamnback, H. (1977). A larval blackfly control field trial using mermithid parasites and its cost implications. *Mosquito News* **37**, 104–108.

Monchadskii, A. S. (1956). Bloodsucking flies in the territory of the USSR and some features of their attacks on man. *Ent. Obzor.* **35**, 547–559. [In Russian].

Monchadskii, A. S. and Radzivilovskaya, Z. A. (1948). A new method of quantitative registration of the attacks of bloodsuckers. *Parazit. Sb.* **9**, 147–166. [In Russian].

Mondet, B., Berl, D. and Bernadou, J. (1977). Étude du parasitisme des Simulies (Diptera) par des Mermithidae (Nematoda) en Afrique de l'Ouest. III. Elevage de *Isomeris* sp. et infestation en laboratoire de *Simulium damnosum s.l. Cah. ORSTOM sér Ent. méd. Parasit.* **15**, 265–269.

Mondet, B., Berl, D. and Prud'hom, J. M. (1978). Infestation de larves de *Simulium damnosum s.l.* (Simuliidae) par *Romanomermis culicivorax* (Mermithidae) parasite de moustique. *ORSTOM/OCCGE No. 16/ONCHO/RAP./78.* Mimeogr. Doc.

Mondet, B., Pendriez, B. and Bernadou, J. (1976). Étude du parasitisme des simulies (Diptera) par des Mermithidae (Nematoda) en Afrique de l'Ouest. I. Observations préliminaires sur un cours d'eau temporaire de savane. *Cah. ORSTOM sér Ent. méd. Parasit.* **14**, 141–149.

Mondet, B., Poinar, G. O., Jr. and Bernadou, J. (1977a). Étude du parasitisme des simulies (Diptera: Simuliidae) par des Mermithidae (Nematoda) en Afrique de l'Ouest. II. Description de deux nouvelles espèces de *Gastromermis. Can. J. Zool.* **55**, 1275–1283.

Mondet, B., Poinar, G. O. Jr. and Bernadou, J. (1977b). Étude du parasitisme des simulies (Diptera: Simuliidae) par des Mermithidae (Nematoda) en Afrique de l'Ouest. IV. Description de *Isomermis lairdi* n.sp., parasite de *Simulium damnosum. Can. J. Zool.* **55**, 2011–2017.

Moore, H. S. and Noblet, R. (1974). Flight range of *Simulium slossonae*, the primary vector of *Leucocytozoon smithi* of turkeys in South Carolina. *Envir. Ent.* **3**, 365–369.

Moore, J. W. (1977). Some factors effecting [sic] algal consumption in subarctic Ephemeroptera, Plecoptera, and Simuliidae. *Oecologia* **27**, 261–273.

Moorhouse, D. E. and Colbo, M. H. (1973).

On the swarming of *Austrosimulium pestilens* Mackerras and Mackerras (Diptera: Simuliidae). *J. Aust. ent. Soc.* **12,** 127–130.

Moreau, R. E. (1933). The food of the red-billed oxpecker, *Buphagus erythrorhynchus* (Stanley). *Bull. ent. Res.* **24,** 325–335.

Mount, D. I. (1967). Consideration for acceptable concentrations of pesticides for fish production. *In* A symposium on water quality criteria to protect aquatic life. (E. Cooper, Ed.). *Spec. Publs Am. Fish. Soc.* **4.**

Muirhead-Thomson, R. C. (1957a). Effect of desiccation on the eggs of *Simulium damnosum* Theobald. *Nature, Lond.* **180,** 1432–1433.

Muirhead-Thomson, R. C. (1957b). The development of *Onchocerca volvulus* in laboratory-reared *Simulium damnosum* Theobald. *Am. J. trop. Med. Hyg.* **6,** 912–913.

Muirhead-Thomson, R. C. (1957c). Laboratory studies on the reactions of *Simulium* larvae to insecticides. I. A laboratory method for studying the effects of insecticide on *Simulium* larvae. *Am. J. trop. Med. Hyg.* **6,** 920–925.

Muirhead-Thomson, R. C. (1966). Blackflies. *In* Insect Colonization and Mass Production, pp. 127–144 (C. N. Smith, Ed.). New York. Academic Press.

Muirhead-Thomson, R. C. (1970). The potentiating effect of pyrethrins and pyrethroids on the action of organophosphorus larvicides in *Simulium* control. *Trans. R. Soc. trop. Med. Hyg.* **64,** 895–906.

Muirhead-Thomson, R. C. (1971). Pesticides and Freshwater Fauna. New York. Academic Press.

Muirhead-Thomson, R. C. (1973). Laboratory evaluation of pesticide impact on stream invertebrates. *Freshwat. Biol.* **3,** 479–498.

Mulla, M. S. and Lacey, L. A. (1976). Feeding rates of *Simulium* larvae on particulates in natural streams (Diptera: Simuliidae). *Envir. Ent.* **5,** 283–287.

Müller, K. (1953). Investigations on the organic drift in north Swedish streams. *Rep. Freshwat. Res. Drottningholm* **35,** 133–148.

Müller, K. (1965). Die Tagesperiodik von Fliesswasserorganismen. *Z. Morph. Ökol. Tiere.* **56,** 93–142.

Müller, K. (1974). Stream drift as a chronobiological phenomenon in running water ecosystems. *A. Rev. Ecol. Syst.* **5,** 309–323.

Mundie, J. H. (1969). Ecological implications of the diet of juvenile Coho in streams. Symp. on Salmon and Trout in Streams. H.

R. MacMillan Lect. in Fish., pp. 135–152. Univ. Brit. Columbia Inst. Fish. (1968).

Murrin, S. F. and Nolan, R. A. (1977). Ultrastructural localization of succinate dehydrogenase in a self-parasitic isolate of *Saprolegnia megasperma. Can. J. Microbiol.* **23,** 491–496.

Muttkowski, R. A. and Smith, G. M. (1929). The food of trout stream insects in Yellowstone National Park. *Bull. NY St Coll. For. Roosevelt wild Life Ann.* **2,** 241–263.

Muu, L. T. (1977). To the identity of the microsporidian parasites of *Odagmia ornata* (Meig). (Diptera: Simuliidae). *Věst. čsl. zool. Spol.* **41,** 45–51.

Nabokov, V. A. (1959). *In* Blackflies and their Control, pp. 1–22. M. Medgiz. [In Russian].

Nakamura, H. (1949). Experimental methods for micro-organisms, Kadokawashoten, Tokyo (cited by Fujita, Y. [1965]. The composition of culture media: Nakamura's culture. *In* Experimental Methods for Algae, pp. 68–104 (H. Tamiya and A. Watanabe, Eds.). Tokyo. Nankoodo. [In Japanese].

Nam, E. A. and Dubitskii, A. M. (1978). A description of a new regulator of the number of bloodsucking blackflies, the fungus *Simuliomyces lairdi* g.n., sp.n. (Entomophthorales). *Vest. Akad. Nauk Kazakh. SSR.* **4,** 73–75.

Needham, P. R. and Usinger, R. L. (1956). Variability in the macrofauna of a single riffle in Prosser Creek, California as indicated by the Surber sampler. *Hilgardia* **24,** 383–409.

Nefedov, D. D. (1964). *In* Blackflies and Measures in Controlling them. M. Izd-vo "Meditzina" 1–163. [In Russian].

Neveu, A. (1973a). Variations biométriques saisonnières chez les adultes de quelques espèces de Simuliidae (Diptera: Nematocera). *Archs. Zool. exp. gén.* **114,** 261–270.

Neveu, A. (1973b). Le cycle de développement des Simuliidae (Diptera: Nematocera) d'une ruisseau des Pyrénées-Atlantiques, le Lessuraga. *Annls. Hydrobiol.* **4,** 51–75.

Newsom, L. D. (1967). Consequences of insecticide use on non-target organisms. *A. Rev. Ent.* **12,** 257–286.

Nicholson, H. P. and Mickel, C. E. (1950). The blackflies of Minnesota (Simuliidae). *Tech. Bull. Minn. agric. Exp. Stn.* **192,** 1–64.

Nilsson, N.-A. (1957). On the feeding habits of trout in a stream of northern Sweden. *Rep. Inst. Freshwat. Res. Drottningholm* **38,** 154–166.

Noamesi, G. K. (1964). The tube bioassay technique in tests to evaluate entomologically the effects of *Simulium* control operations in northwest Ghana. *Ghana med. J.* **3**, 163–165.

Noamesi, G. K. (1971). Dry season survival of *Simulium damnosum* Theobald 1903 (Diptera: Simuliidae) in Northern Ghana, West Africa. *Ann. trop. Med. Parasit.* **65**, 555–565.

Noel-Buxton, M. B. (1956). Field experiments with DDT in association with finely divided inorganic material for the destruction of the immature stages of the genus *Simulium* in the Gold Coast. *Jl W. Afr. Sci. Ass.* **2**, 36–40.

Nolan, R. A. (1970a). The Phycomycete *Catenaria anguillulae*: growth requirements. *J. gen. Microbiol.* **60**, 167–180.

Nolan, R. A. (1970b). Sulfur source and vitamin requirements of the aquatic Phycomycete, *Catenaria anguillulae*. *Mycologia* **62**, 568–577.

Nolan, R. A. (1970c). Carbon source and micronutrient requirements of the aquatic Phycomycete, *Catenaria anguillulae* Sorokin. *Ann. Bot.* (n.s.) **34**, 927–939.

Nolan, R. A. (1975a). Physiological studies with the fungus *Saprolegnia megasperma* isolated from the freshwater nematode *Neomesomermis flumenalis*. *Can. J. Bot.* **53**, 3032–3040.

Nolan, R. A. (1975b). Fungal self-parasitism in *Saprolegnia megasperma*. *Can. J. Bot.* **53**, 2110–2114.

Nolan, R. A. (1977). Physiological studies with the aquatic hyphomycete *Pleuropedium tricladioides* isolated from the freshwater nematode *Neomesomermis flumenalis*. *Mycologia* **69**, 914–926.

Nolan, R. A., Dunphy, G. B. and MacLeod, D. M. (1976). *In vitro* germination of *Entomophthora egressa* resting spores. *Can. J. Bot.* **54**, 1131–1134.

Nolan, R. A. and Lewis, D. J. (1974). Studies on *Pythiopsis cymosa* from Newfoundland. *Trans. Br. mycol. Soc.* **62**, 163–179.

Noyes, A. A. (1914). The biology of the net-spinning Trichoptera of Cascadilla Creek. *Ann. ent. Soc. Am.* **7**, 251–271.

Obeng, L. E. (1967). Life-history and population studies on the Simuliidae of North Wales. *Ann. trop. Med. Parasit.* **61**, 472–487.

Oberlander, H. (1969). Effects of ecdysone, ecdysterone, and inokosterone on *in vitro* initiation of metamorphosis of wing disks of *Galleria mellonella*. *J. Insect Physiol.* **15**, 297–304.

Oberndorfer, R. Y. and Stewart, K. W. (1977). The life cycle of *Hydroperla crosbyi* (Plecoptera: Perlodidae). *Gt. Basin Nat.* **37**, 260–273.

O'Brien, R. D. and Yamamoto, I. (Eds.) (1970). Biochemical Toxicology of Insecticides. New York. Academic Press.

Odetoyinbo, J. A. (1969). Preliminary investigation on the use of "light traps" for day and night time sampling of *Simulium damnosum*, Theobald 1903 (Diptera: Simuliidae) in Ghana. *Brazzaville Regional Office, WHO.* Mimeogr. Doc.

Odintsov, V. S., Tertyshnyi, V. N. and Alekseenko, I. P. (1970). The localization of the acetylcholinesterase activity in the central nervous system of the larvae of bloodsucking blackflies (Diptera: Simuliidae) developing in winter. *Dokl. Akad. Nauk SSSR.* **190**, 224–226.

Oglesby, R. T., Carlson, C. A. and McCann, J. A. (Eds.) (1972). River Ecology and Man. New York. Academic Press.

O'Kane, W. C. (1926). Blackflies in New Hampshire. *Tech. Bull. New Hamps. agric. Exp. Stn.* **32**, 1–23.

Oladimeji, A. A. and Leduc, G. (1974). Effects of dietary methoxychlor on the food requirements of brook trout. *VII Int. Conf. on Wat. Pollut. Res., Paris.*

Omar, M. S. (1976). Histopathologie und Abwehrmechanismen bei Simulien nach starker *Onchocerca volvulus*-Infektion. *Z. angew. Ent.* **82**, 53–57.

Omar, M. S. and Garms, R. (1975). The fate and migration of a Guatemalan strain of *Onchocerca volvulus* in *Simulium ochraceum* and *S. metallicum* and the role of the buccopharyngeal armature in the destruction of microfilariae. *Tropenmed. Parasit.* **26**, 183–190.

Omar, M. S. and Garms, R. (1977). Lethal damage to *Simulium metallicum* following high intakes of *Onchocerca volvulus* microfilariae in Guatemala. *Tropenmed. Parasit.* **28**, 109–119.

Oppenoorth, F. J. (1976). Development of resistance to insecticides. *In* The Future for Insecticides—Needs and Prospects, pp. 41–63 (R. Metcalf and J. McKelvey, Eds.). New York. Wiley.

Orii, M. (1975). Studies on spiders as natural enemies of insect pests. 4. Ecological studies on spiders in the animal-shed and surroundings. *Jap. J. sanit. Zool.* **26**, 93–99.

Orii, T., Kitamura, S., Uemoto, K., Ishino, U. and Kumasawa, N. (1964). Studies on blackfly (Simuliidae) in the Northern suburbs of Kyoto. V. Distribution and seasonal prevalence of larvae and pupae of *Simulium* in Kyoto. *Sanit. injur. Insects* **8**, 36–52. [In Japanese, English summary].

Ovazza, M., Ovazza, L. and Balay, G. (1965). Étude des populations de *Simulium damnosum* Theobald, 1903 (Diptera: Simuliidae) en zones de gîtes non permanents. II. Variations saisonnières se produisant dans les populations adultes et préimaginales. Discussion des différentes hypothèses qui peuvent expliquer le maintien de l'espèce dans les régions sèches. *Bull. Soc. Path. exot.* **58**, 1118–1154.

Ovazza, M., Renard, J. and Balay, G. (1967). Étude des populations de *Simulium damnosum* Theobald, 1903 (Diptera: Simuliidae) en zones de gîtes non permanents. III. Corrélation possible entre certains phénomènes météorologiques et la réapparition des femelles en début de saison des pluies. *Bull. Soc. Path. exot.* **60**, 79–95.

Pacaud, A. (1942). Notes biologiques sur une station de *Simulium aureum* Fries aux environs de Paris. *Bull. biol. Fr.* **76**, 226–238.

Parker, J. D. (1975). The use of aircraft in the WHO Onchocerciasis Control Programme. *5th Ann. Agric. Aviat. Congr., Nat. Agric. Centre, Warwickshire, England. 22–25 September 1975*, 127–136.

Pasternak, J. (1964). Chromosome polymorphism in the blackfly *Simulium vittatum* (Zett.). *Can. J. Zool.* **42**, 135–158.

Patrusheva, V. D. (1966). *In* Blackflies (Simuliidae) in Biological Principles of the Control of Bloodsucking-Diptera in the Ob River Basin, pp. 53–115 (A. I. Cherepanov, Ed.). Siberian Branch, Biological Institute Novosibirsk. Academy of Sciences of the USSR. [In Russian].

Pavlichenko, V. I. (1977a). The natural enemies of blackflies. *Scient. Repts. for Higher Educ., Biol. Sci.* (*Zool.*) **8**, 44–46 [In Russian] (Kafedra Biol. Zaporozhskogo Med. Inst. USSR). [See also *Rev. appl. Ent.* **B66**, 128–129].

Pavlichenko, V. I. (1977b). Role of larvae of *Hydropsyche angustipennis* Curt. (Trichoptera, Hydropsychidae) in the destruction of blackfly larvae in running water of Zaporozhe district. *Ecologiya.* **8**, 104–105. [In Russian, translation *Soviet J. Ecol.*, New York, **8**, 84–85].

Pagel, M. and Rühm, W. (1976). Versuche zur Besiedlung künstlicher Substrate durch präimaginale Stadien von Simuliiden unter besonderer Berücksichtigung von *Boophthora erythrocephala* de Geer (Simuliidae, Dipt.). *Z. angew. Ent.* **82**, 65–71.

Pentelow, F. T. K. (1932). The food of the brown trout (*Salmo trutta* L.). *J. Anim. Ecol.* **1**, 101–107.

Peschken, D. and Thorsteinson, A. J. (1965). Visual orientation of black flies (Simuliidae: Diptera) to colour, shape and movement of targets. *Entomologia exp. appl.* **8**, 282–288.

Petersen, A. (1924). Bidrag til de danske Simuliers Naturhistorie. *K. dansk. Vidensk. Selsk. Skr.* **8**(5), 235–341.

Peterson, B. V. (1959). Observations on mating, feeding, and oviposition of some Utah species of black flies (Diptera: Simuliidae). *Can. Ent.* **91**, 147–155.

Peterson, B. V. (1960). Notes on some natural enemies of Utah black flies (Diptera: Simuliidae). *Can. Ent.* **92**, 266–274.

Peterson, B. V. (1962). *In* Proc. 3rd Conf. on Black Flies (Diptera: Simuliidae), Algonquin Park, Ontario, Canada, p. 77 (B. V. Peterson, Ed.).

Peterson, B. V. (1970). The *Prosimulium* of Canada and Alaska (Diptera: Simuliidae). *Mem. ent. Soc. Can.* **69**, 1–216.

Peterson, B. V. (1977a). A synopsis of the genus *Parasimulium* Malloch (Diptera: Simuliidae), with descriptions of one new subgenus and two new species. *Proc. ent. Soc. Wash.* **79**, 96–106.

Peterson, B. V. (1977b). The black flies of Iceland (Diptera: Simuliidae). *Can. Ent.* **109**, 449–472.

Peterson, B. V. (1978). *Simuliidae. In* An Introduction to the Aquatic Insects of North America, pp. 331–344 (R. W. Merritt and K. W. Cummins, Eds.). Dubuque, Iowa. Kendall/Hunt.

Peterson, B. V. and Davies, D. M. (1960). Observations on some insect predators of black flies (Diptera: Simuliidae) of Algonquin Park, Ontario. *Can. J. Zool.* **38**, 9–18.

Petrishcheva, P. A. and Saf'yanova, V. M. (1961). *In* Blackflies and their Control, pp. 1–34. Moscow. Institute of Hygiene Education of the USSR Ministry of Health. [In Russian].

Phelps, R. J. and DeFoliart, G. R. (1964). Nematode parasitism of Simuliidae. *Res. Bull. agric. Exp. Stn Univ. Wis.* **245**, 1–78.

Philippon, B. (1968). Infestation des femelles de *Simulium damnosum* par *Onchocerca volvulus* dans les conditions naturelles en région de savane. *Rapp. final 8me Conf. techn. OCCGE* **1**, 205–206. Mimeogr. Doc.

Philippon, B. (1976). Étude de la transmission d'*Onchocerca volvulus* (Leuckart, 1893) (Nematoda: Onchocercidae) par *Simulium damnosum* Theobald, 1903 (Diptera: Simuliidae) en Afrique tropicale. *Trav. Doc. OR-STOM* **63**, 1–308.

Philippon, B. and Bain, O. (1972). Transmission de l'onchocerose humaine en zone de savane d'Afrique occidentale. Passage des microfilaires d'*Onchocerca volvulus* Leuck. dans l'hémocèle de la femelle de *Simulium damnosum* Theo. *Cah. ORSTOM sér. Ent. méd. Parasit.* **10**, 251–261.

Philippon, B., Séchan, Y., Rivière, F., Kulzer, H., Pawlich, O. and Krupke, M. (1976). Control of *Simulium damnosum*, vector of human onchocercias in West Africa. VI. Spraying of new insecticide formulations by helicopter in the rainy season in the Guinea-type savanna zone (1973). *WHO/ VBC/76.619*. Mimeogr. Doc.

Phillipson, J. (1956). A study of factors determining the distribution of the larvae of the blackfly *Simulium ornatum* Mg. *Bull. ent. Res.* **47**, 227–238.

Phillipson, J. (1957). The effect of current speed on the distribution of the larvae of the blackflies *Simulium variegatum* (Mg.) and *Simulium monticola* Fried. (Diptera). *Bull. ent. Res.* **48**, 811–819.

Planchon, J. E. (1844). Histoire d'une larve aquatique de genre *Simulium*. Typographie et lithographie de Boehm. Montpellier.

Poehling, H. M. (1977). Die Bildung spezifischer Proteine während der Metamorphose der Speicheldrüsen von *Wilhelmia lineata* (Simuliidae). *J. Insect Physiol.* **23**, 1105–1112.

Poehling, H. M., Wolfrum, D. I. and Neuhoff, V. (1976). Mikro-electrophorese von Proteinen aus Speicheldrüsen und Hämolymphe verschiedener Simuliidenarten in Polyacrylamidgradienten-Gelen. *Entomologia exp. appl.* **19**, 271–286.

Poinar, G. O. Jr. (1975). Entomogenous Nematodes. Leiden. E. J. Brill.

Poinar, G. O. Jr. (1977). A synopsis of the nematodes occurring in blackflies (Diptera: Simuliidae). *Bull. Wld Hlth Org.* **55**, 509–515.

Poinar, G. O. Jr. and Hess, R. (1977). *Romanomermis culicivorax*: morphological evidence of transcuticular uptake. *Expl. Parasit.* **42**, 27–33.

Poinar, G. O. Jr. and Hess, R. (1979).

Mesomermis paradisus sp.n. (Mermithidae: Nematoda), a parasite of *Prosimulium exigens* D. & S. (Simuliidae: Diptera) in California. *Nematologica* **25**, 368–372.

Poinar, G. O. Jr., Hess, R., Hansen, E. and Hansen, J. W. (1979). Laboratory infection of blackflies (Simuliidae) and midges (Chironomidae) by the mosquito mermithid, *Romanomermis culicivorax*. *J. Parasitol.* **65**, 613–615.

Poinar, G. O. Jr., Lane, R. S. and Thomas, G. M. (1976). Biology and redescription of *Pheromermis pachysoma* (v. Linstow) n.gen., n.comb. (Nematoda: Mermithidae), a parasite of yellowjackets (Hymenoptera: Vespidae). *Nematologica* **22**, 360–370.

Poinar, G. O., Jr. and Ono, H. (1979). Parasitic nematodes from *Simulium tobetsuense* Ono and *S. aokii* Takahasi. *Jap. J. sanit. Zool.* **30**, 194.

Poinar, G. O. Jr. and Saito, K. (1979). *Mesomermis japonicus* n.sp. (Mermithidae: Nematoda), a parasite of *Simulium japonicum* (Simuliidae: Diptera). *Jap. J. sanit. Zool.* **30**, 147–149.

Poinar, G. O., Jr. and Takaoka, H. (1979a). Parasitic nematodes from *Simulium japonicum* and *S. bidentatum*. *Jap. J. sanit. Zool.* **30**, 193.

Poinar, G. O., Jr. and Takaoka, H. (1979b). *Isomermis benevolvus* sp.n. (Mermithidae: Nematode), a parasite of *Simulium metallicum* (Diptera: Simuliidae) in Guatemala. *Jap. J. sanit. Zool.* **30**, 305–307.

Poinar, G. O., Jr. and Takaoka, H. (1981). Three new mermithids (Nematoda) from Guatemalan black flies (Diptera: Simuliidae). *Int. J. Systematics* (in press).

Poinar, G. O. Jr. and Welch, H. E. (1978). Nematode parasites of terrestrial invertebrates. *Proc. 4th Int. Congr. Parasit. (ICOPA IV), Warsaw* (in press).

Pomeroy, A. W. J. (1916). Notes on five North American buffalo gnats of the genus *Simulium*. *Bull. U.S. Dep. Agric.* **329**, 1–48.

Pomeroy, A. W. J. (1920). New species of African Simuliidae. *Ann. Mag. nat. Hist.* (9)**6**, 72–81.

Popapov, A. A. and Bogdanova, E. M. (1973). A simple trap for registering the population density of blackflies. *Medskaya Parazit.* **42**, 618–621. [In Russian].

Power, G. (1966). Observations on the speckled trout (*Salvelinus fontinalis*) in Ungava. *Naturaliste can.* **93**, 187–198.

Price, J. W. (1959). A study of the food habits

of some Lake Erie fish. *Final Rep. Res. Fdn.,*
Project 837, Ohio State Univ.

Pritchard, G. (1964). The prey of adult dragonflies in northern Alberta. *Can. Ent.* **96,**
821–825.

Prost, A. (1977). Situation dans un foyer d'onchocercose du Mali après treize ans de
controle anti-simulidien. I. Aspects para-
sitologiques. *Ann. Soc. belge Méd. trop.* **57,**
569–575.

Ptáček, V. and Knoz, J. (1971). Über der
larvale Entwicklung der Art *Simulium* (*S.*)
argyreatum Meigen 1838. *Scripta Fac. Sci.*
Nat. UJEP. Brunensis Biologia **3,** 179–195.

Puri, I. M. (1925). On the life-history and
structure of the early stages of Simuliidae
(Diptera: Nematocera). Part I and II. *Para-*
sitology **17,** 295–369.

Quélennec, G. (1971). Observations sur les
déplacements larvaires des simulies en Afrique de l'Ouest. *Cah. ORSTOM sér. Ent. méd.*
Parasit. **9,** 247–254.

Quélennec, G., Simonkovich, E. and Ovazza,
M. (1968). Recherche d'un type de déver-
soir de barrage défavorable à l'implanta-
tion de *Simulium damnosum* (Diptera:
Simuliidae). *Bull. Wld. Hlth Org.* **38,** 943–
956.

Quillévéré, D., Escaffre, H. Pendriez, B., Grébaut, S., Duchateau, B., Lee, C. W. and
Mouchet, J. (1976a). Control of *Simulium*
damnosum, vector of human onchocerciasis
in West Africa. III. Application by aero-
plane during the wet season—methods of
application—new formulations. *WHO/VBC/*
76.616. Mimeogr. Doc.

Quillévéré, D., Escaffre, H., Pendriez, B.,
Grébaut, S., Ouedraogo, J., Kulzer, H.,
Bellec, C., Philippon, B. and Le Berre, R.
(1976b). Control of *Simulium damnosum,* vector of human onchocerciasis in West Africa.
IV. Evaluation by helicopter of new formulations and simulation of an insecticide
operation. *WHO/VBC/76.617.* Mimeogr.
Doc.

Quillévéré, D., Gouzy, M., Séchan, Y. and
Pendriez, B. (1976c). Étude du complexe
Simulium damnosum en Afrique de l'Ouest.
IV. Analyse de l'eau des gîtes larvaires en
saison sèche. *Cah. ORSTOM, sér. Ent. méd.*
Parasit. **14,** 315–330.

Quillévéré, D., Gouzy, M., Séchan, Y. and
Pendriez, B. (1977a). Étude du complexe
Simulium damnosum en Afrique de l'Ouest.
VI. Analyse de l'eau des gîtes larvaires en
saison des pluies; comparaison avec la

saison sèche. *Cah. ORSTOM, sér. Ent. méd.*
Parasit. **15,** 195–207.

Quillévéré, D. and Pendriez, B. (1975). Étude
du complexe *Simulium damnosum* en Afrique
de l'Ouest. II. Réparation géographique des
cytotypes en Côte d'Ivoire. *Cah. ORSTOM*
sér. Ent. méd. Parasit. **13,** 165–172.

Quillévéré, D., Séchan, Y. and Pendriez, B.
(1977b). Étude du complex *Simulium damno-*
sum en Afrique de l'Ouest. V. Identification
morphologique des femelles en Côte
d'Ivoire. *Tropenmed. Parasit.* **28,** 244–253.

Quillévéré, D. and Séchan, Y. (1978). Morphological identification of females of the
Simulium damnosum complex in West Africa:
differentiation of *S. squamosum* and *S.*
yahense. Trans. R. Soc. trop. Med. Hyg. **72,**
99–100.

Raastad, J. E. (1979). Fennoscandian blackflies
(Diptera: Simuliidae): annotated list of the
species and their gross distribution. *Rhizoc-*
rinus **11,** 1–28.

Raastad, J. E. and Mehl, R. (1972). Night
activity of black flies (Diptera: Simuliidae) in
Norway. *Norsk ent. Tidsskr.* **19,** 173.

Radford, D. S. and Hartland-Rowe, R. (1971).
Subsurface and surface sampling of benthic
invertebrates in two streams. *Limnol.*
Oceanogr. **16,** 114–120.

Raminani, L. N. and Cupp, E. W. (1978). The
male reproductive system of the blackfly,
Simulium pictipes Hagen. *Mosquito News* **38,**
591–594.

Rauther, M. (1930). Nematodes. *In* Hand. d.
Zoologie. Bd. II, Leif 8, Teil 4, 1–402.

Raybould, J. N. (1965). Preliminary experiments with a host-preference trap for *Simu-*
lium. Rep. E. Afr. Inst. Mal. (1 Jul 1963–31
Dec 1964), 32.

Raybould, J. N. (1966). Notes on Simuliidae
visiting flowers and a mercury-vapour
lamp. *Rep. E. Afr. Inst. Mal.* (1 Jan–31 Dec
1965), 41–42.

Raybould, J. N. (1967a). A method of rearing
Simulium damnosum Theobald (Diptera:
Simuliidae) under artificial conditions. *Bull.*
Wld Hlth Org. **37,** 447–453.

Raybould, J. N. (1967b). A study of anthropophilic female Simuliidae (Diptera) at
Amani, Tanzania: The feeding behaviour of
Simulium woodi and the transmission of
onchocerciasis. *Ann. trop. Med. Parasit.* **61,**
76–88.

Raybould, J. N. (1969). Studies on the immature stages of the *Simulium neavei* Roubaud
complex and their associated crabs in the

Eastern Usambara mountains of Tanzania I. Investigations in rivers and large streams. *Ann. trop. Med. Parasit.* **63**, 269–287.

Raybould, J. N. (1979). A new simple technique for rearing F_1 progeny from single females of the *Simulium damnosum* Theobald complex. *WHO/ONCHO/79.148* and *WHO/VBC 79.114*. Mimeogr. Doc.

Raybould, J. N. and Grunewald, J. (1975). Present progress towards the laboratory colonization of African Simuliidae (Diptera). *Tropenmed. Parasit.* **26**, 155–168.

Raybould, J. N., Grunewald, J. and Grunewald, B. (1973). Preliminary investigations of the low temperature storage of *Simulium* eggs. *Rep. E. Afr. Inst. Mal.* (Jan 1972–Dec 1973).

Raybould, J. N., Grunewald, J. and Mhiddin, H. K. (1978). Studies on the immature stages of *Simulium neavei* Roubaud complex and their associated crabs in the Eastern Usambara mountains in Tanzania. IV. Observations on the crabs and their attached larvae under exceptionally dry conditions. *Ann. trop. Med. Parasit.* **72**, 189–194.

Raybould, J. N. and Mhiddin, H. K. (1974). A simple technique for maintaining *Simulium* adults, including onchocerciasis vectors, under artificial conditions. *Bull. Wld Hlth Org.* **51**, 309–310.

Raybould, J. N. and Mhiddin, H. K. (1978). Studies on the immature stages of the *Simulium neavei* Roubaud complex and their associated crabs in the Eastern Usambara mountains in Tanzania. III. Investigations on development and survival and their relevance to control. *Ann. trop. Med. Parasit.* **72**, 177–187.

Raybould, J. N., Vajime, C. G., Quillévéré, D., Barro, T. and Sawadogo, R. (1979). The laboratory maintenance and rearing of *Simulium damnosum* complex species as a research tool for the Onchocerciasis Control Programme in the Volta River Basin. *Tropenmed. Parasit.* **30**, 499–504..

Raybould, J. N. and Yagunga, A. S. K. (1969). Artificial feeding of East African Simuliidae (Diptera) including vectors of human onchocerciasis. *Bull. Wld Hlth Org.* **40**, 463–466.

Reed, E. B. (1962). Limnology and fisheries of the Saskatchewan River in Saskatchewan. *Sask. Dep. Nat. Res., Regina. Fish. Rep. 6.*

Reid, G. D. F. (1978). Cibarial armature of *Simulium* vectors of onchocerciasis. *Trans. R. Soc. trop. Med. Hyg.* **72**, 438.

Reisen, W. K. (1974). The ecology of larval blackflies (Diptera: Simuliidae) in a South Central Oklahoma stream. PhD Thesis. University of Oklahoma, Norman.

Reisen, W. K. (1975). Quantitative aspects of *Simulium virgatum* Coq. and *S.* species life history in a southern Oklahoma stream. *Ann. ent. Soc. Am.* **68**, 949–954.

Reisen, W. K. (1977). The ecology of Honey Creek, Oklahoma: population dynamics and drifting behavior of three species of *Simulium* (Diptera: Simuliidae). *Can. J. Zool.* **55**, 325–337.

Reisen, W. K. and Emory, R. W. (1977). Intraspecific competition in *Anopheles stephensi* (Diptera: Culicidae). II. The effects of more crowded densities and the addition of antibiotics. *Can. Ent.* **109**, 1475–1480.

Rempel, J. G. and Arnason, A. P. (1947). An account of three successive outbreaks of the black fly, *Simulium arcticum*, a serious livestock pest in Saskatchewan. *Scient. Agric.* **27**, 428–445.

Reuter, U. and Rühm, W. (1976). Über die zeitliche Verteilung der anfliegenden Weibchen von *Boophthora erythrocephala* De Geer und *Simulium sublacustre* Davies bei der Eiablage (Simuliidae: Dipt.). *Z. angew Zool.* **63**, 385–391.

Rhame, R. E. and Stewart, K. W. (1976). Life cycle and food habits of three Hydropsychidae (Trichoptera) species in the Brazos river, Texas. *Trans. Am. ent. Soc.* **102**, 65–69.

Richards, O. W. and Richards, M. J. (1951). Observations on the social wasps of South America (Hymenoptera: Vespidae). *Trans. R. ent. Soc. Lond.* **102**, 1–170.

Richman, K. and Oberlander, H. (1971). Effects of fat body on alphaecdysone induced morphogenesis in cultured wing disks of the wax moth, *Galleria mellonella*. *J. Insect Physiol.* **17**, 269–276.

Riley, C. V. (1887). Report of the entomologist. *Rep. U.S. Dep. Agric.* 1886, 459–592.

Ripper, W. E. (1956). Effect of pesticides on balance of arthropod populations. *A. Rev. Ent.* **1**, 403–438.

Rivosecchi, L. (1972). Contributo alla conoscenza dei simulidi Italiani. XXII. Raccolte di simulidi adulti mediante trappole ad anidride carbonica. *Riv. Parassit.* **33**, 293–312.

Rivosecchi, L. (1978). Simuliidae: Diptera,

Nematocera. *Fauna Ital.* **13**. 1–529. [In Italian].

Rivosecchi, L., Cavallini, C., Noccioli, M. and Rubeca, L. (1974). Osservazioni ecologiche sui Simulidi (Diptera: Nematocera) del fiume Arrone e degli affluenti del Lago di Bracciano. *Riv. Parassit.* **35**, 331–356.

Roberts, D. W. (1966). Toxins from the entomogenous fungus *Metarrhizium anisopliae*. I. Production in submerged and surface cultures, and in inorganic and organic nitrogen media. *J. Invert. Pathol.* **8**, 212–221.

Roberts, D. W. (1977). Isolation and development of fungus pathogens of vectors. *In* Biological Regulation of Vectors. The saprophytic and aerobic bacteria and fungi, pp. 85–93 (J. D. Briggs, Ed.). Washington. US Department of Health, Education and Welfare (Pub. No. (NIH) 77–1180).

Roberts, D. W. and Strand, M. A. (Eds.) (1977). Pathogens of Medically Important Arthropods. *Bull. Wld Hlth Org.* **55** (Suppl. 1), 1–419.

Robles, R. (1919). Onchocercose humaine au Guatémala produisant la cécité et l'érysipèle du litoral. *Bull. Soc. Path. exot.* **12**, 442–460.

Rogers, D. and Randolph, S. E. (1978). A comparison of electric-traps and hand-net catches of *Glossina palpalis palpalis* (Robineau-Desvoidy) and *G. tachinoides* Westwood (Diptera: Glossinidae) in the Sudan vegetation zone of northern Nigeria. *Bull. ent. Res.* **68**, 283–297.

Rogers, D. J. and Smith, D. T. (1977). A new electric trap for tsetse flies. *Bull. ent. Res.* **67**, 153–159.

Rohdendorf, B. B. (1964). The Historical Development of Diptera. *Trudy paleont. Inst.* **100**, 1–311 [In Russian]. (For English translation see: The historical development of Diptera, The University of Alberta Press, 1974).

Rolland, A. and Thylefors, B. (1979). Aspects évolutifs de l'onchocercose oculaire en Afrique Occidentale, aprés trois ans de lutte antisimulidienne. *Tropenmed. Parasit.* **30**, 482–488.

Roos, T. (1957). Studies on upstream migration in adult stream-dwelling insects. I. *Rep. Inst. Freshwat. Res. Drottningholm* **38**, 167–193.

Rosenberg, D. M. (1975). Food chain concentration of chlorinated hydrocarbon pesticides in invertebrate communities: A re-evaluation. *Quaest. ent.* **11**, 97–110.

Ross, D. H. and Craig, D. A. (1980). Mechanisms of fine particle capture by larval black flies (Diptera: Simuliidae). *Can. J. Zool.* **58**, 1186–1192.

Ross, D. H. and Merritt, R. W. (1978). The larval instars and population dynamics of five species of black flies (Diptera: Simuliidae) and their responses to selected environmental factors. *Can. J. Zool.* **56**, 1633–1642.

Ross, D. and Service, M. W. (1979). A modified Monks Wood light trap incorporating a flashing light. *Mosquito News* **39**, 610–616.

Rothfels, K. H. (1956). Black flies: siblings, sex, and species grouping. *J. Hered.* **47**, 113–122.

Rothfels, K. H. (1979). Cytotaxonomy of black flies (Simuliidae). *A. Rev. Ent.* **24**, 507–539.

Rothfels, K. H., Feraday, R. and Kaneps, A. (1978). A cytological description of sibling species of *Simulium venustum* and *S. verecundum* with standard maps for the subgenus *Simulium* Davies (Diptera). *Can. J. Zool.* **56**, 1110–1128.

Rothfels, K. H. and Freeman, D. M. (1977). The salivary gland chromosomes of seven species of *Prosimulium* (Diptera: Simuliidae) in the *mixtum* (IIIL–1) group. *Can. J. Zool.* **55**, 482–507.

Rubtsov, I. A. (1936). Eine neue Simuliidenart (*Simulium oligocenicum*, sp.n.) aus dem Bernstein. *C.r. (Dokl.) Acad. Sci. URSS* (n.s.) **2**, 353–355.

Rubtsov, I. A. (1940). Fauna of the USSR. No. 23, Insects. Diptera 6 (6) Simuliidae [in Russian, with English summary and keys].

Rubtsov, I. A. (1948). A biological method of controlling insect pests. M. L. Sel'khozgiv. [In Russian].

Rubtsov, I. A. (1950). Massive propagation of the blackfly and its probable explanation. *Akad. Nauk. SSSR* **2**, 16–22.

Rubtsov, I. A. (1956). Blackflies (Fam. Simuliidae). Ed. 2. *Fauna SSSR* (Ins. Diptera) 6(6). [In Russian].

Rubtsov, I. A. (1957). Toward a biological basis for a system of measures in blackfly control. *Zool. Zh.* **36**, 373–395.

Rubtsov, I. A. (1959). Haemolymph and its function in blackflies (Diptera: Simuliidae). *Ent. Obozr.* **38**, 32–57. [In Russian].

Rubtsov, I. A. (1959–1964). Simuliidae (Melusinidae). *Fliegen palaearkt. Reg.* **14**, 1–689.

Rubtsov, I. A. (1962a). A short guide to the bloodsucking blackflies in the fauna of the USSR. *Opred. faune SSSR* **77**, 1–227. [In Russian; English translation under the title of "Short keys to the bloodsucking Simuliidae of the USSR" by Israel Program for Scientific Translations, 1969].

Rubtsov, I. A. (1962b). On criteria of allopatric species of the group *Odagmia ornata* (Mg.) (Diptera: Simuliidae). *Ent. Obozr.* **41**, 901–920. [In Russian]. (Translated in *Ent. Rev. Wash.* **41**, 559–569).

Rubtsov, I. A. (1962c). Blackflies in the digestion of fish from the basin of the Usa (River). *In* Fish of the Usa Valley and their food. Acad. Sci., USSR, Komi Filial, V.G.O. Vp. 9, 264–268. [In Russian].

Rubtsov, I. A. (1963a). Biological prerequisites for controlling blackflies in Siberia. *Zool. Zh.* **42**, 11–19. [In Russian].

Rubtsov, I. A. (1963b). On mermithids parasitic in blackflies. *Zool. Zh.* **42**, 1768–1784. [In Russian].

Rubtsov, I. A. (1964). Mode and range of blackfly (Diptera: Simuliidae) larval migration. *Ent. Obozr.* **43**, 52–66. [In Russian, translation *Ent. Rev., Wash.* **43**, 27–33].

Rubtsov, I. A. (1966a). Natural enemies of blackflies. *In* The results of Investigations on the Problem in the control of Bloodsucking Diptera. (A. I. Cherepanov, Ed.). *Acad. Sci. USSR*, Siberian Branch. [In Russian].

Rubtsov, I. A. (1966b). Mermithids (Fam. Mermithidae) parasitic in blackflies (Fam. Simuliidae). New species of the genus *Mesomermis* Dad. *In* New Species of the Fauna of Siberia and Adjoining Regions, pp. 109–147. Novosibirsk. [In Russian].

Rubtsov, I. A. (1966c). A new species of parasite from blackflies and "errors" of host instinct. *Dokl. Akad. Nauk. SSSR* **169**, 1236–1238. [In Russian].

Rubtsov, I. A. (1967a). Scheme and organs of the extra-intestinal digestion of mermithids. *Izv. Akad. Nauk. (Biol.)* **10**, 883–891. [In Russian, English summary].

Rubtsov, I. A. (1967b). The natural enemies and biological control of insects of medical importance. *Meditzina*, 1–120. [In Russian].

Rubtsov, I. A. (1967c). Mermithids (Nematoda: Mermithidae), parasitic in blackflies (Diptera: Simuliidae). II. New species of the genus *Gastromermis* Micoletzky 1923. *Trudȳ. zool. Inst. Leningr.* **43**, 59–92. [In Russian].

Rubtsov, I. A. (1967d). Mermithids (Mermithidae), parasites of blackflies (Simuliidae). IV. New species of the genus *Limnomermis* Dad. *Zool. Zh.* **46**, 24–34. [In Russian].

Rubtsov, I. A. (1968). A new species of *Isomermis* (Nematoda: Mermithidae) a parasite of blackflies and its variability. *Zool. Zh.* **47**, 510–524. [In Russian].

Rubtsov, I. A. (1969). Man and the bloodsucking flies. On the characteristics of their interrelations in historical times. *Memorie Soc. ent. Ital.* **48**, 263–268.

Rubtsov, I. A. (1971). Mermithids from the blackflies of Western Europe. *Scr. Fac. Sci. nat. Univ. purkyn. brun. (Biol.)* **2**, 97–132.

Rubtsov, I. A. (1972a). Aquatic Mermithids. Part 1. "Nauka" Leningrad. [In Russian].

Rubtsov, I. A. (1972b). Phoresy of the blackflies (Diptera: Simuliidae) and new phoretic species from larvae of Ephemeroptera. *Ent. Obozr.* **51**, 403–410. [In Russian]. (Translated in *Ent. Rev. Wash.* **51**, 243–247).

Rubtsov, I. A. (1974a). Aquatic Mermithids. Part 2. "Nauka" Leningrad. [In Russian].

Rubtsov, I. A. (1974b). The variability of taxonomic characters in black flies (Diptera: Simuliidae); research tasks and methods. *Ent. Obozr.* **53**, 24–37. [In Russian]. (Translated in *Ent. Rev. Wash.* **53**, 15–24).

Rubtsov, I. A. (1974c). On the evolution, phylogeny and classification of blackflies (Simuliidae: Diptera). *Trudȳ zool. Inst. Leningr.* **53**, 230–282. [In Russian]. (For English translation see: Evolution, phylogeny and classification of the family Simuliidae (Diptera), British Library Lending Division.

Rubtsov, I. A. (1975). Some problems in investigating mermithids for the biological control of blackflies. *Parazitologiya* **9**, 299–300.

Rubtsov, I. A. (1976). An experiment in taxonomic research of the species and intraspecific forms of the family Mermithidae (Braun, 1883) (Nematoda). *In* The physiology and population ecology of animals. *Izdatelstvo Saratovskovo Universiteta* **3**, 19–30.

Rubtsov, I. A. (1977). Mermithids: their Origin, Distribution and Biology. "Nauka", Leningrad. [In Russian].

Rubtsov, I. A. (1978). Mermithids: their Classification, Importance and Use. "Nauka", Leningrad. [In Russian].

Rubtsov, I. A. and Doby, J. M. (1970). Mermithides parasites de simulies (Diptères) en provenance du nord et de l'Ouest de la France. *Bull. Soc. zool. Fr.* **95**, 803–836.

Rühm, W. (1968). Zur Autökologie einiger Simuliidenarten. *Z. angew. Ent.* **61**, 466–471.

Rühm, W. (1970). Zur Dispersion der Larvenstadien und des Puppenstadiums von

Boophthora erythrocephala de Geer (Simuliidae). *Z. angew. Ent.* **66**, 311–321.

Rühm, W. (1971a). Eiablagen einiger Simuliidenarten. *Z. angew. Parasit.* **12**, 68–78.

Rühm, W. (1971b). Über das Kopulationsverhalten bei Kriebelmücken. (Simuliidae: Diptera). *Mitt. dt. ent. Ges.* **30**, 19.

Rühm, W. (1973). Blutsaugen und Eiablage von *Boophthora erythrocephala* de Geer im Labor sowie Kreuzungsexperimente zwischen Boophthora-Populationen Verschiedener Generationen und Herkunft. *Z. angew. Zool.* **60**, 299–320.

Rühm, W. (1975). Freilandbeobachtungen zum Funktionskreis der Eiablage verschiedener Simuliidenarten unter besonderer Berücksichtigung von *Simulium argyreatum* Meig. (Diptera Simuliidae). *Z. angew. Ent.* **78**, 321–334.

Rühm, W. and Hechler, J. (1974). Untersuchungen über die potentielle Natalität verschiedener mammalophiler Kriebelmückenarten unter besonder Berücksichtigung von *Boophthora erythrocephala* de Geer. *Z. angew Ent.* **77**, 19–31.

Rühm, W. and Sander, H. (1975). Die Trennung der präimaginalen Stadien der saisondimorphen *Boophthora erythrocephala* (Diptera: Simuliidae) anhand morphologischer Merkinale. *Z. angew Zool.* **62**, 143–172.

Ruttner, F. (1926). Bemerkungen über den Sauerstoffgehalt der Gewässer und dessen respiratorischen Wert. *Naturwissenschaften* **14**, 1237–1239.

Sailer, R. I. (1953). The blackfly problem in Alaska. *Mosquito News* **12**, 232–235.

Saito, K., Takahashi, M., Nakamura, Y. and Kurihara, T. (1978). Spiders (Araneae) as a natural enemy of blackfly (Simuliidae) adults. *Appl. Ent. Zool.* **22**, 119–121. [In Japanese].

Sanders, H. O. and Cope, O. B. (1968). The relative toxicities of several pesticides to naiads of three species of stoneflies. *Limnol. Oceanogr.* **13**, 112–117.

Sato, S. *et al.* (1978). Immunological and parasitological investigation of onchocerciasis in Guatemala. *Onchocerciasis control project in Guatemala*, First report. *JICA* 107–127.

Savage, J. (1949). Aquatic invertebrates; mortality due to DDT and subsequent reestablishment. *Ont. Dept. Lands and Forests Biol. Bull.* **2**, 39–47.

Schultz, L. P. (1939). Fishing in Pacific coast streams. *In* Book of Fishes, pp. 263–294 (J. O. La Gorce, Ed.). *Natn. geogr. Mag.* 367 pp.

Scott, D. (1958). Ecological studies on the Trichoptera of the river Dean, Cheshire. *Arch. Hydrobiol.* **54**, 340–392.

Séguy, E. (1925). Diptères (Nématocères piqueurs): Ptychopteridae, Orphnephilidae, Simuliidae, Culicidae, Psychodidae, Phlebotominae. *Faune Fr.* **12**, 1–109.

Séguy, E. (1927). Diptères, Brachycères: (Asilidae). *Faune Fr.* **17**, 1–187.

Serra-Tosio, B. (1967). La prise de nourriture chez la larve de *Prosimulium inflatum* Davies 1957 (Diptera: Simuliidae). *Trav. Lab. Hydrobiol. Piscic. Univ. Grenoble* **57–58**, 97–103.

Service, M. W. (1972). Observations on swarming of adults of *Simulium* (*Simulium*) *austeni* Edwards (Dipt. Simuliidae). *Entomologist's mon. Mag.* **107** (1971), 167–168.

Service, M. W. (1976). Mosquito Ecology: Field Sampling Methods. London. Applied Science Publishers.

Service, M. W. (1977). A critical review of procedures for sampling populations of adult mosquitoes. *Bull. ent. Res.* **67**, 343–382.

Service, M. W. (1979). Light trap collections of ovipositing *Simulium squamosum* in Ghana. *Ann. trop. Med. Parasit.* **73**, 487–490.

Service, M. W. and Highton, R. B. (1980). A chemical light trap for mosquitoes and other biting insects. *J. med. Ent.* **17**, 183–185.

Service, M. W. and Lyle, P. T. W. (1975). Detection of the predators of *Simulium damnosum* by the precipitin test. *Ann. trop. Med. Parasit.* **69**, 105–108.

Shapiro, J. (1957). Chemical and biological studies on the yellow organic acids of lake water. *Limnol. Oceanogr.* **2**, 161–179.

Shapiro, M. (1975). *Ann. Rpt. Res. Unit Vec. Pathol., Memorial Univ. Newfoundl.*

Sharleman, E. (1915). On the question of the food of dragonflies. *Lyub. Prir.* **10**, 14–15. [In Russian].

Sheldon, A. L. (1969). Size relationships of *Acroneuria californica* (Perlidae: Plecoptera) and its prey. *Hydrobiologia* **34**, 85–94.

Sheldon, A. L. (1972). Comparative ecology of *Arcynopteryx* and *Diura* (Plecoptera) in a California stream. *Arch. Hydrobiol.* **69**, 521–546.

Sheldon, A. L. and Oswood, M. W. (1977). Blackfly (Diptera: Simuliidae) abundance in a lake outlet: test of a predictive model. *Hydrobiologia* **56**, 113–120.

Shemanchuk, J. A. (1980). Protection of cattle on farms. *In* Control of black flies in the Athabasca River: Technical Report, pp. 215–

216 (W. O. Haufe and G. C. R. Croome, Eds.). Edmonton. Alta. Environ.

Shemanchuk, J. A. and Humber, R. A. (1978). *Entomophthora culicis* (Phycomycetes: Entomophthorales) parasitizing black fly adults (Diptera: Simuliidae) in Alberta. *Can. Ent.* **110**, 253–256.

Shewell, G. E. (1955). Identity of the black fly that attacks ducklings and goslings in Canada (Diptera: Simuliidae). *Can. Ent.* **87**, 345–349.

Shewell, G. E. (1958). Classification and distribution of Arctic and Subarctic Simuliidae. *Proc. Xth Int. Congr. Ent.* (1956) **1**, 635–643.

Shipitsina, N. K. (1963). Parasitic infection of simuliids (Diptera) and its effect on ovarian function. *Zool. Zh.* **42**, 291–294. [In Russian].

Sickmann, F. (1893). Die Hymenopterenfauna von Iburg und seiner nächsten Umgebung, mit biologischen und kritischen Bemerkungen. I. Abteilung die Grabwespen. *Jber. naturw. ver. Osnabr.* **9**, 39–112.

Simmonds, F. J., Nolan, R. A., Briggs, J. D. and Myers, R. F. (1977). Mass production of parasitoids, parasites, and pathogens. *In* Tsetse: The future for biological methods in integrated control, pp. 149–156 (M. Laird, Ed.). Ottawa. IDRC (Publ. No. IDRC–077e).

Simmons, K. R. and Edman, J. D. (1978). Successful mating, oviposition, and complete generation rearing of the multivoltine blackfly *Simulium decorum* (Diptera: Simuliidae). *Can. J. Zool.* **56**, 1223–1225.

Slack, H. D. (1936). The food of caddis fly (Trichoptera) larvae. *J. Anim. Ecol.* **5**, 105–115.

Smart, J. (1934). On the biology of the black fly, *Simulium ornatum* Mg. (Diptera: Simuliidae). *Proc. R. phys. Soc. Edinb.* **22**, 217–238.

Smart, J. (1945). The classification of the Simuliidae (Diptera). *Trans. R. ent. Soc. Lond.* **95**, 463–528.

Smart, J. and Clifford, E. A. (1965). Simuliidae (Diptera) of the Territory of Papua and New Guinea. *Pacif. Insects* **7**, 505–619.

Smith, S. M. (1966). Observations on some mechanisms of host finding and host selection in the Simuliidae and Tabanidae (Diptera). MSc Thesis. McMaster Univ., Hamilton, Ontario, Canada.

Smyly, W. J. P. (1955). On the biology of the stone-loach *Nemacheilus barbatula* (L.). *J. Anim. Ecol.* **24**, 167–186.

Snoddy, E. L. (1967). The common grackle, *Quiscalus quiscula quiscula* L. (Aves: Icter-

idae), a predator of *Simulium pictipes* Hagen larvae (Diptera: Simuliidae). *J. Georgia ent. soc.* **2**, 45–46.

Snoddy, E. L. (1968). Simuliidae, Ceratopogonidae and Chloropidae as prey of *Oxybelus emarginatum*. *Ann. ent. Soc. Am.* **61**, 1029–1030.

Snoddy, E. L. and Hays, K. L. (1966). A carbon dioxide trap for Simuliidae (Diptera). *J. econ. Ent.* **59**, 242–243.

Södergren, A., Svensson, B. and Ulfstrand, S. (1971). DDT and PCB in south Swedish streams. *Envir. Poll.* **3**, 25–36.

Sommerman, K. M. (1962). Notes on two species of *Oreogeton* predaceous on black fly larvae. Diptera: Empididae and Simuliidae. *Proc. ent. Soc. Wash.* **64**, 123–129.

Sommerman, K. M., Sailer, R. I. and Esselbaugh, C. O. (1955). Biology of Alaskan black flies (Simuliidae: Diptera). *Ecol. Monogr.* **25**, 345–385.

Soper, R. S., Holbrook, F. R., Majchrowicz, I. and Gordon, C. C. (1975). Production of *Entomophthora* resting spores for biological control of aphids. *Life Sc. and Agric. Exp. Stn Univ. Me of Orono. Tech. Bull.* 76. Mimeogr. Doc.

Soponis, A. R. and Peterson, B. V. (1976). A preliminary investigation of some morphological characters in adult females of the *Simulium damnosum* complex from Togo. *WHO/VBC/SC/* 76. Mimeogr. Doc.

Speir, J. A. (1976). The ecology and production dynamics of four blackfly species (Diptera: Simuliidae) in western Oregon streams. PhD Thesis, Oregon State Univ.

Stankovitch, S. (1924). Alimentation naturelle de la truite (*Trutta fario*, Linné) dans les cours d'eau alpins. *Trav. Lab. hydrobiol. Piscic. Univ. Grenoble* **14**, 115–191.

Stauffer, J. R. Jr., Wilson, J. H. and Dickson, K. L. (1976). Comparison of stomach contents and condition of two catfish species living at ambient temperatures and in a heated discharge. *Progve Fish Cult.* **38**, 33–35.

Stewart, K. W., Friday, G. P. and Rhame, R. E. (1973). Food habits of hellgrammite larvae, *Corydalus cornutus* (Megaloptera: Corydalidae), in the Brazos River, Texas. *Ann. ent. Soc. Am.* **66**, 959–963.

Steyskal, G. C. (1972). The meaning of the term 'sibling species'. *Syst. Zool.* **21**, 446.

Stone, A. (1964). Guide to the insects of Connecticut. Part VI. The Diptera or true flies of Connecticut. Ninth Fasc. (part).

Family Simuliidae. *Bull. Conn. St. geol. nat. Hist. Surv.* **97**, 1–126.

Stone, A. and Snoddy, E. L. (1969). The black flies of Alabama (Diptera: Simuliidae). *Bull. Ala. agric. Exp. Stn.* **390**, 1–93.

Strand, M. A., Bailey, C. H. and Laird, M. (1977). Pathogens of Simuliidae (Blackflies). *In* Pathogens of medically important arthropods, pp. 213–255 (D. W. Roberts and M. A. Strand, Eds.). *Bull. Wld Hlth Org.* **55**, (Suppl. 1), 1–419.

Straškraba, M., Čihař, J., Franks, S. and Hruška, V. (1966). Contribution to the problem of food competition among the sculpin, minnow and brown trout. *J. Anim. Ecol.* **35**, 303–311.

Strelkov, A. (1964). The biology of a new mermithid from a midge of a piscine reservoir. *Vest. Leningr. Univ. (Biol.)* **3**, 55–69. [In Russian].

Strickland, E. H. (1911). Some parasites of *Simulium* larvae and their effects on the development of the host. *Biol. Bull. mar. biol. Lab. Woods Hole* **21**, 302–338.

Strickland, E. H. (1913). Further observations on the parasites of *Simulium* larvae. *J. Morph.* **24**, 43–105.

Strong, R. P. (1931). *Onchocerca* investigations in Guatemala. Report of progress of the Harvard Expedition. *New Engl. J. Med.* **204**, 916–920.

Strong, R. P., Sandground, J. H., Bequaert, J. C. and Muñoz Ochoa, M. (1934). Onchocerciasis with special reference to the Central American form of the disease. Contr. Harv. Inst. trop. Biol. Med. **6**, 1–234.

Sun, Y. P. (1968). Dynamics of insect toxicology—A mathematical and graphical evaluation of the relationship between insect toxicity and rates of penetration and detoxication of insecticides. *J. econ. Ent.* **61**, 949–955.

Sun, Y. P. (1970). Dynamics of insect toxicology and its relationship to performance, synergism and structure—activity relationship of insecticides. *In* Biochemical Toxicology of Insecticides. (R. O'Brien and I. Yamamoto, Eds.). New York. Academic Press.

Sutcliffe, J. F. and McIver, S. B. (1975). Artificial feeding of simuliids (*Simulium venustum*): factors associated with probing and gorging. *Experientia* **31**, 694–695.

Sutcliffe, J. F. and McIver, S. B. (1976). External morphology of sensilla on the legs of selected black fly species (Diptera: Simuliidae). *Can. J. Zool.* **54**, 1779–1787.

Suzuki, T., Ito, Y. and Harada, S. (1963). A record of blackfly larvae resistant to DDT in Japan. *J. exp. Med.* **33**, 41–46.

Syed Hyder, Ali, Burbutis, P. P., Ritter, W. F. and Lake, R. W. (1974). Blackfly (*Simulium vittatum* Zetterstedt) densities and water quality conditions in Red Clay Creek Pa.-Del. *Envir. Ent.* **3**, 879–881.

Syme, P. D. and Davies, D. M. (1958). Three new Ontario black flies of the genus *Prosimulium* (Diptera: Simuliidae). Part I. Descriptions, morphological comparisons with related species, and distribution. *Can. Ent.* **90**, 697–719.

Tack, E. (1940). Die Ellritz (*Phoxinus laevis* Ag.): eine monographische Bearbeitung. *Arch. Hydrobiol.* **37**, 321–425.

Tada, I. (1974). Report on the visit to Guatemala for studies of onchocerciasis. *OTCA*.

Tada, I. (1976). Onchocerciasis. *J. Kanazawa Med. Univ.* **1**, 227–245.

Tada, I. *et al.* (1977). Onchocerciasis in San Vicente Pacaya, Guatemala. *WHO/ONCHO/77.140.* Mimeogr. Doc.

Tada, I. *et al.* (1978a). Distribution of onchocerciasis cases in San Vicente Pacaya. *Onchocerciasis control project in Guatemala, First report. JICA* 94–106.

Tada, I. *et al.* (1978b). Studies on onchocerciasis in Ethiopia and Guatemala. *Research in filariasis and shistosomiasis.* **3**, 173–194.

Tanabe, T. (1970). Feeding habits of some fishes in the Yoshimo-gawa river. *Nara Rikusui Seibutsu Gakuho* **3**, 9–17.

Tarshis, I. B. (1968). Collecting and rearing black flies. *Ann. ent. Soc. Am.* **61**, 1072–1083.

Tarshis, I. B. (1973). Studies on the collection, rearing, and biology of the blackfly (*Cnephia ornithophilia*). *Fish Wildl. Ser. Wash.*, Special Scientific Report—Wildlife No. 165.

Tarshis, I. B. and Neil, W. (1970). Mass movement of black fly larvae on silken threads (Diptera: Simuliidae). *Ann. ent. Soc. Am.* **63**, 607–610.

Taufflieb, R. (1955). Une campagne de lutte contre *Simulium damnosum* au Mayo Kebbi. *Bull. Soc. Path. exot.* **48**, 564–576.

Thaxter, R. (1888). The Entomophthoreae of the United States. *Mem. Boston Soc. nat. Hist.* **4**, 133–201.

Thomas, J. D. (1962). The food and growth of brown trout (*Salmo trutta* L.) and its feeding relationships with the salmon par (*Salmo salar* L.) and the eel (*Anguilla anguilla* (L.)) in the River Teify, West Wales. *J. Anim. Ecol.* **31**, 175–205.

Thomas, K. A. (1977). Enzyme studies in the *Simulium (Edwardsellum) damnosum* complex and *Aedes (Stegomyia) scutellaris* group. MSc Thesis. University of Liverpool.

Thompson, B. H. (1976a). The intervals between the bloodmeals of man-biting *Simulium damnosum* (Diptera: Simuliidae). *Ann. trop. Med. Parasit.* **70**, 329–342.

Thompson, B. H. (1976b). Studies on the flight range and dispersal of *Simulium damnosum* (Diptera: Simuliidae) in the rainforest of Cameroon. *Ann. trop. Med Parasit.* **70**, 343–354.

Thompson, B. H. (1976c). Studies on the attraction of *Simulium damnosum s.l.* (Diptera: Simuliidae) to its hosts. I. The relative importance of sight, exhaled breath, and smell. *Tropenmed. Parasit.* **27**, 455–473.

Thompson, B. H. (1976d). Studies on the attraction of *Simulium damnosum s.l.* (Diptera: Simuliidae) to its hosts. II. The nature of substances on the human skin responsible for attractant olfactory stimuli. *Tropenmed. Parasit.* **27**, 83–90.

Thompson, B. H. (1977). Studies on the attraction of *Simulium damnosum s.l.* (Diptera: Simuliidae) to its hosts. III. Experiments with animal-baited traps. *Tropenmed. Parasit.* **28**, 226–228.

Thompson, B. H. and Adams, B. G. (1979). Laboratory and field trials using Altosid® insect growth regulator against blackflies (Diptera: Simuliidae) of Newfoundland Canada. *J. med. Ent.* **16**, 536–546.

Thorsteinson, A. J., Bracken, G. K. and Hanec, W. (1965). The orientation behaviour of horse flies and deer flies (Tabanidae: Diptera). III. The use of traps in the study of orientation of tabanids in the field. *Entomologia exp. appl.* **8**, 189–192.

Thorup, J. (1974). Occurrence and size distribution of Simuliidae (Diptera) in a Danish spring. *Arch. Hydrobiol* **74**, 316–335.

Thylefors, B. and Rolland, A. (1977). Situation dans un foyer d'onchocercose du Mali après treize ans de controle anti-simulidien. 2. Aspects oculaires. *Annls. Soc. belge Méd. trop.* **57**, 577–582.

Timofeeva, L. V., Mitrofanov, A. M., Markovich, N. Ya., Murav'yeva, T. V., Shvan'kov, M. E. and Tupitzin, L. F. (1962). A successful experiment in controlling blood-feeding blackflies (Diptera: Simuliidae) by treating their hatching sites. Preliminary communication. *Med. Parazitol.* **1**, 3–9.

Tippets, W. E. and Moyle, P. B. (1978). Epibenthic feeding by rainbow trout (*Salmo gairdneri*) in the McCloud River, California. *J. Anim. Ecol.* **47**, 549–559.

Townson, H. and Meredith, S. E. O. (1979). Identification of the Simuliidae in relation to onchocerciasis. *In* Problems in the identification of parasites and their vectors. 17th Symposium of the British Society for Parasitology, pp. 145–174 (A. E. R. Taylor and R. Muller, Eds.). Oxford. Blackwell Scientific.

Traver, J. R. (1939). Himalayan mayflies (Ephemeroptera). *Ann. Mag. nat. Hist.* (11)**4**, 32–56.

Travis, B. V. and Guttman, D. (1966). Additional tests with blackfly larvicides. *Mosquito News,* **26**, 157–160.

Travis, B. V. and Schuchman, S. M. (1968). Tests (1967) with blackfly larvicides. *J. econ. Ent.* **61**, 843–845.

Travis, B. V. and Wilton, D. P. (1965). A progress report on simulated stream tests of black fly larvicides. *Mosquito News* **25**, 112–118.

Trpis, M. (1977). Autogeny in diverse populations of *Aedes aegypti* from East Africa. *Tropenmed. Parasit.* **28**, 77–82.

Trpis, M. (1978). Genetics of hematophagy and autogeny in the *Aedes scutellaris* complex (Diptera: Culicidae). *J. med. Ent.* **15**, 73–80.

Tsacas, L. and Disney, R. H. L. (1974). Two new African species of *Drosphila* (Diptera: Drosophilidae) whose larvae feed on *Simulium* larvae. *Tropenmed. Parasit.* **25**, 360–377.

Twinn, C. R. (1939). Notes on some parasites and predators of blackflies (Simuliidae: Diptera). *Can. Ent.* **71**, 101–105.

Uemoto, K. (1971). Blackfly control in Kyoto City. *Prog. sanit. Zool.* **1**, 137–159.

Ulfstand, S. (1967). Microdistribution of benthic species (Ephemeroptera, Plecoptera, Trichoptera, Diptera: Simuliidae) in Lapland streams. *Oikos,* **18**, 293–310.

Undeen, A. H. and Berl, D. (1979). Laboratory studies on the effectiveness of *Bacillus thuringiensis* var. *israelensis* de Barjac against *Simulium damnosum* (Diptera: Simuliidae) larvae. *Mosquito News* **39**, 742–745.

Undeen, A. H. and Nagel, W. L. (1978). The effect of *Bacillus thuringiensis* ONR–60A strain (Goldberg) on *Simulium* larvae in the laboratory. *Mosquito News* **38**, 524–527.

Undeen, A. H. and Nolan, R. A. (1977). Ovarian infection and fungal spore oviposition in the blackfly *Prosimulium mixtum*. *J. Invert. Pathol.* **30,** 97–98.

Undeen, A. H., Takaoka, H. and Hansen, K. (1980). The evaluation of *Bacillus thuringiensis* var. *israelensis* de Barjac as a larvicide for use against *Simulium ochraceum*, the Central American vector of onchocerciasis. *Mosquito News* **41,** 37–40.

Ussova, Z. V. (1961). Flies of the Karelia and the Murmansk region (Diptera: Simuliidae). *Izdatel'stvo Akad. Nauk.,* SSSR [In Russian]. [Translated by Israel Program for Scientific Translations (1964)].

Vaillant, F. (1951). Un empidide destructeur de simulies. *Bull. Soc. zool. Fr.* **76,** 371–379.

Vaillant, F. (1953). *Hemerodromia seguyi*, nouvel empidide d'Algérie destructeur de simulies. *Hydrobiologia*, **5,** 180–188.

Vajime, C. G. and Dunbar, R. W. (1975). Chromosomal identification of eight species of the subgenus *Edwardsellum* near and including *Simulium* (*Edwardsellum*) *damnosum* Theobald (Diptera: Simuliidae). *Tropenmed. Parasit.* **26,** 111–138.

Vajime, C. G. and Quillévéré, D. (1978). The distribution of the *Simulium* (*Edwardsellum*) *damnosum* complex in West Africa with particular reference to the Onchocerciasis Control Programme area. *Tropenmed. Parasit.* **29,** 473–482.

Vale, G. A. (1974a). New field methods for studying the responses of tsetse flies (Diptera: Glossinidae) to hosts. *Bull. ent. Res.* **64,** 199–208.

Vale, G. A. (1974b). The responses of tsetse flies (Diptera: Glossinidae) to mobile and stationary baits. *Bull. ent. Res.* **64,** 545–588.

Van Den Bosch, R. (1971). Biological control of insects. *A. Rev. Ecol. Syst.* **2,** 45–66.

Van Den Heuvel, M. J. (1963). The effect of rearing temperature on the wing length, thorax length, leg length and ovariole number of the adult mosquito, *Aedes aegypti* (L.). *Trans. R. ent. Soc. Lond.* **115,** 197–216.

Van Someren, V. D. and McMahon, J. P. (1950). Phoretic association between *Afronurus* and *Simulium* species, and the discovery of the early stages of *Simulium naevei* on freshwater crabs. *Nature, Lond.* **166,** 350–351.

Vargas, L. (1945). Simúlidos del Nuevo Mundo. *Monografias Inst. Salubr. Enferm trop. Méx.* **1,** 1–241.

Vargas, L. and Díaz Nájera, A. (1957). Simúlidos mexicanos. *Revta. Inst. Salubr. Enferm trop. Méx.* **17,** 143–399.

Vargas, M., Rubtsov, I. A. and Fallas, B. (1980). Bionomics of black flies (Diptera: Simuliidae) in Costa Rica. V. Description of *Neomesomermis travisi* sp.n. (Nematode: Mermithidae). *Rev. Biol. Trop.* **28.** 73–89.

Vaught, G. L. and Stewart, K. W. (1974). The life history and ecology of the stonefly *Neoperla clymene* (Newman) (Plecoptera: Perlidae). *Ann. ent. Soc. Am.* **67,** 167–178.

Vaux, W. G. (1962). Interchange of stream and intergravel water in a salmon spawning riffle. *Rep. Fish Wildl. Serv. US* **405.**

Verdat, F. J. (1822). Mémoire pour servir à l'historie des simulies, genre d'insectes de l'ordre des diptères, famille des tipulaires; lu à la réunion de la Société helvétique des Sciences naturelles à Bâle, le 25 Juillet 1821. *Naturw. Anz.* **9,** 65–70.

Vidy, G. (1976). Étude du regime alimentaire de quelques poissons insectivores dans les rivières de Côté d'Ivoire. *ORSTOM, Bouaké Rap. 2.* Mimeogr. Doc.

Vladimirova, V. V. and Popapov, A. A. (1963). New types of traps for horse flies and blackflies. *Medskaya Parazit.* **32,** 83–87.

Von Linstow, O. F. B. (1898). Das Genus *Mermis. Arch. mikrosk. Anat.* **53,** 149–168.

Von Linstow, O. F. B. (1905). Helminthologische Beobachtungen. *Arch. mikrosk. Anat.* **66,** 355–366.

Vulcano, M. A. (1967). Family Simuliidae. *In* A catalogue of the Diptera of the Americas south of the United States, Fasc. 16. São Paulo. Departamento de Zoologia, Secretaria da Agricultura.

Wächtler, K., Rühm, W. and Welsch, U. (1971). Histologische und histochemische Untersuchungen an den Speicheldrüsen der Imagines von *Boophthora erythrocephala* de Geer und *Odagmia ornata* (Meig.). (Diptera: Simuliidae). *Z. angew. Ent.* **67,** 189–201.

Waddy, B. B. (1963). A project for the control of Onchocerciasis in the Volta Basin. *WHO/ PA/53.63.* Mimeogr. Doc.

Wade, J. O. (1976). A new design of membrane feeder incorporating an electrical blood stirring device. *Ann. trop. Med. Parasit.* **70,** 113–120.

Waiwood, K. G. and Johansen, P. H. (1974). Oxygen consumption and activity of the white sucker (*Catostomus commersoni*), in lethal and nonlethal levels of the organochlorine insecticide, methoxychlor. *Wat. Res.* **8**, 401–406.

Wallace, R. R. (1971). The effects of several insecticides on blackfly larvae and on other stream-dwelling aquatic invertebrates. MSc Thesis. Queen's Univ., Kingston, Ontario.

Wallace, R. R. (1973). The effect of methoxychlor on, and the accumulation of methoxychlor in, some insects of running waters. PhD Thesis, Univ. Waterloo, Ontario.

Wallace, R. R. and Hynes, H. B. N. (1975). The catastrophic drift of stream insects after treatments with methoxychlor (1,1,1-Trichloro-2,2-Bis (p-methoxyphenyl)ethane). *Envir. Poll.* **8**, 255–268.

Wallace, R. R., Hynes, H. B. N. and Kaushik, N. K. (1975). Laboratory experiments on factors affecting the activity of *Gammarus pseudolimnaeus* Bousfield. *Freshwat. Biol.* **5**, 533–546.

Wallace, R. R., Hynes, H. B. N. and Merritt, W. (1976). Laboratory and field experiments with methoxychlor as a larvicide for Simuliidae (Diptera). *Environ. Poll.* **10**, 251–269.

Wallace, R. R., West, A. S., Downe, A. E. R. and Hynes, H. B. N. (1973). The effects of experimental blackfly (Diptera: Simuliidae) larviciding with abate, dursban, and methoxychlor on stream invertebrates. *Can. Ent.* **105**, 817–831.

Walsh, F. [sic, J. F. W.] (1979). Perils of pesticiding or virtue unrewarded. *Antenna* **3**, 128–129.

Walsh, J. F. (1970a). The control of *Simulium damnosum* in the River Niger and its tributaries in relation to the Kainji Lake research project, covering the period 1961 to 1969. *WHO/PD/70.4.* Mimeogr. Doc.

Walsh, J. F. (1970b). Evidence of reduced susceptibility to DDT in controlling *Simulium damnosum* (Diptera: Simuliidae) on the River Niger. *Bull. Wld Hlth Org.* **43**, 316–318.

Walsh, J. F. (1972). Observations on the resting of *Simulium damnosum* in trees near a breeding site in the West African savanna. *WHO/ONCHO/72.99.* Mimeogr. Doc.

Walsh, J. F. (1978). Light trap studies on *Simulium damnosum s.l.* in northern Ghana. *Tropenmed. Parasit.* **29**, 492–496.

Walsh, J. F., Davies, J. B., Le Berre, R. and Garms, R. (1978). Standardization of criteria for assessing the effect of *Simulium* control in onchocerciasis control programmes. *Trans. R. Soc. trop. Med. Hyg.* **72**, 675–676.

Wang, W., Lee, G. F. and Spyridakis, D. (1972). Adsorption of parathion in a multi-component solution. *Wat. Res.* **6**, 1219–1228.

Wanson, M. and Henrard, C. (1945). Habitat et comportement larvaire du *Simulium damnosum* Theobald. *Recl Trav. Sci. méd. Congo belge* **4**, 113–121.

Wanson, M., Henrard, C. and Peel, E. (1945). *Onchocerca volvulus* Leuckart, indices d'infection des simulies agressives pour l'homme. Cycle de dévelopement chez *Simulium damnosum* Theobald. *Recl Trav. Sci. méd. Congo belge* **4**, 122–137.

Wanson, M. and Lebied, B. (1948). Note sur le cycle gonotrophique de *Simulium damnosum. Revue Zool. Bot. afr.* **41**, 66–82.

Warren, C. and Davies, G. (1971). Laboratory stream research: Objectives, possibilities and constraints. *A. Rev. Ecol. Syst.* **2**, 111–144.

Watanabe, M. (1977). Observations on nectar-sucking behaviour and parous rates of five species of blackflies. *Jap. J. sanit. Zool.* **28**, 401–407.

Watanabe, M. (1978). Observations on the gonotrophic cycle of *Simulium ochraceum.* Onchocerciasis control project in Guatemala, First report. *JICA*, 61–68.

Waters, T. F. (1964). Recolonization of denuded stream bottom areas by drift. *Trans. Am. Fish. Soc.* **93**, 311–315.

Waters, T. F. (1965). Interpretation of invertebrate drift in streams. *Ecology* **46**, 327–334.

Waters, T. F. (1969). Invertebrate drift-ecology and significance to stream fishes. *In* Symposium-salmon and trout in streams, p. 121–134. H. R. MacMillan Lectures in Fisheries (T. G. Northcote, Ed.). Vancouver. Univ. British Columbia.

Waters, T. F. (1972). The drift of stream insects. *A. Rev. Ent.* **17**, 253–272.

Webb, F. E. and MacDonald, D. R. (1958). Studies of aerial spraying against the spruce budworm in New Brunswick. *Forest. Biol. Lab. Fredericton, N.B. Can. Dept. Agric. Sci. Service, Forest Biol. Sec.* Mimeogr. Doc.

Webster, F. M. (1902). Winds and storms as agents in the diffusion of insects. *Am. Nat.* **36**, 795–808.

Webster, F. M. (1914). Natural enemies of *Simulium*: notes. *Psyche* **21**, 95–99.

Weed, C. W. (1904). Experiments in destroying blackflies. *Bull. New Hamps. agric. Exp. Stn* **12**, 133–136.

Wegesa, P. (1970a). The present status of onchocerciasis in Tanzania. A review of the distribution and prevalence of the disease. *Trop. geogr. Med.* **22**, 345–351.

Wegesa, P. (1970b). *Simulium nyasalandicum* (Amani form) and *S. adersi*, two new potential vectors of *Onchocerca volvulus* in the Eastern Usambaras, north-eastern Tanzania. *E. Afr. med. J.* **47**, 364–367.

Weiser, J. (1946). Studie o mikrosporidiich z larev hyzn nasich vod. *Věst. čsl. zool. Spol.* **10**, 245–272.

Weiser, J. (1947). Klič k určování Mikrosporidií. *Acta. Soc. Scient. nat. moravo-siles* **18**, 1–64.

Weiser, J. (1960). *Thelohania columbaczense* n.sp., a new microsporidian of *Simulium reptans columbaczense* (Diptera: Simuliidae). *Věst. čsl. zool. Spol.* **24**, 196–198.

Weiser, J. (1966). Nemoci hymzu, Academia.

Weiser, J. (1968). Iridescent virus from the blackfly *Simulium ornatum* Meigen in Czechoslovakia. *J. Invert. Pathol.* **12**, 36–39.

Weiser, J. (1977). An Atlas of Insect Diseases (2nd revised Ed.). The Hague, The Netherlands. Dr W. Junk B.V. Publishers (Co-edition with Academia, Prague).

Weiser, J. (1978). A new host *Simulium argyreatum* for the cytoplasmic polyhedrosis virus of blackflies in Czechoslovakia. *Folia Parasitol.* **25**, 361–365.

Weiser, J. and Vankova, J. (1978). Toxicity of *Bacillus thuringiensis israelensis* for blackflies and other freshwater invertebrates. *Proc. Int. Colloq. Invert. Pathol. Prague*, 11–17 Sept., 243–244.

Weiser, J. and Zizka, Z. (1978). Thick walled sporangia of *Coelomycidium simulii*. *Česka. mykologie* (in press).

Weitz, B. (1952). The antigenicity of sera of man and animals in relation to the preparation of specific precipitating antisera. *J. Hyg.* **50**, 275–294.

Weitz, B. (1956). Identification of blood meals of blood-sucking arthropods. *Bull. Wld Hlth Org.* **15**, 473–490.

Weitz, B. and Burton, P. A. (1953). The rate of digestion of blood meals of various haematophagous arthropods as determined by the precipitin test. *Bull. ent. Res.* **44**, 445–450.

Welch, H. E. (1958). *Agamomermis pachysoma* (Linstow, 1905). n. comb. (Mermithidae: Nematoda), a parasite of social wasps. *Insectes soc.* **5**, 353–355.

Welch, H. E. (1962). New species of *Gastromermis*, *Isomermis* and *Mesomermis* (Nematoda: Mermithidae) from black fly larvae. *Ann. ent. Soc. Am.* **55**, 535–542.

Welch, H. E. (1964). Mermithid parasites of blackflies. *Bull. Wld Hlth Org.* **31**, 857–863.

Welch, H. E. and Rubtsov, I. A. (1965). Mermithids (Nematoda: Mermithidae) parasitic in blackflies (Insecta: Simuliidae). I. Taxonomy and bionomics of *Gastromermis boophthorae* sp.n. *Can. Ent.* **97**, 581–596.

Welling, W. (1977). Dynamic aspects of insecticide interactions. *A. Rev. Ent.* **22**, 53–78.

Welsch, U., Wächtler, K. and Rühm, W. (1968). Die Feinstruktur der Speicheldrüse von *Boophthora erythrocephala* de Geer (Simuliidae: Diptera) vor und nach der Blutaufnahme. *Z. Zellforsch. microsk. Anat.* **88**, 340–352.

Wenk, P. (1962). Anatomie des Kopfes von *Wilhelmia equina* L. ♀ (Simuliidae syn. Melusinidae, Diptera). *Zool. Jb.* (Anat.) **80**, 81–134.

Wenk, P. (1965a). Über die Biologie blutsaugender Simuliiden (Diptera). I. Besamungsrate der ♀♀ beim Blütenbesuch und Anflug auf den Blutwirt. *Z. Morph. ökol. Tiere* **55**, 656–670.

Wenk, P. (1965b). Über die Biologie blutsaugender Simuliiden (Diptera). II. Schwarmverhalten, Geschlechterfindung und Kopulation. *Z. Morph. ökol. Tiere* **55**, 671–713.

Wenk, P. (1965c). Über die Biologie blutsaugender Simuliiden (Diptera). III. Kopulation, Blutsaugen und Eiablage von *Boophthora erythrocephala* de Geer im Laboratorium *Z. Tropenmed. Parasit.* **16**, 207–226.

Wenk, P. (1976). Koevolution von Überträger und Parasit bei Simuliiden und Nematoden. *Z. angew Ent.* **82**, 38–44.

Wenk, P. and Raybould, J. (1972). Mating, blood feeding and oviposition of *Simulium damnosum* Theobald in the laboratory. *Bull. Wld Hlth Org.* **47**, 627–634.

Wenk, P. and Schlörer, G. (1963). Wirtsorientierung und Kopulation bei blutsaugenden Simuliiden (Diptera). *Z. Tropenmed. Parasit.* **14**, 177–191.

Wenk, P. and Schulz-Key, H. (1974). Versuche zur quantitativen Wirkungsbestimmung von Larviziden gegen Simuliiden in Westafrika. *Tropenmed. Parasit.* **25**, 381–394.

West, A. S. (1961). Biting fly control on the Quebec north shore. *Proc. New Jers. Mosq. Exterm. Ass.* **48**, 87–96.

West, A. S. (1971). Canada Region. *Mosquito News* **31**, 345–346.

West, A. S., Baldwin, W. F. and Gomery, J. (1971). A radioisotopic-sticky trap-autoradiographic technique for studying the dispersal of black flies. *WHO/ONCHO/ 71.84. WHO/VBC/71.280.* Mimeogr. Doc.

White, H. C. (1936). The food of salmon fry in eastern Canada. *J. Fish. Res. Bd Can.* **2**, 499–506.

Whitton, B. A. (Ed.) (1975). River Ecology. Studies in Ecology 2. Oxford. Blackwell Scientific.

Wilhelmi, J. In Edwards, F. W. (1920). On the British species of *Simulium*. II. early stages, with corrections and additions to Part I. *Bull. ent. Res.* **11**, 211–246.

Wilhm, J. (1972). Graphical and mathematical analyses of biotic communities in polluted streams. *A. Rev. Ent.* **17**, 223–252.

Wilkes, F. G. and Weiss, C. M. (1971). The accumulation of DDT by the dragonfly nymph, *Tetragoneuria*. *Trans. Am. Fish. Soc.* **100**, 222–236.

Williams, C. B. (1962). Studies on black flies (Diptera: Simuliidae) taken in a light trap in Scotland. III. The relation of night activity and abundance to weather conditions. *Trans. R. ent. Soc. Lond.* **114**, 28–47.

Williams, C. B. (1964). Patterns in the Balance of Nature and Related Problems in Quantitative Ecology. London. Academic Press.

Williams, C. B. and Davies, L. (1957). Simuliidae attracted at night to a trap using ultra-violet light. *Nature, Lond.* **179**, 924–925.

Williams, D. D. and Hynes, H. B. N (1974). The occurrence of benthos deep in the substratum of a stream. *Freshwat. Biol.* **4**, 233–255.

Williams, T. R., Connolly, R. C., Hynes, H. B. N. and Kershaw, W. E. (1961). Size of particles ingested by *Simulium* larvae. *Nature, Lond.* **189**, 76.

Williams, T. R., Connolly, R. C. Hynes, H. B. N. and Kershaw, W. E. (1961). The size of particulate material ingested by *Simulium* larvae. *Ann. trop. Med. Parasit.* **55**, 125–127.

Williams, T. R. and Hynes, H. B. N. (1971). A survey of the fauna of streams on Mount Elgon, East Africa, with special reference to the Simuliidae (Diptera). *Freshwat. Biol.* **1**, 227–248.

Williams, T. R. and Obeng, L. (1962). A comparison of two methods of estimating changes in *Simulium* larval populations with a description of a new method. *Ann. trop. Med. Parasit.* **56**, 359–361.

Wilson, R. and Snow, D. (1972). Non-target effects of abate, an organophosphorus insecticide. Unpub. rept. of a field program to evaluate blackfly control methods at Baie Verte, Nfld. Ms. Rept. 72–8, Environment Canada, St John's, Newfoundl. Mimeogr. Doc.

Wilton, D. P. and Travis, B. V. (1965). An improved method for simulated stream tests for blackfly larvicides. *Mosquito News* **25**, 118–123.

Winner, R. A., Steelman, C. D. and Schilling, P. E. (1978). Effects of selected insecticides on *Romanomermis culicivorax*, a mermithid nematode parasite of mosquito larvae. *Mosquito News* **38**, 546–553.

Wirth, W. W. and Stone, A. (1956). Aquatic Diptera. *In* Aquatic Insects of California, pp. 372–482 (R. L. Usinger, Ed.). Univ. Calif. Press.

Wirtz, H. P. (1976). Untersuchungen über den Einfluss der Blutnahrung auf die Eientwicklung von *Boophthora erythrocephala* de Geer und *Wilhelmia lineata* Meigen (Diptera: Simuliidae). Tübingen. Diplomarbeit.

Wolfe, L. S. and Peterson, D. G. (1959). Black flies (Diptera: Simuliidae) of the forests of Quebec. *Can. J. Zool.* **37**, 137–159.

Wolfe, L. S. and Peterson, D. G. (1960). Diurnal behavior and biting habits of black flies (Diptera: Simuliidae) in the forests of Quebec. *Can. J. Zool.* **38**, 489–497.

Wood, D. M. (1978). Taxonomy of the Nearctic species of *Twinnia* and *Gymnopais* (Diptera: Simuliidae) and a discussion of the ancestry of the Simuliidae. *Can. Ent.* **110**, 1297–1337.

Wood, D. M. and Davies, D. M. (1964). The rearing of simuliids (Diptera). *Proc. XII Int. Congr. Ent. London*, 821–823.

Wood, D. M. and Davies, D. M. (1966). Some methods of rearing and collecting black flies (Diptera: Simuliidae). *Proc. ent. Soc. Ont.* **96**(1965), 81–90.

Wood, D. M., Peterson, B. V., Davies, D. M. and Gyorkos, H. (1963). The black flies (Diptera: Simuliidae) of Ontario. Part II. Larval identification. With descriptions and illustrations. *Proc. ent. Soc. Ont.* **93** (1962), 99–129.

World Health Organization (1969). Joint USAID/OCCGE/WHO technical meeting on the feasibility of onchocerciasis control. Tunis, 1–8 July 1968. *WHO/ONCHO/69.75.* Mimeogr. Doc.

World Health Organization (1973). Onchocerciasis control in the Volta river basin area. Annex III–1: The distribution and biology of the vector and transmission of onchocerciasis. *OCP/73.1.* Mimeogr. Doc.

World Health Organization (1976). Epidemiology of Onchocerciasis. Report of a WHO Expert Committee. Techn. Rep. Series No. 597, 94 pp.

World Health Organization (1977). Informal Consultation. Species complexes in insect vectors of disease (Blackflies, mosquitoes, tsetse flies). *WHO/VBC/77.656; WHO/ONCHO/77.131.* Mimeogr. Doc.

World Health Organization (1978). Species complexes in the Simuliidae. *Bull. Wld Hlth Org.* **56**, 53–61.

Wotton, R. S. (1976). Evidence that blackfly larvae can feed on particles of colloidal size. *Nature, Lond.* **261**, 697.

Wotton, R. S. (1977). The size of particles ingested by moorland stream blackfly larvae (Simuliidae). *Oikos* **29**, 332–335.

Wotton, R. S. (1978a). The feeding-rate of *Metacnephia tredecimatum* larvae (Diptera: Simuliidae) in a Swedish lake outlet. *Oikos* **30**, 121–125.

Wotton, R. S. (1978b). Growth, respiration and assimilation of blackfly larvae (Diptera: Simuliidae) in a lake-outlet in Finland. *Oecologia* **33**, 279–290.

Wotton, R. S. (1978c). Life-histories and production of blackflies (Diptera: Simuliidae) in moorland streams in Upper Teesdale, Northern England. *Arch. Hydrobiol.* **83**, 232–250.

Wotton, R. S. (1979). The influence of a lake on the distribution of blackfly (Diptera: Simuliidae) species along a river. *Oikos* **32**, 368–372.

Wotton, R. S., Friberg, F., Hermann, J., Malmqvist, B., Nilsson, L. M. and Sjöström, P. (1979). Drift and colonization of three coexisting species of blackfly larvae in a lake outlet. *Oikos* **33**, 290–296.

Wright, F. N. (1957). Rearing of *Simulium damnosum* Theobald (Diptera: Simuliidae) in the laboratory. *Nature, Lond.* **180**, 1059.

Wright, R. E. and DeFoliart, G. R. (1970). Some hosts fed upon by ceratopogonids and simuliids. *J. med. Ent.* **7**, 600.

Wu, Y. F. (1931). A contribution to the biology of *Simulium* (Diptera). *Pap. Mich. Acad. Sci.* **13** (1930), 543–599.

Wülker, W. (1964). Parasite-induced changes of internal and external sex characters in insects. *Expl. Parasit.* **15**, 561–597.

Wülker, W. (1975). Parasite-induced castration and intersexuality in insects. *In* Intersexuality in the Animal Kingdom, pp. 121–134 (R. Reinboth, Ed.). Springer-Verlag. Berlin Heidelberg New York.

Wygodzinsky, P. and Coscarón, S. (1973). A review of the Mesoamerican and South American blackflies of the tribe Prosimuliini (Simuliinae: Simuliidae). *Bull. Am. Mus. nat. Hist.* **151**, 129–199.

Yang, Y. J. and Davies, D. M. (1968a). Digestion, emphasizing trypsin activity, in adult simuliids (Diptera) fed blood, blood-sucrose mixtures and sucrose. *J. Insect Physiol.* **14**, 205–222.

Yang, Y. J. and Davies, D. M. (1968b). Occurrence and nature of invertase activity in adult black flies (Simuliidae). *J. Insect Physiol.* **14**, 1221–1232.

Yang, Y. J. and Davies, D. M. (1968c). Amylase activity in black flies and mosquitoes (Diptera). *J. med. Ent.* **5**, 9–13.

Yang, Y. J. and Davies, D. M. (1974). The saliva of adult female blackflies (Simuliidae: Diptera). *Can. J. Zool.* **52**, 749–753.

Yang, Y. J. and Davies, D. M. (1977). The peritrophic membrane in adult simuliids (Diptera) before and after feeding on blood and blood-sucrose mixtures. *Entomologia. exp. appl.* **22**, 132–140.

Zahar, A. R. (1951). The ecology and distribution of blackflies (Simuliidae) in south-east Scotland. *J. Anim. Ecol.* **20**, 33–62.

Zivkovic, V. (1955). Recherches morphologiques et écologiques sur les Simulies du Danube, avec une étude particulière de *S. colombaschense* (Fabr.). *Monogr. Acad. Serbe. Sci.* **245**, 1–95. [In Serbian, French summary].

Zivkovic, V. (1958). Neuere Beiträge zur Ökologie der Kolumbatscher Mücke (*Simulium columbaczense*, Fabricius, 1787). *Z. Tropenmed. Parasit.* **9**, 193–200.

Zivkovic, V. and Kacanski, D. (1965). Der Anteil der Kriebelmücken (Diptera, Simuliidae) inder Ernährung der Fische. *Zeit. Fischerei Daren Hilfwiss.* **13**, 261–268.

Zvyagintzev, S. N. (1965). Materials on the biology of blackflies (Diptera: Simuliidae) in reservoirs. The distribution of larvae in the environment of the controlled Volga. *Med. Parasitol.* **1**, 32–37. [In Russian].

Zwick, H. (1974). Faunistisch-ökologische und taxonomische Untersuchungen an Simuliidae (Diptera) unter besonderer Berücksichtigung der Arten des Fulda-Gebietes. *Abh. Senckenb. naturforsch. Ges.* **533**, 1–116.

Zwick, H. (1978). Simuliidae. *In* Limnofauna europaea. Eine Zusammenstellung aller die europäischen Binnengewässer bewohnenden mehrzelligen Tierarten mit Angaben über ihre Verbreitung und Ökologie, pp. 396–403 (Ed. 2) [J. Illies, Ed.]. Stuttgart and New York. Gustav Fischer Verlag; Amsterdam. Swets and Zeitlinger B.V.

Index

NOTE: Page numbers in italics indicate a reference to an illustration.

JCC = Joint Coordinating Committee, WHO

JICA = Japan International Cooperation Agency

NTO = non-target organism

OCCGE = Organisation de Coordination et de Coopération pour la lutte contre les Grandes Endémies

OCP = Onchocerciasis Control Programme, WHO

ORSTOM = Office de la Recherche Scientifique et Technique Outre-Mer

PAG MISSION = The Preparatory Assistance to Governments Mission, WHO

PCP = pest control product

RNA = ribonucleic acid

RUVP = Research Unit on Vector Pathology, Memorial University of Newfoundland

SNEM = Servicio Nacional de Erradicación de Malaria

STAC = Scientific and Technical Advisory Committee, WHO

UN = United Nations

UNDP = United Nations Development Programme

UNEP = United Nations Environment Programme

USAID = United States Agency for International Development

UV = ultraviolet

WB = World Bank

WHO = World Health Organization

Glossary

ABR = Annual Biting Rate

ADP = adenosine diphosphate

ARBFRP = Athabasca River Black Fly Research Program

ATP = Annual Transmission Potential

BOD = Biological Oxygen Demand

CDC = Center for Disease Control, United States Department of Agriculture

CPV = Cytoplasmic Polyhedrosis Virus

DDD = 2, 2-bis (p-chlorophenyl)-1, 1-dichloroethane

DDE = 2, 2-bis (p-chlorophenyl)-2-(p-chlorophenyl)-1, 1-dichloroethane

DDT = 2, 2-bis (p-chlorophenyl)-1, 1, 1-trichloroethane

DEET = diethyl toluamide

DMP = dimethyl phthalate

DNA = deoxyribonucleic acid

DV = Densonucleosis Virus

EC = emulsifiable concentrate

EDAP = Economic Development Advisory Panel, WHO

EM = electron microscopy

EP = Ecological Panel, WHO

FAO = Food and Agricultural Organization of the United Nations

FED = Fonds Européen de Developpement, organe de la Communauté Economique Européene

IBRD = International Bank for Regional Development

IDRC = International Development Research Centre, Ottawa

IGR = Insect Growth Regulator

IRO = Institut de Recherches sur l'Onchocercose

ITCZ = Intertropical Convergence Zone

IV = Iridescent Virus